AF527737

Potassium Solubilizing Microorganisms for Sustainable Agriculture

Vijay Singh Meena •
Bihari Ram Maurya •
Jay Prakash Verma •
Ram Swaroop Meena
Editors

Potassium Solubilizing Microorganisms for Sustainable Agriculture

Editors
Vijay Singh Meena
Crop Production Division
ICAR – Vivekananda Institute of Hill Agriculture
Almora, Uttarakhand, India

Bihari Ram Maurya
Institute of Agricultural Sciences
Banaras Hindu University
Varanasi, Uttar Pradesh, India

Jay Prakash Verma
Institute of Environment and Sustainable Development
Banaras Hindu University
Varanasi, Uttar Pradesh, India

Ram Swaroop Meena
Institute of Agricultural Sciences
Banaras Hindu University
Varanasi, Uttar Pradesh, India

ISBN 978-81-322-2774-8 ISBN 978-81-322-2776-2 (eBook)
DOI 10.1007/978-81-322-2776-2

Library of Congress Control Number: 2016941176

Printed on acid-free paper

This Springer imprint is published by Springer Nature
The registered company is Springer (India) Pvt. Ltd.

Foreword

Institute of Environment and Sustainable Development

After nitrogen and phosphorus, potassium is one of the major nutrients required by all the plants. It plays a major role in the activation of several metabolic processes including protein synthesis, photosynthesis, enzyme activation, the transport of water and nutrients in the plants through xylem and synthesis of starch. It also plays a role in improving shelf life of crops and disease resistance. It is often observed that without adequate potassium, the plants have poorly developed roots, grow slowly, produce small seeds and have lower yields.

Potassium in soil exists in different forms but the crop can absorb what is present in soil solution only. The readily available potassium in soil solution is just 2 % out of the total, and the remaining part of soil potassium is locked up in soil minerals and hence most of it is unavailable for plant uptake. Also, in most of the soils, potassium decreases easily due to crop uptake, runoff, leaching and soil erosion. Soil potassium management strategies rely on inorganic chemical-based fertilizers cause a serious threat to human health and environment. In many developing countries, the entire requirement of potassium fertilizer is met by imports and distributed to farmers at subsidized price. But at places, where soils are naturally rich in potassium, the application of potassium solubilizing bacteria as biofertilizer for agriculture can reduce the use of chemical fertilizer and support sustainable agriculture. A large number of saprophytic bacterial and fungal strains have been identified which solubilize potassium from insoluble forms like mica, fledspar and others by microbial production of protons and organic ligands, hydroxyl anions and extracellular polysaccharides (EPS), which can either directly dissolve rock potassium or chelate silicon ions to bring potassium into the solution.

I commend this book to researchers and students interested in the area of PGPM and sustainable agriculture. The editors of this book deserve credit for compiling articles on this interesting topic of potassium solubilizing microorganisms. Authors of the chapters present a fascinating and stimulating collection of articles focusing on isolation and screening of potassium solubilizing microorganisms, mechanism of solubilization and evaluation of their effectiveness in greenhouse as well as in field. This book is a comprehensive treatment of past and current knowledge on the subject. I am sure that this book will provide a rapid update and truly inspire further research in this fascinating area relevant to sustainable agriculture

Varanasi
11 November, 2015

Opp. LD Guesthouse
Banaras Hindu University
Varanaisi

Akhilesh Singh Raghubanshi
Professor and Director

Contents

About the Editors

Vijay Singh Meena is currently working as a soil scientist in the Indian Council of Agricultural Research–Vivekananda Institute of Hill Agriculture, Almora, Uttarakhand. He obtained his B.Sc. (Ag) in 2009 and M.Sc. (Ag) in 2011 with specialization in soil science and agricultural chemistry from Banaras Hindu University, Varanasi, Uttar Pradesh, India. Subsequently, he submitted his Ph.D. in 2016. He has accomplished leading work on potassium-solubilizing microbes, soil biological fertility, rhizospheric chemistry, and conservation agriculture. Consequently, he has also published more than 30 original research articles in national and international peer-reviewed journals. In addition, he has published 11 book chapters. He is serving as managing editor in the journal *Agriculture for Sustainable Development* published by the Farmers' Social Welfare Society and reviewers in various peer-reviewed national and international journals of publishers.

Jay Prakash Verma is an assistant professor in the Institute of Environment and Sustainable Development, Banaras Hindu University, Varanasi, w.e.f. February 26, 2011, to continue. He completed his B.Sc. (Hons) in botany in 2003 and M.Sc. in biotechnology in the year 2006. Also he completed his Ph.D. (botany) in 2010 entitled "Studies on new bioformulation of plant growth-promoting rhizobacteria to develop a novel biofertilizer for chickpea (*Cicer arietinum* L.)." He has accomplished leading work on soil microbiology, biofertilizers, PGPR, PGPF, plant–microbe–soil interaction, soil fertility and health management, and sustainable agriculture. Also he has expertise in the research and development of pesticide-degrading microbes and cellulose-degrading microbes for bioethanol production from cellulosic material. He has a total of 9 years of experience in research and 4 years in teaching B.Sc., M.Sc. Tech, M.Phil., and Ph.D. courses. He has published 39 articles having more than 70 impact factors, 28 abstracts in national and international seminars and conferences, and four book chapters. He has also visited Spain, Paris, and Shanghai, China. He is serving as technical editor in *Biotechnology*, *Journal of Applied Sciences*, *International Journal of Soil Science*, *Asian Journal of Biological Sciences*, and *Bacteriology Journal* and is a member of the editorial board in *Agriculture for Sustainable Development* published by the Farmers' Social Welfare Society and reviewers in various peer-reviewed national and international journals of publishers, e.g., Nature Publishing Group, Elsevier, Springer, Wiley, and Taylor and Francis. He is also a project evaluator for the Czech Academy of Sciences, Science Support Division in the Czech Republic, and European Commission.

Bihari Ram Maurya is a professor and the head of the Department of Soil Science and Agricultural Chemistry, Institute of Agricultural Sciences, Varanasi, Uttar Pradesh, India. He obtained his B.Sc. (Hons) in 1976 and M.Sc. in 1978 from Banaras Hindu University in Uttar Pradesh, India. Subsequently, he was awarded a Ph.D. in 1984. His current areas of research are soil microbiology and biofertilizers. His Ph.D. thesis focuses on topics related to soil microbiology particularly on agriculturally important microorganisms which are used as biofertilizers. He supervised two Ph.D. and 18 M.Sc. (Ag.) students working on various aspects of soil microbiology. For more than 35 years, he has been doing teaching and research work in this specific area. He is a nodal officer of a biofertilizer production unit under an experiential learning program. He has accomplished leading work on soil microbiology, soil fertility, and soil chemistry. Consequently, he has also published more than 50 original research articles in national and international peer-reviewed journals. In addition, he has published 16 book chapters.

Ram Swaroop Meena is an assistant professor in the Department of Agronomy, Institute of Agricultural Sciences, Varanasi, Uttar Pradesh, India. He obtained his B.Sc. (Hons) and M. Sc. in agronomy from SKRAU, Bikaner, India. Subsequently, he was awarded a Ph.D. in 2011. He has accomplished leading work on agronomic practices for enhancing soil fertility and health. Consequently, he has also published more than 50 original research articles in national and international peer-reviewed journals. In addition, he has published 9 book chapters. He is serving as an editor in *Agriculture for Sustainable Development* published by the Farmers' Social Welfare Society and reviewers in various peer-reviewed national and international journals of publishers.

Potassium-Solubilizing Microorganism in Evergreen Agriculture: An Overview

1

Vijay Singh Meena, Indra Bahadur, Bihari Ram Maurya, Ashok Kumar, Rajesh Kumar Meena, Sunita Kumari Meena, and Jay Prakash Verma

Abstract

Increasing cost of the fertilizers with lesser nutrient use efficiency necessitates alternate means to fertilizers. Soil is a storehouse of nutrients and energy for living organisms under the soil-plant-microorganism system. These rhizospheric microorganisms are crucial components of sustainable agricultural ecosystems. They are involved in sustaining soil as well as crop productivity under organic matter decomposition, nutrient transformations, and biological nutrient cycling. The rhizospheric microorganisms regulate the nutrient flow in the soil through assimilating nutrients, producing biomass, and converting organically bound forms of nutrients. Soil microorganisms play a significant role in a number of chemical transformations of soils and thus, influence the availability of macro- and micronutrients. Use of plant growth-promoting microorganisms (PGPMs) helps in increasing yields in addition to conventional plant protection. The most important PGPMs are *Azospirillum*, *Azotobacter*, *Bacillus subtilis*, *B. mucilaginosus*, *B. edaphicus*, *B. circulans*, *Paenibacillus* spp., *Acidithiobacillus ferrooxidans*, *Pseudomonas*, *Burkholderia*, potassium, phosphorous, zinc-solubilizing

V.S. Meena (✉)
Department of Soil Science and Agricultural Chemistry, Institute of Agricultural Sciences, Banaras Hindu University, Varanasi 221005, Uttar Pradesh, India

Indian Council of Agricultural Research – Vivekananda Institute of Hill Agriculture, Almora 263601, Uttarakhand, India
e-mail: vijayssac.bhu@gmail.com; vijay.meena@icar.gov.in

I. Bahadur • B.R. Maurya
Department of Soil Science and Agricultural Chemistry, Institute of Agricultural Sciences, Banaras Hindu University, Varanasi 221005, Uttar Pradesh, India

A. Kumar
Department of Botany, MMV, Banaras Hindu University, Varanasi 221005, India

R.K. Meena
Department of Plant Sciences, School of Life Sciences, University of Hyderabad, Hyderabad 500046, TG, India

S.K. Meena
Division of Soil Science and Agricultural Chemistry, Indian Agriculture Research Institute, New Delhi 110012, India

J.P. Verma
Institute of Environment and Sustainable Development, Banaras Hindu University, Varanasi 22100, Uttar Pradesh, India

V.S. Meena et al. (eds.), *Potassium Solubilizing Microorganisms for Sustainable Agriculture*,
DOI 10.1007/978-81-322-2776-2_1

microorganisms, or SMART microbes; these are eco-friendly and environmentally safe. The rhizosphere is the important area of soil influenced by plant roots. It is composed of huge microbial populations that are somehow different from the rest of the soil population, generally denominated as the "rhizosphere effect." The rhizosphere is the small region of soil that is immediately near to the root surface and also affected by root exudates.

Keywords
Potassium-solubilizing microorganisms (KSMs) • Nutrient use efficiency • Soil-plant-microorganism system • Rhizosphere • Evergreen agriculture

1.1 Introduction

Among nitrogen (N), phosphorus (P), and potassium (K), potassium is the third important macronutrient element of plant nutrient that plays significant roles in the activation of several metabolic processes, including photosynthesis, protein synthesis, and enzymes, as well as in resistance to diseases, insects, etc. (Rehm and Schmitt 2002). Potassium though is present as an abundant element in the soil or is applied to fields as natural or synthetic fertilizers, only 1–2 % of this is available to plants, the rest being bound to other minerals and therefore unavailable to plants. The most common soil components of potassium, 90–98 %, are feldspar and mica (McAfee 2008). Soil microorganisms influence the availability of soil minerals, playing a central role in ion cycling and soil fertility (Lian et al. 2008). Their uses as efficient biofertilizer agents for agriculture improvement, environmental protection, and soil fertility have been a focus of recent research. Certain bacteria are capable of decomposing aluminosilicate minerals and releasing a portion of the potassium contained therein (Biswas and Basak 2009). A detailed understanding of how microorganisms affect mineral dissolution rates is essential to quantify mineral weathering on global element cycling (Xiufang et al. 2006). Plants absorb potassium only from the soil; its availability in the soil is dependent upon the K dynamics and total K content. Out of the three forms of potassium present in the soil, only soil minerals make up more than 90–98 % of soil potassium (Sparks 1987) and most of it is unavailable for plant uptake. The second non-exchangeable form of potassium found approximately 1–10 % of soil potassium and consists predominantly under the interlayer K of non-expanded clay minerals such as illite and lattice K in K-feldspars that play an important role to the plant uptake (Sharpley 1989). Release of non-exchangeable K to the third exchangeable form occurs when the exchangeable level and solution K is decreased by crop removal, runoff, erosion, or leaching (Sheng and Huang 2002; Meena et al. 2013; Maurya et al. 2014). With the introduction of high-yielding crop varieties and the progressive intensification of sustainable agriculture, the soils are generally getting depleted in the potassium reserve at a faster rate. Moreover, due to imbalanced fertilizer application, potassium deficiency is becoming one of the major constraints in the crop production (Meena et al. 2015a, b; Singh et al. 2015).

This emphasized the search to find an alternative and effective indigenous source of K for plant uptake and also to maintain K status in soils for sustaining crop production (Meena et al. 2015b). Identification of efficient microbial strains that have the capability to solubilize potassium minerals quickly that can conserve our existing resources to avoid environmental pollution hazards caused by the heavy use of chemical fertilizers. A wide range of bacteria, namely, *Acidithiobacillus ferrooxidans*, *Pseudomonas*, *Burkholderia*, *Bacillus mucilaginosus*, *B. circulans*, *B. edaphicus*, and *Paenibacillus* spp., have been reported to release potassium in accessible form from potassium-bearing minerals in soils (Sheng 2005; Liu et al. 2012; Meena et al. 2014a, b; Kumar et al. 2015). These potassium-solubilizing bacteria (KSB) microorganisms were found to dissolve potassium, aluminum, and silicon from insoluble K-bearing minerals such as illite, micas, and orthoclases through excreting organic acids, by either directly dissolved rock K or via chelated silicon ions to bring K into the solution (Zhang and Kong 2014). Inoculation with potassium-solubilizing microorganisms (KSMs) has been reported to exert beneficial effects on growth of cotton and rape (Sheng 2005), cucumber and pepper (Han et al. 2006), wheat (Sheng and He 2006), and sudan grass (Basak and Biswas 2010). Similarly, inoculation of wheat and maize plants with *Azotobacter chroococcum*, *Bacillus mucilaginosus*, and *Rhizobium* resulted in significantly higher mobilization of potassium from waste mica, which in turn acted as a source of potassium for plant growth (Singh et al. 2010). Therefore, potassium-solubilizing bacteria are extensively applicable as biofertilizers in China and Korea as significant cultivated areas of soils in these countries the soil-available K are deficient (Zhang and Kong 2014). Thus, application of K-solubilizing bacteria as biofertilizer for agricultural improvement can reduce the use of agrochemicals and support eco-friendly crop production (Sindhu et al. 2010).

Currently, very little information is available on potassium solubilization through bacteria, mechanisms of solubilization, and the effect of KSM inoculation on nutrient availability in soils and growth of different crops. Potassium is available in soil in four forms, which are an exchangeable cation form, as K ions (K^+) in the soil solution, tightly bound on the surfaces of clay minerals and organic matter, tightly held or fixed by weathered micaceous minerals, and present in the lattice of few K-containing primary minerals. There are various processes which help to the availability of potassium in the soil. Soil solution potassium is present in the available form in the soil for plant uptake; however the concentration of potassium is affected by cropping history, soil weathering, and fertilizer use. Thus, the amount present is not sufficient to meet the crop requirement. Then, an indicator of soil potassium status, such as exchangeable potassium, has a rapid equilibrium with the soil solution potassium and is considered as readily available. Fixed and lattice potassium that can be grouped together make up the pool of non-exchangeable inorganic potassium in the soil.

1.1.1 Potassium and Its Importance

Potassium is an essential macroelement for all living organisms. In plant physiology it is the crucial cation in regard to its content in plant tissues and with respect to its biochemical and physiological functions (Zhang and Kong 2014). The quantity of potassium absorbed by roots is only second after that of nitrogen for most cultivated plants; however the demand for potassium is higher than that for nitrogen in banana, cotton, and some other species (Mora et al. 2012). As the cation K^+, it is dissolved in the soil solution as well as adsorbed to clay and organic colloids. But it can also be part of more complex chemical compounds (Sheng et al. 2003; Zheng et al. 2009; Zandonadi et al. 2010).

Potassium is absorbed by roots and translocated inside the plant as the positive cation K^+. It is characterized by a high mobility at all levels inside the plant in individual cells, tissues, and in long-distance transport through the xylem and phloem, although no structural role of potassium has been found. This is in contrast to calcium and magnesium, which have important structural functions, but only limited mobility in plants. Control of plant water status is an important potassium function. It promotes water absorption by the roots, keeps osmotic tension and turgor in cells and plant tissues, and regulates the activity of stomata cells to prevent unnecessary water loss by transpiration. Potassium has a role in photosynthesis and in the production and translocation of carbohydrate to areas of meristematic growth, fruit development, and storage.

The carbohydrate production and transport function is very important in vegetables and fruit production as it directly affects sugar and starch accumulation. Early stages of potassium deficiency are reflected to decreases in the yield production. This stage of potassium deficiency is called *hidden hunger* as no specific symptoms appear on the plants. As the intensity of the deficiency increases, symptoms consisting in yellowing and eventual necrosis of the border in older leaves do appear, beginning from the tip and progressing backward over the leaf. Because it is a mobile element, when the potassium deficiency occurs, the element is transferred from old leaves to the young growing points. A slowdown of the growth rate is also present at this level of potassium deficiency. The main internal consequences of potassium deficiency are general reductions of the strength of plant structures, loss of vigor; slowdown of carbohydrate transport, and less resistance to low water availability and to diseases.

Potassium deficiency decreases the stunting of growth with shortening of internodes. Its deficiency caused a reduction in photosynthesis; blackening of tubers in the case of potato and tips or margin of lower leaves of legumes, cotton, maize, and tobacco; and either scorching or burning of all small grains (Ashley et al. 2005). Potassium plays a significant role in the maintenance of cellular organizations for regulating permeability of cell membranes and also keeping the protoplasm in a proper degree of hydration. It activates the certain essential enzymes in protein and translocation of carbohydrates and carbohydrate metabolism and imparts resistance to plants against bacterial and fungal disease. Potassium is an integral part for the development of chlorophyll. It plays an important role in photosynthesis, which is converting carbon dioxide and hydrogen into sugars, for translocation of sugars, and for starch formation, and also metabolic activities of plants (www.ikisan.com).

Adequate quantities of potassium are essential for a plant to achieve its full yield potential and also for various aspects of product quality such as grain size and appearance, oil content, tuber size, dry matter and percentage sugar, starch content, and fruit ripening and quality. The different functions of potash in the plant are also related to physiological conditions and various stresses. These functions are diverse and include efficient nitrogen and drought tolerance, water use, frost resistance, and resistance to pests and diseases. It is therefore not surprising that the lack of plant-available potash in the soil results in weaker, less vigorous crops that suffer major growing seasons. In years when growing conditions and yield production are good, the response to potash may be modest, especially for cereal crops, but during adverse years, its contribution to optimum yields will be substantial. Adequate potassium is crucial for which it provides some "insurance" against adverse conditions in difficult growing seasons.

1.1.2 Global Potassium Demand

Potash fertilization started during in the nineteenth century, when Justus v. Liebig discovered that the plants required it in different proportions and quantities. During the past 40 years, world potassium consumption has increased up to 2.5-fold. Between 1960 and 2000, the world use of potassium fertilizers rose from 9 to 22 Mt and potassium fertilizer use accounts for 16 % of total fertilizer usage. In the developed countries, potash consumption has increased 1.25-fold in the past 40 years, while in the developing countries its demand expended 22-fold from 0.5 Mt in 1960 to 11.3 Mt in 2000. The world potash production had been increased up to 37 Mt (USGS Mineral Commodity Summary 2012) as well as the price of potash $470/per tons since 2011 (www.infomine.com, 2013). However, K fertilizer cost has not stopped enhancing every year; this has led to the increased cost of rice production and should reduce farmer's income.

1.1.3 Potassium Status of Soils in India

In India, nutrient removal continues to exceed 10 Mt of N + P_2O_5 + K_2O every year. Clearly, expansion in fertilizer application (input) continues to fall short of nutrient removals

Table 1.1 Nutrient balance (NPK) under major states of India

States	Additions	Removal	Balance
Andhra Pradesh	411.3	708.6	−297.3
Karnataka	330.3	734.4	−404.0
Madhya Pradesh	168.9	943.9	−775.0
Kerala	72.3	278.6	−206.3
Uttar Pradesh	863.1	1842.1	−979.0
Gujarat	146.1	1137.4	−991.3
Haryana	16.6	669.1	−652.5
Maharashtra	420.8	1482.9	−1062.1
Rajasthan	20.9	1014.8	−993.9
Punjab	38.4	1022.8	−984.5
Jharkhand	9.8	169.6	−159.9
Assam	56.0	255.8	−199.8
Tamil Nadu	304.2	726.1	−421.9
Orissa	63.0	383.6	−320.6
West Bangal	304.4	972.8	−668.4
Total (15 states)	3226.1	12,342.5	−9116.5

Adapted from Srinivasrao et al. 2011

(output) resulting in the depletion of soil fertility and negative nutrient balance sheet. Out of 371 districts for which more than 11 million soil test data are available, 76 districts are in low, 190 in medium, and 105 districts are in the high category. Thus, 21 % districts are in the low, 51 % in the medium, and 28 % in the high category of potassium fertility status (Table 1.1). All the low and medium K soils which constitute 72 % of the total need K fertilization for optimum yield and balanced soil fertility (Hasan 2002; Ramamurthy and Bajaj 1969).

About 70–75 % of the K absorbed is retained by leaves, straw, and stover. The remainder is found in harvested portions such as grains, fruits, nuts, etc. Whenever the soil cannot adequately supply the K required to produce high yields, farmers must supplement soil reserves with fertilizer K. Improvements in both quantity and quality will add to export earnings. The information presented here is based on more than 11 M soil samples made available by soil testing laboratories run by state departments of agriculture and the fertilizer industry. Though 11 M soil tests are not sufficient to comprehensively cover a country which has >140 Mha cultivated land, they reflected changing K-fertility status of soils in different parts of the country and provide some measure of the need for scientific use of K fertilizers (Fig. 1.1).

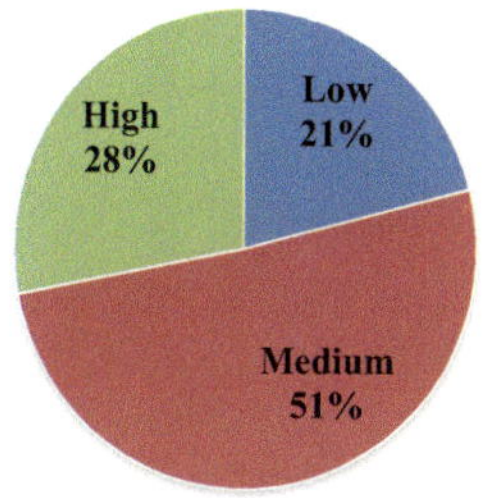

Fig. 1.1 The graphical representation of soil potassium status of India

1.1.4 Potassium Fixation in Soil

In addition to releasing K, soil minerals can also fix K, significantly affecting K availability. This involves the adsorption of K ions on two sites in the interlayers of weathered sheet silicates, such as illite and vermiculite. The degree of K fixation in soils depends on the type of clay mineral and its charge density, moisture content, competing ions, and soil pH. Montmorillonite, vermiculite, and weathered micas are the major clay minerals that tend to fix K (Sparks 1987). Additionally, soil wetting and drying also significantly affect the K fixation. The fixation process of K is relatively fast, whereas the release of fixed K is very slow due to the strong binding force between K and clay minerals (Oborn et al. 2005). Whether a soil fixes or releases K highly depends on the K concentration in the soil solution (Schiavon et al. 2010). In addition to organic acids, the H+ concentration in the soil solution (via soil pH) seems to play a key role in K release from clay minerals.

Optimization of soil pH may be a means of enhancing K release. For optimized K fertilizer management practices, it is crucial to understand the factors that regulate K release from soil non-exchangeable pool. Recent investigations have raised awareness of the impact of K on the soil structure and its ability to capture water. It has been reported that the application of mineral K fertilizers enhances the water-holding capacity of soils and also improves the structural stability of sandy soil in particular (Holthusen et al. 2010). Higher water retention plays a key role in securing the soil productivity in water-limited areas. Therefore, more information is needed in order to understand the effect of K fertilization on the soil's physical properties and soil water-holding capacity (Fig. 1.1).

1.2 Need for Fertilizers

Diffusion of fertilizer consumption in Indian agriculture field has been quite widespread. The uses of imbalanced NPK in the agriculture field have become highly conspicuous. The efficiency of fertilizer application has gradually gone up to 3 kg/ha. It is universally accepted that the application of chemical fertilizers is an integral part of raising the agricultural production to a higher place. The Food and Agricultural Organization of the United Nations (FAO) studied and have established beyond doubt that there is a close relationship between the fertilizer consumption level and average crop yields. The advantages of using nutrients or inorganic fertilizers are immediately available to plants and the exact amount of a given element that can be measured before feeding plants. However, commercial fertilizer, particularly nitrogen, has been easily leached out through rain or irrigation.

1.2.1 Potassic Fertilizers

The potassium content of potassic fertilizers is usually expressed as potassium oxide, K_2O, also referred to as potash. These fertilizers are manufactured from minerals and ores. The commercial fertilizers are salts of potassium usually chlorides and sulfates which are soluble, hence readily available to the plants (www.incitecpivot.com).

The entire requirement of potash is met via imports as there are no exploitable reserves of potassic minerals in the country. The total import of MOP during the current year was 30.40 lakh Mt (www.sobip.com). Studies also have shown that building up readily available potassium reserves in the soil ensures the best opportunity for plants to achieve their optimum economic yield. The application of large amounts of potassium fertilizer to agriculture soil with little readily available potassium will not always enhance yields to the equal amount as those in enriched soil. This is due to the enriched soil reserves that potassium is equally and uniformly distributed throughout the layer of soil in which most of the roots grow. The crop uptake of the depleted potassium in the soil solution as a result is rapidly replenished from the reserves. The benefits are somewhat more with those crops that have a short growing season. Such crops do not have extensive root systems and as a result they must acquire quickly nutrients to optimize growth. But the emergence of potassium fertilizers has not given the ultimate solution for developing countries like India because it lays a major economic constraint since large sums are spent on potassium fertilizers alone (The Economic Times 2006).

1.2.2 Fertilizers and Environmental Pollution

Fertilizers are generally safer in comparison to the pesticides that exhibit toxic properties of living systems. However, all doses of fertilizers that have been applied to the soil are not fully utilized by plants. About 50 % of applied fertilizers to plants are left behind as residues. Generally, inorganic fertilizers are not directly toxic to humans and other life forms; they have been present to upset the existing ecological balance. The nutrients escape from the agriculture or other fields that have been found in excessive quantities in lakes, rivers, and coastal waters. Algae blooms found where the nutrient load is high and these smother other aquatic vegetation in the water bodies. These phenomena may lead to death of the many aquatic dwellers. Environmental contamination arises because not all the fertilizer applied is taken up by the crop and removed at harvest.

1.2.3 Availability of Potassium in Soil

Only little quantities of K are maintained in the soil solution 5–20 kg/ha K (6–24 kg K_2O). The majority of the reserves of potassium in the soil are held through negative charges in organic matter and clay minerals. The potassium may be held weakly or strongly depending upon the position in the clay lattice. Plant uptake K^+ which is loosely bound (exchangeable K) is quickly released, while the strongly bound reserve ("non-exchangeable K") is very slowly released. Individual soils have various intensities to hold potassium according to the clay type and content and also the amount of the soil organic matter. The sandy soils have a very limited amount of the reserves of exchangeable K. Clay minerals themselves contain potassium (Fig. 1.2).

1.2.4 Forms and Availability of Potassium

Potassium is the seventh most abundant element in the Earth's crust and makes up about 2.4 % by weight of the earth's crust. Yet only 1–2 % is available to plants. Potassium exists in the soil as exchangeable K, dissolved K^+ ions (solution K), mineral K, and non-exchangeable K. Plants can take up solution K and exchangeable K from the soil, but the non-exchangeable K and mineral K are unavailable to plants. Depending on the soil type, ~98 % of total soil K is found in unavailable form. Most potassium minerals are insoluble. Feldspars and micas are minerals that contain most of the K and common parent materials for most soils (Foth and Ellis 1997). The crops cannot use the K in this crystalline (insoluble form). Over very long periods of time (Fig. 1.2), these minerals weather (break down) and K is released; this process is too slow to supply the full K needs of field crops.

1.3 Environmental Factors Affecting Potassium Solubilization

In soil, K mobilization is affected by many biotic and abiotic environmental factors like soil properties (physicochemical characteristics, aeration, and pH), the presence of mycorrhizae fungi and rhizosphere bacteria, and the composition of plant root exudates. These parameters may also influence the K mobilization ability of bacteria. Gahoonia et al. (1997) explained three

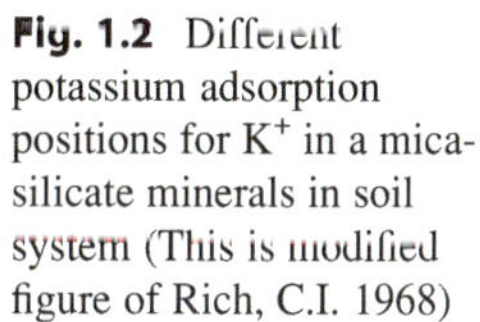

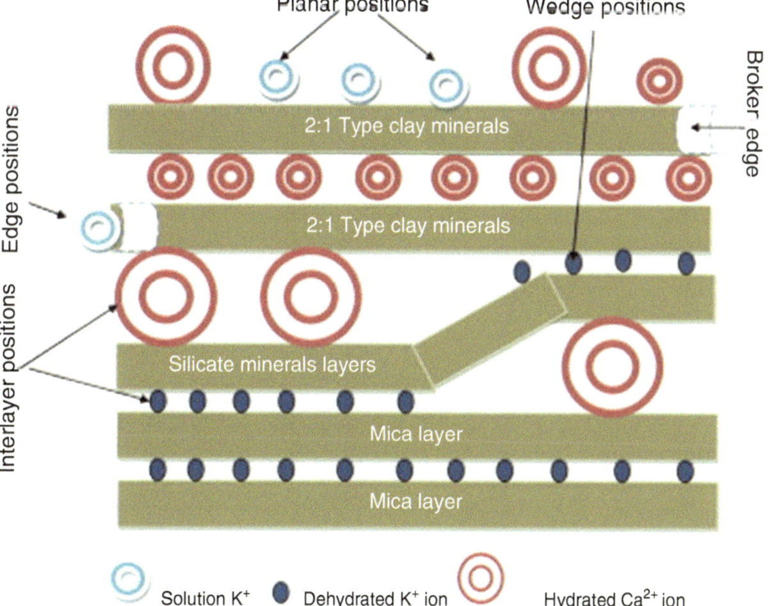

Fig. 1.2 Different potassium adsorption positions for K^+ in a mica-silicate minerals in soil system (This is modified figure of Rich, C.I. 1968)

hypothetical processes:(1) Mineral fragmentation caused by the activity of the root increases the bacterial effect on mineral mobilization due to increased surface area for reactivity. (2) Root exudates help by indirectly providing the substrates for the production of weathering metabolites by bacteria. (3) Besides producing weathering agents, bacteria produce phytohormones which stimulate the development of root, modify root exudation and physiology, and help improve nutrient uptake and mobilization of minerals. Potassium concentration in various plant parts is not similar as it may vary from different climate conditions. In case of feldspar, illite, and muscovite, more K is released under aerobic conditions as compared to anaerobic conditions (Badar 2006). In liquid medium *Bacillus edaphicus* showed better growth and greater K-releasing ability with illite as compared to feldspar (Sheng and He 2006).

1.3.1 Potassium Functions in Plants

Potassium has many functions in plant growth, such as to smooth the progress of cell division and growth, to increase disease resistance and drought tolerance, and to regulate the opening and closing of stomata required for osmotic regulation. Besides, this potassium is essential for photosynthesis process and acts as key to activate enzymes to metabolize carbohydrates for the manufacture of amino acids and proteins. Furthermore, potassium assimilates transport during plant ontogeny, and one of the most important influences is improving the oil content of plants. Since the use of potassium has covered a lot of plant activities, depletion in potassium uptake can cause problems for plant growth. A symptom of potassium deficiency is chlorosis along the leaf margins. Then, in severe cases, the leaf will turn into yellow color and eventually will fall off. It also affects plant growth and canopy photosynthesis process. There are several factors that lead to this problem, for instance, low soil potassium-supplying capacity, insufficient application of mineral potassium fertilizer and biofertilizer, complete removal of plant straw, leaching losses, and phosphorus and nitrogen deficiency (Das and Sen 1981). Fundamentally, K^+ is highly water soluble and highly mobile and transported in the plants' xylem (Lack et al. 2005). Membrane transport of the potassium can be mediated either through potassium channels, utilizing the membrane potential to easy transport of potassium down its electrochemical gradient, or through secondary transporters. In plants, potassium acts as a regulator since it is constitutes 60 different enzyme systems of drought tolerance and water use efficiency.

Plants' K concentrations of crops vary widely with the site, year, crop species, and fertilizer input. It ranges from 0.4 % to 4.3 % (Askegaard et al. 2004). Oborn et al. (2005) concluded that crop K concentrations are often well below (~3.5 %). For many crops, the critical K concentration is in the range of 0.5–2 % in dry matter (Leigh and Jones 1984). With the exception of the cytosol and the vacuole, the subcellular distribution of K is largely uncharacterized. The concentration of K in the cytoplasm is kept relatively constant at around 50–150 mM, while its concentration in the vacuole varies substantially depending on supply status. Together with accompanying anions (NO^{-3}, Cl^-, malate$^-$), vacuolar K largely determines the osmotic potential of the cell sap. In the agronomic literature, high K concentration in crops has often been termed *luxury consumption*. However, as outlined below ("Potassium Nutrition and Crop Stress Resistance" section), high accumulation of K by crops during optimal growing conditions may be considered as an *insurance strategy* to enable the plant to better survive a sudden environmental stress (Kafkafi 1990).

Plant species are known to differ in their K requirement and in their ability to take up K. The differences in absorption of K among different plant species are attributed to variations in root structure, such as root density, rooting depth, and root hair length (Nieves-Cordones et al. 2014). All crops require potassium, especially high-carbohydrate plants such as maizes, bananas, and potatoes (Hillel 2008). KSB is a heterotrophic bacterium which obtains their energy and carbon source from the buildup of dead organic materials. Besides these, KSBs are aerobic bacteria which play a significant role in maintaining the soil structure through contribution in the formation and stabilization of water-stable soil aggregates. In addition, this gram-positive bacterium can produce a substance that stimulates plant growth or inhibits root pathogens (Egamberdiveya 2006). Moreover, KSB specifically are well known for its capability to solubilize rock potassium mineral such as illite, orthoclases, and micas. This is done through the production and excretion of different organic acids (Subhashini and Kumar 2014; Maurya et al. 2014). Therefore, KSB increases potassium availability in soils and increases mineral contents in plants.

1.3.2 Quantification of Potassium Solubilizatlon

Studies on the dynamics of potassium solubilization by microorganisms are best carried out in vitro based on their measurement of K released into culture broth from cultures developed by using an insoluble mineral compound as the only K source. The rate of potassium solubilization is often estimated by the subtracting of the final K concentration (minus that of a non-inoculated control) from the initial theoretical K supplied through the K-bearing substrates. This estimate has the limitation of not taking into account the K utilized by the cells during growth. The efficiency of potassium solubilization by different microorganisms varies with the nature of potassium-bearing minerals and more specifically it depends on the structure and chemical composition of minerals. It was observed that bacteria degraded kietyote and pegatolite and released 47 and 44.4 mg soluble potassium, respectively, after 38 h of incubation (Yakhontova et al. 1987). The easy release of the potassium from the minerals follow in order as illite > feldspar > muscovite (Sheng et al. 2002; Badar 2006).

Besides, the releases of K from minerals also get affected by pH of the medium, dissolved oxygen, and the microbial strain used. It was estimated that with the increase in pH from 6.5-to-8.0, the soluble K content increased from 490 to 758 mg/L. Instability of the potassium-solubilizing character of a few strains after several times of inoculation has been reported. However, the traits seem to remain stable in most of the isolates. Although no exact quantitative comparison can be made from experiments with different sources of insoluble K, the review of literature suggests that fungi such as *Aspergillus* spp. and *Penicillium* spp. are more effective solubilizers than bacteria such as *Bacillus* spp., *Paenibacillus* spp., *Pseudomonas* spp., etc.

1.4 Potassium-Solubilizing Microorganisms (KSMs)

Diverse groups of soil microflora were reported to be involved in the solubilization of insoluble and fixed forms of K into available forms of K which are easily absorbed by plants (Zarjani et al. 2013; Gundala et al. 2013). Microbial inoculants which are able to dissolve K from mineral and rocks that enhanced plant growth and yield are also economically viable and eco-friendly. The first evidence of solubilization of rock potassium by microbial involvement had been shown by Muentz (1890). A wide range of KSMs, namely, *Bacillus mucilaginosus*, *B. edaphicus*, *B. circulans*, *Paenibacillus* spp., *Acidithiobacillus ferrooxidans*, *Pseudomonas*, and *Burkholderia* (Sheng et al. 2008; Singh et al. 2010, Basak and Biswas 2012), have been reported to release potassium in an accessible form from K-bearing minerals in soils. Several fungal and bacterial species, popularly called KSMs that assist plant growth by mobilization of insoluble forms of K. KSMs are ubiquitous whose numbers vary from soil to soil. The rhizosphere microorganisms widely contribute in the solubilization of bound form of soil minerals in the soil (Supanjani et al. 2006; Sindhu et al. 2009). A variety of soil microorganisms have been found to solubilize silicate minerals (Sheng et al. 2001).

Many microorganisms like fungi, bacteria, mycorrhizae, and actinomycetes colonized even on the surface of mountain rocks (Groudev 1987;

Gundala et al. 2013) and it has been reported that the *B. mucilaginosus* sub spp. *siliceus* is silicate-solubilizing bacteria that liberates K from aluminosilicates and feldspar. According to Aleksandrov et al. (1967) that were isolated from agricultural land at different locations and found that various bacterial species like silicate bacteria were found to dissolve K, silicates, and aluminum from insoluble minerals, it also helps in the decomposition of organic matter, crop residues etc., and suggested that they play a major role in nutrient cycling in the soil-plant system. *B. mucilaginosus* stain CS1 is a silicate bacterium which exhibited inhibitory effect on the growth of gram-negative bacteria. It has also been reported as silicate-solubilizing bacteria present in rhizosphere as well as non-rhizosphere soil (Lin et al. 2002; Liu 2001). K-solubilizing rhizosphere bacteria were isolated from the roots of cereal crops which are grown in potassium and silicate-amended soil (Mikhailouskaya and Tcherhysh 2005).

1.4.1 Potassium-Solubilizing Bacteria (KSB)

A wide range of rhizospheric microorganisms reported as K-solubilizers include *B. mucilaginosus* (Zarjani et al. 2013), *B. edaphicusand* (Sheng 2002), *B. circulans* (Lian et al. 2002), *Burkholderia*, *Acidithiobacillus ferrooxidans*, *B. mucilaginosus* (Zhang and Kong 2014), *Bacillus edaphicus* (Sheng and He 2006), *Arthrobacter* spp. (Zarjani et al. 2013), *Enterobacter hormaechei* (Prajapati et al. 2013), *Paenibacillus mucilaginosus* (Liu et al. 2012; Hu et al. 2006), *P. frequentans*, *Cladosporium* (Argelis et al. 1993), *Aminobacter*, *Sphingomonas*, *Burkholderia* (Uroz et al. 2007), and *Paenibacillus glucanolyticus* (Sangeeth et al. 2012). These microbial strains have the ability to solubilize K from K-bearing minerals, but only a few bacteria, such as *B. edaphicus* and *B. mucilaginosus*, have high capacity for mobilizing and solubilizing of K from minerals (Zhao et al. 2008; Rajawat et al. 2012). Bacteria have wide applications in mining, metallurgy, microbial fertilizer, and feed (Maurya et al. 2014; Meena et al. 2014a; Zhang and Kong 2014).

1.4.2 Potassium-Solubilizing Fungi (KSF)

Arbuscular mycorrhiza can increase the solubility of the mineral form of potassium by releasing protons, H^+, or CO_2 and organic acid anions such as citrate, oxalate, and malate. This also increased the nitrogen, potassium, calcium, and iron in the plant leaves and fruits (Veresoglou et al. 2011; Yousefi et al. 2011). The inoculant of the two arbuscular mycorrhizal fungi (AMF) species *G. intraradices* and *G. mosseae* were applied in soil on a weight basis and recorded the increasing potassium uptake by maize crop (Wu et al. 2005) Information on the uptake of the macronutrient cations through AM plants has been relatively inconsistent in that increases, decreases, and no effects have been reported (Clark and Zeto 1996), and in the case of potassium, it depends on the soil condition as well as nature of plant growth and other conditions (Clark and Zeto 2000). Alves et al. (2010) reported that after 90 days, the plant height, root length, shoot dry weight potassium and phosphorus contents, and mycorrhizae colonization were increased in comparison to control. Potassium uptake compared to Mg and Ca was especially enhanced in AM switch grass grown in acid soil (Clark et al. 1999). Ectomycorrhizal fungi particularly isolated UFSC-Pt22 and UFSC-Pt186 and contributed to the increase of the efficiency of alkaline breccias as a source of P and K to the plant growth of *Eucalyptus dunnii* seedlings, respectively (Alves et al. 2010).

Prajapati et al. (2012) reported that potassium-solubilizing fungi (KSF) strains such as *Aspergillus terreus* and *Aspergillus niger* were isolated from various K-rich soil samples and observed that *A. terreus* and *A. niger* could solubilize insoluble potassium and showed the highest available potassium in liquid medium by using two various insoluble sources of potassium, i.e., feldspar and potassium aluminum silicate, based on their colonies and morphology characters.

A. terreus shows highest solubilization as well as acid production on both insoluble potassium sources. The concentration of trace elements is another relevant factor in the context of rock solubilization by fungi (*A. niger*) also reported by production of acids (Mirminachi et al. 2002). Furthermore, symbiotic nitrogen-fixing rhizobia and *Pseudomonas*, which fix atmospheric nitrogen into ammonia and that export the fixed nitrogen to the host plants, have also shown K- and P-solubilizing activity. For instance, *Aspergillus* spp., *Aspergillus terreus* (Prajapati et al. 2013), *Aspergillus niger* (Prajapati et al. 2012), and *Penicillium* spp. (Sangeeth et al. 2012) enhanced K-solubilization by mobilizing inorganic and organic K and release of structural K from rocks and minerals.

1.4.3 Mechanisms of K-Solubilization

Currently, little information is available on K-solubilization by rhizospheric microorganism, in which the mechanisms of K-solubilization by production of organic acid to provide potassium nutrients as well as other nutrients for enhancing crop growth. Sheng and Huang (2002) found that K release from the minerals was affected by oxygen, pH, and the bacterial strains used. The efficiency of the K-solubilization by various microorganisms was found to vary according to the nature of potassium-bearing minerals and aerobic conditions (Uroz et al. 2009). The extent of potassium solubilization of *B. edaphicus* in the liquid media was more growth on illite than feldspar (Sheng and He 2006). Indigenous rhizospheric microorganisms have the greater to absorb and mobilize the fixed form of nutrients (potassium) from trace mineral sources. Silicate bacteria were found to dissolve aluminum, potassium, and silica from insoluble minerals. Hydrogen ion of soil or soil solution is directly related to releases of K from minerals. The content of potassium solubilization was enhanced 84.8–127.9 % in microbial inoculated than un-inoculated treatment. The extent of potassium solubilization was reported higher in illite by *B. edaphicus* in the broth culture compared to feldspar (Sheng and He 2006). Badar (2006) reported that the extent of potassium solubilization by silicate-solubilizing bacteria were recorded as 4.90 mg/L at pH 6.5–8.0. *B. mucilaginosus* was solubilized the 4.29 mg/L K-solubilization in media supplemented with muscovite mica (Sugumaran and Janarthanam 2007). The K-releasing affected by pH, soil mineral properties, and aerobic conditions (Chen et al. 2008; Bin et al. 2010).

Mechanism of potassium solubilization means by which the insoluble potassium and structurally unavailable form of potassium compounds are mobilized and solubilized due to the production of various types of organic acids which are accompanied by acidolysis and complexolysis exchange reactions, and these are key processes attributed to the conversion in a soluble form (Uroz et al. 2009). The organic and inorganic acids convert insoluble K (mica, muscovite, biotite feldspar) to the soluble form of K (soil solution form) with the net result of increasing the availability of the nutrients to the plants. The various types of organic acid produced by KSMs were differing with different organisms. Organic acids were detected in the microbial suspension (Verma et al. 2014; Zhang and Kong 2014; Maurya et al. 2014). KSMs have the ability to weather phlogopite through acidic dissolution and aluminum chelation of the crystal network (Leyval and Berthelin 1989; Abou-el-Seoud and Abdel-Mageed 2012; Meena et al. 2014b).

The release of various types of organic acids were reflected by microorganisms to solubilized the insoluble K to an available form of K which is easily uptaken by the plant. Researchers suggested that the plant growth promotion activities were related to K-solubilization as well as the release of organic acids by the K-solubilizing strains. Sheng and He (2006) reported that solubilization of feldspar and illite and via rhizospheric microorganisms is due to the production of organic acids like citric acid, tartaric acids, 2-ketogluconic acid, oxalic acid, gluconic acid, malic acid, propionic, fumaric, glycolic, and succinic acid seems to be the most frequent agent of K-solubilization mineral (Zarjani et al. 2013; Prajapati et al. 2012; Prajapati and Modi 2012; Wu et al. 2005).

The solubilization of structural K compounds by naturally-abundant KSMs is common under in vitro (Meena et al. 2013; Maurya et al. 2014), field, and greenhouse conditions (Prajapati et al. 2013; Parmar and Sindhu 2013). Indigenous, rhizospheric microorganisms are very effective in releasing K from structural K through solubilization and from exchangeable pools of total soil K by acidolysis, chelation, and solubilization by KSMs (Uroz et al. 2009). Biomass of rhizospheric microorganism in the soil also contains a major quantity of fixed K which is potentially available to plants (Girgis 2006; Subhashini and Kumar 2014).

Mechanisms for KSMs to solubilization of K are by: (i) lowering the pH or (ii) through increasing the chelation of the cations bound to K and (iii) acidolysis of the surrounding area of microorganism. The lowering in pH of the medium suggests the release of different organic acids and protons by the K-solubilizing microorganisms (Zarjani et al. 2013; Parmar and Sindhu 2013). Such acidolysis by organic acids produced by the rhizospheric microorganisms can either directly dissolve the mineral K as a result of slow releases of exchangeable K or readily available exchangeable K or can chelate by both Al and Si ions associated with K mineral (Romheld and Kirkby 2010). Thus, the synthesis and discharge of organic acids through microorganisms into the surrounding environment acidify the microbe's cells and their surrounding environment that ultimately lead to the release of K ions from the mineral K by protonation and acidification (Goldstein 1994). Of the various organic acids involved in the solubilization of insoluble K, gluconic, oxalic acids a-ketogluconic and succinic citric are the most efficient acids released by microbial strains (Table 1.2). Figure 1.3 showed the direct and indirect mechanisms of plant growth-promoting properties of potassium-solubilizing microorganism (KSMs) and their K-solubilizing ability of mica (K- bearing mineral) on Aleksandrov medium (this figure modified from Meena et al. 2014b).

Organic acids produced by KSMs can be detected through enzymatic and high-performance liquid chromatography methods (Archana et al. 2013; Zhang et al. 2013). However, the acidification does not seem to be the only mechanism of solubilization, as they have ability to reduce the pH in some cases which did not correlate with the ability to solubilize mineral K (Zhang and Kong 2014; Subhashini and Kumar 2014). Furthermore, the chelating ability of the different organic acids is also very important, as it has been shown that the addition of 0.05 M EDTA into the medium which has the same solubilizing effect as compared to the inoculation with *Penicillium bilaii* (Sheng and Huang 2002; Liu et al. 2006).

1.5 Morphological and Biochemical Characterization

Cell morphologies of the KSB isolate were determined through using an optical microscope after stained with the phenol red. Physiological tests, including Proskauer (VP) and Voges-Methyl red (MR) reactions, utilization of organic acids, anaerobic growth, production of acid from carbohydrates, were observed by the method of Claus and Berkeley (1986). Subhashini and Kumar (2014) studied the substrate utilization patterns along with salt tolerance and temperature, pH, and salt.

1.6 Molecular Biology of Potassium-Solubilizing Microorganism

For maintaining turgid pressure of microbial and plant cells, potassium (K^+) is one of the important elements. Stimulation of potassium acquisition is one of the most rapid responses to an osmotic up-shock in bacteria. Potassium (K) is mostly present as intracellular cations, which has

Table 1.2 The categorization of districts as K-fertility status of India

States	Potassium status (kg K_2O/ha)		
	<130	130–335	>335
Andhra Pradesh	2	14	3
Delhi	0	1	0
Himachal Pradesh	6	4	3
Bihar and Jharkhand	1	24	2
Assam	7	3	0
Dadra and Nagar Haveli	0	1	0
Chandigarh	0	1	0
Goa	1	0	0
Gujarat	0	3	16
Manipur	1	0	0
Kerala	4	6	0
Jammu and Kashmir	5	5	0
Meghalaya	1	0	0
Haryana	0	2	9
Nagaland	5	0	0
Maharashtra	0	12	13
Karnataka	3	10	7
Punjab	0	9	3
Mizoram	1	0	0
Rajasthan	0	23	0
Orissa	2	11	0
Tamil Nadu	0	6	7
Pondicherry	1	0	0
Madhya Pradesh and Chhattisgarh	3	10	31
Sikkim	0	4	0
Arunachal Pradesh	2	3	0
Tripura	3	0	0
Uttar Pradesh and Uttaranchal	26	23	7
West Bengal	2	13	1
Total districts, (%)	76 (21)	190 (51)	105 (28)

a major role in maintaining the turgor pressure of the cells and also plays an important role in bacterial osmoadaptation, gene expression, pH regulation, and activation of cellular enzymes (Epstein 2003). Three different types of K transporters (Trk, Kdp, and Kup) have been involved for the uptake of K. Trk is a multi-component complex widespread in bacteria and archea and it has a moderate affinity to the K uptake. Trk consist of a trans-membrane protein named TrkH or TrkG, which is the actual K-trans-locating subunit, and TrkA is a cytoplasmic membrane surface protein, which binds to NAD-binding protein that is required for the system's activity (Sleator and Hill 2002). Kdp is an inducible system with greater affinity and specificity for K, found in *Escherichia coli* and other bacteria. Kdp is the only bacterial K acquisition system whose expression is significantly regulated at the transcriptional level through KdpE response regulator and KdpD sensor kinase (Epstein 2003; Domínguez-Ferreras et al. 2009). *Bacillus subtilis* has the Ktr gene which is involved in K uptake. HAK/KT/KUP family, the genes of these families are homologous to bacterial KUP (TrkD) potassium transporters. The KUP transporter from *E. coli* is characterized by a midrange (0.37 mM) KM for K^+ and a similar affinity to the Cs^+ and Rb^+. Complementation of TK2463 cells by AtKUP1. The *E. coli* TK2463 mutant is defective in three K1 uptake transporters (Trk, Kdp and Kup)

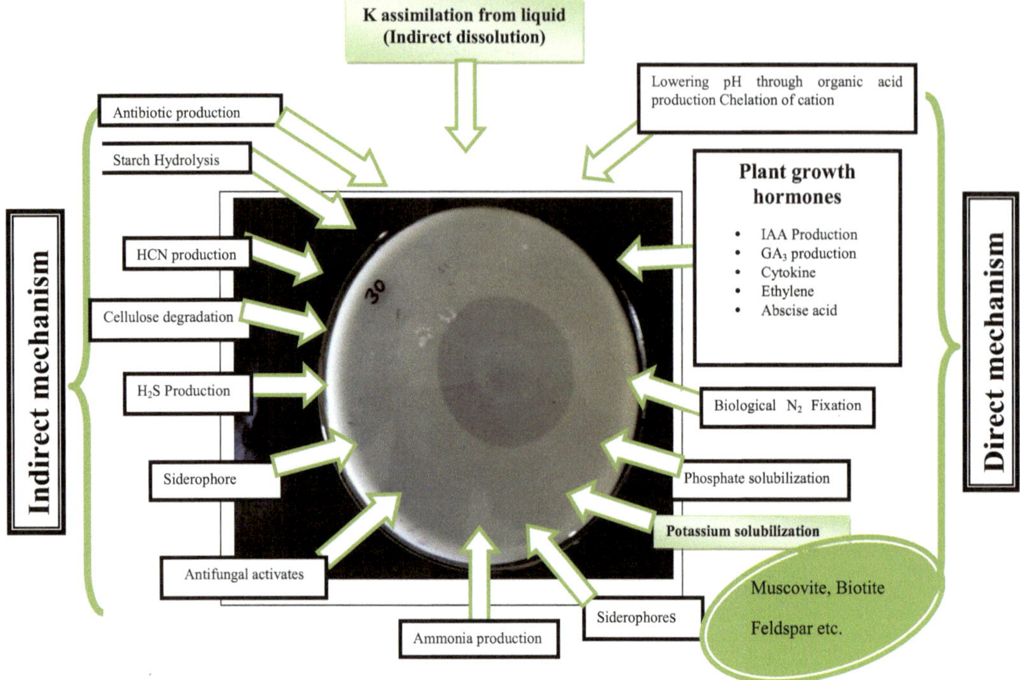

Fig. 1.3 Direct and indirect mechanisms of potassium solubilizing microorganism (KSMs) and their K-Solubilizing ability of mica (K-bearing mineral) on Aleksandrov medium

(Epstein and Kim 1971) and was transformed through empty vector or with plasmids containing the AtKUP1 gene. The production of different organic acids is considered as the principal mechanism for solubilization of mineral phosphate and potassium by microorganism, this assumption has been corroborated by the cloning of two genes which is involved in gluconic acid production, viz., pqq, and gab Y. Gluconic acid is the principal organic acid produced by *Bacillus* spp., *Pseudomonas* spp., *Arthrobacter* spp., *Aspergillus* spp., and *Penicillium* spp. Chelating substances and inorganic acids such as sulphydric, carbonic, and nitric acid are considered as other mechanisms for potassium and phosphate solubilization. Xiufang et al. (2006) identified and characterized potassium-solubilizing bacteria by 16S rDNA gene sequencing analysis using universal primers, the forward primer and the reverse primer; the recombinant plasmids were developed from the cloned products and the relatedness between the strains and were estimated from phylogenetic analysis (Xiufang et al. 2006).

1.7 Effect of Potassium-Solubilizing Microorganisms on Plant Growth and Yield

Inoculation of seeds and seedling treatments of plants with KSMs generally showed significant enhancement of germination percentage, seedling vigor, plant growth, yield, and K uptake by plants in particular, under glasshouse conditions (Singh et al. 2010; Awasthi et al. 2011; Zhang et al. 2013; Zhang and Kong 2014; Subhashini and Kumar 2014). The application of organominerals with a combination of silicate bacteria for enhancing plant growth and yield of maize and wheat was first reported by Aleksandrov (1958). More importantly, research investigation conducted under field level test crops such as wheat, forage crop, maize, and sudan grass crops has revealed that KSMs could drastically reduce the usage of chemical or organic fertilizers (Xie 1998). As reported by previous researchers (Singh et al. 2010; Sindhu et al. 2012; Zeng et al. 2012), the enhancement of plant K nutrition might be due to the stimulation of root growth or the elongation of root hairs by specific microorganisms, thus no direct increase in the availability of soil solution K is expected.

KSMs have been isolated from rhizospheric soil of various plants and from K-bearing mineral (Parmar and Sindhu 2013; Zhang et al. 2013), feldspar (Sheng et al. 2008), potato-soybean cropping sequence (Biswas 2011), Iranian soils (Zarjani et al. 2013), ceramic industry soil (Prajapati and Modi 2012), mica core of Andhra Pradesh (Gundala et al. 2013), common bean (Kumar et al. 2012), biofertilizers (Zakaria 2009), sorghum, maize, bajra, chili (Archana et al. 2013), cotton, tomato, soybean, groundnut, and banana (Archana et al. 2012), soil of Tianmu Mountain, Zhejiang Province (China) (Hu et al. 2006), rice (Muralikannan 1996), tea (Bakyalakshmi et al. 2012), Valencia orange (Shaaban et al. 2012), black pepper (Sangeeth et al. 2012), potato (Abdel-Salam and Shams 2012), thyme (Yadegari et al. 2012), eggplant (Han and Lee 2005), peanut and sesame (Youssef et al. 2010), and tobacco (Subhashini and Kumar 2014). Better crop performance was reported to be achieved from several horticultural plants, vegetables, and cereals, which were successfully inoculated with KSMs (Singh et al. 2010; Basak and Biswas 2012; Prajapati et al. 2013). K-use efficiency in agricultural lands could effectively be improved through the inoculation of relevant KSMs, which is, in fact, an integration and sustainable means of nutrient management of crop production. Enhancement of plant growth by improving N-fixer and P and K-solubilizers is another beneficial effect of microorganisms with K-solubilizing potential (Verma et al. 2010; Basak and Biswas 2012; Meena et al. 2014a; Subhashini and Kumar 2014). A hydroponics study was carried out by Singh et al. (2010) to evaluate the effect of *B. mucilaginosus*, *Rhizobium* spp., and *A. chroococcum* on their capacity to mobilize K from waste mica using wheat and maize as the test crops under a phytotron growth chamber. The significant K assimilation was recorded in wheat and maize, where waste mica was the sole source of K; this has been translated into higher biomass accumulation, K content, and acquisition through plants as well as crude protein and chlorophyll content in plant tissue. Among the rhizobacteria, *B. mucilaginosus* showed significantly greater mobilization of potassium over *Rhizobium* and *A. chroococcum* inoculation. According to Sheng and He (2006), in the investigation of K mobilization by the wild-type strain NBT of *B. edaphicus*, in a pot experiment, wheat has been grown in a yellow-brown soil that contained low available K. After inoculation with bacterial strains, the root growth and shoot growth of wheat were significantly increased and higher NPK contents of plant components as compared to un-inoculated.

Inoculation with KSMs have been reported to exert beneficial effects on growth of cotton and rape (Sheng 2005), pepper and cucumber (Han et al. 2006), khella (Hassan et al. 2010), sorghum (Badr et al. 2006), wheat (Sheng and He 2006), tomato (Lin et al. 2002), chili (Ramarethinam and Chandra 2005), sudan grass (Basak and Biswas 2010), and tobacco (Zhang and Kong 2014). Similarly, Zahra et al. (1984) reported that the effect of soil inoculation of the silicate bacteria *B. circulans* for solubilization of Si and K from various minerals and soil showed significant increase of organic matter and 17 % yield of rice (Muralikannan 1996). Increased wheat yield up to 1.04 t/ha reported by Mikhailouskaya and Tcherhysh (2005) with inoculation of KSMs on several eroded soils and comparable with yields on moderately eroded soil without bacterial inoculation and dry matter production also increased. According to Badar et al. (2006), the co-inoculation of KSMs with P- and K-bearing minerals on sorghum was recorded to enhance dry matter yield (48 %, 65 %, and 58 %), P (71 %, 110 %, and 116 %) and K (41 %, 93 %, and 79 %) uptake in three different soils: sandy, calcareous, and clay soils, respectively. Archana et al. (2008) reported that the KSMs was isolated from rock and rhizosphere soils of greengram (*Vigna radiata*) and reported that these KSMs enhanced the solubilization of K in acid-leached soil as well as increased seedling growth and yield of greengram. Sugumaran and Janarthanam (2007) reported that increased in the oil content

35.4 % and dry matter by 25 % in groundnut crop and available K and P increased from 86.57 to 99.60 mg/kg and 6.24 and 9.28 mg/kg, respectively, in soil by inoculation of *B. mucilaginosus* as compared to un-inoculated control.

According to Archana et al. (2012), the efficient K-solubilizing bacteria *Bacillus* spp. showed increase in growth and yield of maize. It indicates that the KSMs significantly increased yield, plant growth, and nutrient uptake component over absolute fertilizer control. Supanjani et al. (2006) reported that integration of P and K rocks with inoculation of K- and P-solubilizing bacteria increased K availability from 13 % to 15 % and P availability from 12 % to 21 %, respectively. Soil application of KSMs plant has ~16 % photosynthesis and 35 % higher leaf area to control. The overall results of this experiment is the treatment of P and K rocks with P- and K-solubilizing bacterial strain were sustainable and alternative of chemical fertilizer for crop production. Bagyalakshmi et al. (2012) reported that the K-solubilizing stains were isolated from rhizosphere of tea and used as biofertilizers of K in tea that have a solubilizing capacity of muriate of potash (MOP) was increased as compared to mineral K sources. Supplementation of glucose and ammonium nitrate was found to be highly effective in solubilization of MOP as compared to the other sources which should be considered prior to the application of these strains in tea soils as bio-inoculants.

K-solubilizing bacteria are extensively used as biofertilizers in China and Korea in the significant areas of cultivated soils in both of the countries are deficient in plant-available K is considered to be a major limiting factor for food production in many agricultural soils (Xie 1998). Thus the application of K-solubilizing bacteria as biofertilizers for agriculture improvement that can reduce the application of agrochemicals and support eco-friendly crop production (Prajapati et al. 2013; Maurya et al. 2014; Meena et al. 2014a; Zhang and Kong 2014). Therefore, it is imperative to isolate more species of mineral-solubilizing bacteria to enrich the pool of microbial species and genes as microbial fertilizers, which will be of great benefit to the ecological development of agriculture (Liu et al. 2012).

1.8 Future Prospect of Potassium-Solubilizing Microorganisms

Potassium-solubilizing microorganisms play an important role in plant nutrition that enhance the K acquisition of plants through soil which increase plant growth promotion activities, these KSMs contribution important role as to bio-fertilization of agricultural crops. Accordingly, further investigation is required to improve the performance and use of potassium-solubilizing microorganism as efficient microbial bio-inoculants. The greater attention is needed for studies and application of new efficient combinations of potassium-solubilizing microorganisms and other plant growth-promoting microorganisms for improved results. The mechanisms explaining the synergistic interaction among KSMs required further research to elucidate the biochemical basis of these interactions. On the other hand, the application of biotechnological tools for genetic manipulation of potassium-solubilizing microorganism that increase their potassium-solubilizing efficiency/ability/capabilities and/or the insertion of this trait into other strains of plant growth-promoting effects is not only crucial but also causes it to be practically feasible. In addition to application of classical genetic methods to mutants, the strains that increased the organic acids production which could constitute an effective approach that cannot be underestimated. The genetic manipulation through recombinant DNA technology use to a feasible approach for enhanced the strains efficiency. Gene cloning approaches involved in mineral potassium solubilization, such as those influencing the synthesis of different organic acids, would be the first step in such a genetic manipulation program.

Future research work should also investigate or improve the performance and stability of the potassium solubilization trait once the microorganisms have been inoculated in soil as in both genetically modified and natural strains. The survival efficiency and their establishment of the inserted strains can be affected by little competitiveness as a result limiting the effectiveness of application. However, the putative risk rose during the release of genetically modified microorganisms in soil which is an important matter of controversy, in this regard to the possibility of horizontal transfer of the introduced DNA to many other soil microorganisms. For these reasons, the application of genetic reporter systems, such as green fluorescent protein genes or bioluminescence genes, are important in studying the survival and fate of the strain in soil. The genetic engineering of the potassium-solubilizing character must be eventually required for the chromosomal integration of the gene for greater stability of the character and also to avoid horizontal transfer of the introduced gene in soil. This strategy would also decrease/prevent the risk of metabolic load that showed the presence of the plasmid in the bacterial cell. The chromosomal integration may also have disadvantage for little expression activity, due to the small copy number of the gene as compared to the plasmid-harbored genes. An alternative approach to this situation might be due to the integration of multicopies of the target gene. In addition to these, the application of major and species-specific promoters, which have been activated under certain specific environmental conditions of soil and another interesting approach for successful gene expression, is engineered strain.

Acknowledgment Authors are thankful to the Head of the Department of Soil Science and Agricultural Chemistry, Institute of Agricultural Sciences, Banaras Hindu University, Varanasi, for providing all the necessary facility to conduct this experiment. VSM is thankful to the University Grants Commission (UGC), New Delhi, for the fellowship during his Ph.D. work and SKM is thankful to the Indian Council of Agricultural Research (ICAR), New Delhi, and Government of India (GOI) for the junior research fellowship (JRF) during her study.

References

Abou-el-Seoud II, Abdel-Mageed A (2012) Impact of rock materials and biofertilization on P and K availability for maize (*Zea maize*) under calcareous soil conditions. Saudi J Biol Sci 19:55–63

Aleksandrov VG (1958) Organo-mineral fertilizers and silicate bacteria. Dokl Akad Nauk 7:43–48

Aleksandrov V, Blagodyr R, Ilev I (1967) Liberation of phosphoric acid from apatite by silicate bacteria. Mikrobiol Z 29:111–114

Alves L, Oliveira VL, Filho GNS (2010) Utilization of rocks and ectomycorrhizal fungi to promote growth of eucalypt. Braz J Microbiol 41:76–84

Archana DS, Nandish MS, Savalagi V, Alagawadi A (2013) Characterization of potassium solubilizing bacteria (KSB) from rhizosphere soil. Bioinfolet 10:248–257

Argelis DT, Gonzala DA, Vizcaino C, Gartia MT (1993) Biochemical mechanism of stone alteration carried out by filamentous fungi living in monuments. Biogeochemistry 19:129–147

Ashley DL, Blount B, Singer PC, Depaz E, Wilkes C, Gordon S (2005) Changes in blood trihalomethane concentrations resulting from differences in water quality and water-use activities. Arch Environ Occup Health 60(1):7–15

Askegaard M, Eriksen J, Johnston AE (2004) Sustainable management of potassium. In: Schjorring P, Elmholt S, Christensen BT (eds) Managing soil quality: challenges in modern agriculture. CABI Publishing, Wallingford, pp 85–102

Awasthi R, Tewari R, Nayyar H (2011) Synergy between plants and P-solubilizing microbes in soils: effects on growth and physiology of crops. Int Res J Microbiol 2: 484–503

Badar MA (2006) Efficiency of K feldspar combined with organic material and silicate dissolving bacteria on tomato yield. J Appl Sci Res 2:1191–1198

Bagyalakshmi B, Ponmurugan P, Marimuthu S (2012) Influence of potassium solubilizing bacteria on crop productivity and quality of tea (*Camellia sinensis*). Afr J Agric Res 7:4250–4259

Basak BB, Biswas DR (2009) Influence of potassium solubilizing microorganism (*Bacillus mucilaginosus*) and waste mica on potassium uptake dynamics by sudan grass (*Sorghum vulgare* Pers.) grown under two Alfisols. Plant and Soil 317:235–255

Basak BB, Biswas DR (2010) Co-inoculation of potassium solubilizing and nitrogen fixing bacteria on solubilization of waste mica and their effect on growth promotion and nutrient acquisition by a forage crop. Biol Fertil Soils 46:641–648

Basak B, Biswas D (2012) Modification of waste mica for alternative source of potassium: evaluation of potassium release in soil from waste mica treated with potassium solubilizing bacteria (KSB). LAMBERT Academic Publishing, Germany. ISBN 978-3659298424

Bin L, Bin W, Mu P, Liu C, Teng HH (2010) Microbial release of potassium from K-bearing minerals by thermophilic fungus *Aspergillus fumigatus*. Geochim Cosmochim Acta 72:87–98

Biswas DR (2011) Nutrient recycling potential of rock phosphate and waste mica enriched compost on crop productivity and changes in soil fertility under potato–soybean cropping sequence in an Inceptisol of Indo-

Gangetic Plains of India. Nutr Cycl Agroecosyst 89: 15–30
Chen S, Lian B, Liu CQ (2008) *Bacillus mucilaginosus* on weathering of phosphorite and primary analysis of bacterial proteins during weathering. Chin J Geochem 27:209–216
Clark RB, Zeto SK (1996) Growth and root colonization of mycorrhizal maize grown on acid and alkaline soil. Soil Biol Biochem 28:1505–1511
Clark RB, Zeto SK (2000) Mineral acquisition by arbuscular mycorrhizal plants. J Plant Nutr 23: 867–902
Clark RB, Zobel RW, Zeto SK (1999) Effects of mycorrhizal fungus isolate on mineral acquisition by *Panicum virgatum* in acidic soil. Mycorrhiza 9:167–176
Claus D, Berkeley CW (1986) The genus *Bacillus*. In: PHA Sneath (ed) Bergey's manual of systematic bacteriology, vol 2. Williams, Wilkins, Baltimore. 34, 1105–1139
Das BK, Sen SP (1981) Effect of nitrogen, phosphorus and potassium deficiency on the uptake and mobilization of ions in Bengal gram (*Cicer arietinum*). J Biosci 3:249–258
Domínguez-Ferreras A, Munoz S, Olivares J, Soto MJ, Sanjuan J (2009) Role of potassium uptake systems in *Sinorhizobium meliloti* adaptation and symbiotic performance. J Bacteriol 21:33–43
Egamberdiveya D (2006) Enhancement of wheat performance with plant growth promoting bacteria in different soils. In: Mukerji KG, Manoharachary C (eds) Current concepts in botany. International Publishing House Ltd, New Delhi, pp 417–425
Epstein W (2003) The roles and regulation of potassium in bacteria. Prog Nucleic Acid Res Mol Biol 75: 293–320
Epstein W, Kim BS (1971) Potassium transport loci in Escherichia coli K-12. J Bacteriol 108:639–644
Foth HD, Ellis BG (1997) Soil fertility. CRC Press, Boca Raton, p 290
Gahoonia TS, Care D, Nielsen NE (1997) Root hairs and phosphorus acquisition of wheat and barley cultivars. Plant and Soil 191:181–188
Goldstein AH (1994) Involvement of the quinoprotein glucose dehydrogenase in the solubilization of exogenous phosphates by gram-negative bacteria. Phosphate in microorganisms: cellular and molecular biology. ASM Press, Washington, DC, pp 197–203
Groudev SN (1987) Use of heterotrophic microorganisms in mineral biotechnology. Acta Biotechnol 7:299–306
Gundala PB, Chinthala P, Sreenivasulu B (2013) A new facultative alkaliphilic, potassium solubilizing, *Bacillus* spp. SVUNM9 isolated from mica cores of Nellore district, Andhra Pradesh, India. J Microbiol Biotechnol 2(1):1–7
Han HS, Supanjani E, Lee KD (2006) Effect of co-inoculation with phosphate and potassium solubilizing bacteria on mineral uptake and growth of pepper and cucumber. Plant Soil Environ 52(3): 130–136
Hasan R (2002) Potassium status of soils in India. Better Crops 16(2):3–5
Hassan EA, Hassan EA, Hamad EH (2010) Microbial solubilization of phosphate-potassium rocks and their effect on khella (*Ammi visnaga*) growth. Annu Agric Sci 55:37–53
Hillel M (2008) balanced crop nutrition: fertilizing for crop and food quality. Turk J Agric For 32:183–193
Holthusen D, Peth S, Horn R (2010) Impact of potassium concentration and matric potential on soil stability derived from rheological parameters. Soil Tillage Res 111:75–85
Hu X, Chen J, Guo J (2006) Two phosphate and potassium solubilizing bacteria isolated from Tianmu Mountain, Zhejiang, China. World J Microbiol Biotechnol 22:983–990
Kafkafi U (1990) The functions of plant K in overcoming environmental stress situations. In: Development of K-fertilizer recommendations: proceedings 22nd colloquium of the International Potash Institute, Bern, Switzerland, pp 81–93
Kumar P, Dubey R, Maheshwari D (2012) *Bacillus* strains isolated from rhizosphere showed plant growth promoting and antagonistic activity against phytopathogens. Microbiol Res 167:493–499
Kumar A, Bahadur I, Maurya BR, Raghuwanshi R, Meena VS, Singh DK, Dixit J (2015) Does a plant growth-promoting rhizobacteria enhance agricultural sustainability? J Pur Appl Microbiol 9(1):715–724
Lack A, Evans J, David E (2005) Bios instant notes plant biology. Taylor & Francis, New York/Abingdon, pp 351
Leyval C, Berthelin J (1989) Interaction between *Laccaria laccata*, *Agrobacterium radiobacter* and beech roots: influence on P, K, Mg and Fe mobilization from minerals and plant growth. Plant and Soil 117:103–110
Lian B, Fu PQ, Mo DM, Liu CQ (2002) A comprehensive review of the mechanism of potassium release by silicate bacteria. Acta Mineral Sin 22:179–183
Lian B, Wang B, Pan M, Liu C, Teng HH (2008) Microbial release of potassium from K-bearing minerals by thermophilic fungus *Aspergillus fumigatus*. Geochim Cosmochim Acta 72:87–98
Lin Q, Rao Z, Sun Y, Yao J, Xing L (2002) Identification and practical application of silicate-dissolving bacteria. Agric Sci China 1:81–85
Liu GY (2001) Screening of silicate bacteria with potassium releasing and antagonistic activity. Chin J Appl Environ Biol 7:66–68
Liu D, Lian B, Dong H (2012) Isolation of *Paenibacillus* spp. and assessment of its potential for enhancing mineral weathering. Geomicrobiol J 29(5):413–421
Maurya BR, Meena VS, Meena OP (2014) Influence of inceptisol and Alfisol's potassium solubilizing bacteria (KSB) isolates on release of K from waste mica. Vegetos 27(1):181–187

McAfee J (2008) Potassium, a key nutrient for plant growth. Department of Soil and Crop Sciences: http://jimmcafee.tamu.edu/files/potassium

Meena OP, Maurya BR, Meena VS (2013) Influence of K-solubilizing bacteria on release of potassium from waste mica. Agric Sustain Dev 1(1):53–56

Meena VS, Maurya BR, Bahadur I (2014a) Potassium solubilization by bacterial strain in waste mica. Bangladesh J Bot 43(2):235–237

Meena VS, Maurya BR, Verma JP (2014b) Does a rhizospheric microorganism enhance K+ availability in agricultural soils? Microbiol Res 169:337–347

Meena RK, Singh RK, Singh NP, Meena SK, Meena VS (2015a) Isolation of low temperature surviving plant growth-promoting rhizobacteria (PGPR) from pea (*Pisum sativum* L.) and documentation of their plant growth promoting traits. Biocatal Agric Biotechnol. doi:10.1016/j.bcab.2015.08.006

Meena VS, Maurya BR, Verma JP, Aeron A, Kumar A, Kim K, Bajpai VK (2015b) Potassium solubilizing rhizobacteria (KSR): isolation, identification, and K-release dynamics from waste mica. Ecol Eng 81: 340–347

Mikhailouskaya N, Tcherhysh A (2005) K-mobilizing bacteria and their effect on wheat yield. Latv J Agron 8:154–157

Mirminachi F, Zhang A, Roehr M (2002) Citric acid fermentation and heavy metal ions. Acta Biotechnol 22:363–373

Mora V, Baigorri R, Bacaicoa E, Zamarreñob AM, García-Mina JM (2012) The humic acid-induced changes in the root concentration of nitric oxide, IAA and ethylene do not explain the changes in root architecture caused by humic acid in cucumber. Environ Exp Bot 76:24–32

Muentz (1890) Surla décomposition desroches etla formation de la terrearable. CR Acad Sci 110:1370–1372

Muralikannan N (1996) Biodissolution of silicate, phosphate and potassium by silicate solubilizing bacteria in rice ecosystem. M.Sc. (Ag) thesis submitted to Tamil Nadu Agricultural University, Coimbatore. p 125

Nieves-Cordones M, Aleman F, Martinez V, Rubio F (2014) K^+ uptake in plant roots. The systems involved their regulation and parallels in other organisms. J Plant Physiol 171:688–695

Oborn I, Andrist-Rangel Y, Askekaard M, Grant CA, Watson CA, Edwards AC (2005) Critical aspects of potassium management in agricultural systems. Soil Use Manag 21:102–112

Prajapati K, Modi H (2012) Isolation and characterization of potassium solubilizing bacteria from ceramic industry soil. CIB Technol J Microbiol 1:8–14

Prajapati K, Sharma MC, Modi HA (2013) Growth promoting effect of potassium solubilizing microorganisms on *Abelmoscus esculantus*. Int J Agric Sci 3:181–188

Rajawat MVS, Singh S, Singh G, Saxena AK (2012) Isolation and characterization of K-solubilizing bacteria isolated from different rhizospheric soil. In: Proceeding of 53rd annual conference of association of microbiologists of India, p 124

Ramamurthy B, Bajaj JC (1969) Soil fertility map of India. Indian Agricultural Research Institute, Annual report New Delhi

Ramarethinam S, Chandra K (2005) Studies on the effect of potash solubilizing/mobilizing bacteria *Frateuria aurantia* on brinjal growth and yield. Pestology 11: 35–39

Rehm G, Schmitt M (2002) Potassium for crop production. University of Minnesota Extension, www.extension.umn.edu/distribution/cropsystems. 46, pp 229–236. doi 10.1139/cjm-46-3-229

Rich CI (1968) Mineralogy of soil potassium. In: Kilmer VJ et al (eds) The role of potassium in agriculture. ASA, CSSA, SSSA, Madison, pp 79–108

Romheld V, Kirkby EA (2010) Research on potassium in agriculture: needs and prospects. Plant and Soil 335: 155–180

Sangeeth KP, Bhai RS, Srinivasan V (2012) *Paenibacillus glucanolyticus*, a promising potassium solubilizing bacterium isolated from black pepper (*Piper nigrum* L.) rhizosphere. J Spice Aromat Crops 21:118–124

Schiavon M, Pizzeghello D, Muscolo A, Vaccoro S, Francioso O, Nardi S (2010) High molecular size humic substances enhance phylpropanoid metabolism in maize (*Zea mays* L.). J Chem Ecol 36:662–669

Shaaban EA, El-Shamma IMS, El Shazly S, El-Gazzar A, Abdel-Hak RE (2012) Efficiency of rock-feldspar combined with silicate dissolving bacteria on yield and fruit quality of valencia orange fruits in reclaimed soils. J Appl Sci Res 8:4504–4510

Sharpley AN (1989) Relationship between soil potassium forms and mineralogy. Soil Sci Soc Am J 52: 1023–1028

Sheng XF (2002) Study on the conditions of potassium release by strain NBT of silicate bacteria scientia. Agric Sin 35(6):673–677

Sheng XF (2005) Growth promotion and increased potassium uptake of cotton and rape by a potassium releasing strain of *Bacillus edaphicus*. Soil Biol Biochem 37:1918–1922

Sheng XF, He LY (2006) Solubilization of potassium bearing minerals by a wild type strain of *Bacillus edaphicus* and its mutants and increased potassium uptake by wheat. Can J Microbiol 52:66–72

Sheng XF, Huang WY (2002) Mechanism of potassium release from feldspar affected by the strain NBT of silicate bacterium. Acta Pedol Sin 39(6):863–871

Sheng X, He L, Huang W (2001) The conditions of releasing potassium by a silicate dissolving bacterial strain NBT. Agric Sci China 1(6):662–666

Sheng XF, He LY, Huang W (2002) The conditions of releasing potassium by a silicate dissolving bacterial strain NBT. Agric Sci China 1:662–666

Sheng XF, Xia JJ, Chen J (2003) Mutagenesis of the *Bacillus edaphicus* strain NBT and its effect on growth of chilli and cotton. Agric Sci China 2: 400–412

Sheng XF, Zhao F, He H, Qiu G, Chen L (2008) Isolation, characterization of silicate mineral solubilizing *Bacillus globisporus* Q12 from the surface of weathered feldspar. Can J Microbiol 54:1064–1068

Sindhu SS, Dua S, Verma MK, Khandelwal A (2010) Growth promotion of legumes by inoculation of rhizosphere bacteria. In: Khan MS, Zaidi A, Musarrat J (eds) Microbes for legume improvement. Springer-Wien, New York, pp 195–235

Sindhu SS, Parmar P, Phour M (2012) Nutrient cycling: potassium solubilization by microorganisms and improvement of crop growth. In: Parmar N, Singh A (eds) Geomicrobiology and biogeochemistry: soil biology. Springer-Wien, New York

Singh G, Biswas DR, Marwah TS (2010) Mobilization of potassium from waste mica by plant growth promoting rhizobacteria and its assimilation by maize (*Zea mays*) and wheat (*Triticum aestivum* L.). J Plant Nutr 33: 1236–1251

Singh NP, Singh RK, Meena VS, Meena RK (2015) Can we use maize (*Zea mays*) rhizobacteria as plant growth promoter? Vegetos 28(1):86–99

Sleator RD, Hill C (2002) Bacterial osmoadaptation: the role of osmolytes in bacterial stress and virulence. FEMS Microbiol Rev 26:49–71

Sparks DL (1987) Potassium dynamics in soils. Adv Soil Sci 6:1–63

Srinivasrao CH, Satyanarayana T, Venkateswarulu B (2011) Potassium mining in Indian agriculture: input and output balance. Karnataka J Agric Sci 24:20–28

Subhashini DV, Kumar AV (2014) Phosphate solubilizing *Streptomyces* spp. obtained from the rhizosphere of *Ceriops decandra* of Corangi mangroves. Indian J Agric Sci 84(5):560–564

Sugumaran P, Janarthanam B (2007) Solubilization of potassium containing minerals by bacteria and their effect on plant growth. World J Agric Sci 3(3):350–355

Supanjani, Han HS, Jung SJ, Lee KD (2006) Rock phosphate potassium and rock solubilizing bacteria as alternative sustainable fertilizers. Agron Sustain Dev 26:233–240

Uroz S, Calvaruso C, Turpault MP, Pierrat JC, Mustin C, Frey-Klett P (2007) Effect of the mycorrhizosphere on the genotypic and metabolic diversity of the bacterial communities involved in mineral weathering in a forest soil. Appl Environ Microbiol 73:3019–3027

Uroz S, Calvaruso C, Turpault MP, Frey-Klett P (2009) Mineral weathering by bacteria: ecology, actors and mechanisms. Trends Microbiol 17:378–387

Veresoglou SD, Mamolos AP, Thornton B, Voulgari OK, Sen R, Veresoglou S (2011) Medium-term fertilization of grassland plant communities masks plant species-linked effects on soil microbial community structure. Plant and Soil 344:187–196

Verma JP, Yadav J, Tiwari KN, Lavakush, Singh V (2010) Impact of plant growth promoting rhizobacteria on crop production. Int J Agric Res 5: 954–983

Verma JP, Yadav J, Tiwari KN, Jaiswal DK (2014) Evaluation of plant growth promoting activities of microbial strains and their effect on growth and yield of chickpea (*Cicer arietinum* L.) in India. Soil Biol Biochem 70:33–37

Wu SC, Cao ZH, Li ZG, Cheung KC, Wong MH (2005) Effects of biofertilizer containing N-fixer, P and K solubilizers and AM fungi on maize growth: a greenhouse trial. Geoderma 125:155–166

Xie JC (1998) Present situation and prospects for the world's fertilizer use. Plant Nutr Fertil Sci 4:321–330

Xiufang H, Jishuang C, Jiangfeng G (2006) Two phosphate and potassium solubilizing bacteria isolated from Tianmu Mountain Zhejiang, China. World J Microbiol Biotechnol 22:983–990

Yadegari M, Farahani GHN, Mosadeghzad Z (2012) Biofertilizers effects on quantitative and qualitative yield of Thyme (*Thymus vulgaris*). Afr J Agric Res 7:4716–4723

Yousefi AA, Khavazi K, Moezi AA, Rejali F, Nadian NH (2011) Phosphate solubilizing bacteria and arbuscular mycorrhizal fungi impacts on inorganic phosphorus fractions and wheat growth. World Appl Sci J 15(9): 1310–1318

Youssef GH, Seddik WMA, Osman MA (2010) Efficiency of natural minerals in presence of different nitrogen forms and potassium dissolving bacteria on peanut and sesame yields. Am J Sci 6:647–660

Zahra MK, Monib MS, Abdel-Al I, Heggo A (1984) Significance of soil inoculation with silicate bacteria. Zentralbl Mikrobiol 139(5):349–357

Zandonadi DB, Santos MP, Dobbss LB, Olivares FL, Canellas LP, Binzel ML, Okorokova-Façanha AL, Façanha AR (2010) Nitric oxide mediates humic acids-induced root development and plasma membrane H^+-ATPase activation. Planta 231:1025–1036

Zarjani JK, Aliasgharzad N, Oustan S, Emadi M, Ahmadi A (2013) Isolation and characterization of potassium-solubilizing bacteria in some Iranian soils. Arch Agron Soil Sci 59:1713–1723

Zhang C, Kong F (2014) Isolation and identification of potassium-solubilizing bacteria from tobacco rhizospheric soil and their effect on tobacco plants. Appl Soil Ecol 82:18–25

Zhang A, Zhao G, Gao T, Wang W, Li J, Zhang S (2013) Solubilization of insoluble potassium and phosphate by *Paenibacillus kribensis* CX-7: a soil microorganism with biological control potential. Afr J Microbiol Res 7(1):41–47

Zheng C, Jiang D, Liub F, Dai T, Liu W, Jing Q, Cao W (2009) Exogenous nitric oxide improves seed germination in wheat against mitochondrial oxidative damage induced by high salinity. Environ Exp Bot 67(1): 222–227

Potassium as an Important Plant Nutrient in Sustainable Agriculture: A State of the Art

2

Durgesh Kumar Jaiswal, Jay Prakash Verma, Satya Prakash, Vijay Singh Meena, and Ram Swaroop Meena

Abstract

The current scenario of potassium (K) depletion in soil is slowly increasing due to K fixation or the unavailable form of K in soil. Presently, farmers are faced with a problem of higher price of K fertilizer or other fertilizers in market so farmers are unable to fulfill the demand of potassium in soil for plant growth. Potassium deficiency affects the nutritional quality, mechanical stability, and also pathogen resistance of crops. Therefore, that times needs to fallow the sustainable technology for sustainable agricultural production through use of microbial consortia of potassium-solubilizing microbes or biofertilizer/PGPR under organic farming system. The potassium-solubilizing microorganism is one of the best sustainable technologies, which solubilizes the fixed form of K available for plant uptake. Thus, the bio-formula of the potassium-solubilizing microorganism as biofertilizer offers environmentally sustainable approach and also fulfills the requirement of potassium for crop production.

Keywords

Potassium • Potassium-solubilizing microorganisms (KSMs) • Microbial diversity • Sustainable agriculture

D.K. Jaiswal • J.P. Verma (✉)
Institute of Environment and Sustainable Development, Banaras Hindu University, Varanasi 22100, Uttar Pradesh, India
e-mail: verma_bhu@yahoo.co.in; jpv.iesd@bhu.ac.in

S. Prakash
Geological Survey of India, Northern Region, Aliganj, Lucknow, India

V.S. Meena
Crop Production Division, Vivekananda Institute of Hill Agriculture, (ICAR), Almora 263 601, India

R.S. Meena
Department of Agronomy, Institute of Agricultural Sciences, Banaras Hindu University, Varanasi 221005, U.P., India

2.1 Introduction

Potassium is the third major essential macronutrient and most abundantly absorbed cation that plays an important role in the growth, metabolism, and development of plants. Without adequate potassium, the plants will have poorly developed roots, grow slowly, produce small seeds, and have lower yields. The concentrations of soluble potassium in the soil are usually very

V.S. Meena et al. (eds.), *Potassium Solubilizing Microorganisms for Sustainable Agriculture*,
DOI 10.1007/978-81-322-2776-2_2

low and more than 90 % of potassium in the soil exists in the form of insoluble form of silicate (Meena et al. 2015b). In the present study, soil K is divided into water soluble, exchangeable, HNO_3-extractable, and mineral K forms. Water-soluble K is taken up directly by plant, although amounts found in soils are generally small. Exchangeable K, determined as that replaceable by NH_4 from NH_4OAc, is held by negative charges on organic matter and clay particles and is readily available to plants (Arnold 1958; Conyers and McLean 1969). Igneous rocks of the Earth's crust have higher K contents as compared to sedimentary rocks. Of the igneous rocks, granites and syenites contain 46–54 g kg^{-1} K, basalts contain 7 g kg^{-1}, and peridotites contain 2.0 g kg^{-1}. Among the sedimentary rocks, clayey shales contain 30 g of K per kilogram, whereas limestones have an average of only 6 g kg^{-1} (Malavolta 1985). Mineral soils generally range between 0.04 and 3 % K. Total K contents in soils range between 3,000 and 100,000 kg ha^{-1} in the upper 0.2 m of the soil profile. Of this total K content, 98 % is bound in the mineral form, whereas approximately 2 % K is in soil solution as exchangeable phases (Schroeder 1979; Bertsch and Thomas 1985). Potassium is involved in numerous biochemical and physiological processes in plants like stomatal regulation for plants depending upon K to regulate the opening and closing of stomata. Proper functioning of stomata is essential for photosynthesis (Shanware et al. 2014). Potassium also plays a major role in the transport of water and nutrients in the plant through the xylem. The enzyme responsible for synthesis of starch (starch synthetase) is activated by K; hence, it plays a crucial role in water and nutrient transport (Pettigrew 2008). Due to deficiency of K, severe loss was faced in yield and quality of crop production. Its role is known in improving shelf life of crops and disease resistance (Khawilkar and Ramteke 1993).

The current scenario of the K level in soil is depleted in faster rate, due to the introduction of high-yielding crop varieties during green revolution and intensification of agriculture (Sheng et al. 2008). Consequently, potassium deficiency is becoming one of the major constraints in crop production, especially in coarse-textured soils and even in fine-textured soils. Therefore, crops do respond to K fertilization in soils (Meena et al. 2013; Maurya et al. 2014; Sindhu et al. 2014). Therefore, application of potassium fertilizer to the soil gives positive response. In India, there is no reserve of K-bearing minerals for production of commercial K fertilizers, so that the whole consumption of K fertilizers is imported in the form of muriate of potash (KCl) and sulfate of potash (K_2SO_4) (Sindhu et al. 2014). Due to increasing demand of potassium fertilizer, India ranks fourth in consumption of potassium fertilizers after the USA, China, and Brazil as far as the total consumption of K fertilizers in the world is concerned (FAI 2007). As a result of the extensive use of chemical fertilizer, it causes a negative impact on economic and environmental sustainability (Xie 1998). This leads to the focus and to the exploration of an alternative indigenous resource of low-grade K minerals (such as feldspar and waste mica) for plant uptake and to maintain the K status in soils for sustaining crop production. The use of plant growth-promoting microorganisms as biofertilizers was reported as an alternative solution for improving plant growth and nutrition (Vessey 2003). There are some bacteria in the soil which are capable of solubilizing K-bearing minerals to bring K into an available form. These bacteria are specially known as K-solubilizing bacteria (KSB). Some of the selected plant growth-promoting microorganisms such as *Pseudomonas* spp., *Acidithiobacillus ferrooxidans*, *Bacillus mucilaginosus*, *B. edaphicus, and B. megaterium* have shown potassium solubilization of K-bearing minerals (Sheng et al. 2008; Meena et al. 2014b; Zhang and Kong 2014).

2.2 Forms of Soil Potassium

The accessibility of K depends on soil type and is affected by physicochemical properties of the soil. Soil K exists in soils as in four forms: solution, exchangeable, fixed or non-exchangeable, and structural or mineral (Fig. 2.1). Exchangeable K and non-exchangeable K levels comprise a small

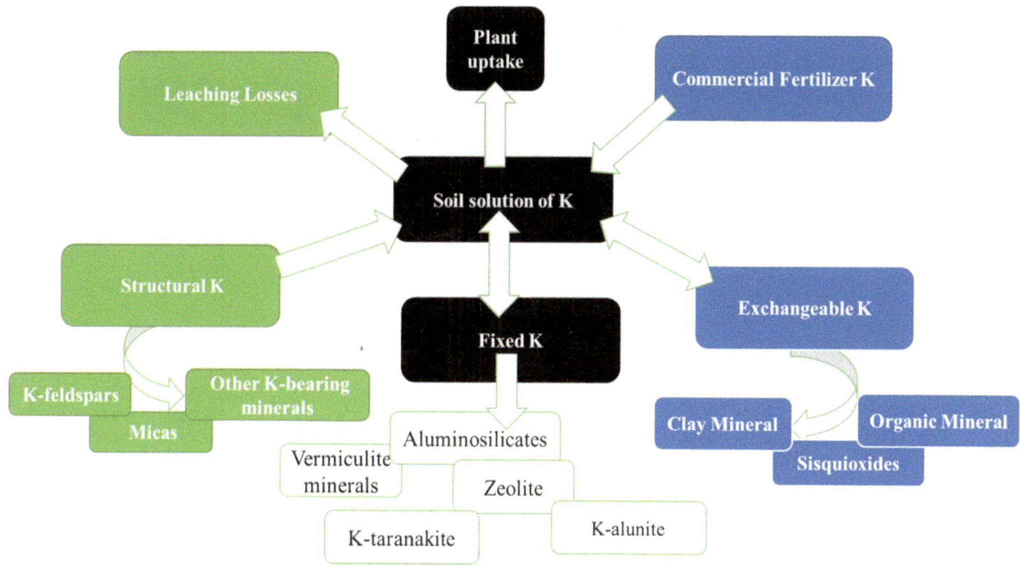

Fig. 2.1 Interrelationships of various forms of soil K (Modified from Sparks and Huang 1985)

portion of the total K. The bulk of total soil K is in the mineral fraction (Sparks and Huang 1985). The forms of soil K in the order of their availability to plants and microbes are solution > exchangeable > fixed (non-exchangeable) > mineral (Sparks and Huang 1985; Sparks 2000). Soil solution K is the form of K that is directly taken up by plants and microbes and also is the form most subject to leaching in soils (Sparks et al. 1996). The quantity of K in the soil solution varies from 2 to 5 mg K L^{-1} for normal agricultural soils of humid regions and is an order of magnitude higher in arid region soils (Haby et al. 1990). Levels of solution K are affected by the equilibrium and kinetic reactions that occur between the forms of soil K, the soil moisture content, and the concentrations of bivalent cations in solution and on the exchanger phase (Sparks and Huang 1985; Sparks 2000).

2.3 Potassium-Solubilizing Microorganisms (KSMs)

Some plant growth-promoting microorganism has been reported to play a key role in the natural potassium cycle. A wide range of bacteria, namely, *Pseudomonas*, *Burkholderia*, *Acidithiobacillus ferrooxidans*, *Bacillus mucilaginosus*, *Bacillus edaphicus*, *B. circulans*, and *Paenibacillus* sp., have been found to release potassium in an accessible form from potassium-bearing minerals in soils. These potassium-solubilizing bacteria (KSB) were found to dissolve potassium, silicon, and aluminum from insoluble K-bearing minerals such as micas, illite, and orthoclases, by excreting organic acids which either directly dissolved rock K or chelated silicon ions to bring K into the solution (Sheng and He 2006). Therefore, the application of K-solubilizing microorganisms is a promising approach for increasing K availability in soils. Scientists from China and India have reported that pH near neutral is the best chemical state for specific bacteria to solubilize micaceous minerals, making K available to the plant root for absorption (Meena et al. 2015). In a study by Indian scientists, Parmar and Sindhu (2013) and Meena et al. (2015a) isolated different K-solubilizing bacteria (KSB) from rhizosphere soil samples collected from rabi and kharif crop. They found that the bacteria were more efficient and able to solubilize K at 28 °C with decreasing pH. This indicates that soils must be warmed up considerably by summer temperatures to have bacteria active and increase in population for effectively solubilized minerals. Therefore, the use of K-solubilizing bacteria as biofertilizer for agriculture improvement and environmental protection has been focused in recent research (Sheng et al. 2008; Meena et al. 2014a; Singh et al. 2015).

2.4 Mechanisms Involved in K Solubilization

Solubilization of K-bearing minerals (micas, illite, and orthoclases) by microorganisms was reported due to the lowering of pH, by enhancing chelation of the cations bound to K; the production of organic acids like citric, oxalic, and tartaric; and also due to the production of capsular polysaccharides which helped in dissolution of minerals to release potassium (Fig. 2.2) (Sheng and He 2006; Liu et al. 2006; Parmar and Sindhu 2013). Organic acids can directly enhance dissolution by either a proton- or ligand-mediated mechanism. They can also indirectly enhance dissolution by the formation of complexes in solution with reaction products and as a consequence increase the chemical affinity for the overall dissolution (Ullman and Welch 2002; Meena et al. 2014b; Kumar et al. 2015). Production of carboxylic acids like citric, tartaric, and oxalic acids was associated with feldspar solubilization by *B. mucilaginosus* and *B. edaphicus* (Malinovskaya et al. 1990; Sheng and Huang 2002a). Silicate bacteria were found to dissolve potassium, silica, and aluminum from insoluble minerals by liberation of phosphoric acids that solubilized apatite and released available form of nutrients from apatite (Heinen 1960). Burford et al. (2003) isolated several fungal isolates having the potential to release metal ions and silicate ions from minerals, rocks, and soils. These fungal isolates were found to produce citric acid and oxalic acid that are mainly known to decompose or solubilize natural silicates and help in removal of metal ions from the rocks and soils (Gadd 1999). Thus, production of

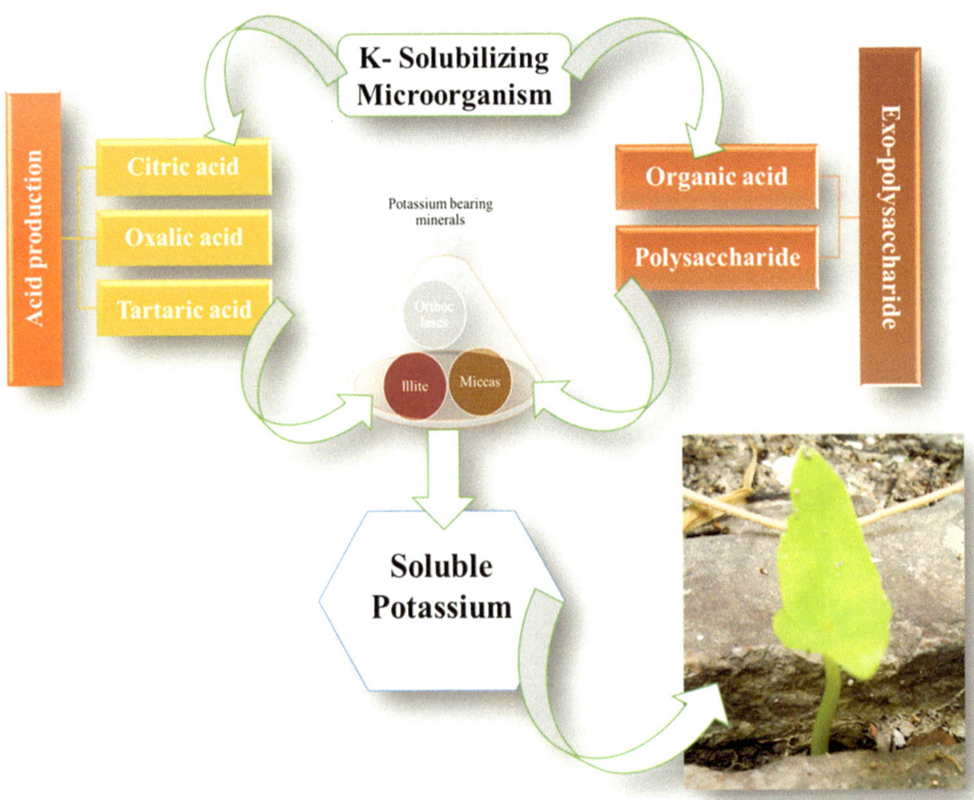

Fig. 2.2 Mechanism of K – solubilization by the potassium-solubilizing microorganisms

organic acids such as acetate, citrate, and oxalate by microorganisms was found to increase mineral dissolution rate (Barker et al. 1998). In another study it was found that production of slime or acidic exopolysaccharides (EPS) contributed to the mechanism of releasing potassium from silicates (Groudev (1987). Liu et al. (2006) found that polysaccharides strongly adsorbed the organic acids and attached to the surface of the mineral, resulting in an area of high concentration of organic acids near the mineral. It was suggested that the extracellular polysaccharides adsorbed SiO_2, and this affected the equilibrium between the mineral and fluid phases and led to the reaction toward SiO_2 and K^+ solubilization.

2.5 Function of Potassium in Plant

Unlike N, P, and most of the other essential nutrients, K does not become a part of the chemical structure of the plant. Thus, its mobility in the plant allows it to influence almost all aspects of plant growth. Many questions arise during the study on role of potassium in plants, regarding how potassium works to increase crop yield through the following fundamental physiological processes such as increasing the root growth, activating many enzyme system, maintaining turgor, reducing water loss and wilting, photosynthesis and food formation, reducing respiration, preventing energy losses, enhancing translocation of sugars and starch, increasing grain rich in starch, building cellulose, reducing lodging, and helping retard crop diseases (Clarkson and Hanson 1980; Carraretto et al. 2013; Mengel 1980; Pettigrew 2008; Zörb et al. 2014). Consequently, a number studies found that a sufficient level of K supply in plants by application of potassium-solubilizing microorganism or natural resources may enhance tolerance to various environmental stresses (salt, drought, cold, high light, and pathogen) (Wang and Wu 2015) (Fig. 2.3). Worldwide, one third of the irrigated lands are facing a salinity problem (abiotic stress) because the salt stress not only exerts osmotic stress on plant cells but

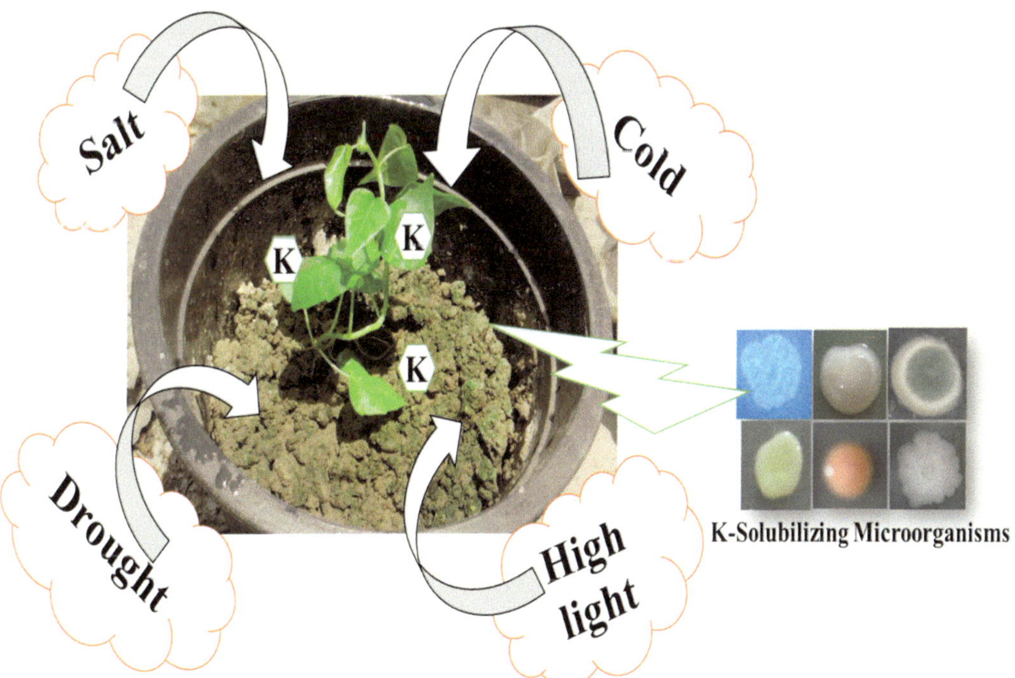

Fig. 2.3 Application of potassium solubilizing microorganism (K-SM) enhance plant tolerance to diverse environmental stress

simultaneously reduces K^+ uptake by plant root cells (Shabala and Cuin 2008). Huertas et al. (2013) found that overexpression of a K^+ transporter *LeNHX2* (tomato Na^+/H^+ Exchanger 2) increases the salt tolerance in transgenic tomato. Therefore, the application of K in agriculture field, as in the form biofertilizer or commercial available K fertilizer, frequently improves the crop performance on saline soils (Sheng and Huang 2002b).

2.6 Factors Affecting Potassium Solubilization

Different environmental factors such as pH, temperature, exo-polysaccharide, and nature of the K-bearing minerals affected the rate of potassium solubilization (Sheng and Huang 2002a). Yakhontova et al. (1987) showed that efficiency solubilization of K-bearing minerals by the bacterium was dependent on the nature of mineral used. Welch et al. (1999) found that a variety of extracellular polysaccharides significantly enhanced the solubilization of K-bearing minerals. Sheng et al. (2002) observed 35.2 mg/L potassium release from strains of potassium-solubilizing bacteria in 7 days at 28 °C at a pH range from 6.5 to 8.0. Lian et al. (2008) found that the K-solubilizing rate showed direct relation with pH when a thermophile fungus (*Aspergillus fumigatus*) and minerals were mixed directly. Lopes-Assad et al. (2010) found that the percentage of solubilization of K decreased at higher volumetric scales (Richards and Bates 1989; Supanjani et al. 2006).

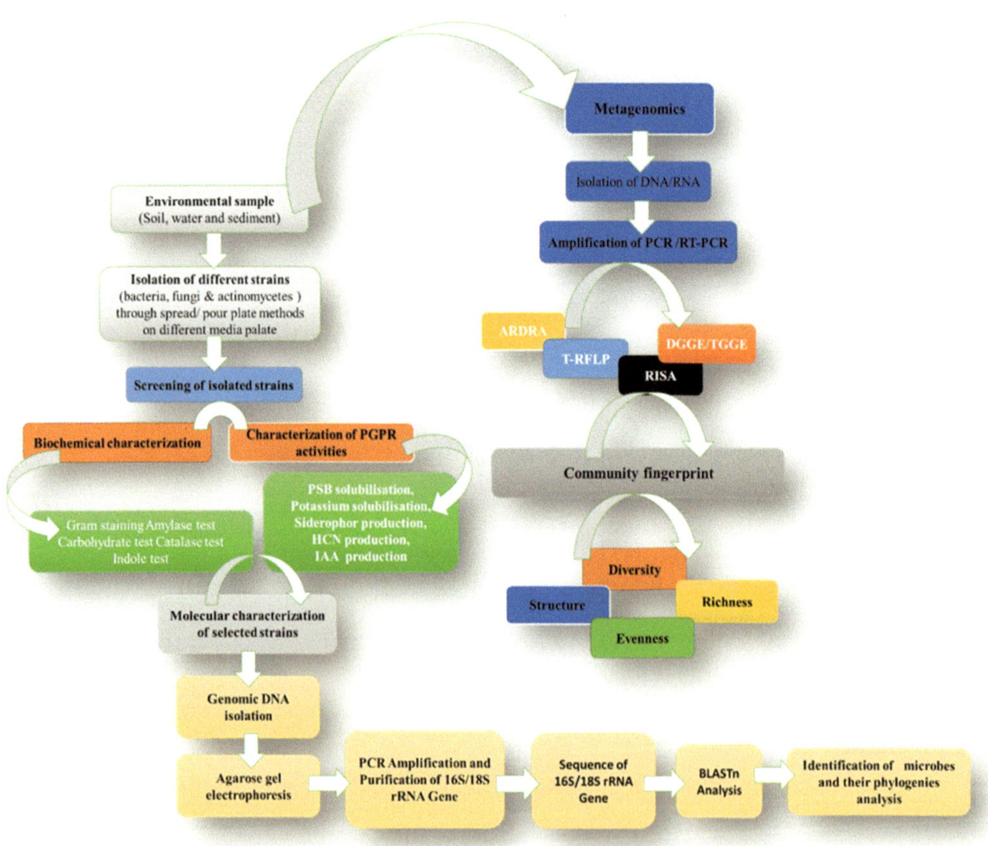

Fig. 2.4 Molecular characterization of isolated strains and collected environmental samples

2.7 Molecular Characterization of KSMs

Formerly, in the application of potassium-solubilizing microorganisms in crop fields, they are characterized through biochemical tests (staining, amylase, carbohydrate, and IMViC) and PGPR activities and then used as KSM biofertilizer (Fig. 2.4). Consequently, molecular techniques (like metagenomics) have provided the tools for analyzing the entire bacterial community including those which we are not able to grow in the laboratory. Because conventional methods, isolating and identifying microbes have been a limiting factor for exploring to new enzymes and plant growth-promoting activity due to very limited variability. Therefore, microbiology has focused on two areas of study: (i) microbial diversity, including the isolation, identification, and quantification of microorganisms in various habitats and (ii) microbial activity, that is, what microorganisms are doing in their habitats and how their activities contribute to the observed microbial diversity (Fenchel. 2005; Torsvik et al. 1996; Meena et al. 2015a, b). Thus, time is needed to focus on soil microbial diversity for developing effective microbial consortia of potassium-solubilizing microorganisms (KSMs) in a way of sustainability.

2.8 Future Approaches and Development Strategies for Sustainable Agriculture

Soil microorganisms play a pivotal role in soil quality (through various biogeochemical cycles) and crop productivity. Potassium-solubilizing microorganism (KSM) and other plant growth-promoting rhizobacteria (PGPR) are important for releasing nutrients in soil, which exert beneficial effects on plant growth. These microbes influence aboveground ecosystems by contributing to plant nutrition, plant health, and soil structure. Therefore, microorganisms offer an environment-friendly sustainable system (Meena et al. 2015b). However, these microorganisms (KSM and PSM) often fail to confer these beneficial effects when applied in the field; it is due to its deficiency in root colonization (Lugtenberg et al. 2001). Lin et al. (2002) found that silicate-dissolving bacteria increased 70 % in the rhizosphere soil and 20 % in the non-rhizosphere soil, respectively. According to Sugumaran and Janarthanam (2007), the number of K-solubilizing bacteria increased to 10^6–10^7 cfu/g in soil after 90 days of inoculation, whereas the count of these bacteria was only 10^3 cfu/g in the control soil. Therefore, currently, the research needs to focus on exploring the microbial diversity and genetic manipulation of potassium-solubilizing bacteria to increase efficiency and to confer strain survival in the rhizosphere. Hence, this approach allows to strain for better functioning in the rhizosphere system and also helps in leading to substantial improvement in the sustainability of an agriculture system (Ryan et al. 2009; Sindhu et al. 2014). Consequently, KSM will be very useful for higher yield of agricultural production under adverse environmental condition. It will be also an eco-friendly, cost-effective, and socially acceptable technology.

2.9 Conclusions

Potassium is a macronutrient for plant growth and yield. Microbial consortia of KSMs is a sustainable technology for solubilizing the fixed form of potassium into an available form of potassium in soil for survival of plant growth as well as microbial growth. This is only an alternative practice to fulfill the potassium deficiency in soils.

References

Arnold PW (1958) Potassium uptake by cation-exchange resins from soils and minerals. Nature 182:1594–1595

Barker WW, Welch SA, Chu S, Banfield JF (1998) Experimental observations of the effects of bacteria on alu minosilicate weathering. Am Mineral 83:1551–1563

Bertsch PM, Thomas GW (1985) Potassium status of temperate region soils. In: Munson RD (ed) Potassium in agriculture. Soil Science Society of America, Madison, pp 1:131–162

Burford EP, Fomina M, Gadd GM (2003) Fungal involvement in bioweathering and biotransformation of rocks and minerals. Mineral Mag 67(6):1127–1155

Carraretto L, Formentin E, Teardo E, Checchetto V, Tomizioli M, Morosinotto T, Szabó I (2013) A thylakoid-located two-pore K+ channel controls photosynthetic light utilization in plants. Science 342 (6154):114–118

Clarkson DT, Hanson JB (1980) The mineral nutrition of higher plants. Annu Rev Plant Physiol 31(1):239–298

Conyers ES, McLean EO (1969) Plant uptake and chemical extractions for evaluating potassium release characteristics of soils. Soil Sci Soc Am J 33(2): 226–230

FAI (2007) Fertiliser statistics 2006–2007. The Fertilizer Association of India, New Delhi

Fenchel T (2005) Cosmopolitan microbes and their cryptic' species. Aquat Microb Ecol 41(1):49–54

Gadd GM (1999) Fungal production of citric and oxalic acid: importance in metal speciation, physiology and biogeochemical processes. Adv Microb Physiol 41: 47–92

Groudev SN (1987) Use of heterotrophic microorganisms in mineral biotechnology. Acta Biotechnol 7(4): 299–306

Haby VA, Russelle MP, Skogley EO (1990) Testing soil for potassium, calcium, and magnesium. In: Westerman RL (ed) Soil testing and plant analysis, 3rd edn, SSSA book series 3. Soil Science Society of America, Madison, America, pp 181–228

Heinen W (1960) Silicon metabolism in microorganisms. Arch Microbiol 37:199–210

Huertas R, Rubio L, Cagnac O, García-Sánchez MJ, Alché JDD, Venema K, Rodríguez-Rosales MP (2013) The K+/H+ antiporter LeNHX2 increases salt tolerance by improving K+ homeostasis in transgenic tomato. Plant Cell Environ 36(12):2135–2149

Khawilkar SA, Ramteke JR (1993) Response of applied K in cereals in Maharashtra. Agriculture 11:84–96

Kumar A, Bahadur I, Maurya BR, Raghuwanshi R, Meena VS, Singh DK, Dixit J (2015) Does a plant growth-promoting rhizobacteria enhance agricultural sustainability? J Pure Appl Microbiol 9(1):715–724

Lian B, Wang B, Pan M, Liu C, Teng HH (2008) Microbial release of potassium from K-bearing minerals by thermophilic fungus *Aspergillus fumigatus*. Geochim Cosmochim Acta 72(1):87–98

Lin QM, Rao ZH, Sun YX, Yao J, Xing LJ (2002) Identification of a silicate-dissolving bacterium and its effect on tomato. Sci Agric Sin 35:59–62

Liu W, Xu X, Wu X, Yang Q, Luo Y, Christie P (2006) Decomposition of silicate minerals by *Bacillus mucilaginosus* in liquid culture. Environ Geochem Health 28(1–2):133–140

Lopes-Assad ML, Avansini SH, Rosa MM, De Carvalho JR, Ceccato-Antonini SR (2010) The solubilization of potassium-bearing rock powder by *Aspergillus niger* in small-scale batch fermentations. Can J Microbiol 56(7):598–605

Lugtenberg BJJ, Dekkers L, Bloemberg CV (2001) Molecular determinants of rhizosphere colonization by *Pseudomonas*. Annu Rev Phytopathol 39:461–490

Malavolta E (1985) Potassium status of tropical and subtropical region soils. In: R.E. Munson (ed.) Potassium in Agriculture. American Society of Agronomy, Crop Science Society of America, and Soil Sci Soc Am, Madison, WI, pp. 163–200.

Malinovskaya IM, Kosenko LV, Votselko SK, Podgorskii VS (1990) Role of *Bacillus mucilaginosus* polysaccharide in degradation of silicate minerals. Mikrobiologie 59:49–55

Maurya BR, Meena VS, Meena OP (2014) Influence of Inceptisol and Alfisol's potassium solubilizing bacteria (KSB) isolates on release of K from waste mica. Vegetos 27(1):181–187

Meena OP, Maurya BR, Meena VS (2013) Influence of K-solubilizing bacteria on release of potassium from waste mica. Agric Sustain Dev 1(1):53–56

Meena VS, Maurya BR, Bahadur I (2014a) Potassium solubilization by bacterial strain in waste mica. Bangladesh J Bot 43(2):235–237

Meena VS, Maurya BR, Verma JP (2014b) Does a rhizospheric microorganism enhance K+ availability in agricultural soils? Microbiol Res 169:337–347

Meena RK, Singh RK, Singh NP, Meena SK, Meena VS (2015a) Isolation of low temperature surviving plant growth-promoting rhizobacteria (PGPR) from pea (*Pisum sativum* L.) and documentation of their plant growth promoting traits. Biocatal Agric Biotechnol. doi:10.1016/j.bcab.2015.08.006

Meena VS, Maurya BR, Verma JP, Aeron A, Kumar A, Kim K, Bajpai VK (2015b) Potassium solubilizing rhizobacteria (KSR): isolation, identification, and K-release dynamics from waste mica. Ecol Eng 81: 340–347

Mengel K (1980) Effect of potassium on the assimilate conduction to storage tissue. Ber Deut Bot Gesch 93(1):353–362

Parmar P, Sindhu SS (2013) Potassium solubilization by rhizosphere bacteria: influence of nutritional and environmental conditions. J Microbiol Res 3(1):25–31

Pettigrew WT (2008) Potassium influences on yield and quality production for maize, wheat, soybean and cotton. Physiol Plant 133:670–681

Richards JE, Bates TE (1989) Studies on the potassium supplying capacities of southern Ontario soils. Measurement of available K. Can J Soil Sci 69: 597–610

Ryan PR, Dessaux Y, Thomashow LS, Weller DM (2009) Rhizosphere engineering and management for sustainable agriculture. Plant Soil 321:363–383

Schroeder D (1979) Structure and weathering of potassium containing minerals. Proc Congr Int Potash Inst 2:43–63

Shabala S, Cuin TA (2008) Potassium transport and plant salt tolerance. Physiol Plant 133(4):651–669

Shanware AS, Kalkar SA, Trivedi MM (2014) Potassium solubilisers: occurrence, mechanism and their role as competent biofertilizers. Int J Curr Microbiol Appl Sci 3(9):622–629

Sheng XF, He LY (2006) Solubilization of potassium-bearing minerals by a wild-type strain of Bacillus edaphicus and its mutants and increased potassium uptake by wheat. Can J Microbiol 52(1):66–72

Sheng XF, Huang WY (2002a) Mechanism of potassium release from feldspar affected by the strain NBT of silicate bacterium. Acta Pedol Sin 39:863–871

Sheng XF, Huang WY (2002b) Study on the conditions of potassium release by strain NBT of silicate bacteria. Sci Agric Sin 35:673–677

Sheng XF, He LY, Huang WY (2002) The conditions for releasing potassium by a silicate dissolving bacterial strain NBT. Agric Sci China 1:662–666

Sheng XF, Zhao F, He LY, Qiu G, Chen L (2008) Isolation and characterization of silicate mineral-solubilizing *Bacillus globisporus* Q12 from the surfaces of weathered feldspar. Can J Microbiol 54(12): 1064–1068

Sindhu SS, Parmar P, Phour M (2014). Nutrient cycling: potassium solubilization by microorganisms and improvement of crop growth. In: Geomicrobiology and biogeochemistry (p 175–198). Springer, Berlin.

Singh NP, Singh RK, Meena VS, Meena RK (2015) Can we use maize (*Zea mays*) rhizobacteria as plant growth promoter? Vegetos 28(1):86–99

Sparks DL (2000) Bioavailability of soil potassium, D-38-D-52. In: Sumner ME (ed) Handbook of soil science. CRC Press, Boca Raton

Sparks DL, Huang PM (1985) Physical chemistry of soil potassium. Potassium Agric 16:238–249

Sparks DL, Page AL, Helmke PA, Loeppert RH, Soltanpour PN, Tabatabai MA, Johnston CT, Summer ME (1996) Methods of soil analysis. Part 3: Chemical methods, 3rd edn. Soil Science Society of America and American Society of Agronomy, Madison, pp 46–64

Sugumaran P, Janarthanam B (2007) Solubilization of potassium containing minerals by bacteria and their effect on plant growth. World J Agric Sci 3(3):350–355

Supanjani Han HS, Jung SJ, Lee KD (2006) Rock phosphate potassium and rock solubilizing bacteria as alternative sustainable fertilizers. Agron Sustain Dev 26:233–240

Torsvik V, Sørheim R, Goksøyr J (1996) Total bacterial diversity in soil and sediment communities—a review. J Ind Microbiol 17(3–4):170–178

Ullman WJ, Welch SA (2002) Organic ligands and feldspar dissolution. Geochem Soc 7:3–35

Vessey JK (2003) Plant growth promoting rhizobacteria as biofertilizers. Plant Soil 255(2):571–586

Wang Y, Wu WH (2015) Genetic approaches for improvement of the crop potassium acquisition and utilization efficiency. Curr Opin Plant Biol 25:46–52

Welch SA, Barker WW, Banfield JF (1999) Microbial extracellular polysaccharides and plagioclase dissolution. Geochim Cosmochim Acta 63:1405–1419

Xie JC (1998) Present situation and prospects for the world's fertilizer use. Plant Nutr Fertil Sci 4: 321–330

Yakhontova LK, Andreev PI, Ivanova MY, Nesterovich LG (1987) Bacterial decomposition of smectite minerals. Dokl Akad Nauk USSR 296:203–206

Zhang C, Kong F (2014) Isolation and identification of potassium-solubilizing bacteria from tobacco rhizospheric soil and their effect on tobacco plants. Appl Soil Ecol 82:18–25

Zörb C, Senbayram M, Peiter E (2014) Potassium in agriculture–status and perspectives. J Plant Physiol 171(9):656–669

Growth-Promoting Effect of Potassium-Solubilizing Microorganisms on Some Crop Species

3

Hossein Zahedi

Abstract

Potassium (K) is an essential substance for plant growth. With the fast development of world agriculture, attainable soils K levels have dropped due to crop removal, leaching, runoff, and erosion. The potassium is a fundamental macronutrient for plant growth and plays important roles in making active of several metabolic processes including protein synthesis, photosynthesis, and enzyme activation. Whole soil K reserves are usually large despite the fact that the distribution of potassium forms differs from soil to soil as a role of the dominant soil minerals present. Many microorganisms in the soil are capable to solubilize "unavailable" forms of K-bearing minerals, such as micas, illite, and orthoclases by expelling organic acids which either directly dissolve rock K or chelate silicon ions to bring the K into solution. In this article, we are focusing on the enhancement of potassium to the plant by the help of efficient potassium-solubilizing microorganisms (KSMs) with specific crop species examples.

Keywords

Potassium-solubilizing microorganisms (KSMs) • Crop species • Stress resistance • Sustainable agriculture

3.1 Introduction

After nitrogen (N) and phosphorus (P), potassium (K) is the third important plant nutrient, and it is a fundamental macronutrient for plant growth and plays influential roles in activation of several metabolic processes including protein synthesis and photosynthesis and enzymes, as well as in resistance to diseases, insects, etc. (Rehm and Schmitt 2002). The potassium though presents in as an abundant element in soil or is applied to fields as natural or synthetic fertilizers, only 1–2 % of this is available to plants, the rest being bound with other minerals and therefore unavailable to plants. The most common soil

H. Zahedi (✉)
Department of Agronomy and Plant Breeding, Eslamshahr Branch, Islamic Azad University, P.O.Box: 33135-369, Tehran, Iran
e-mail: hzahedi2006@gmail.com

V.S. Meena et al. (eds.), *Potassium Solubilizing Microorganisms for Sustainable Agriculture*,
DOI 10.1007/978-81-322-2776-2_3

components of potassium 90–98 % are feldspar and mica (McAfee 2008). The efficient soil microorganisms influence the availability of soil minerals, playing a central role in ion cycling and soil fertility (Lian et al. 2010). They are utilized as biofertilizer factors for agriculture improvement and environmental protection and have been a concentration of recent research. Specific bacteria are capable of disintegrating aluminosilicate minerals and liberating a portion of the potassium contained therein (Biswas and Basak 2009). A detailed understanding of how bacteria influence mineral dissolution rates is needed to quantify mineral weathering on global element cycling (Xiufang et al. 2006). Plants take in potassium only from the soil and its availability in soil is dependent on the potassium dynamics as well as on sum potassium content. Without the three forms of potassium found in the soil, soil minerals make up more than 90–98 % of soil K (Sparks 1987) and the majority of it is unavailable for plant uptake. The second nonexchangeable form of K makes up approximately 1–10 % of soil potassium and consists generally of an interlayer K of non-expanded clay minerals such as illite, which contribute meaningfully to the plant absorption (Sharpley 1989). The release of non-interchangeable K to the third exchangeable form happens when the level of exchangeable and solution K is lessened by crop removal, runoff, or leaching (Sheng and Huang 2002a). With the introduction of high-yielding crop varieties and the advanced strengthening of agriculture, the soils that are getting drained in potassium retain at a faster rate. In addition, due to imbalance in fertilizer application, potassium lack is becoming one of the major restrictions in crop production.

This emphasized the search to find an alternative native source of potassium for plant uptake and to sustain K status in soils for maintaining crop production (Sindhu et al. 2012). Recognition of microbial strains able of solubilizing potassium minerals rapidly can preserve our existing resources and avoid environmental pollution hazards caused by heavy application of chemical fertilizers. A wide extent of bacteria, namely, *Pseudomonas*, *Burkholderia*, *A. ferrooxidans*, *B. mucilaginosus*, *B. edaphicus*, *B. circulans*, and *Paenibacillus* spp., has been reported to release potassium in an easily obtainable form from potassium-bearing minerals in soils (Sheng 2005; Liu et al. 2012; Maurya et al. 2014). These potassium-solubilizing bacteria (KSB) were found to dissolve potassium, silicon, and aluminum from insoluble potassium-bearing minerals such as micas, illite, and orthoclases, by discharging organic acids which either directly dissolved rock potassium or chelated silicon ions to carry with potassium into the solution (Bennett et al. 1998; Meena et al. 2014). Inoculation with potassium-solubilizing bacteria have been reported to use beneficial effects on growth of cotton and rape (Sheng 2005), pepper and cucumber (Han et al. 2006), wheat (Sheng and He 2006), and sudan grass (Basak and Biswas 2010). In a similar manner, inoculation of corn and wheat plants with *B. mucilaginosus*, *A. chroococcum,* and *Rhizobium* brings about significant higher movement of potassium from waste mica, which in order acted as a source of potassium for plant growth (Singh et al. 2010). Hence, potassium-solubilizing bacteria are widely used as biofertilizers in Korea and China as significant areas of cultivated soils in these countries are deficient in soil-available potassium (Xie 1998). Therefore, application of K-solubilizing bacteria as biofertilizer for agriculture improvement can lessen the use of agrochemicals and support eco-friendly crop production (Sindhu et al. 2010).

Presently, some information is available on potassium solubilization by bacteria, their mechanisms of solubilization, and influence of KSB inoculation on nutrient availability in soils and growth of different crops. Potassium is attainable in four forms in the soil which are K ions in the soil solution, as an interchangeable cation, firmly held on the surfaces of clay minerals and organic matter, and tightly held or fixed by weathered micaceous minerals and present in the lattice of certain K-containing basic minerals. Fig. 3.1 shows the potassium's cycle in the soil-plant-animal system. There are several processes that contribute to the availability of potassium in the soil. Soil solution potassium is already available in the soil for plant absorbing; however, the

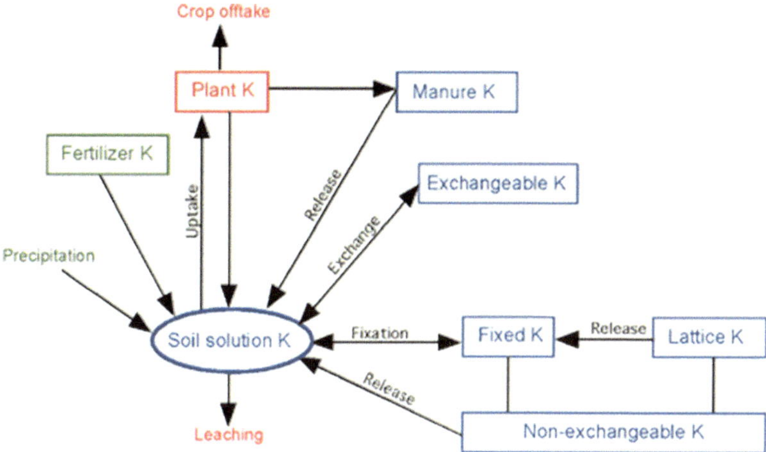

Fig. 3.1 Potassium cycle in plant-soil-animal system (Syers 1998)

concentration of potassium is influenced by soil weathering, cropping history, and fertilizer use. Therefore, the amount present is insufficient to meet the crop requirement. Then, an indicator of soil potassium status such as exchangeable potassium has rapid equilibrium with the soil solution potassium, and it is considered as readily available. Fixed and lattice potassium can be grouped simultaneously and make up the pool of noninterchangeable inorganic potassium in the soil.

3.2 Potassium Fixation in Soil

In addition to releasing K, soil minerals can also fix K, significantly affecting K availability. This involves the adsorption of potassium ions onto sites in the interlayers of weathered sheet silicates, such as illite and vermiculite. The extent of K fixation in soils depends on the type of clay mineral and its charge density, moisture content, competing ions, and soil pH. Montmorillonite, vermiculite, and weathered micas are the major clay minerals that tend to fix K (Sparks 1987). Additionally, soil wetting and drying also significantly affects the K fixation. The fixation process of K is relatively fast, whereas the release of fixed K is very slow due to the strong binding force between K and clay minerals (Oborn et al. 2005). Whether soil fixes or releases K highly depends on the K concentration in the soil solution (Schneider et al. 2013).

As mentioned above, in addition to organic acids, the H^+ concentration in soil solution (via soil pH) seems to perform a key role in K release from clay minerals. Therefore, optimization of soil pH may be a means of enhancing K release. For optimized K fertilizer management practices, it is crucial to understand the factors that regulate K release from soil nonexchangeable pool, with the effects of potassium fertilization on soil physical properties. Recent investigations have raised awareness of the impact of K on the soil structure and its ability to capture water. It has been informed that the applications of mineral K fertilizers enhance the water-holding capacity of soils and also improve the structural stability of sandy soil in particular (Holthusen et al. 2010). Higher water retention plays a key role in securing the soil productivity in water-limited areas. Therefore, more information is needed in order to comprehend the effect of K fertilization on the soil's physical properties and soil water-holding capacity.

3.3 Determination and Adjustment of the Potassium Supply of Crops

In agricultural production, fertilizer application is required not only to ensure but also to sustain an adequate supply of soluble K to crops.

However, application rates and timing of organic or inorganic fertilizers are frequently based on an optimal nitrogen supply, but unfortunately often not on K requirements. This may lead to an excess or a deficit of potassium, depending on both crop and soil characteristics (Oborn et al. 2005).

3.3.1 Soil Analysis

A monitoring of soil K reserves is, therefore, extremely important in order to make precise fertilizer recommendations. Simple soil extraction methods for measuring exchangeable potassium are widely used to estimate the K fertilizer demand of a crop. An estimation of exchangeable soil K by extracting the soil with neutral ammonium acetate, ammonium chloride, calcium chloride, or ammonium fluoride (Mehlich 3) from air-dried or oven-dried soil samples is the most widely used soil test for K and provides the basis for most K fertilizer recommendations worldwide (Barbagelata and Mallarino 2012). Here, the preparation procedure includes the drying of soil samples to a maximum of 40 °C and crushing them to pass a 2 mm sieve to provide a homogenous mixture for analysis. However, it has long been identified that drying of soil samples can influence the amount of K extracted by the traditional extract. Some researchers have reported an increase in extractable potassium upon drying (Luebs et al. 1956). Haby and Ruselle (Haby et al. 1990) concluded that the impact of sample drying on K extracted by tests that estimate exchangeable K depends on the deviation from the equilibrium K concentration at sampling time and on soil mineralogy. Illite appeared to be the source of K released by drying, while vermiculite and montmorillonite were associated with K fixation in selected Kentucky soils (Dowdy and Hutcheson 1963).

In field and greenhouse studies, another K extraction method, called the *wet-extraction method* (ammonium acetate-extraction of field-moist soil), was reported to yield a better correlation with crop K-uptake than methods extracting K from air-dried soil (Luebs et al. 1956). A recent study by Barbagelata and Mallarino (2012) also demonstrated that moist extraction K tests had a superior capacity to predict crop responses to K fertilization, as compared to the commonly used dry extraction method. However, few laboratories have adopted wet-extraction methods due to impractical procedures, such as sieving moist soil. The extraction methods discussed above may provide sufficient information for fertilizer recommendations in light-textured soils that do not contain 2:1 clay minerals (Mengel and Kirkby 2001). In soils that contain 2:1 clay minerals, however, the nonexchangeable K pool usually contributes largely, sometimes to over 50 %, to crop K supply (Mengel and Busch 1982). Measuring plant-available soil K that is released from nonexchangeable K reserves is very difficult due to the complexity of the dynamic equilibrium among the various forms of soil K during crop growth.

Thus, no routine methods are available to measure this parameter. Nevertheless, various methods have been established to assess the slowly or potentially available K in soils, e.g., extraction by 1 M HCl, boiling in 0.5 M or 1 M HNO_3, electro ultrafiltration, exchange resins, Jackson's test (sodium tetra-phenyl-boron; NaTPB), and field balances (Andrist-Rangel et al. 2006). However, common acid extraction methods only remove a proportion of the reserves of K present in the nonexchangeable pool. In a grassland study, Ogaard and Krogstad (2005) compared several methods to estimate the plant-available K and concluded that, in light-textured soils, boiling of soil in 0.5 M HNO_3 was a better predictor of K-uptake than other acid extraction methods. Additionally, they reported that boiling in 1 M HNO_3 extracts more K than would normally become available to a crop.

3.3.2 Bacterial Culture and Inoculant Preparation

Potassium-solubilizing bacteria were cultured in Tryptone Yeast medium (Vincent 1970) and sucrose-minimal salts medium (Sheng and Huang 2002a), respectively, and incubated on an orbital shaker at 150 rpm for 48 h at 27 °C. The cells in cultured bacterial broth were collected by centrifugation at 2,822 × g for 15 min at 4 °C and washed with sterilized tap water. The pelleted cells were re-suspended in sterilized tap water and then the cells were modified to about 10^8 cells mL^{-1}, based on optical density 620 = 0.08 (Bhuvaneswari et al. 1980).

3.4 Potassium Functions in Plants

Potassium has many functions in plant growth such as to smooth the progress of cell division and growth, to increase disease resistance and drought endurance, to regulate the opening and closing of the stomata and it is required for osmotic regulation. Besides, potassium is essential for photosynthesis process and acts as a key to activate enzymes to metabolize carbohydrates for the manufacture of amino acids and proteins. Furthermore, potassium assimilates transport during plant ontogeny and one of the most important influences is to improve oil content in plants. Since the use of potassium is covered in a lot of plants' activities, depletion potassium uptake can cause a problem for plant growth, a deficiency of this element is also known as potash deficiency exhibiting chlorosis (loss of green color) along the leaf edges. Then, in severe cases, the leaf will turn into yellow color and eventually will fall off. It also affects plant growth and canopy photosynthesis process. There are several factors that lead to this problem, for instance, low-soil potassium-supplying capacity, insufficient application of mineral potassium fertilizer and biofertilizer, complete removal of plant straw, leaching losses, and phosphorus and nitrogen deficiency (Das and Sen 1981). Fundamentally, K^+ is very water soluble and highly mobile and transported in the plants' xylem (Lack and Evans 2005). Membrane transport of K can be mediated either by K channels, utilizing the membrane potential to make easy transport of potassium down its electrochemical gradient, or by secondary transporters.

In plants, potassium acts as a regulator because it is constituent of ~60 different enzyme systems of drought tolerance and water use efficiency. Plant's K concentrations of crops vary widely with site, year, crop species, and fertilizer input; concentrations in the range of 0.4–4.3 % have been reported (Askegaard et al. 2004). Oborn et al. (2005) concluded in a literature survey that crop potassium concentrations are often well below (<2.5–3.5 %). For many crops, the critical potassium concentration is in the range 0.5–2 % in dry matter (Leigh and Wyn Jones 1984). With the exception of the cytosol and the vacuole, the subcellular distribution of K is largely uncharacterized. The concentration of potassium in the cytoplasm is kept relatively constant at around 50–150 mM, while the concentration in the vacuole varies substantially depending on supply status. Together with accompanying anions (NO_3^-, Cl^-, malate$^-$), vacuolar K largely determines the osmotic potential of the cell sap. In the agronomic literature, high potassium concentrations in crops have often been termed "luxury consumption." However, as outlined in the "Potassium nutrition and crop stress resistance" section, high gathering of potassium by crops during optimal growing conditions may be considered as an "insurance strategy" to enable the plant to better survive a sudden environmental stress (Kafkafi 1990).

Plant species are known to differ in their K requirement and in their ability to take up K. The differences in absorption of K among different plant species are attributed to variations in root structure, such as root density, rooting depth, and root hair length (Nieves-Cordones et al. 2014) All crops require potassium especially high carbohydrate plants such as bananas and potatoes (Hillel 2008). In addition, the current study has showed that for optimum growth, crops need more potassium than needed (Simonsson et al. 2007). This means that the uptake of potassium depends on the rate at which it is supplied through roots rather than amounts of potassium availability in the soil. Although it is not an integral constituent part of cell structure, potassium regulates many metabolic processes required for growth and fruit and seed development. Furthermore, the presence of extreme amounts of reduced substance in poorly drained soils will cause retarded root growth and reduced K-uptake. There are plenty of deficiency symptoms associated with potassium. Some of them are older leaves changing from yellow to brown, leaf tips and margins dry up, poor root oxidation power, discolorization of younger leaf, unhealthy root system that may cause reduction other nutrient uptake, and reduced cytokinin production in roots. Based on the specific crop's type of potassium deficiency, grain crops such as corn become small in size and have low yield and tomatoes exhibit uneven fruit ripening.

Besides, cotton leaves turn reddish-brown, appear burned, become bronze then black, and finally fall while yield of fodder crops is low and its quality is indigent. Recent studies have proved that biofertilizer can increase crop yield. For example, by introducing KSB and phosphate-solubilizing bacteria (PSB), the primary macronutrient of nitrogen, phosphorus, and potassium uptake increases on eggplant, pepper, and cucumber and lead to higher yield (Han et al. 2006). KSB are a heterotrophic bacteria which are achieving all their energy and cellular carbon of buildup from dead organic materials. Besides, KSB are aerobic bacteria which play a significant role in sustaining soil structure by their contribution in the formation and stabilization of water-stable soil aggregates. In addition, this gram-positive bacterium can produce a substance that stimulates plant growth or inhibits root pathogens (Egamberdiveya 2006). Moreover, KSB specifically are well known for its capability to solubilize rock potassium mineral such as micas, illite, and orthoclases. This is done through the production and excretion of organic acids (Han et al. 2006). Thus, KSB are functioning to increase potassium availability in soils besides increase mineral contents in plants.

3.5 Potassium and Crop Stress Resistance

It has become clear in the preceding section that the physiological functions of potassium are critical for yield formation and product quality under otherwise optimal and undisturbed growth conditions. In the field, however, crops are faced with a multitude of factors that affect metabolism, growth, and thus yield. Abiotic stresses, such as drought, salinity, cold, or high light and biotic stresses, such as bacterial and fungal diseases or insect pests, reduce crop yield to a large extent. Climate models predict that incidences and duration of drought and heat stress events will raise in some parts of the world, while other parts will suffer from heavy storms and periodic flooding. These conditions will have a surprising effect on agricultural production and farming practices (Brouder and Volenec 2008). Even in Europe, future climate changes are anticipated to be problematic, resulting more often in dry spring sand rainy summers in northern Europe and longer dry periods in the South. The impacts of such climate changes could be monitored in the European heat wave of 2003, which reduced crop production by around 30 % (Ciais et al. 2005).

On a worldwide scale, abiotic stress often leads to massive, often complete and crop failures. Due to its fundamental roles in the plant, K nutrition impacts on the resistance of crops to virtually all abiotic and biotic stresses, in both direct and indirect ways. This part of the review highlights selected agronomic aspects of this relationship, while some of the fundamental mechanisms are covered in-depth in other articles of this topical issue (Demidchik 2014). Potassium fertilizer application lessened frequency of diseases in most cases, but occasionally had no effect or even the opposite effect. The variable effects of potassium on disease incidence could be affected by the amount and source of potassium, plant and pathogen species, and trial type. Nam et al. (2006) discovered that strawberries that were grown with excess potassium were very vulnerable to infection by the anthracnose pathogen, *C. gloeosporioides*, but its resistance was extremely improved when no potassium was supplied. This outcome was discerned due to the fact that the low plant potassium status causes the synthesis of molecules, including reactive oxygen species and phytohormones, such as auxin, ethylene, and jasmonic acid, as a result of its improved plant stress tolerance (Amtmann et al. 2008).

Potassium is also necessary to the execution of multiple plant enzyme functions, and it controls the metabolite pattern of higher plants, ultimately changing metabolite concentrations. In a potassium-sufficient plant, the synthesis of high-molecular-weight compounds (such as proteins, starches, and cellulose) was markedly raised, thereby depressing the concentrations of low-molecular-weight compounds, such as soluble sugars, organic acids, amino acids, and amides, in the plant tissues. These low-molecular-weight compounds are significant for the development of infections and insect invasions, so lower concentrations, by means of that, leave plants less susceptible to disease and pest attacks in potassium-sufficient plants (Marschner 2012). Sufficient potassium raises phenol concentrations, which play a crucial role in plant resistance (Prasad et al. 2010).

3.6 Effect of KSB on Some Plants

The roles of KSB and the specific requirement of K supply differ with crop types. Here we exemplary review some different crops:

3.6.1 Eggplant

A research was directed to evaluate the potential of phosphate-solubilizing bacteria (PSB) *B. megaterium* and potassium-solubilizing bacteria (KSB) *B. mucilaginosus* inoculated in nutrient-limited soil planted with eggplant. Results revealed that the shoot and root dry weight in eggplant increase ($P < 0.05$) in the combined fertilizer treatment with rock phosphate or potassium materials, in the inoculation with a single respective bacterial strain (PSB or KSB) treatment and in the combined bacteria plus rock material treatment, as compared with the control (Table 3.1) (Han and Lee 2005).

The application of rock P and K materials alone did not significantly enhance plant height, dry weight, or photosynthesis. Despite the fact that inoculation with a single bacterial strain in combination with its respective phosphate source consistently increased shoot and root dry weight as compared with the control, the treatment which combined both bacteria and mineral rocks further increased shoot dry weight by ~27 % and root dry weight by 30 % over the control 30 days following planting. Photosynthetic responses of eggplant showed a similar trend to plant dry weight responses (Table 3.1).

The combined treatment of the two strains co-inoculated with each other along with the applied insoluble rock material significantly raised leaf photosynthesis ~12 % over the control. In our case, co-inoculation of PSB and KSB strains synergistically solubilized the rock P and K materials which were added into the soil and made them more available to the plant. This led to the promotion of their uptake and plant growth. Growth improvement by *Bacillus* may

Table 3.1 Effects of PSB and KSB strains on dry matter and photosynthesis of eggplant

Treatments	Plant height (cm)	Dry weight (g plant^{-1}) Shoot	Root	Total	Photosynthetic rate (μmol cm^{-2} s^{-1})
Control	18.9	1.33	0.60	1.93	12.7
Rock P	19.6	1.38	0.68	2.06	13.1
Rock K	19.5	1.30	0.61	1.91	12.9
Rock (P+K)	20.0	1.46	0.70	2.16	13.4
PSB	19.5	1.41	0.70	2.11	12.9
KSB	19.7	1.35	0.66	2.01	12.5
(P+K)SB	20.8	1.53	0.75	2.28	13.5
Rock P+PSB	21.5	1.68	0.72	2.40	13.9
Rock K+KSB	20.6	1.49	0.71	2.20	13.0
Rock (P+K)+(P+K)SB	21.9	1.69	0.78	2.47	14.2
$LSD_{0.05}$	2.8	0.14	0.11	0.25	0.8

also relate to its ability to produce hormones, particularly IAA (Sheng and Huang 2002b). Briefly, co-inoculation of PGPR with different advantageous properties may be the future trend for biofertilizer application to enable sustainable crop production. In conclusion, co-inoculation of PSB and KSB in conjunction with directs the application of rock phosphate and potassium materials into the soil increased N-, P-, and K-uptake, photosynthesis, and the yield of eggplant grown on P- and K-limited soils.

3.6.2 Pepper and Cucumber

Experiments were performed to assess the potential of phosphate-solubilizing bacteria (PSB) *B. megaterium* var. *phosphaticum* and potassium-solubilizing bacteria (KSB) *B. mucilaginosus* inoculated in nutrient-limited soil planted with pepper and cucumber (Han et al. 2006). Results demonstrated that rock P and K applied either singly or in combination did not significantly enhance soil availability of phosphate and potassium, indicating their improperty for direct application. PSB were more influential P-solubilizers than KSB, and co-inoculation of PSB and KSB resulted in regularly higher phosphate and potassium obtains ability than in the control without bacterial inoculum and without rock material fertilizer. Integrated rock phosphate with inoculation of PSB increased the availability of P and K in soil; the uptake of nitrogen, phosphorus, and potassium by shoot and root; and the growth of pepper and cucumber. Similar but less significant results were obtained when rock K and KSB were added concomitantly.

Combined together, rock materials and both bacterial strains consistently increased further mineral availability, uptake, and plant growth of pepper and cucumber, suggesting its potential use as fertilizer. Single or double application of rock phosphate and rock potassium did not significantly improve plant growth (shoot and root dry weight) and photosynthesis in pepper and cucumber plants (Table 3.2). Double inoculation of PSB and KSB increased shoot and root dry weight; PSB inoculation increased shoot dry weight, whereas KSB did not. Although combined PSB inoculation with application of rock phosphate consistently increased shoot and root dry weight as compared to control, a treatment which joints together both bacteria and mineral rocks further increased plant growth: ~26 % in shoot and ~29 % in root dry weight for pepper and 22 % in shoot and 27 % in root dry weight for cucumber plant over the controls during 30 days following planting. Photosynthesis responses of pepper and cucumber to the treatments demonstrated similar trends to shoot dry weight responses. The integrated treatment of co-inoculation with two strains and application of insoluble rock materials significantly increased leaf photosynthesis: 20 % in pepper and 16 % in cucumber plants over the controls.

3.6.3 Sorghum

A study was done to determine the effect of the bacteria on releasing potassium and phosphate as well as their effect on sorghum growth (Badr et al. 2006). The dry matter of sorghum plants inoculated with silicate-dissolving bacteria (SBS strain) and supplied with minerals (feldspar and rock phosphate) increased by ~48 %, 65 %, and 58 % for clay, sandy, and calcareous soil, respectively, as compared to the plants supplied with minerals alone (Table 3.3). Moreover, the dry weight of sorghum inoculated with bacteria alone increased by ~34 %, 28 %, and 26 %, respectively, as compared to non-inoculated treatment. Potassium and phosphorus uptake improved markedly in both cases with inoculation of bacteria in the tested soils as compared to corresponding controls. However, these bacteria were found to develop in the rhizosphere of the sorghum plants.

The release of dissolved potassium- and phosphate-bearing minerals indicates that the silicate-dissolving bacteria could grow under the soil conditions and have a potential of mineral dissolution. After 30 days of inoculation, the cells in rhizosphere of sorghum plants treated with minerals increased to 38.6×10^4,

Table 3.2 Effects of PSB and KSB strains on dry weight (g/plant) and photosynthetic rate (μmole/cm²/s) of pepper and cucumber; the values are the mean of four replications (± SD)

Treatments	Pepper			Cucumber		
	Dry weight			Dry weight		
	Shoot	Root	Photosynthetic rate	Shoot	Root	Photosynthetic rate
Control	0.67 ± 0.033	0.268 ± 0.023	18.2 ± 0.5	0.757 ± 0.039	0.291 ± 0.016	11.1 ± 0.59
R-P	0.692 ± 0.035	0.288 ± 0.016	18.7 ± 0.67	0.81 ± 0.046	0.307 ± 0.02	11.3 ± 0.61
R-K	0.685 ± 0.037	0.283 ± 0 019	18.5 ± 0.47	0.80 ± 0.042	0.302 ± 0.023	11.5 ± 0.54
R-(P+K)	0.70 ± 0.037	0.289 ± 0.021	18.8 ± 1.13	0.781 ± 0.044	0.317 ± 0.007	11.6 ± 0.49
PSB	0.762 ± 0.035	0.294 ± 0.027	19.4 ± 0.48	0.845 ± 0.043	0.314 ± 0.021	12.1 ± 0.57
KSB	0.723 ± 0.033	0.286 ± 0.013	18.9 ± 0.22	0.819 ± 0.04	0.310 ± 0.023	11.6 ± 0.79
(P+K) SB	0.786 ± 0.055	0.30 ± 0.016	19.7 ± 0.87	0.832 ± 0.069	0.338 ± 0.018	12.1 ± 0.70
R-P+PSB	0.790 ± 0.039	0.310 ± 0.030	20.5 ± 1.02	0.871 ± 0.044	0.339 ± 0.024	12.5 ± 0.32
R-K+KSB	0.765 ± 0.042	0.295 ± 0.024	19.2 ± 0.40	0.828 ± 0.038	0.317 ± 0.014	12.0 ± 12.0
R-(P+K)+(P+K) SB	0.846 ± 0.039	0.345 ± 0.021	21.8 ± 0.54	0.920 ± 0.047	0.369 ± 0.011	12.9 ± 0.74
$LSD_{0.05}$	0.056	0.031	0.99	0.066	0.027	0.085

Table 3.3 Effect of silicate-dissolving bacteria (SBS strain) on sorghum growth and nutrient availability in different soils

Treatments	D.W.	K-uptake mg/pot	P-uptake mg/pot	mg/kg	Avail. P mg/kg	No. of cells × 10^4/g
Clay soil						
Minerals A	26.2	878	96	135	37	38.6
Minerals B	17.7	513	68	59	18	4.3
No minerals A	20.6	562	75	76	25	31.2
No minerals B	15.4	327	58	45	16	2.5
Sandy soil						
Minerals A	21.8	740	73	82	32	30.5
Minerals B	13.2	383	36	24	13	0.8
No minerals A	14.7	422	43	30	15	22.3
No minerals B	11.5	283	12	21	9	0.5
Calcareous soil						
Minerals A	19.8	657	61	73	12	26.8
Minerals B	12.5	304	34	18	7	0.6
No minerals A	12.8	345	25	27	8	20.1
No minerals B	10.2	232	24	18	6	0.3

A: Inoculated with bacteria; B: non- inoculated; LSD at 5% 3.24, 2.47 and 2.15 for clay, sandy and calcareous soil respectively.

30.5 × 10^4, and 26.8 × 10^4 for clay, sandy, and calcareous soil, respectively. In comparison to non-inoculated treatments, the release of potassium and phosphate from the added minerals was considerably enhanced with bacterial treatments. Soil available potassium increased to 135 (clay), 82 (sandy), and 73 (calcareous) mg/kg soil, respectively. Soil available phosphate also increased to 37 (clay), 32 (sandy), and 12 (calcareous) mg/kg soil, respectively. The release of potassium was about 129 %, 242 %, and 306 % much higher than that of non-inoculated treatment, while the release of P was ~106 %, 146 %, and 71 % in the clay, sandy, and calcareous soil, respectively. On the other hand, inoculation of silicate-dissolving bacteria without mineral additions appreciably activates the initial K and P content in all soils. The available potassium content in the rhizosphere soil of sorghum was ~69 %, 43 %, and 50 % much higher than that of non-inoculated treatments for clay, sandy, and calcareous soil, respectively, while the available P content increased by about ~56 %, 67 %, and 33 % in the same soil, respectively. The most possible reason was that the inoculation of silicate-dissolving bacteria accelerated the transformation of non-available forms of K and P into an available one.

3.7 Concluding Remarks and Future Prospects

The issue of sustainable management of potassium in soil has partly been ignored during the last decades when the focus was aimed at the potent environmental impact on use of nitrogen and phosphorus. There are many reports in the recent past that the soils do show potassium deficiency because available soil potassium levels have dropped due to the rapid development of agriculture without replenishing it and the application of potassium fertilizer to those soils gives positive response. Consequently, for maximal crop growth, soil solution, and exchangeable potassium need to be replenished continually with potassium through the release of noninterchangeable potassium through the weathering of potassium reserves (i.e., micas and feldspars) or the addition of potassium fertilizers. Many microorganisms in the soil are able to solubilize "unavailable" forms of potassium-bearing minerals. Thus, the application of potassium-solubilizing microorganisms (KSM) is a promising approach for increasing potassium availability in KSM-amended soils. Their use in agriculture can improve of plant growth, plant nutrition, root growth pattern, plant competitiveness, and responses to external stress factors. Use of these bacteria as bio-inoculants will increase the available potassium in soil, help to minimize the chemical fertilizer application, reduce environmental pollution, and promote sustainable agriculture. Further studies are necessary to see the beneficial effects of these bacterial strains on the crops.

References

Amtmann A, Troufflard S, Armengaud P (2008) The effect of potassium nutrition on pest and disease resistance in plants. Physiol Planta 133:682–691

Andrist-Rangel Y, Simonsson M, Andersson S, Öborn I, Hillier S (2006) Mineralogical budgeting of potassium in soil: a basis for understanding standard measures of reserve potassium. J Plant Nutr Soil Sci 169:605–615

Askegaard M, Eriksen J, Johnston AE (2004) Sustainable management of potassium. In: Schjorring P, Elmholt S, Christensen BT (eds) Managing soil quality: challenges in modern agriculture. CABI Publishing, Wallingford, pp 85–102

Badr MA, Shafei AM, Sharaf El-Deen SH (2006) The dissolution of K and P-bearing minerals by silicate dissolving bacteria and their effect on Sorghum growth. Res J Agric Biol Sci 2(1):5–11

Barbagelata PA, Mallarino AP (2012) Field correlation of potassium soil test methods based on dried and field-moist soil samples for corn and soybean. Nutr Manag Soil Plant Anal 77:318 327

Basak BB, Biswas DR (2010) Co-inoculation of potassium solubilizing and nitrogen fixing bacteria on solubilization of waste mica and their effect on growth promotion and nutrient acquisition by a forage crop. Biol Fertil Soils 46:641–648

Bennett PC, Choi WJ, Rogera JR (1998) Microbial destruction of feldspars. Mineral Manag 8(62A): 149–150

Bhuvaneswari TV, Turgeon BG, Bauer WD (1980) Early events in the infection of soybean (*Glycine max* L. Merr.) by Rhizobium japonicum I. location of infectable root cells. Plant Physiol 66:1027–1031

Biswas DR, Basak BB (2009) Influence of potassium solubilizing microorganism (Bacillus mucilaginosus) and waste mica on potassium uptake dynamics by sudan grass (Sorghum vulgare) grown under two Alfisols. Plant Soil Environ J 317:235–255

Brouder SM, Volenec JJ (2008) Impact of climate change on crop nutrient and water use efficiencies. Physiol Plant 133:705–724

Ciais P, Reichstein M, Viovy N, Granier A, Ogee J, Allard V et al (2005) Europe-wide reduction in primary productivity caused by the heat and drought in 2003. Nature 437:529–533

Das BK, Sen SP (1981) Effect of nitrogen, phosphorus and potassium deficiency on the uptake and mobilization of ions in Bengal gram (*Cicer arietinum*). J Biosci 3:249–258

Demidchik V (2014) Mechanisms and physiological roles of K^+ efflux from root cells. J Plant Physiol 171: 696–707

Dowdy RJ, Hutcheson TB (1963) Effects of exchangeable potassium level and drying on release and fixation of potassium by soils as related to clay mineralogy. Soil Sci Soc Am Proc 27:31–34

Egamberdiyeva D (2006) Enhancement of wheat performance with plant growth promoting bacteria in different soils. In: Mukerji KG, Manoharachary C (eds) Current concepts in botany. International Publishing House Ltd, New Delhi, India, pp 417–425

Haby VA, Ruselle MP, Skogley EO (1990) Testing soils for potassium, calcium, and magnesium. In: Westerman RL (ed) Soil testing and plant analysis. Soil Science Society of America, Madison, pp 181–227

Han HS, Lee KD (2005) Phosphate and potassium solubilizing bacteria effect on mineral uptake, soil availability and growth of eggplant. Res J Agric Biol Sci 1(2):176–180

Han HS, Supanjani, Lee KD (2006) Effect of co-inoculation with phosphate and potassium solubilizing bacteria on mineral uptake and growth of pepper and cucumber. Plant Soil Environ 52: 130–136

Hillel M (2008) Balanced crop nutrition: fertilizing for crop and food quality. Turk J Agric For 32:183–193

Holthusen D, Peth S, Horn R (2010) Impact of potassium concentration and matric potential on soil stability derived from rheological parameters. Soil Tillage Res 111:75–85

Kafkafi U (1990) The functions of plant K in overcoming environmental stress situations. In: Proceedings 22nd Colloquium, International Potash Institute, Bern, Switzerland, pp. 81–93.

Lack AJ, Evans DE (2005) Instant notes in plant biology, 2nd edn. Taylor and Francis, Oxford, p 351

Leigh RA, Wyn Jones RG (1984) A hypothesis relating critical potassium concentrations for growth to the distribution and functions of this ion in the plant cell. New Phytol 97:1–13

Lian B, Wang B, Pan M, Liu C, Henry H (2010) Microbial release of potassium from K-bearing minerals by thermophilic fungus *Aspergillus fumigatus*. Geochim Cosmochim Acta 72:87–98

Liu D, Lian B, Dong H (2012) Isolation of *Paenibacillus* sp. and assessment of its potential for enhancing mineral weathering. Geomicrob J 29:413–421

Luebs RE, Stanford G, Scott AD (1956) Relation of available potassium to soil moisture. Soil Sci Soc Am Proc 20:45–50

Marschner P (2012) Marschner's mineral nutrition of higher plants, 3rd edn. Academic, London, pp 178–189

Maurya BR, Meena VS, Meena OP (2014) Influence of Inceptisol and Alfisol's potassium solubilizing bacteria

(KSB) isolates on release of K from waste mica. Vegetos 27(1):181–187
McAfee J (2008) Potassium, a key nutrient for plant growth. Department of Soil and Crop Sciences. http://jimmcafee.tamu.edu/files/potassium
Meena VS, Maurya BR, Bahadur I (2014) Potassium solubilization by bacterial strain in waste mica. Bangladesh J Bot 43(2):235–237
Mengel K, Busch R (1982) The importance of the potassium buffer power on the critical potassium level in soils. Soil Sci 133:27–32
Mengel K, Kirkby EA (2001) Principles of plant nutrition. Kluwer Academic Publishers, Dordrecht
Nam MH, Jeong SK, Lee YS, Choi JM, Kim HG (2006) Effects of nitrogen, phosphorus, potassium and calcium nutrition on strawberry anthracnose. Plant Pathol 55: 246–249
Nieves-Cordones M, Aleman F, Martinez V, Rubio F (2014) K^+ uptake in plant roots. The systems involved, their regulation and parallels in other organisms. J Plant Physiol 171:688–695
Öborn I, Andrist-Rangel Y, Askekaard M, Grant CA, Watson CA, Edwards AC (2005) Critical aspects of potassium management in agricultural systems. Soil Use Manag 21:102–112
Ogaard AF, Krogstad T (2005) Ability of different soil extraction methods to predict potassium release from soil in ley over three consecutive years. J Plant Nutr Soil Sci 168:186–192
Prasad D, Singh R, Singh A (2010) Management of sheath blight of rice with integrated nutrients. Indian Phytopathol 63:11–15
Rehm G, Schmitt M (2002) Potassium for crop production. University of Minnesota. http://www.extension.umn.edu/distribution/cropsystems/dc6794.html
Schneider A, Tesileanu R, Charles R, Sinaj S (2013) Kinetics of soil potassium sorption-desorption and fixation. Commun Soil Sci Plant Anal 44:837–849
Sharpley AN (1989) Relationship between soil potassium forms and mineralogy. Soil Sci Soc Am J 52:1023–1028
Sheng XF (2005) Growth promotion and increased potassium uptake of cotton and rape by a potassium releasing strain of Bacillus edaphicus. Soil Biol Biochem 37: 1918–1922
Sheng XF, He LY (2006) Solubilization of potassium bearing minerals by a wild type strain of Bacillus edaphicus and its mutants and increased potassium uptake by wheat. Can J Microbiol 52:66–72
Sheng XF, Huang WY (2002a) Study on the conditions of potassium release by strain NBT of silicate bacteria. Sci Agric Sin 35:673–677
Sheng XF, Huang WY (2002b) Mechanism of potassium release from feldspar affected by the strain NBT of silicate bacterium. Acta Pedol Sin 39(6):863–871
Simonsson M, Andersson S, Andrist-Rangel Y, Hillier S, Mattsson L, Oborn I (2007) Potassium release and fixation as a function of fertilizer application rate and soil parent material. Geoderma 140:188–198
Sindhu SS, Dua S, Verma MK, Khandelwal A (2010) Growth promotion of legumes by inoculation of rhizosphere bacteria. In: Khan MS, Zaidi A, Musarrat J (eds) Microbes for legume improvement. Springer-Wien, New York/Heidelberg, pp 195–235
Sindhu SS, Parmar P, Phour M (2012) Nutrient cycling: potassium solubilization by microorganisms and improvement of crop growth. In: Parmar N, Singh A (eds) Geomicrobiology and biogeochemistry: soil biology. Springer-Wien, New York/Heidelberg, in press
Singh G, Biswas DR, Marwah TS (2010) Mobilization of potassium from waste mica by plant growth promoting rhizobacteria and its assimilation by maize (Zea mays) and wheat (Triticum aestivum L.). J Plant Nutr 33: 1236–1251
Sparks DL (1987) Potassium dynamics in soils. Adv Soil Sci 6:1–63
Syers JK (1998) Soil and plant potassium in agriculture. The International Fertilizer Society, York, p 32
Vincent JM (1970) A manual for the practical study of the root-nodule bacteria. Burgess and Son Ltd, Great Britain, p 45
Xie JC (1998) Present situation and prospects for the world's fertilizer use. Plant Nutr Fert Sci 4: 321–330
Xiufang H, Jishuang C, Jiangfeng G (2006) Two phosphate and potassium solubilizing bacteria isolated from Tianmu Mountain Zhejiang, China. World J Microbiol Biotechnol 22:983–990

Role of Rhizosphere Microflora in Potassium Solubilization

4

M.P. Raghavendra, S. Chandra Nayaka, and B.R. Nuthan

Abstract

The K^+ is not just the essential nutrient required to support optimal plant growth and yield, it is also an important signaling agent mediating a wide range of plant adaptive responses to abiotic and biotic stresses such as drought, salinity, oxidative stress, and apoptosis. The main source of K^+ for plant is soil which is taken up by the plant roots through the epidermal and cortical cells and transported to the shoot and distributed to the leaves. Movement of K^+ is facilitated by transport systems present in the cell membrane and the availability of which strongly determines crop yield. It is important to note that only a small percentage of the total K^+ of the soil exists in a form available for plant uptake. The remainder is in complex with other elements and organic matter, making it unavailable and even intensive agricultural practices is adding to K^+ deficiency in soil. It is already reported that large agricultural areas of world are deficient in K^+ availability which is a major concern today. In this connection, efforts to understand the K^+ uptake by plants and its solubilization from the K-bearing minerals such as waste muscovite, biotite, feldspars, orthoclase, illite, and mica have been undertaken. Recent investigations have shown that organic exudates of some microorganisms such as *Pseudomonas* spp., *Burkholderia* spp., *Acidithiobacillus ferrooxidans*, *Bacillus mucilaginosus*, *B. edaphicus*, *B. megaterium*, and *Aspergillus* spp., and even plant roots play a key role in releasing non-available K^+ from the minerals. The list of rhizosphere microflora with potassium solubilization is increasing, and hence the present chapter discusses the mechanism of K^+ solubilization and its role in signaling its uptake system in plants. Plant species effective in K^+ uptake and K^+-solubilizing microbial populations may be further key factors that control the K^+ release from soil minerals.

M.P. Raghavendra (✉) • B.R. Nuthan
Postgraduate Department of Microbiology, Maharani's Science College for Women, JLB Road, Mysore 570 005, Karnataka, India
e-mail: mpraghavendra@gmail.com

S. Chandra Nayaka
Department of Studies in Biotechnology, University of Mysore, Mysore 570 005, Karnataka, India

V.S. Meena et al. (eds.), *Potassium Solubilizing Microorganisms for Sustainable Agriculture*,
DOI 10.1007/978-81-322-2776-2_4

Keywords
Potassium solubilization • Rhizosphere microflora • K^+ uptake • Minerals

4.1 Introduction

Sustainable agriculture is taking front seat for the next green revolution, and it is considered to be the best alternative for conventional agriculture, which is considered to be slow and failing to yield to the expectations of human beings in agricultural needs. Conventional agriculture also demands more inputs in terms of chemical fertilizers and pesticides which is converting the agricultural practice nonprofit for farmers (Kumar et al. 2015; Singh et al. 2015). Along with these, the steady increase in corporate farming-based conventional methods in the last few decades, which is primarily profit driven, has increased the destabilization of rural communities as well as increased the detrimental effects on both the farmland ecology and surrounding natural environments. The sustainable agriculture in turn is eco-friendly with fewer inputs to agricultural fields, and supply of important nutrients required for plant growth will be taken care by microbial activity. Microorganisms are supporting plant growth from the time immemorial as a component of soil and in continuous association with roots. This type of agriculture uses a special farming technique wherein the environmental resources can be fully utilized and at the same time ensuring that no harm was done to it. Thus the technique is environment friendly and ensures safe and healthy agricultural products.

Soil microorganisms in particular and microbial populations in general are instrumental in fundamental processes such as biogeochemical cycles that drive stability and productivity of agroecosystems. Even today there is a continuous investigation on the diversity, dynamics, and importance of these microbes in agricultural productivity (Singh et al. 2011).

Plants require several nutrients for its growth. These nutrients are usually supplied through chemical fertilizers which contain inorganic salts of the macronutrients such as nitrogen, phosphorus, and potassium. Fertilizers having only one of these three nutrients are termed as straight fertilizers such as superphosphate, ammonium nitrate, and muriate of potash (a source of potassium), where as fertilizers that contain two or more mineral nutrients are referred to as compound fertilizers or mixed fertilizers used in combinations such as 10:14:10, which refers to the effective percentages of N, P_2O_5, and K_2O, respectively, in the fertilizer. Among these, potassium is an essential plant nutrient required in large amounts by plants and hence referred to as a macronutrient. The common fertilizer sources of K^+ are KCl, K_2SO_4-2$MgSO_4$, KNO_3, and K_2SO_4.

K is available mainly in minerals such as feldspar, mica, and clays such as montmorillonite, vermiculite, and illite. Biotite is also known as a mica-type mineral containing K, Fe, and Mg. However, of immediate concern to crop production is the portion of this nutrient that is in an exchangeable (available) form available for plant use. ~72 % of India's agricultural area needs immediate K fertilizers for increased crop production. However, K fertilizer cost has increased every year. This has led to an increase in the cost of rice production and thus farmer's income should decrease (Meena et al. 2014a).

K and N are absorbed by plants in larger amounts than either magnesium or calcium. Among these K plays an important role in numerous plant functions such as nutrient absorption, respiration, transpiration, and enzyme activity. K is unique because it does not become part of plant compounds, but remains in ionic form in the plant; it remains in plant residues after harvest and in manure which is quickly returned to the soil by water leaching through both the components. Once released to the soil, understanding its availability to plants is actually a mystery even to modern science. Even though effects of K on plants growth and yield are well known, but exactly why and how it happens is not completely understood (Xie 1998). Hence deficiency in plant-available P and K is still considered to be a major limiting factor to food production in many agricultural soils.

4.2 Sources of K in Soil

Soil is a primary source of several mineral nutrients including potassium. The continental crust comprises of potassium 2.1 % along with other elements. It is a cation generally bound to the surface of soil particles which is negatively charged. Addition of a cation such as potassium K^+ can displace another cation such as calcium (Ca^{2+}) from its binding on the surface of the soil particle and make it available for uptake by the root.

The total K^+ content of soils frequently exceeds 20,000 ppm. Most of K component of soil minerals is not available for plant growth. Due to large differences in soil parent materials and the effect of weathering of these minerals in different agricultural fields, the amount of K^+ supplied by soils to plants also varies. Therefore, the need of K^+ in a fertilization program varies for different agriculture production system.

K is available in soil such as unavailable, slowly available or fixed, and readily available or exchangeable. Several factors such as soil type, temperature, wetting and drying cycles, pH, aeration, moisture and even complex interactions of soil mineralogical and textural factors, and biological processes determine how readily structural or fixed K^+ in soil minerals may become available for crop uptake (Hinsinger and Jaillard 1993; Hinsinger et al. 1993; Wang et al. 2000). Typically 98 % of all soil K^+ is found in K-feldspars and phyllosilicates (Andrist-Rangel et al. 2010). The exact mechanism by which some of these factors influence the reactions required for release of K in soil is not clearly understood.

The imbalance in the potassium is generally caused by crop removal, fixation, and leaching. Aside from the contributions of rainwater, atmospheric deposits, and the recycling of elements contained in falling leaves and dead roots, soil minerals (primary and secondary minerals) constitute the main reservoir of inorganic nutrients for the long-lasting functioning of these ecosystems. In fact a fertile soil is distinguished from an infertile soil not by the fact that it contains more NPK but by the fact that the nutrients present in the soil are liberated with greater rapidity in the fertile than in the unfertile soil.

In most virgin soils, plants rely totally on the soluble and exchangeable K pools for their requirements. The unavailable form is contained in unweathered or slightly weathered minerals and is not available to plants. This accounts for 90–98 % of the total potassium in soils. Along with this after prolonged periods of agricultural production and removal of produce, pools of potassium in soil are slowly depleted. Once depletion of exchangeable potassium occurs, plants rely on the release of fixed potassium to recharge the exchangeable and soluble pools. However in many soils, the size of the fixed potassium pool, or the rate at which it is released, is insufficient to meet plant demand. This is particularly the case where intensive, high-yielding production systems are established.

4.3 Rhizosphere Microflora

Rhizosphere is the zone of soil surrounding a plant root where the biology and chemistry of the soil are influenced by the root exudates such as amino acids, organic acids, carbohydrates, sugars, mucilage, and proteins (Lugtenberg and Kamilova 2009). The ability of microorganisms in particular and rhizobacteria in particular to use these exudates as carbon source correlates with rhizosphere competence and also the microbial diversity around roots. The specific chemotoxicity between the root and its surroundings selectively decides the group of microbial flora to be colonized around the root, and hence these selected bacteria in turn may play an important role in plant growth promotion (Drogue et al. 2012).

In turn, rhizosphere microorganisms play a significant role on plant growth by various mechanisms, namely, fixation of atmospheric nitrogen, utilization of 1-aminocyclopropane-1-

carboxylic acid (ACC) as the sole N source, production of siderophores, and production of plant growth regulators (hormones). Further, many microorganisms in serpentine soil are able to solubilize "unavailable" forms of heavy metal-bearing minerals by excreting organic acids (Rajkumar et al. 2009).

The intensification of mineral weathering in the rhizosphere can be attributed, in part, to root-associated fungal and bacterial communities. There is increasing evidence that the mineral weathering by soil fungi and bacteria affects ion cycling and plant nutrition (Toro et al. 1997; Wallander 2000; Calvaruso et al. 2006). Until recently, mineral-weathering ability to release P and K from apatite and biotite has been mainly documented in mycorrhizal fungi (Wallander 2000; Blum et al. 2002; Glowa et al. 2003). However, the presence of weathering bacteria was reported in the rhizosphere or rhizoplane of several plants, including desert plants (*Helianthus annuus jaegeri* and annual sunflower) (Goldstein et al. 1999), mangrove trees (Vazquez et al. 2000), and cactus (Puente et al. 2004).

These bacteria isolated from rhizosphere of different plants are characterized for their ability to solubilize minerals. *Sphingomonas* spp. and *Collimonas* spp. isolated from oak-*Scleroderma citrinum* mycorrhizosphere, *Pseudomonas* spp. isolated from forest soil (Calvaruso et al. 2009; Uroz et al. 2007), *Rahnella* spp. isolated from rhizosphere of soybean, *Rhizobium* spp. isolated from agricultural soil (Kim et al. 1998), *Azospirillum* spp. isolated from rhizosphere of sugarcane (Rodriguez et al. 2004), *Acinetobacter* spp. isolated from rhizosphere of soybean, *Burkholderia* spp., and *Enterobacter* spp. (Vassilev et al. 2006) are reported to solubilize the soil minerals (Meena et al. 2013, 2014a, b Maurya et al. 2014).

It is well documented that the production of siderophores in the rhizosphere greatly contributes to the growth-promoting effect of bacteria on plants. Siderophore-producing bacteria *Variovorax paradoxus*, *Flavobacterium* spp., *Rhodococcus* spp., *Ralstonia* spp., *Arthrobacter* spp., *Stenotrophomonas* spp., and *Pseudomonas* spp. isolated from rhizosphere of *Brassica juncea* (Belimov et al. 2005); *Pseudomonas* spp., *Serratia marcescens*, and *Streptomyces* spp. isolated from different rhizosphere soil (Kuffner et al. 2008); *Microbacterium* spp., *S. liquefaciens*, *P. tolaasii*, *P. fluorescens*, *Ralstonia taiwanensis*, *Agrobacterium tumefaciens*, *Paracoccus* spp., and *Cellulomonas* spp. isolated from rhizosphere of perennial Gramineae grasses (Dell'Amico et al. 2005); and *Methylobacterium mesophilicum*, *M. extorquens*, *Methylobacterium* spp., *Burkholderia terricola*, *Okibacterium fritillariae*, *Rhodococcus fascians*, *Rhodococcus* spp., and *Microbacterium* spp. isolated from rhizosphere of *Thlaspi goesingense* (Idris et al. 2004) need to be evaluated for its capacity to solubilize K as they are already reported to be associated with metal-contaminated rhizosphere.

4.4 Potassium-Solubilizing Microorganisms (KSMs)

Evolution studies have proved that stones can be considered as a primary ecosystem which are rich in several minerals and are supplied to the plants through effective microbial mineral-leaching abilities due to its extraordinary adaptive capacity through simple gene regulation. These pioneer microbes will release the inorganic nutrients/minerals required for the establishment of other organisms including plants (Banfield et al. 1999). Even to date, it has been accepted that lichens are the first weathering pioneer organisms in both types of ecological successions, whereas now the list also includes bacteria from different genera such as *Anabaena* spp., *Bradyrhizobium* spp., *Burkholderia* spp., and *Collimonas* spp. (Mannisto and Haggblom 2006; Seneviratne and Indrasena 2006; Gorbushina 2007).

Nowadays, several bacteria including KSMs play an essential role in the environment by contributing to the release of key nutrients from primary minerals that are required not only for their own nutrition but also for that of plants. Fundamentally, KSB is a heterotrophic bacterium which obtains their energy and cellular carbon from preexisting organic material. Thus, they are important in the formation of humus in soil, the cycling of other minerals tied up in

Table. 4.1 Potential potassium-solubilizing microorganisms

Potassium-solubilizing microorganisms	References
Penicillium frequentans, *Cladosporium* spp.	Argelis et al. (1993)
Paenibacillus mucilaginosus	Liu et al. (2012) and Hu et al. (2006)
Aspergillus niger, *Penicillium* spp.	Sperberg (1958)
B. megaterium, *Pseudomonas* spp., *B. subtilis*	Taha et al. (1969)
B. megaterium, *E. freundii*	Taha et al. (1969)
Arthrobacter spp., *Bacillus* spp., *B. firmus*	Bajpai and Sundara (1971)
Aspergillus fumigatus, *Aspergillus candidus*	Banik and Dey (1982)
Pseudomonas aeruginosa	Sheng et al. (2003) and Badar et al. (2006)
B. mucilaginosus	Vandevivere et al. (1994), Welch and Ullman (1999), Sheng and He (2006), and Zakaria (2009)
Pseudomonas spp.	Krishnamurthy (1989)
Pseudomonas spp., *Burkholderia* spp., *Acidothiobacillicus ferrooxidans*, *Bacillus mucilaginosus*, *Bacillus edaphicus*, and *Bacillus megaterium*	Sheng et al. (2002)
Bacillus edaphicus	Sheng and He (2006)
A. fumigatus	Teng and Lian (2007)
Bacillus globisporus	Sheng et al. (2008)
Pseudomonas, *Burkholderia*, *Acidithiobacillus ferrooxidans*, *Bacillus mucilaginosus*, *Bacillus edaphicus*, *B. circulans* and *Paenibacillus* spp.	Sheng (2005), Lian et al. (2002), Li et al. (2006), and Liu et al. (2012)
Pseudomonas chlororaphis and *Bacillus megaterium*	Yu et al. (2012)
Bacillus altitudinis	Huang et al. (2013)
Bacillus spp.	Gundala et al. (2013)
Buttiauxella izardii, *Enterobacter cancerogenus*, *Burkholderia ubonensis*, *E. hormaechei*, and *Burkholderia pyrrocinia*	Ruangsanka (2014)
Klebsiella variicola, *Enterobacter cloacae*, *E. asburiae*, *E. aerogenes*, *Pantoea agglomerans*, *Agrobacterium tumefaciens*, *Microbacterium foliorum*, *Myroides odoratimimus*, and *Burkholderia cepacia*	Zhang and Kong (2014)

organic matter, and the prevention of buildup of dead organic materials. Besides, the KSB are aerobic bacteria which play an important role in maintaining soil structure by their contribution in the formation and stabilization of water-stable soil aggregates (WSA). In addition, Gram-positive bacteria can produce substance that stimulates plant growth or inhibit root pathogens (Zakaria 2009). Moreover, KSB are able to solubilize rock K mineral powder, such as micas, illite, and orthoclases (feldspar) through production and excretion of organic acids or chelate silicon ions to bring the K into solution (Ullman et al. 1996; Bennett et al. 1998; Meena et al. 2015a, b; Singh et al. 2015). Reports on KSMs are increasing and Table 4.1 summarizes the list of microorganisms in potassium mineral leaching and its solubilization.

4.4.1 Role of Earthworms' Gut Microflora in Mineralization of Soil Minerals

Even though earthworms are considered as main component of vermicompost, the role of its gut microflora is the degradation of complex polysaccharides and also helps in mineralizing soil minerals. Earthworms are an ecosystem's engineers, contributing to a wide range of nutrient cycling and geochemical processes in the ecosystem. Their activities can increase rates of silicate mineral weathering. Their intestinal microbes usually are thought to be one of the key drivers of mineral degradation mediated by earthworms, but the diversities of the intestinal microorganisms which were relevant with mineral weathering are unclear. The phylogenetic tree based on the 16S rRNA gene sequences of the clones obtained from the soil in which earthworms fed for 10 days revealed that the library contained sequences homologous to *Verrucomicrobia* (26.86 %); *Bacteroidetes* (24.30 %); *Proteobacteria* (12.8 %) including *Alphaproteobacteria* (0.26 %), *Betaproteobacteria* (0.77 %), *Gammaproteobacteria* (7.42 %), and *Deltaproteobacteria* (4.35 %); *Firmicutes* (2.05 %); *Actinobacteria* (0.52 %); and two unidentified OTUs. The results of the phylogenetic tree and homologous sequences indicated the presence of *Acidobacteriaceae* (*Acidobacteria*, *Acidobacteriales*), *Opitutae* group (*Verrucomicrobia*), *Flavobacterium* spp. (*Bacteroidetes*, *Flavobacteria*, *Flavobacteriales*, *Flavobacteriaceae*), *Sphingobacteriales* (*Bacteroidetes*, *Sphingobacteria*), and *Aquicella* spp. (*Proteobacteria*, *Gammaproteobacteria*, *Legionellales*, *Coxiellaceae*) (Liu et al. 2011). These group organisms need to be evaluated further for its efficacy in mineral weathering in general and K solubilization in particular along with siderophore-producing bacteria as already mentioned.

4.4.2 Transformation of Potassium by Soil Microbes

K is considerably less susceptible to effects of microorganisms in its relationships to plant development in soils. It belongs to a large group of elements in soil which are not affected in a great variety of ways, on account of the fact that they do not enter into organic combination to the extent of such elements as nitrogen or sulfur. Its entrance into organic compounds is more generally confined to replacing the hydrogen of the acid groups where it forms the salts of these acids. In plant residues a large portion of the K is present in inorganic form, occurring in the various fluids of the cells. Upon ignition of the organic substances, the potassium remains in the ash residue. Another factor appreciably limiting the capacities of microorganisms to affect the element is its stable character. It is not oxidized and reduced and consequently cannot be carried through the extensive series of changes common to nitrogen, sulfur, and iron. The influence of microorganisms on such elements as potassium is confined to causing its solution from organic and inorganic substances and its assimilation in growth of the microbial cells. Considerable potassium becomes added to soils in the form of various organic and inorganic compounds. Stable manures, green manures, plant stubble, and microbial cells all contain K (Wang et al. 2000).

When these organic substances are decomposed by microorganisms in the soil, the potassium is liberated in forms available to higher plants. A small part of this K may be re-assimilated by the microbes which bring about the decomposition processes, thus part of the potassium may be temporarily removed from circulation. The ash of bacteria and fungi usually contains between 5 and 40 % of potassium calculated as K_2O. The inorganic forms of K added to the soil as fertilizers are generally soluble, but the minerals in the soil which contain potassium are quite insoluble. Solution of potassium from these minerals is accelerated by interaction with various acids produced by microorganisms, as shown in the case of orthoclase:

$$\begin{aligned} & Al_2O_3\tilde{n}K_2O\tilde{n}6SiO_2 + 8HNO_3 \\ & \quad = 2Al(NO_3)_3 + 2KNO_3 + 6SiO_2 + 4H_2O \end{aligned}$$

The products of nitrification and sulfur oxidation, as well as the carbonic acid produced by the microbial population in general, aid in such solution of potassium.

4.5 Role of Potassium in Plant Growth and Developments

The nature provides both organic and inorganic nutrients required for living organisms through soil. Some of these inorganic elements, such as sodium, potassium, calcium, and magnesium, are present in bulk because it is available in large concentrations.

Potassium, present in the form of cation K^+ within plants, plays an important role in the regulation of the osmotic potential of plant cells. Enzymes involved in respiration and photosynthesis are activated by K^+. Mottled or marginal chlorosis is the first observable symptom of potassium deficiency which is then develops into necrosis primarily at the leaf tips, at the margins, and between veins. These necrotic lesions may initially form at the leaf tips and margins and then extend toward the leaf base in many monocots. These symptoms will appear first on mature leaves than younger leaves because potassium can be mobilized more to the younger leaves. The leaves may also curl and crinkle. The stems of potassium-deficient plants may be slender and weak, with abnormally short internodal regions. In potassium-deficient corn, the roots may have an increased susceptibility to root-rotting fungi present in the soil, and this susceptibility, together with effects on the stem, results in an increased tendency for the plant to be easily bent to the ground (lodging).

When relating acute deficiency symptoms to a particular essential element, an important clue is the extent to which an element can be recycled from older to younger leaves. Some elements, such as nitrogen, phosphorus, and potassium, can readily move from leaf to leaf; others, such as boron, iron, and calcium, are relatively immobile in most plant species. If an essential element is mobile, deficiency symptoms tend to appear first in older leaves.

Maintenance of osmotic balance and for the generation of ionic gradients across membranes, Na^+ and K^+ (together with H^+ and Cl^-) which bind weakly to organic ligands, are ideally suited. In contrast, Mg^{2+} and Ca^{2+} with intermediate-binding strengths to organic ligands can play important structural roles and, in the particular case of Ca^{2+}, serve as a charge carrier and a trigger for signal transmission. Zn^{2+} not only plays a structural role but can also fulfill a very important function as a Lewis acid. Organic ligands which bind tightly with redox metal ions like iron and copper play an important role in oxygen transport besides participating in innumerable redox reactions.

The regulation of the flow of ions across cell membranes is absolutely essential for the functioning of living cells. Because of the hydrophobicity of cellular membranes, the energetically driven preference of ionic species such as Na, K, Cl, H, and Ca to cross, never mind to find them preferentially on one side or the other of a biological membrane, would be impossible. Without ionic gradients, which maintain high concentrations of K^+ within the cell and low concentrations of Na^+, cells would not be able to carry out their normal metabolic activities. This means, in simplistic terms, that some molecular machines must be able to distinguish between Na^+ and K^+ ions (presumably unhydrated, since the degree of hydration could make for difficulties in discrimination).

Another significant function of potassium is in the construction of the cuticle layer. This cuticle layer is the plant's first line of defense to disease and insect attack. Plants receiving adequate potassium nutrition have a stronger enzyme activity and are capable of withstanding more fungal attack. Increasing potassium will reduce the amount of disease both in the root and in the above ground parts of the plant.

In plants, hence potassium plays a vital role in a wide range of both biophysical and biochemical processes. It exists as a monovalent cation and does not participate in covalent binding; it functions to maintain charge balance. The preservation of cell turgor pressure is very sensitive to a limited K^+ supply. Indeed, due to its high mobility, K^+ is usually the principle cation that contributes to vacuole and cell expansion (Hamamoto and Uozumi 2014). Nonetheless, over a longer time scale, it can be replaced by Na^+ (Jeschke and Wolf 1988) and/or organic solutes, explaining the observed highly variable (10–200 mM) vacuolar K^+ levels. In contrast, cytoplasmic levels are relatively stable, near 100 mM. It is suggested that as total tissue K^+ concentration declines, the cytoplasm maintains a homeostatic concentration of K^+ to enable K^+-dependent processes. Consequently, any initial changes in tissue K^+ concentrations are likely to be at the expense of vacuolar K^+, with other solutes being diverted to the vacuole to maintain the osmotic potential.

Extent of turgor pressure (usually around 4 atm for *E. coli*) is determined by the concentration of intracellular solutes (glutamate, potassium). To establish cell turgor and maintain cell electroneutrality, K^+ will act as a principle cation and > 70 enzymes in plants will utilize K^+ as a cofactor.

At the biochemical level, K^+ plays an important role in the activation of many enzymes, especially in protein and starch synthesis, as well as in respiratory and photosynthetic metabolism. Starch synthesis, for instance, has a requirement of about 50 mM K^+ for normal activity. In line with this level, most K^+-requiring enzymes need about 10–50 mM of K^+ for optimal activity. Interestingly, these levels may be reduced by the substitution of other monovalent cations for K^+. For example, Rb^+, Cs^+, and NH^{4+} are about 80 % as effective as K^+ at maintaining starch synthetase activity, while Na^+ is approximately 20 % as effective. Hence, in both biophysical and biochemical processes, K^+ may be the most efficient monovalent cation, although to a certain extent, it can be replaced.

4.5.1 Potassium Uptake in Plants

There are three families of plant K^+ transport systems that are active at the plasma membrane: the Shaker K^+ channel family, comprised of voltage-gated channels that dominate the plasma membrane conductance to K^+ in most environmental conditions, and two families of transporters, the HAK/KUP/KT K^+ transporter family, which includes some high-affinity transporters, and the HKT K^+ and/or Na^+ transporter family, in which K^+-permeable members seem to be present in monocots only (Very et al. 2014).

A number of regulatory partners have been identified that control the targeting or the activity of plant Shaker channels at the plasma membrane. These include β-subunits (Zhang et al. 1999), 14-3-3 proteins (Sottocornola et al. 2006), different types of kinases (Xu et al. 2006; Zhao et al. 2013), phosphatase (Chérel et al. 2002; Lee et al. 2007), and SNAREs (Sutter et al. 2006; Honsbein et al. 2009).

4.5.2 Potassium Deficiencies in Plants

Proteomic study of the plant response to potassium deficiency has revealed the overexpression of several proteins compared to normal K availability. HAK/KUP transporters at the tonoplast (Hamamoto and Uozumi 2014) are one such example, which are energized via coupling to the transtonoplast motive force. This transporter helps in the active transport of the K from the vacuole to the cytoplasm against concentration gradient (Walker et al. 1996). This indicates that the prolonged K starvation imparts additional energy burden on plants directly affecting the crop yield.

Sensing of and response to K^+ by the plant appear to be mediated by a number of different means. These include the cell membrane potential, reactive oxygen species (ROS), Ca^{2+}, hormones (ethylene, jasmonic acid, or cytokinins), and direct sensing of the environmental K^+ concentrations by K^+ channels. During K^+ starvation, there is a transcriptional upregulation of *Arabidopsis* HAK5 and its homologues from other plant species (Alemán et al. 2009; Martínez-Cordero et al. 2004; Nieves-Cordones et al. 2007; Wang et al. 2002) and repressed after K^+ resupply (Ahn et al. 2004; Armengaud et al. 2004; Gierth et al. 2005; Nieves-Cordones et al. 2008).

Hyperpolarization of the membrane potential is considered to be the first effects of root cells to K^+ deprivation (Amtmann et al. 2006). This suggests a role for the membrane potential in the regulation of this type of genes. In addition, AtHAK5 expression is dependent on reactive oxygen species (ROS) production and K^+-deficient plants rapidly accumulate ROS (Hernandez et al. 2012; Shin et al. 2005; Shin and Schachtman 2004; Demidchik 2014). Initiation of the signal cascades which activates the K^+ uptake epidermal cells in the root tip is by a rapid accumulation of ROS at 24 h of K^+ starvation. After the accumulation of ROS, the activity of antioxidative enzymes such as superoxide dismutase, peroxidase, and ascorbate peroxidase, is increased, and oxidative damage resulted in

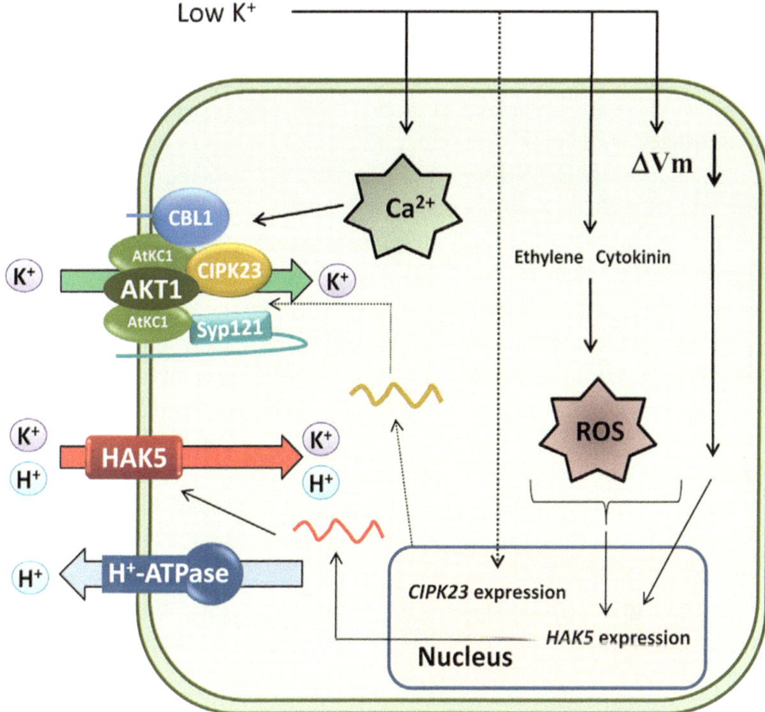

Fig. 4.1 K^+ sensing in plants: elements involved in low K^+ response of root cells (Nieves-Cordones et al. 2014)

prolonged K^+ starvation (Hernandez et al. 2012). ROS production in response to K^+ deficiency and modulation of AtHAK5 expression after K^+ starvation is mainly contributed by *Arabidopsis* class III peroxidase RCI3 (Kim et al. 2010). Ethylene inhibitors will block the production of ROS in K^+ -deficient plants, indicating that this hormone is involved in low K^+ signaling upstream ROS (Fig. 4.1).

4.6 Roles of KSMs in Agriculture

Soil erosion, crop uptake, runoff, and leaching will decrease the potassium content in soils (Sheng and Huang 2002). In these instances, potassium fertilizer must be applied to ensure healthy plant growth. The direct application of rock phosphate (RP) and potassium (rock K) materials is more agronomically useful and even environmentally feasible than soluble P and K (Rajan et al. 1996; Ranawat et al. 2009). It is also considered as cheaper sources of P and K, but they are not easily available to plants because of its slow release of the mineral, and its use seems to be insignificant in present-day agriculture due to overemphasis on more yields at short time. In contrast, plant growth, nutrition, root growth pattern, plant competitiveness, and responses to external stress factors are improved by the application of P and K solubilizers (Vessey 2003; Sheng 2005; Setiawati and Handayanto 2010; Ekin 2010).

Increased availability and uptake of P and K by the maize plants grown on P and K-limited soil by co-inoculating the PSB (*B. megaterium* var. *phosphaticum*) and KSB (*B. mucilaginosus* and *B. subtilis*) with direct application of RP and K materials [R+ (P+ K)] into the soil was reported by Abou-el-Seoud and Abdel-Megeed (2012). It was also observed that plant growth is increased by the co-inoculation of beneficial microorganisms.

Increasing the bioavailability of P and K in the soils may also be achieved by adding rock materials along with co-inoculation (Lin et al. 2002; Han and Lee 2005; Han et al. 2006), and it will also lead to increased P and K uptake and plant growth (Han et al. 2006; Chen et al. 2006; Eweda et al. 2007;

Jorquera et al. 2008; Sabannavar and Lakshman 2009).

The highest biomass and seedling height of *Zea mays* were observed by the use of biofertilizer containing arbuscular mycorrhizal fungus (*G. mosseae*), N fixer (*A. chroococcum*), P solubilizers (*B. megaterium*), and K solubilizers (*B. mucilaginous*) (Wua et al. 2005). Maize and wheat plants which are inoculated with *B. mucilaginosus*, *A. chroococcum*, and *Rhizobium* spp. made a good source of K availability for plant growth from waste mica, which in turn acted as a source of potassium (Singh et al. 2010). Therefore, KSB are extensively used as biofertilizers in Korea and China as significant areas of cultivated soils in these countries are deficient in soil-available K (Xie 1998). Thus, application of KSB as biofertilizer for agriculture improvement can reduce the use of agrochemicals and support eco-friendly crop production under sustainable agriculture (Sheng et al. 2003; Sindhu et al. 2010).

According to Simonsson et al. (2007), a study was conducted on the release and fixation rates of K (kg ha^{-1} $year^{-1}$) estimated in five long-term field experiments located on varying parent materials in South and Central Sweden, each having an array of K fertilizer application rates. The resulted estimates ranged from 65 ± 07 kg ha^{-1} $year^{-1}$ of K released from soil reserves to 85 ± 10 kg ha^{-1} $year^{-1}$ of K trapped by fixation. Release and fixation rates were highly dependent on the soil K balance, confirming that these are reversible processes that depend on plant uptake and fertilizer input. Soils, with a texture from sandy loam to clay, showed release rates approaching the net K output, when not K fertilized. These calculations carried out by them were based on accumulated K inputs and outputs from the soil during the 30 years and on changes in topsoil and subsoil exchangeable K.

It is estimated that 1 ha of soil may have anywhere between several tonnes to several hundred tonnes of K held in the structure of micas and K feldspars within the rooting zone (Sparks and Huang 1985). Of this, exchangeable K is traditionally believed to form a reserve that is available for uptake by plants in the short term. The size of this pool may be a few hundred kg ha^{-1}, i.e., corresponding to the demand of just a few years of intense cropping. It has long been known that crops may also extract K from non-exchangeable sources on a time scale of only a few years (Gholston and Hoover 1948). Nonetheless, negative K balances (harvest > fertilization) may be a concern in agriculture, if release of fixed or structural mineral K from soil minerals does not compensate for nutrient offtake and losses. Indeed, negative field K balances are common in mixed systems with both livestock and arable farming, especially in organic farming systems, which mainly rely on on-farm nutrient recycling and internal sources (Oborn et al. 2005).

Recently *Frateuria aurantia*, a potassium-mobilizing bacteria (KMB) isolated from banana rhizosphere, has the ability to mobilize the elementary or mixture of potassium which can be easily absorbed by plants. It is estimated that 50–60 % of potash chemical fertilizers usage can be reduced by using *F. aurantia* as a biological inoculant. It belongs to the family *Pseudomonadaceae* and shown to have the extra ability to mobilize K in almost all types of soils especially, low K content soils, soils of pH 5–11, and it survived in the temperature up to 42 °C. Pindi and Satyanarayana (2012) proved that this KMB can be used as a biofertilizer and can be applied in combination with *Rhizobium*, *Azospirillum*, *Azotobacter*, PSM, etc.

4.7 Mechanism of K Solubilization by Soil Microflora

The soil is influenced by the biological weathering of minerals. Mineral weathering is more rapid in the rhizosphere than in bulk soil without roots. This intensification of the mineral-weathering process can be linked to the pH status of the rhizosphere, which is generally different to that of bulk soil (Fig. 4.2). This variation in pH and the mineral-weathering intensification can be due to the plants themselves: their growing roots might physically disrupt the mineral particles, exposing new surface areas to weathering,

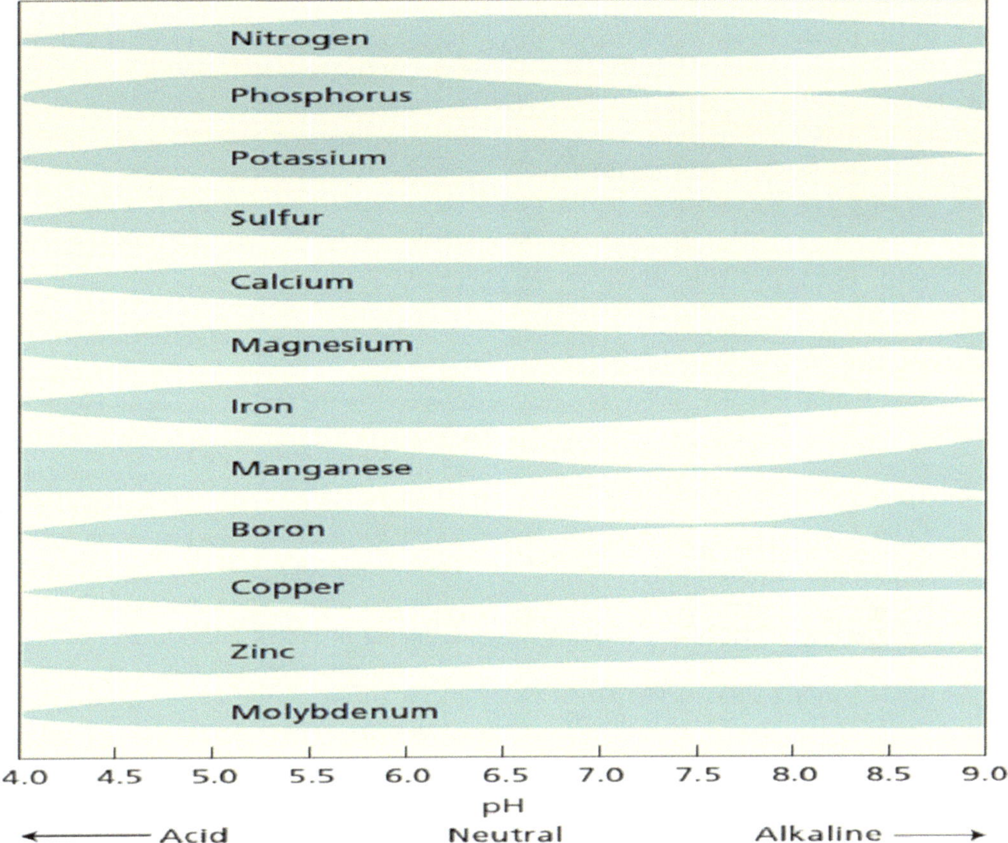

Fig. 4.2 Influence of soil pH on the availability of nutrient elements in organic soils. The width of the shaded areas indicates the degree of nutrient availability to the plant root. All of these nutrients are available in the pH range of 5.5–6.5 (Lucas and Davis 1961)

and/or the exuded metabolites (such as organic acids and protons) might act directly on mineral-weathering kinetics (Barker et al. 1997).

The protons associated with organic acid molecules decrease the pH of the solution and, therefore, induce the releasing of cations such as iron, potassium, and magnesium (Uroz et al. 2009). Major factors that lower the soil pH are the decomposition of organic matter and the amount of rainfall. Carbon dioxide is produced as a result of the decomposition of organic material and equilibrates with soil water in the following reaction:

$$CO_2 + H_2O \rightarrow H^+ + HCO_3^-$$

This reaction releases hydrogen ions (H^+), lowering the pH of the soil. Microbial decomposition of organic material also produces ammonia and hydrogen sulfide that can be oxidized in the soil to form the strong acids such as nitric acid (HNO_3) and sulfuric acid (H_2SO_4)y. Hydrogen ions also displace K^+, Mg^{2+}, Ca^{2+}, and Mn^{2+} from the cation-exchange complex in a soil. Leaching then may remove these ions from the upper soil layers, leaving a more acid soil. By contrast, the weathering of rock in arid regions releases K^+, Mg^{2+}, Ca^{2+}, and Mn^{2+} to the soil, but because of the low rainfall, these ions do not leach from the upper soil layers, and the soil remains alkaline.

The monovalent cations such as potassium (K^+) can form electrostatic bonds with the carboxylic groups of many organic acids. Nonetheless, much of the potassium that is accumulated by plant cells and functions in osmotic regulation and enzyme activation remains in the cytosol and the vacuole as the free ion.

The chemical processes involved in the weathering of rocks are those of hydrolysis, oxidation, hydration, solution, and carbonation or carbonate formation. The following reactions illustrate the chemical changes involved in the weathering of orthoclase and olivine, two rock-forming minerals:

$$\underset{\text{Orthoclase}}{2KAlSi_3O_8} + \underset{\text{Water}}{2H_2O} + \underset{\text{Carbon dioxide}}{CO_2} \rightarrow \underset{\text{Kaolinite}}{H_4Al_2Si_2O_9} + \underset{\text{Potassium carbonate}}{K_2CO_3} + \underset{\text{Silica}}{4SiO_2}$$

$$\underset{\text{Olivine}}{12MgFeSiO_4} + 26H_2O + 3O_2 \rightarrow \underset{\text{Serpentine}}{4H_4Mg_3Si_2O_9} + \underset{\text{Silica}}{4SiO_2} + \underset{\text{Limonite}}{6Fe_2O_3.3H_2O}$$

4.7.1 Role of Acids in Mineral Solubilization

B. mucilaginosus and *B. edaphicus* can generate polysaccharide and carboxylic acids, such as tartaric acid and citric acid, to solubilize K compounds (Lian et al. 2002). Using K-solubilizing microbes to increase the concentration of available K ions in the soil may mitigate K deficiency (Barker et al. 1998). New facultative alkaliphilic *Bacillus* spp. are isolated from mica mines of Nellore District of Andhra Pradesh, India. The analysis of the culture medium by high-pressure liquid chromatography identified gluconic acid as the main organic acid released by *Bacillus* spp. SVUNM9. This study is the first report on the isolation and characterization of indigenous KSB from mica ore (Gundala et al. 2013).

The release of potassium in accessible form from K-bearing minerals in soils has been reported by a wide range of bacteria, namely, *Pseudomonas*, *Burkholderia*, *Acidithiobacillus ferrooxidans*, *B. mucilaginosus*, *B. edaphicus*, *B. circulans*, and *Paenibacillus* spp. (Sheng 2005; Lian et al. 2002; Li et al. 2006; Liu et al. 2012). These KSB were found to dissolve potassium, silicon, and aluminum from insoluble K-bearing minerals such as micas, illite, and orthoclases, by excreting organic acids which either directly dissolved rock K or chelated silicon ions to bring K into the solution (Bennett et al. 1998).

Solubilization of K and silicon from the silicate minerals by the KSMs resulted mostly from the action of organic acids. Gluconic acid seemed to be the most active agent for the solubilization of the three silicate minerals. Gluconic and acetic acids were likely involved in the solubilization of feldspar. The strain could be acid or alkali and salt tolerant and temperature resistant (Sheng et al. 2008).

The *B. altitudinis* strain could accelerate weathering of potash feldspar, change mineral surface morphology, and induce the formation of new mineral complex. The strain dissolved potash feldspar and significantly released more Si, Al, and Fe elements by producing more organic acids. Many bacteria and some spherical minerals were observed on the surfaces of the feldspar and the energy-dispersive spectrometry analysis showed that the new minerals contained more Fe. After 30 days, siderite might be the newly formed mineral identified by X-ray diffraction (XRD) in the mineral-weathering process (Huang et al. 2013).

KSB, such as *B. mucilaginosus*, increased K availability in agricultural soils and increased mineral content in plant (Sheng et al. 2002; Zakaria 2009). *B. mucilaginosus* increases the rate of dissolution of silicate and alumino-2 silicate minerals, and it releases the K^+ and SiO_2 from the crystal lattice primarily by generating organic acids (Vandevivere et al. 1994) and accelerates the dissolution of a variety of silicates by the production of extracellular polysaccharides (EPS) (Welch and Ullman 1999). The dispute about the mechanism by which *B. mucilaginosus* decomposes silicate minerals and releases K^+ and SiO_2 may have severely limited the use of the organism in agriculture as a form of biological K fertilizer. Polysaccharides strongly adsorbed the organic acids and attached to the surface of the mineral, resulting in an area of high concentration of organic acids near the mineral (Liu et al. 2006).

4.8 Effect of KSMs on K Availability in Soils

Organic acids as explained earlier are considered to play an important role in bringing rock K (K-containing primary minerals) to solution through dissolution (Basak and Biswas 2009). In recent studies application of rock K along with the inoculation of KSMs to soil has yielded good results. Cotton, oilseed rape, pepper, cucumber, and Sudan grass resulted beneficial effects when KSMs along with mica were applied to the soil (Sheng et al. 2002; Han and Lee 2005, 2006). The exudates from the KSMs can also effectively increase the release of K from clay minerals. After a series of incubation trials carried out in the laboratory, it revealed that the application of inoculated feldspars into the soil enhances about 40–60 % of K solubility and plant K uptake (Han and Lee 2006; Basak and Biswas 2009; Abou-el-Seoud and Abdel-Megeed 2012). A less information is available on the field application of such methods, which is most likely due to difficulties in soil inoculation under field conditions. In order to evaluate the potential of such applications for agricultural production systems, we need more field studies that evaluate their effect on both soil as well as control conditions.

4.9 Future Approach for KSMs

The possible alternative to solve K deficiency in soil could be to fully exploit the reservoir of K present in the soil. Because minerals are present in a myriad of environments, more bacterial species with mineral-weathering ability, as well as additional molecular mechanisms involved in this process, will continue to be discovered in the future. The use of KSMs has several advantages over conventional chemical fertilizers in agriculture such as the following (a) microbial products are eco-friendly compared to chemical fertilizers which have a residual effect in the soil; (b) microbial products or microbes will not accumulate in the environment; (c) due to less generation time, repeated inoculation to field is not required; and (d) KSMs are not considered harmful to ecological processes or the environment.

There are many questions which need to be addressed in the future about the application of microorganisms to enhance K solubility in soil. Along with this more studies are needed to identify additional microbial genes, enzymes, and metabolites involved in mineral weathering such as metagenomic analysis of the different agricultural soil. It is also possible to develop methods combining molecular and isotopic approaches, such as stable-isotope probing of nucleic acids, to identify mineral-weathering organisms and quantify their relative contributions.

References

Abou-el-Seoud II, Abdel-Megeed A (2012) Impact of rock materials and biofertilizations on P and K availability for maize (*Zea maize*) under calcareous soil conditions. Saudi J Biol Sci 19:55–63

Ahn SJ, Shin R, Schachtman DP (2004) Expression of KT/KUP genes in *Arabidopsis* and the role of root hairs in K^+ uptake. Plant Physiol 134:1135–1145

Alemán F, Nieves-Cordones M, Martínez V, Rubio F (2009) Differential regulation of the HAK5 genes encoding the high-affinity K^+ transporters of *Thellungiella halophila* and *Arabidopsis thaliana*. Environ Exp Bot 65:263–269

Amtmann A, Hammond JP, Armengaud P, White PJ, Callow JA (2006) Nutrient sensing and signalling in plants: potassium and phosphorus. Adv Bot Res (Academic Press, London) 43:209–257

Andrist-Rangel Y, Hillier S, Öborn I, Lilly A, Towers W, Edwards AC, Paterson E (2010) Assessing potassium reserves in northern temperate grassland soils: a perspective based on quantitative mineralogical analysis and aqua-regia extractable potassium. Geoderma 158:303–314

Argelis DT, Gonzala DA, Vizcaino C, Gartia MT (1993) Biochemical mechanism of stone alteration carried out by filamentous fungi living in monuments. Biogeochemistry 19:129–147

Armengaud P, Breitling R, Amtmann A (2004) The potassium-dependent transcriptome of *Arabidopsis* reveals a prominent role of jasmonic acid in nutrient signaling. Plant Physiol 136:2556–2576

Badar MA, Shafei AM, Sharaf El-Deen SH (2006) The dissolution of K and phosphorus bearing minerals by silicate dissolving bacteria and their effect on sorghum growth. Res J Agric Biol Sci 2:5–11

Bajpai PD, Sundara R (1971) Phosphate solubilizing bacteria, solubilization of phosphate in liquid culture by selected bacteria as affected by different pH values. Soil Sci Plant Nutr 17:41–43

Banfield JF, Barker WW, Welch SA, Taunton A (1999) Biological impact on mineral dissolution: application of the lichen model to understanding mineral weathering in the rhizosphere. Proc Natl Acad Sci U S A 96:3403–3411

Banik S, Dey BK (1982) Available phosphate content of an alluvial soil as influenced by inoculation of some isolated phosphate solubilizing microorganisms. Plant Soil 69:353–364

Barker WW, Welch SA, Banfield JF (1997) Biogeochemical weathering of silicate minerals. In: Banfield JF, Nealson KH (eds) Geomicrobiology: interactions between microbes and minerals, vol 35. Minerological Society of America, Washington, DC, pp 391–428

Barker WW, Welch SA, Chu S, Banfield J (1998) Experimental observations of the effects of bacteria on aluminosilicate weathering. Am Mineral 83:1551–1563

Basak BB, Biswas DR (2009) Influence of potassium solubilizing microorganism (*Bacillus mucilaginosus*) and waste mica on potassium uptake dynamics by sudan grass (*Sorghum vulgare* Pers.) grown under two Alfisols. Plant Soil 317:235–255

Belimov AA, Hontzeas N, Safronova VI, Demchinskaya SV, Piluzza G, Bullitta S, Glick BR (2005) Cadmium-tolerant plant growth-promoting bacteria associated with the roots of Indian mustard (*Brassica juncea* L. Czern.). Soil Biol Biochem 37:241–250

Bennett PC, Choi WJ, Rogers JR (1998) Microbial destruction of feldspars. Miner Mag 8:149–150

Blum JD, Klaue A, Nezat CA, Driscoll CT, Johnson CE, Siccama TG, Eagar C, Fahey TJ, Likens GE (2002) Mycorrhizal weathering of apatite as an important calcium source in base-poor forest ecosystems. Nature 417:729–731

Calvaruso C, Turpault MP, Frey-Klett P (2006) Root-associated bacteria contribute to mineral weathering and to mineral nutrition in trees: a budgeting analysis. Appl Environ Microbiol 72:1258–1266

Calvaruso C, Mareschal L, Turpault MP, Leclerc E (2009) Rapid clay weathering in the rhizosphere of Norway spruce and oak in an acid forest ecosystem. Soil Sci Soc Am J 73:331–338

Chen YP, Rekha PD, Arun AB, Shen FT, Lai WA, Young CC (2006) Phosphate solubilizing bacteria from subtropical soil and their tricalcium phosphate solubilizing abilities. Appl Soil Ecol 34:33–41

Chérel I, Michard E, Platet N, Mouline K, Alcon C, Sentenac H, Thibaud JB (2002) Physical and functional interaction of the *Arabidopsis* K^+ channel AKT2 and phosphatase AtPP2CA. Plant Cell 14:1133–1146

Dell'Amico E, Cavalca L, Andreoni V (2005) Analysis of rhizobacterial communities in perennial Graminaceae from polluted water meadow soil, and screening of metal-resistant, potentially plant growth-promoting bacteria. FEMS Microbiol Ecol 52:153–162

Demidchik V (2014) Mechanisms and physiological roles of K^+ efflux from root cells. J Plant Physiol 171:696–707

Drogue B, Doré H, Borland S, Wisniewski-Dyé F, Prigent-Combaret C (2012) Which specificity in cooperation between phytostimulating rhizobacteria and plants. Res Microbiol 163:500–510

Ekin Z (2010) Performance of phosphate solubilizing bacteria for improving growth and yield of sunflower (*Helianthus annuus* L.) in the presence of phosphorus fertilizer. Afr J Biotechnol 9(25):3794–3800

Eweda WE, Selim SM, Mostafa MI, Abd El-Fattah DA (2007) Use of *Bacillus circulans* as bio-accelerator enriching composted agricultural wastes identification and utilization of the microorganism for compost production. In: Proceedings of the 12th conference of the microbiology. The Egyptian Soc of App Micro (ESAM), Giza, pp 43–65

Gholston LE, Hoover CD (1948) The release of exchangeable and nonexchangeable potassium from several Mississippi and Alabama soils upon continuous cropping. Soil Sci Soc Am Proc 13:116–121

Gierth M, Maser P, Schroeder JI (2005) The potassium transporter AtHAK5 functions in K^+ deprivation-induced high-affinity K^+ uptake and AKT1 K^+ channel contribution to K^+ uptake kinetics in *Arabidopsis* roots. Plant Physiol 137:1105–1114

Glowa KR, Arocena JM, Massicotte HB (2003) Extraction of potassium and/or magnesium from selected soil minerals by *Piloderma*. Geomicrobiol J 20:99–111

Goldstein AH, Braverman K, Osorio N (1999) Evidence for mutualism between a plant growing in a phosphate-limited desert environment and a mineral phosphate solubilizing (MPS) rhizobacterium. FEMS Microbiol Ecol 30:295–300

Gorbushina AA (2007) Life on the rocks. Environ Microbiol 9:1613–1631

Gundala PB, Chinthala P, Sreenivasulu B (2013) A new facultative alkaliphilic, potassium solubilizing, *Bacillus* Sp. SVUNM9 isolated from mica cores of Nellore District, Andhra Pradesh, India. Res Rev: J Microbiol Biotechnol 2(1):1–7

Hamamoto S, Uozumi N (2014) Organelle-localized potassium transport systems in plants. J Plant Physiol 171:743–747

Han HS, Lee KD (2005) Phosphate and potassium solubilizing bacteria effect on mineral uptake, soil availability and growth of eggplant. Res J Agric Biol Sci 1:176–180

Han HS, Supanjani E, Lee KD (2006) Effect of co-inoculation with phosphate and potassium solubilizing bacteria on mineral uptake and growth of pepper and cucumber. Plant Soil Environ 52(3):130–136

Hernandez M, Fernandez-Garcia N, Garcia-Garma J, Rubio-Asensio JS, Rubio F, Olmos E (2012)

Potassium starvation induces oxidative stress in *Solanum lycopersicum* L. roots. J Plant Physiol 169: 1366–1374

Hinsinger P, Jaillard B (1993) Root-induced release of interlayer potassium and vermiculitization of phlogopite as related to potassium depletion in the rhizosphere of rye grass. J Soil Sci 44:525–534

Hinsinger P, Elsass F, Jaillard B, Robert M (1993) Root-induced irreversible transformation of a trioctahedral mica in the rhizosphere of rape. J Soil Sci 44:535–545

Honsbein A, Sokolovski S, Grefen C, Campanoni P, Pratelli R, Chen Z, Paneque-Corralles M, Johansson I, Blatt MR (2009) A tri-partite SNARE-K^+ channel complex mediates in channel-dependent K^+ nutrition in *Arabidopsis*. Plant Cell 21:2859–2877

Hu X, Chen J, Guo J (2006) Two phosphate and potassium solubilizing bacteria isolated from Tianmu Mountain, Zhejiang, China. World J Microbiol Biotechnol 22:983–990

Huang Z, He L, Sheng X, He Z (2013) Weathering of potash feldspar by *Bacillus* sp. L11. Wei Sheng Wu Xue Bao 53(11):1172–1178

Idris R, Trifinova R, Puschenreiter M, Wenzel WW, Sessitsch A (2004) Bacterial communities associated with flowering plants of the Ni hyperaccumulator *Thlaspi goesingense*. Appl Environ Microbiol 70: 2667–2677

Jeschke D, Wolf O (1988) External potassium supply is not required for root growth in saline conditions: experiments with *Ricinus communis* L. grown in a reciprocal split-root system. J Exp Bot 39(9): 1149–1167

Jorquera MA, Hernâandez MT, Rengel Z, Marschner P, Mora ML (2008) Isolation of culturable phosphobacteria with both phytate-mineralization and phosphate-solubilization activity from the rhizosphere of plants grown in a volcanic soil. Biol Fertil Soils 44:1025–1034

Kim KY, Jordan D, Krishnan HB (1998) Expression of genes from *Rahnella aquatilis* that are necessary for mineral phosphate solubilization in *Escherichia coli*. FEMS Microbiol Lett 159:121–127

Kim MJ, Ciani S, Schachtman DP (2010) A peroxidase contributes to ROS production during *Arabidopsis* root response to potassium deficiency. Mol Plant 3: 420–427

Krishnamurthy HA (1989) Effect of pesticides on phosphate solubilizing microorganisms, M.Sc. (Agri.) thesis. University of Agricultural Sciences, Dharwad

Kuffner M, Puschenreiter M, Wieshammer G, Gorfer M, Sessitsch A (2008) Rhizosphere bacteria affect growth and metal uptake of heavy metal accumulating willows. Plant Soil 304:35–44

Kumar A, Bahadur I, Maurya BR, Raghuwanshi R, Meena VS, Singh DK, Dixit J (2015) Does a plant growth-promoting rhizobacteria enhance agricultural sustainability? J Pure Appl Microbiol 9(1):715–724

Lee SC, Lan WZ, Kim BG, Li L, Choeng YH, Pandey GK, Lu G, Buchanan B, Luan S (2007) A protein phosphorylation/dephosphorylation network regulates a plant potassium channel. Proc Natl Acad Sci U S A 104:15959–15964

Li FC, Li S, Yang YZ, Cheng LJ (2006) Advances in the study of weathering products of primary silicate minerals, exemplified by mica and feldspar. Acta Pet Mineral 25:440–448

Lian B, Fu PQ, Mo DM, Liu CQ (2002) A comprehensive review of the mechanism of potassium release by silicate bacteria. Acta Mineral Sin 22:179–183

Lin Q, Rao Z, Sun Y, Yao J, Xing L (2002) Identification and practical application of silicate-dissolving bacteria. Agric Sci China 1:81–85

Liu W, Xu X, Wu X, Yang Q, Luo Y, Christie P (2006) Decomposition of silicate minerals by *Bacillus mucilaginosus* in liquid culture. Environ Geochem Health 28:133–140

Liu D, Lian B, Wang B, Jiang G (2011) Degradation of potassium rock by earthworms and responses of bacterial communities in its gut and surrounding substrates after being fed with mineral. PLoS One 6(12):28803–28820

Liu D, Lian B, Dong H (2012) Isolation of *Paenibacillus* sp. and assessment of its potential for enhancing mineral weathering. Geomicrobiol J 29(5):413–421

Lucas RE, Davis JF (1961) Relationships between pH values of organic soils and availabilities of 12 plant nutrients. Soil Sci 92:177–182

Lugtenberg B, Kamilova F (2009) Plant-growth-promoting rhizobacteria. Annu Rev Microbiol 63:541–556

Mannisto MK, Haggblom MM (2006) Characterization of psychrotolerant heterotrophic bacteria from Finnish Lapland. Syst Appl Microbiol 29:229–243

Martínez-Cordero MA, Martínez V, Rubio F (2004) Cloning and functional characterization of the high-affinity K^+ transporter HAK1 of pepper. Plant Mol Biol 56: 413–421

Maurya BR, Meena VS, Meena OP (2014) Influence of Inceptisol and Alfisol's potassium solubilizing bacteria (KSB) isolates on release of K from waste mica. Vegetos 27(1):181–187

Meena OP, Maurya BR, Meena VS (2013) Influence of K-solubilizing bacteria on release of potassium from waste mica. Agric Sustain Dev 1(1):53–56

Meena VS, Maurya BR, Bahadur I (2014a) Potassium solubilization by bacterial strain in waste mica. Bangladesh J Bot 43(2):235–237

Meena VS, Maurya BR, Verma JP (2014b) Does a rhizospheric microorganism enhance K+ availability in agricultural soils? Microbiol Res 169:337–347

Meena RK, Singh RK, Singh NP, Meena SK, Meena VS (2015a) Isolation of low temperature surviving plant growth-promoting rhizobacteria (PGPR) from pea (*Pisum sativum* L.) and documentation of their plant growth promoting traits. Biocatal Agric Biotechnol. doi:10.1016/j.bcab.2015.08.006

Meena VS, Maurya BR, Verma JP, Aeron A, Kumar A, Kim K, Bajpai VK (2015b) Potassium solubilizing rhizobacteria (KSR): isolation, identification, and

K-release dynamics from waste mica. Ecol Eng 81: 340–347

Nieves-Cordones M, Martinez-Cordero MA, Martinez V, Rubio F (2007) An NH_4^+ sensitive component dominates high-affinity K^+ uptake in tomato plants. Plant Sci 172:273–280

Nieves-Cordones M, Miller A, Alemán F, Martínez V, Rubio F (2008) A putative role for the plasma membrane potential in the control of the expression of the gene encoding the tomato high-affinity potassium transporter HAK5. Plant Mol Biol 68:521–532

Nieves-Cordones M, Alemán F, Martínez V, Rubio F (2014) K^+ uptake in plant roots. The systems involved, their regulation and parallels in other organisms. J Plant Physiol 171:688–695

Oborn I, Andrist-Rangel Y, Askegaard M, Grant CA, Watson CA, Edwards AC (2005) Critical aspects of potassium management in agricultural systems. Soil Use Manag 21:102–112

Pindi PK, Satyanarayana SDV (2012) Liquid microbial consortium a potential tool for sustainable soil health. J Biofertil Biopestici 3:124–133

Puente ME, Bashan Y, Li CY, Lebsky VK (2004) Microbial populations and activities in the rhizoplane of rock weathering desert plants in root colonization and weathering of igneous rocks. Plant Biol 6: 629–642

Rajan SSS, Watkinson JH, Sinclair AG (1996) Phosphate rocks for direct application to soils. Adv Agron 57: 77–159

Rajkumar M, Vara Prasad MN, Freitas H, Ae N (2009) Biotechnological applications of serpentine soil bacteria for phytoremediation of trace metals. Crit Rev Biotechnol 29(2):120–130

Ranawat P, Kumar KM, Sharma NK (2009) A process for making slow-release phosphate fertilizer from low-grade rock phosphate and siliceous tailings by fusion with serpentinite. Curr Sci 96(6):843–848

Rodriguez H, Gonzalez T, Goire I, Bashan Y (2004) Gluconic acid production and phosphate solubilization by the plant growth-promoting bacterium *Azospirillum* spp. Naturwissenschaften 91:552–555

Ruangsanka S (2014) Identification of phosphate-solubilizing bacteria from the bamboo rhizosphere. Sci Asia 40:204–211

Sabannavar SJ, Lakshman HC (2009) Effect of rock phosphate solubilization using mycorrhizal fungi and phosphobacteria on two high yielding varieties of *Sesamum indicum* L. World J Agric Sci 5(4):470–479

Seneviratne G, Indrasena IK (2006) Nitrogen fixation in lichens is important for improved rock weathering. J Biosci 31:639–643

Setiawati A, Handayanto E (2010) Role of phosphate solubilizing bacteria on availability phosphorus in oxisols and tracing of phosphate in corn by using 32P. In: 19th world congress of soil science, soil solutions for a changing world, Brisbane

Sheng XF (2005) Growth promotion and increased potassium uptake of cotton and rape by a potassium releasing strain of *Bacillus edaphicus*. Soil Biol Biochem 37:1918–1922

Sheng XF, He LY (2006) Solubilization of potassium-bearing minerals by a wild type strain of *Bacillus edaphicus* and its mutants and increased potassium uptake by wheat. Can J Microbiol 52:66–72

Sheng XF, Huang WY (2002) Mechanism of potassium release from feldspar affected by the strain NBT of silicate bacterium. Acta Pedol Sin 39:863–871

Sheng XF, He LY, Huang WY (2002) The conditions of releasing potassium by a silicate-dissolving bacterial strain NBT. Agric Sci China 1:662–666

Sheng XF, Xia JJ, Chen J (2003) Mutagenesis of the *Bacillus edphicaus* strain NBT and its effect on growth of chili and cotton. Agric Sci China 2:40–51

Sheng XF, Zhao F, He LY, Qiu G, Chen L (2008) Isolation and characterization of silicate mineral-solubilizing *Bacillus globisporus* Q12 from the surfaces of weathered feldspar. Can J Microbiol 54(12):1064–1068

Shin R, Schachtman DP (2004) Hydrogen peroxide mediates plant root cell response to nutrient deprivation. Proc Natl Acad Sci U S A 101:8827–8832

Shin R, Berg RH, Schachtman DP (2005) Reactive oxygen species and root hairs in *Arabidopsis* root response to nitrogen, phosphorus and potassium deficiency. Plant Cell Physiol 46:1350–1357

Simonsson M, Andersson S, Andrist-Range Y, Hillier S, Mattsson L, Öborn I (2007) Potassium release and fixation as a function of fertilizer application rate and soil parent material. Geoderma 140:188–198

Sindhu SS, Dua S, Verma MK, Khandelwal A (2010) Growth promotion of legumes by inoculation of rhizosphere bacteria. In: Khan MS, Zaidi A, Musarrat J (eds) Microbes for legume improvement. Springer-Wien, New York, pp 195–235

Singh G, Biswas DR, Marwah TS (2010) Mobilization of potassium from waste mica by plant growth promoting rhizobacteria and its assimilation by maize (*Zea mays*) and wheat (*Triticum aestivum* L.). J Plant Nutr 33: 1236–1251

Singh JS, Pandey VC, Singh DP (2011) Efficient soil microorganisms: a new dimension for sustainable agriculture and environmental development. Agric Ecosys Environ 140:339–353

Singh NP, Singh RK, Meena VS, Meena RK (2015) Can we use maize (*Zea mays*) rhizobacteria as plant growth promoter? Vegetos 28(1):86–99

Sottocornola B, Visconti S, Orsi S, Gazzarrini S, Giacometti S, Olivari C, Camoni L, Aducci P, Marra M, Abenavoli A, Thiel G, Moroni A (2006) The potassium channel KAT1 is activated by plant and animal 14-3-3 proteins. J Biol Chem 281:35735–35741

Sparks DL, Huang PM (1985) Physical chemistry of soil potassium. In: Munson RD (ed) Potassium in agriculture. American Society of Agronomy, Madison, pp 201–276

Sperberg JI (1958) The incidence of apatite-solubilizing organisms in the rhizosphere and soil. Aust J Agric Res Econ 9:778

Sutter JU, Campanoni P, Tyrrell M, Blatt MR (2006) Selective mobility and sensitivity to SNAREs is exhibited by the *Arabidopsis* KAT1 K^+ channel at the plasma membrane. Plant Cell 18:935–954

Taha SM, Mahmod SAZ, Halim El-Damaty A, Hafez AM (1969) Activity of phosphate dissolving bacteria in Egyptian soils. Plant Soil 31:149–160

Teng H, Lian B (2007) Potassium solubilization in fungal degradation of aluminosilicate minerals. American Geophysical Union, Fall meeting 2007, abstract #B13C-1384

Toro M, Azcon R, Barea J (1997) Improvement of arbuscular mycorrhiza development by inoculation of soil with phosphate-solubilizing rhizobacteria to improve rock phosphate bioavailability (sup32) P and nutrient cycling. Appl Environ Microbiol 63:4408–4412

Ullman WJ, Kirchman DL, Welch SA (1996) Laboratory evidence by microbiologically mediated silicate mineral dissolution in nature. Chem Geol 132:11–17

Uroz S, Calvaruso C, Turpault MP, Pierrat JC, Mustin C, Frey-Klett P (2007) Effect of the mycorrhizosphere on the genotypic and metabolic diversity of the bacterial communities involved in mineral weathering in a forest soil. Appl Environ Microbiol 73:3019–3027

Uroz S, Calvaruso C, Turpault MP, Frey Klett P (2009) Mineral weathering by bacteria: ecology, actors and mechanisms. Trends Microbiol 17(8):378–387

Vandevivere P, Welch SA, Ullman WJ, Kirchman DL (1994) Enhanced dissolution of silicate minerals by bacteria at near neutral pH. Microbiol Ecol 27:241–251

Vassilev N, Vassileva M, Nikolaeva I (2006) Simultaneous P-solubilizing and biocontrol activity of microorganisms: potentials and future trends. Appl Microbiol Biotechnol 71:137–144

Vazquez P, Holguin G, Puente ME, Lopez-Cortes A, Bashan Y (2000) Phosphate-solubilizing microorganisms associated with the rhizosphere of mangroves in a semiarid coastal lagoon. Biol Fertil Soils 30:460–468

Very A-A, Nieves-Cordones M, Daly M, Khan I, Fizames C, Sentenac H (2014) Molecular biology of K+ transport across the plant cell membrane: what do we learn from comparison between plant species? J Plant Physiol 171:748–769

Vessey KJ (2003) Plant growth promoting rhizobacteria as biofertilizers. Plant Soil 255:571–586

Walker DJ, Leigh RA, Miller AJ (1996) Potassium homeostasis in vacuolate plant cells. Proc Natl Acad Sci U S A 93:10510–10514

Wallander H (2000) Uptake of P from apatite by *Pinus sylvestris* seedlings colonised by different ectomycorrhizal fungi. Plant Soil 218:249–256

Wang JG, Zhang FS, Cao YP, Zhang XL (2000) Effect of plant types on release of mineral potassium from gneiss. Nutr Cycl Agroecosyst 56:37–44

Wang Y-H, Garvin DF, Kochian LV (2002) Rapid induction of regulatory and transporter genes in response to phosphorus, potassium, and iron deficiencies in tomato roots. Evidence for cross talk and root/rhizosphere mediated signals. Plant Physiol 130:1361–1370

Welch SA, Ullman WJ (1999) The effect of microbial glucose metabolism on bytownite feldspar dissolution rates between 5 and 35 °C. Geochem Cosmochim Acta 63:3247–3259

Wua SC, Cao ZH, Li ZG, Cheunga KC, Wonga MH (2005) Effects of biofertilizer containing N-fixer, P and K solubilizers and AM fungi on maize growth: a greenhouse trial. Geoderma 125:155–166

Xie JC (1998) Present situation and prospects for the world's fertilizer use. Plant Nutr Fertil Sci 4:321–330

Xu J, Li HD, Chen LQ, Wang Y, Liu LL, He L, Wu WH (2006) A protein kinase, interacting with two calcineurin B-like proteins, regulates K^+ transporter AKT1 in *Arabidopsis*. Cell 125:1347–1360

Yu X, Liu X, Zhu TH, Liu GH, Mao C (2012) Co-inoculation with phosphate-solubilizing and nitrogen-fixing bacteria on solubilization of rock phosphate and their effect on growth promotion and nutrient uptake by walnut. Eur J Soil Biol 50:112–117

Zakaria AAB (2009) Growth optimization of potassium solubilizing bacteria isolated from biofertilizer. Bachelor of Chem. Eng. (Biotech.), Fac. of Chem., Natural Resources Eng Univ, Malaysia Pahang p 40

Zhang C, Kong F (2014) Isolation and identification of potassium-solubilizing bacteria from tobacco rhizospheric soil and their effect on tobacco plants. Appl Soil Ecol 82:18–25

Zhang S, Zhou Z, Gong Q, Makielski JC, January CT (1999) Mechanism of block and identification of the verapamil binding domain to HERG potassium channels. Circ Res 84:989–998

Zhao L-N, Shen L-K, Zhang W-Z, Zhang W, Wang Y, Wu W-H (2013) Ca^{2+} dependent protein kinase11 and 24 modulate the activity of the inward rectifying K^+ channels in *Arabidopsis* pollen tubes. Plant Cell 25: 649–661

5 The Molecular Mechanisms of KSMs for Enhancement of Crop Production Under Organic Farming

Ashok Kumar, Jai Singh Patel, Indra Bahadur, and Vijay Singh Meena

Abstract

The continuous use of chemical fertilizers and pesticides for the enhancement of crop yield and instant action of pest control causes harmful and hazardous effect on the environment due to the accumulation of minerals and biomagnifications at higher hierarchical level. Therefore, the current need is alternative and eco-friendly technology as integrated pest management (IPM) and plant growth-promoting microorganisms (PGPMs) for enhancing organic farming practices. One of the promising microorganisms is potassium-solubilizing microorganisms (KSMs) as PGPMs are applicable for sustainable agriculture. Plant growth promotion (PGP) is a complex phenomenon rarely attributable to a single mechanism as most PGP microbes influence plant growth through multiple mechanisms. However, any microbial agent added to the rhizosphere has to interact not only with the plant but also with other organisms around the microenvironment. The KSMs have the ability for IAA production, K solubilization, antifungal, HCN, and siderophore production. Due to secretion of organic acids, KSMs solubilize various forms of K in soil to available forms which helps enhance plant growth, yield, and fertility status of soil. This book chapter is a critical summary of the efforts of scientist in efficient use of KSMs, mechanism of K solubilization, and use of these microorganisms for increasing the crop production. They also help plant to combat against pathogenic microbes and other environmental stresses. The indigenous microbes proven their effectiveness; such microbes suit the environmental conditions in the cropping system for which they are intended. This chapter covers the studies of KSMs, their sources, mechanism of K solubilization, and their effect on crops.

A. Kumar (✉) • I. Bahadur
Department of Botany, MMV, Banaras Hindu University, Varanasi 221005, India
e-mail: ashokbhu2010@gmail.com

J.S. Patel
Department of Botany, Banaras Hindu University, Varanasi 221005, India

V.S. Meena
Department of Soil Science and Agricultural Chemistry, Institute of Agricultural Sciences, Banaras Hindu University, Varanasi 221005, Uttar Pradesh, India

ICAR-Vivekananda Institute of Hill Agriculture, Almora 263 601, Uttarakhand, India

V.S. Meena et al. (eds.), *Potassium Solubilizing Microorganisms for Sustainable Agriculture*, DOI 10.1007/978-81-322-2776-2_5

Keywords
Growth • K solubilizing • Microbes • Potassium • Yield

5.1 Introduction

Potassium (K) is one of the most important mineral elements found in soil. It is the seventh most abundant element on the Earth's crust. First report of K was done by Sir Humphry Davy in 1807. He isolated this element from caustic potash (KOH, potassium hydroxide) by electrolysis. However, this element is first to be isolated by using electrolysis and collected on the cathode (Holmes 2011). Its actual concentrations vary from 0.04 % to 3 % (Sparks and Huang 1985a). Concerning its availability in the soil to plant, four different forms are available: (i) soil solution, (ii) exchangeable K, (iii) fixed K, and (iv) lattice K (Syers 1998). Only one form, i.e., solution K, could be taken directly by plants. In the soil its occurrence is either in exchangeable or in non-changeable minerals. In the Indian context, its occurrence in soil ranges from 0.5 % to 3.0 % (Mengel and Kirkby 1987). Only ~2 % soluble K is found in Indian soil. However, ~98 % are in the form of minerals like biotite, feldspar, mica, muscovite, and vermiculite (Goldstein 1994). A number of studies reported deficiency of available K in the large area of agricultural fields, including three-fourth of the paddy soils of China and two-third of the wheat belt of Southern Australia (Mengel and Kirkby 2001; Römheld and Kirkby 2010).

Soil deficient in K is generally found to be sandy, waterlogged, saline, and acidic. A report by Sparks and Huang (1985b) showed the importance of K in plant growth and development. Only K^+ in solution can be taken by plants and its availability is dependent upon the nutrient dynamics as well as on total K content. Conversion of K from one form to other form largely depends on the concentration of macronutrients like nitrate in the soil solution (Yanai et al. 1996). Few studies (Pretty and Stangel 1985; Johnston 2005) showed that uptake of K by plant is faster than its exchange from non-available to available form. Higher level of few monovalent cations like Na^+ and $NH4^+$ can interfere with the potassium uptake (Spalding et al. 1999; Qi and Spalding 2004; Rus et al. 2004).

Potassium (K) plays an essential role in various plants' growth, metabolism, and development. In addition to plant metabolism, potassium improves crop quality because it helps in grain filling and kernel weight, strengthens straw, increases disease resistance against pest and diseases, and also helps the plant to withstand during stress. It activates plant enzymes, enhances photosynthesis, reduces respiration, maintains cell turgor, helps in transport of sugars and starches, facilitates in nitrogen uptake, and is also important for protein synthesis. Without adequate potassium, the plants will have poorly developed roots, grow slowly, produce small seeds, and have lower yields. In the introduction of high-yielding and hybrid crop varieties during green revolution with the progressive intensification of agriculture, the soils are getting depleted in potassium reserve at a faster rate, and available soil K levels have also dropped due to leaching, runoff, and erosion (Sheng and Huang 2002a).

5.1.1 Solubilization/Mobilization of Potassium from Rocks or Mica

The low availability of potassium present in significant areas of cultivated soils in India, Korea, and China results in low crop productivity (Xie 1998). The use of plant growth-promoting rhizobacteria including potassium-solubilizing bacteria (KSB) as a biofertilizer could work as a sustainable solution to improve plant growth, nutrient uptake, and production (Vessey 2003). Furthermore, the direct use of potassium-containing rock materials may be agronomically more useful and environmentally more feasible than soluble form of K (Rajan et al. 1996). Rock K materials are cheaper sources and most of them are readily available to a plant because the minerals are slowly released by potassium-solubilizing microorganisms (KSMs) and their use as biofertilizer causes significant yield enhances of the various crops (Meena et al. 2015b; Singh et al. 2015).

5.1.2 Efficient Potassium-Solubilizing Microorganisms (KSMs) and Their Occurrence

Potassium-solubilizing bacteria are heterotrophic bacteria which are obtaining their energy and cellular carbon from preexisting organic material. Thus, they are participating in the humus formation in soil, the cycling of other minerals tied up in organic matter, and the prevention of buildup of dead organic materials (Zakaria 2009). In addition, these aerobic Gram-positive bacteria can maintain soil structure and produce substance that stimulates plant growth or inhibits root pathogens (Zakaria 2009). However, KSB are able to solubilize rock K mineral powder, such as micas, illite, and orthoclases (feldspar) and also to produce and excrete different organic acids (Friedrich et al. 1991; Ullman et al. 1996) or chelate silicon ions to bring the K into solution (Groudev 1987; Friedrich et al. 1991; Ullman et al. 1996; Bennett et al. 1998; Meena et al. 2013; Maurya et al. 2014). It was shown that the potassium-solubilizing strain such as *Bacillus mucilaginosus* enhanced the K availability in soils and increased mineral content in plant (Sheng et al. 2002; Zakaria 2009). In addition, Vandevivere et al. (1994) proposed that *B. mucilaginosus* increases the dissolution rate of silicate and alumino-2 silicate minerals and releases the K^+ and SiO_2 from the crystal lattice primarily by generating of various organic acids. However, this hypothesis is controversial because *B. mucilaginosus* is also thought to accelerate the dissolution of a variety of silicates with the help of extracellular polysaccharide (EPS) production (Welch and Ullman 1993).

The dispute about the mechanism by which *B. mucilaginosus* decomposes silicate minerals and releases K^+ and SiO_2 may have severely limited the use of the organism in agriculture as a form of living biological K^+ fertilizer. Recently, Liu et al. (2006) reported that the polysaccharides strongly adsorbed the various organic acids and attached to the surface of the mineral, resulting in an area of high concentration of organic acids near the mineral. Experiments revealed that *Bacillus* species enhanced the soluble content of K^+ in the culture medium. Microorganisms like *Aspergillus niger*, *Bacillus extorquens*, and *Clostridium pasteurianum* were found to grow on potassium-solubilizing media, muscovite, biotite, orthoclase microclase, and mica in vitro (Archana et al. 2013). The varieties of bacterial species like silicate bacteria were found to dissolve potassium, silicates, and aluminum from insoluble minerals (Aleksandrov et al. 1967). Zhang and Kong (2014) reported that the 27 KSB strains were isolated and identified through the 16S rDNA. Among these strains, 17 strains belonged to *Klebsiella variicola*, 2 strains belonged to *Enterobacter cloacae*, 2 strains belonged to *Enterobacter asburiae*, and the remaining 6 strains belonged to *Enterobacter aerogenes*, *Pantoea agglomerans*, *Agrobacterium tumefaciens*, *Microbacterium foliorum*, *Myroides odoratimimus*, and *Burkholderia cepacia*, respectively. All isolated KSB strains were capable of solubilizing K-feldspar powder in solid and liquid media.

5.1.3 Potassium Solubilizers as Alternate Method for Biofertilizers

The biofertilizer, an alternative use of chemical fertilizers, is a substance containing microorganisms, having the ability to colonize the plant rhizosphere (Shen 1997). In terms of these microorganisms, they have the ability to solubilize the nutritionally important macronutrients, nitrogen, phosphorus, and potassium (NPK), from their unavailable form by the biological process.

Reports exist to ensure the effect of beneficial microbes in agriculture practices, and nowadays it was shown that these microbes can enhance tolerance in plant against environmental stresses (Shen 1997). Different potassium-solubilizing bacteria were isolated from soil, e.g., *Pseudomonas, Burkholderia, Acidithiobacillus ferrooxidans, Bacillus mucilaginosus, Bacillus edaphicus, B. circulans, Paenibacillus* sp., and few fungal microorganisms like *Aspergillus* and *Penicillium* (Sheng 2005; Lian et al. 2002; Li et al. 2006; Liu et al. 2012; Meena et al. 2014a, b). Another recent report by Singh et al. (2010) showed the mobilization of K from waste mica significantly by *Bacillus mucilaginosus, Azotobacter chroococcum*, and *Rhizobium*, which can be the source of K for the plants.

Uses of potassium- and phosphate-solubilizing bacteria (KSB and PSB) as biofertilizers in the cultivated P- and K-deficient soil have the ability to reduce the chemical fertilizers and control pollution caused by these chemical fertilizers (Glick 1995; Sindhu et al. 2010). Another report (Sheng et al. 2003) suggests the use of K-solubilizing bacteria as biofertilizer in agriculture as well as environmental protection from pollution. Muentz (1890) reported the first evidence of solubilization of K rocks by microbes. A number of microbes such as *Aspergillus niger, Bacillus extorquens*, and *Clostridium pasteurianum* were found to grow on orthoclase, muscovite, microclase, biotite, and micas under in vitro conditions (Reitmeir 1951). Another report by Bunt and Rovira (1955) showed the use of a variety of soil microorganisms to solubilize silicate minerals. Similar report by Gromov (1957) showed the growth of microorganisms like bacteria, fungi, and actinomycetes on the surface of mountain rocks. A bacterium *B. mucilaginous* subsp. *siliceous* is a silicate-solubilizing bacterium (SSB) that has the ability to liberate potassium from feldspar and alumino-silicates (Norkina and Pumpyanskaya 1956). Silicate-dissolving action of Gram-negative bacteria such as *Erwinia herbicola* and *Pseudomonas* strains was reported by Duff and Webley (1959). Another study by Heinen (1960) showed clearing zone in silicate medium on growth of *Pseudomonas* strain isolated from soil. Metabolic products of microorganisms can attack on the siliceous minerals in rocks (Webley et al. 1963). Different bacterial species isolated by Aleksandrov et al. (1967) were found to dissolve potassium, silica, and aluminum from insoluble minerals. A study suggested the involvement of SSB in the cycling of silicon in sea water and distribution of these microbes in marine environment (Purushothaman et al. 1974). Similarly *B. mucilaginosus* reported to solubilize insoluble silicates (Avakyan et al. 1981). Another report suggested cleavage of siloxane bond in quartz by *B. mucilaginosus* (Belkanova et al. 1985). Among the K-bearing silicate minerals, mica was found to weather readily (Tandon and Sekhon 1988). Li (1994) isolated K-solubilizing bacteria identified as *B. mucilaginosus* from soil, rock, and mineral samples. *Bacillus* sp. isolated from rice rhizosphere has silicate solubilization ability (Muralikannan 1996). Kannan and Raj (1998) studied enumeration of silicate and phosphate-solubilizing bacteria from soil's tank sediments based on biochemical characteristics; 3 out of 17 isolates were identified as *Bacillus* spp. Two strains of bacteria, CS1 and CS2, as silicate bacteria, have an inhibitory activity on the growth of Gram-negative bacteria, *E. coli*; one is identified as *B. mucilaginosus* (Liu 2001). Lin et al. (2002) reported the slime production by *B. mucilaginosus* strain, having the ability to dissolve silicate minerals and also colonize in the rhizospheric region. Twenty-seven strains of heterotrophic bacteria, isolated from feldspar-rich soil, were studied by Hutchen et al. (2003) and reported dissolution of silicate mineral in liquid and solid minimal media. Raj (2004) reported isolation and characterization of SSB in a medium containing 0.25 % insoluble magnesium trisilicate; the microorganism identified as *Bacillus* sp. solubilizes silicate minerals more efficiently under in vitro conditions (Kumar et al. 2015; Meena et al. 2015a).

KSB were also isolated from rhizosphere of cereal crops by using specific potassium-bearing minerals (Mikhailouskaya and Tcherhysh 2005). Silicate solubilizers were isolated by using modified Bunt and Rovira medium soil samples collected from coconut palms (Murali et al. 2005). Mostly, the silicate solubilizers were identified as *Pseudomonas* spp. and *Bacillus* spp. Another study by Hu et al. (2006) isolated phosphate solubilizers from the soil of Tianmu Mountain, Zhejiang Province (China), strains KNP413 and KNP414. These two isolates reported to be effective phosphate solubilizers even greater than *Bacillus mucilaginosus*, the most widely used biofertilizer in China. The two strains KNP413 and KNP414 were rod-shaped spore formers with large-capsule bacteria and form slimy translucent colonies on Aleksandrov medium. Further by molecular study, these two strains were classified as *Paenibacillus mucilaginosus*. Zhou et al. (2006) characterized Gram-positive, rod-shaped endospore former with thick capsule *Bacillus mucilaginosus* which solubilized silicon from illite at 30 °C. Researchers isolated KSB from soil, rocks, and mineral samples, viz., microcline, orthoclase, and muscovite mica (Sugumaran and Janarthanam 2007). Among the isolates *B. mucilaginosus* MCRCp1 solubilized more potassium by producing slime in muscovite mica; however nutritional status of soil in context to phosphorus and potassium was markedly improved through inoculation of this bacterium. Another study reported KSB with multiple activities like potassium solubilization, production of indole acetic acid (IAA), and siderophore production (Zhao et al. 2008). Seventy bacterial isolates were isolated from rhizosphere of wheat by using modified Aleksandrov medium plates (consisting of glucose, 5.0 g; $MgSO_4 \cdot 7H_2O$, 0.5 g; $CaCO_3$, 0.1 g; $FeCl_3$, 0.006 g; Ca_3PO_4, 2.0 g; mica powder, 2.0 g; and agar, 20.0 g) (Parmar 2010). These 70 bacterial strains were tested for potassium solubilization ability using 67 standard reference strains on Aleksandrov medium using mica as potassium source. Sindhu et al. (1999) spotted different rhizobacterial isolates on medium plates, incubated at 28 °C for 3 days, and detection of potassium solubilization was based upon the ability of solubilization zone formation, and 20 rhizobacterial isolates were found to solubilize potassium from mica powder. Prajapati et al. (2012) isolated 20 - potassium-solubilizing fungi from ceramic industry soils and four fungal isolates showed the high ratio of clear zone on Aleksandrov agar supplemented with 0.5 % potassium aluminum silicate. Two fungal strains (KF1 and KF2) showed the highest available potassium in liquid medium containing potassium aluminum silicate. Further molecular study confirmed the isolates as *Aspergillus niger* and *Aspergillus terreus* which have greater potassium solubilization activity in the presence of 1 % rock potassium (feldspar). In a recent study, Liu et al. (2012) isolated mineral-solubilizing *Paenibacillus* strain KT from a soil in Henan Province, China. In another study potassium-solubilizing bacterium isolated from soil was characterized as *B. circulans* Z1–3, having the ability to solubilize potassium from feldspar (Xiaoxi et al. 2012).

5.1.4 Potassium-Solubilizing Mechanisms of KSMs

Solubilization of illite and feldspar by microorganisms and also due to the production a number of organic acids, like citric, oxalic, and tartaric acids, and of capsular polysaccharides which helped in dissolution of minerals to release potassium (Sheng and He 2006; Liu et al. 2006). The main mechanism of potassium solubilization is discussed below.

5.1.4.1 Acid Production

Production of carboxylic acids like citric, tartaric, and oxalic acids has been reported as a predominant mechanism contributing to K solubilization (Styriakova et al. 2003; Sheng and He 2006). The mechanism of K-solubilizing silicate bacteria has the ability to dissolve potassium, silicon, and aluminum from insoluble K-bearing minerals such as micas, illite, and orthoclases by excreting organic acids that either dissolve rock K directly or chelate Si ions to bring K in the solution (Aleksandrov et al. 1967; Friedrich et al. 1991; Ullman et al. 1996; Bennet et al. 1998). Organic acids dissolve K by a proton- or ligand-mediated mechanism. Ullman and Welch (2002) suggested that organic acids indirectly enhance solubilization of K by formation of complexes in solution with reaction products and increase chemical affinity for the overall dissolution. In soil the potassium present in sufficient amount or source of potassium provided in K-deficient soil, KSMs have potential ability to solubilize it and release to plant for plant growth promotion (Fig. 5.1).

Weathering of minerals can also release K for the plants (Bertsch and Thomas 1985). Heinen (1960) studied silicate bacteria and found that they can dissolve potassium, silicon, and aluminum from insoluble minerals by liberation of phosphoric acids that solubilize apatite and release available form of nutrients from apatite. Several fungal isolates isolated by Moira et al. (1963) have the ability to release metal ions and silicates from minerals (saponite and vermiculite), rocks, and soils. The fungal isolates also have the ability to produce citric acid and oxalic acid, which are known to solubilize silicate minerals and also help in the removal of metal ions from rock and soil. Another report showed similar mechanism, i.e., dissolution of minerals by formation of organic acids (Vainberg et al. 1980). Inorganic acid production by *Thiobacillus*, *Clostridium*, and *Bacillus* also reported to dissolve potassium (Berthelin 1983) from minerals. Organic acids such as citric, tartaric, and oxalic acids are involved in K solubilization from feldspar secreted by *B. mucilaginosus* and *B. edaphicus* (Malinovskaya et al. 1990; Sheng and Huang 2002b). Mineral dissolution rates were found to be increased by organic compounds produced by microorganisms such as acetate, citrate, and oxalate (Palmer et al. 1991). Production of polysaccharide like gluconate promoted dissolution of silicates like albite, quartz, and kaolinite by

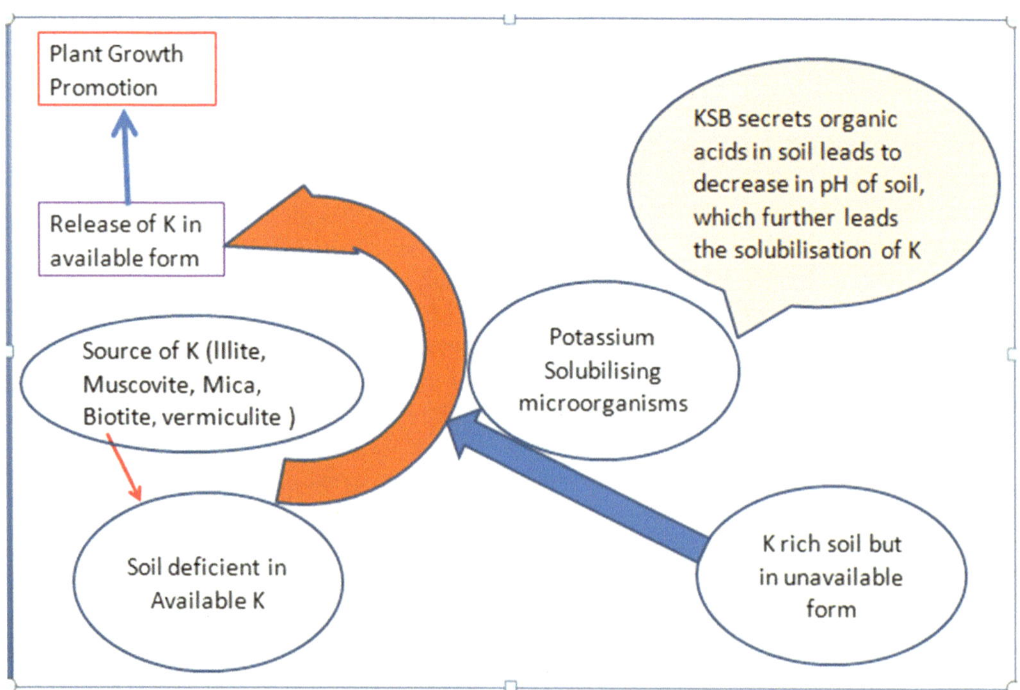

Fig. 5.1 Potassium-solubilizing microorganisms solubilize K released to plants

subsurface bacteria (Duff et al. 1963; Surdam and MacGowan 1988; Vandevivere et al. 1994). Mineral dissolution was found to be increased by production of organic acids such as acetate, citrate, and oxalate, which are readily produced by microorganisms (Hazen et al. 1991). However, production of organic acid favors K dissolution compared to inorganic acids produced by microorganisms (Welch and Ullman 1993), and polysaccharide produced by KSB during their reproduction can combine with minerals to form complex, which causes dissolution of that mineral.

5.1.4.2 Exopolysaccharides of KSM Can Also Dissolve K-Based Minerals

First report about production of slime acidic exopolysaccharides by microorganisms contributes to K dissolution from minerals (Groudev 1987). Polysaccharides can strongly absorb the organic acids attached to the surface of silicate minerals, which results high concentration of acids around the mineral (Liu et al. 2006). They have suggested the activity of extracellular polysaccharides to absorb SiO_2, which affects the equilibrium between the mineral and fluid phases and led to the reaction toward SiO_2 and K^+ solubilization. A study by Welch and Vandevivere (2009) suggested that naturally occurring polymers can affect the mineral dissolution. Solution containing fresh microbial EPS increases the dissolution rate of feldspars probably by forming complexes with framework ions in solution.

5.2 Inoculation Effect of Potassium Solubilizers or with Other Bacteria on Different Crops

Plant growth-promoting bacteria associated with plant roots may exert their beneficial effects on the uptake of essential nutrients and plant growth through a number of mechanisms such as N_2 fixation, P solubilizers, K solubilizers, production of phytohormones and siderophores, and transformation of macroelements such as phosphorus, potassium, and iron, when they are either applied to seeds or incorporated into the soil (Herridge et al. 2008; Sindhu et al. 2010). Khudsen et al. (1982) isolated KSB from rock and mineral samples which caused higher activity in potassium release from acid-leached soil and improved green gram's seedling growth. Bacteria persisted for a long time and increase population densities were detected after 14 months particularly in soils containing higher levels of organic matter (Argelis et al. 1993). Bacterial inoculation could improve phosphorus and potassium availability in the soils by secretion of different organic acid and other chemicals and thereby stimulated growth and mineral uptake of plants (Park et al. 2003).

The inoculation of *B. edaphicus* resulted in enhanced root and shoot growth and potassium content in plants (Sheng 2005). Christophe et al. (2006) reported that *Burkholderia glathei* in association with pine roots significantly enhances weathering of biotite which affects growth and root morphology, which was attributed to the release of K from the mineral. Chandra et al. (2005) reported that the yield of yam and tapioca was enhanced due to the KSB application in combination with other biofertilizers like *Rhizobium*, *Azospirillum*, *Azotobacter*, and *Acetobacter* and phosphate-solubilizing microorganism (PSM). The efficient bacterial inoculation improved the soil health such as organic matter content and total N in soil. Han and Lee (2005) found that co-inoculation of PSB (*B. megaterium*) and KSB (*B. mucilaginosus* strain KCTC3870) mixed with direct application of rock P and K materials into the soil enhanced N, P, and K uptake, improved photosynthesis, and promoted growth of eggplant in nutrient-limited soil.

The enhancement of plant growth by combining together rock materials and both bacterial strains suggested their potential use as biofertilizer. Nitrogen fixers and P- and K-solubilizing bacterial strains act as plant growth-promoting rhizobacteria (PGPR) which increases the availability of essential nutrients in soil and also enhances their uptake by plant. Thus, inoculation with PGPR including phosphate- and potassium-solubilizing bacteria (PSB and KSB) acting as potential biofertilizers could be a sustainable solution to improve plant

Table 5.1 Inoculation effect of potassium-solubilizing microorganisms (KSMs) or with other beneficial bacteria on crops

Crops	Microorganisms	Increased parameters	References
Cotton	*Azotobacter* and silicate bacteria	Yield	Ciobanu (1961)
Rice	*Bacillus* spp.	Yield	Muralikannan (1996)
Egg plant	*Frateuria aurantia*	Uptake and biomass	Nayak (2001)
Tomato	*Bacillus mucilaginosus*	Uptake and biomass	Lin et al. (2002)
Chilly-cotton	*Bacillus edaphicus*	P and K contents in plant biomass	Sheng et al. (2003)
Rice	*Bacillus* spp.	Yield	Raj (2004)
Cotton rape	*Bacillus edaphicus*	Growth, uptake	Sheng (2005)
Maize	N fixer, P and K solubilizers, and AM-fungi	Growth and uptake	Wu et al. (2005)
Brinjal	*Frateuria aurantia*	Growth, yield	Ramarethinam and Chandra (2005)
Yam-tapioca	Potash solubilizers and PSMs	Yield	Chandra et al. (2005)
Egg plant	*Bacillus megaterium* and *B. mucilaginosus*	Uptake, photosynthesis and yield	Han and Lee (2005)
Wheat	*Bacillus edaphicus*	Growth and nutrient contents	Sheng and He (2006)
Sorghum	*Aspergillus* and *Bacillus*	Dry matter yield and uptake	Badr et al. (2006)
Pepper-cucumber	*Bacillus megaterium* and *Bacillus mucilaginosus*	Growth, uptake	Han et al. (2006)
Chilly	*Bacillus* spp.	Uptake	Supanjani et al. (2006)
Maize-wheat	*Bacillus mucilaginosus*, *Azotobacter chroococcum*	Growth, yield	Singh et al. (2010)
Groundnut	*Bacillus mucilaginosus* MCRCp1	Growth, yield	Sugumaran and Janarthanam (2007)
Sorghum	*Bacillus mucilaginous*	Uptake	Basak and Biswas (2009)
Wheat	K-solubilizing isolate HWP47	Growth, yield	Parmar (2010)
Forage	*Bacillus mucilaginosus*; *Azotobacter chroococcum* A-41	Growth and acquisition	Basak and Biswas (2010)

nutrition and crop production (Vessey 2003). Table 5.1 shows the effect of potassium-solubilizing microorganisms (KSMs) alone or with combination of other beneficial bacteria on different crops.

The KSMs have the ability to enhance direct and indirect plant growth and yield as well as protect under stress conditions (Fig. 5.2). Potassium is one of the crucial nutrients for plant growth and development. Most of the K in soil exists in different insoluble forms like rocks, minerals, and sedimentary materials (Goldstein 1994). Soil K may be solubilized and absorbed by plants with the help of KSMs which secretes organic acids and decreased the pH consequently solubilized various form of K in soil which have uptake by plants. These microbes decompose silicate minerals such as K-feldspar and mica which transform solid K in the soil into available K that can be directly absorbed by plants, and they release various active substances that promote plant growth (Sheng et al. 2001). The use of KSB as a live biofertilizer is a hotspot in the study of agriculture and environmental conservation (Deng et al. 2003). KSB can be obtained from the various rhizosphere crops easily (Altamare et al. 1999). Most of the KSB obtained from the rhizosphere are *Bacillus* spp. and *Pseudomonas* spp., but *Bacillus mucilaginosus* is the most abundant studied species of KSB (Liu 2001; Zhou et al. 2006; Sugumaran and Janarthanam 2007).

Potassium ions activate certain enzymes especially those involved in photosynthesis,

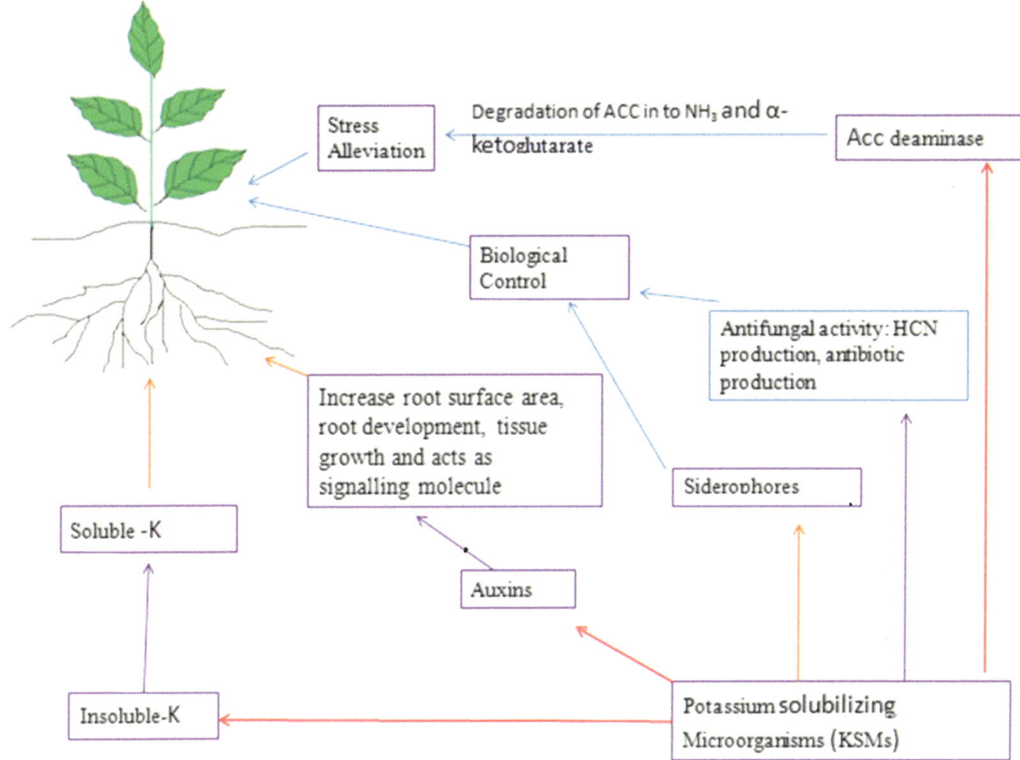

Fig. 5.2 Mechanism of plant growth promotion by potassium-solubilizing microorganisms

respiration, and starch and protein synthesis (Hopkins 1995). Moreover, opening and closure of stomata guard cells or daily changes in the orientation of leaves occurred by potassium concentration (Shehata and El-Khawas 2003).

5.3 Effect of Potassium-Solubilizing Bacteria on Availability of Potassium in Soils

Some soil microorganisms, viz., *Pseudomonas* spp., *Burkholderia* spp., *Acidithiobacillus ferrooxidans*, *Bacillus mucilaginosus*, *Bacillus edaphicus*, and *Bacillus megaterium*, are able to release K from K-bearing minerals by secreting different organic acids (Sheng et al. 2002). These organic acids either directly solubilize rock K (K-containing primary minerals) or chelate the primary mineral's silicon ions to bring the K into solution form (Bennett et al. 1998; Basak and Biswas 2009). Therefore, combining inoculation of K-solubilizing microorganisms with application of rock K to soil has recently gained great attention. Beneficial effects of inoculated-mica mixer to soil have been reported for enhancement of K uptake in different crops (Sheng et al. 2002; Han and Lee 2005).

This indicates that exudate of these microorganisms can efficiently increase the release of K from clay minerals. Similarly, a number of in vitro incubation trials have shown that the use of inoculated feldspars into soil increases about 40–60 % of K solubility and K uptake by plant (Han et al. 2006; Basak and Biswas 2009; Abou-el-Seoud and Abdel-Megeed 2012). Currently, little field experimental information is available, which is most likely due to difficulties in soil inoculation under field conditions. In order to increase the potential of such applications for agricultural production systems, we need more field experiment studies that evaluate their effect on both soil properties and crop growth.

5.4 Effect of Root Exudates on Potassium Availability in Soils

The utilization of non-exchangeable K sources is a crucial factor for the K uptake capacity in different crops (Claassen and Steingrobe 1999), and plant species or genotypes within species have been evaluated to differ in their efficiency to use this resource (Wang et al. 2011). El Dessougi et al. (2002) found sugar beet took up to six times more K per unit of root length than barley and wheat grown on K-fixing soils. The efficiency of K uptake by different crops, differences is generally attributed not only to the capacity in absorption, but also to the mobilization of non-exchangeable K by root exudates. Many plants released different organic acids such as citric and oxalic acids by maize (Kraffczyk et al. 1984), tartaric acid by pak-choi and radish (Chen et al. 2000), and malic acid by oilseed rape (Zhang et al. 1997). Similarly, various amino acids were detected in root exudates of wheat and sugar beet which help in the release of K from clay minerals (Rengel and Damon 2008). The depletion of K in rhizosphere below a threshold level from soil solution has been reported to be a key signal that activates the root exudation mechanism (Hosseinpur et al. 2012; Schneider et al. 2013).

The organic acids are well known to facilitate the weathering of soil minerals through the metal–organic complexes formation, and by increasing the exchange of H^+ for K^+ (Hinsinger and Jaillard 1993; Wang et al. 2011). Better understandings of the mechanisms behind the K release from soil minerals are new approaches to develop for sustainable agriculture. However, the K absorbed by the crop and the contribution of non-exchangeable K are very difficult to estimate due to the lack of suitable methods under field experiments or even in microcosm conditions. Therefore, in the K-limited areas, the selection of few species or varieties that are better in K-solubilizing efficiency via exudates should have a great potential to enhance the resource use efficiency.

5.5 Potassium Fixation in Soil

In addition to releasing K, soil minerals can also fix K that significantly affects K availability. This involves the adsorption of K ions onto interlayer sites of weathered sheet silicates, such as illite and vermiculite. The degree of K fixation in soils mainly depends on the type of clay mineral and its moisture content, charge density, competing ions, and soil pH. The major clay minerals responsible for K fixation are montmorillonite, vermiculite, and weathered micas (Sparks 1987). In addition to this, soil wetting and drying also significantly affect the K fixation.

The fixation process of K is relatively fast, while the release of fixed K is very slow because the strong binding force exists between K and clay minerals (Öborn et al. 2005). On the other hand, a soil that fixes or releases K highly depends on the concentration of K in the soil solution (Schneider et al. 2013). In addition to organic acids, the H^+ concentration in soil solution (via soil pH) seems to play an important role in K release from clay minerals. Therefore, the major K release may depend upon the optimization of soil pH. For optimized K fertilizer management practices, it is essential to understand the different factors that regulate K release from soil non-exchangeable pool.

5.6 Potassium Nutrition and Crop Stress Resistance

Environmental stresses limited the crops' productivity worldwide. The potassium-solubilizing microorganisms are multifunctional agents that not only provide nutrients to the plants but also help in alleviation of different stresses. The stresses affect crops and also lead to wastage of farmer's work for good production. The eco-friendly approach to deal with the situation where K-solubilizing microorganisms can be used as stress releasing agents. The physiological functions of potassium are critical for growth and yield formation and also product quality under otherwise optimal and undisturbed growth conditions. However, the crops are faced with a multitude of factors that affect metabolism, growth, and yield in field experiment. The biotic stresses, such as bacterial and fungal diseases or insect pests, and abiotic stresses, like drought, salinity, cold, or heat, reduce crop yield to a large extent.

Climate models predict that incidences and duration of drought and heat stress will affect some parts of the world, whereas other parts will suffer from heavy storms and periodic flooding. These fluctuated conditions will have a dramatic impact on agricultural production and farming practices (Brouder and Volenec 2008). In Europe, future climate changes are expected to be creates very problematic in dry springs and rainy summers in northern Europe and longer dry periods in the South. The effects of such climate variations could be observed in the European heat wave of 2003, which decreased crop production by around 30 % (Ciais et al. 2005). Abiotic stress often leads to massive crop failures in worldwide scale.

5.7 Future Prospectus of KSMs

Keeping in mind all these beneficial roles of KSMs present in soil rhizosphere, it can be concluded that integrating the K source with efficient K- inoculants with less fertilizer based on integrated nutrient management (INM) system should be considered in many situations as it promises high crop productivity and agricultural sustainability. Interventions and technologies for the ultimate transfer of novel, genetically modified soil and region-specific KSMs to the fields have to be developed, pilot tested, and transferred to farmers in a relatively short time. Search for new strains of beneficial KSM for biofertilizer and development of microbial diversity map for any region just like nutrient mapping may be helpful too. There is an urgent need for research to clear definition of what KSMs traits are useful and necessary for various environmental conditions and plants, so that optimal KSM strains can either be selected and/or improved. Combinations of other beneficial bacterial strains that interact synergistically are being devised, and numerous recent studies show a promising trend in the field of inoculation technology.

Many beneficial KSMs isolated from plants growing on nonsalinated soil are perfectly able to promote plant growth and control plant diseases in salinated soil. In other words, salination caused by a possible future climate change does not seem to be a threat for the use of presently applicable plant-beneficial bacteria. Future work is planned to test mixtures of the selected KSM for biocontrol against multiple plant pathogens in bioassays. KSMs can be used for the efficient restoration of sodic soil. Presently, the scope of potassium solubilization microorganism application is limited because it colonizes the rhizosphere only in certain plants and cannot survive in harsh environments, such as high concentrations of heavy metals and salts. Genotypic study of the potassium-solubilizing strains and molecular characterization of the plant parts is necessary to understand plant mechanisms for potassium absorption and its need in plant. Furthermore, scientists need to address certain issues, like how to improve the efficacy of biofertilizers, what should be an ideal and universal delivery system, how to stabilize these microbes in soil systems, and how nutritional and root exudation aspects could be controlled in order to get maximum benefits from KSM application.

5.8 Concluding Remarks

It is obvious that the use of chemical fertilizers and pesticides cannot be eliminated at the moment, which may cause increase in food production and harmful environmental side effects. Hence, the potassic biofertilizers should target agricultural inputs and lower the adverse environmental impacts of chemical fertilizers under sustainable agriculture. Better understanding of the interactions between microbe and plants is very important. The potassium-solubilizing microorganisms have efficient plant growth-promoting properties. The combined inoculation of KSM strain formulations will be better for plant growth, yield of crops, soil fertility, and their health under sustainable agriculture.

References

Abou-el-Seoud, Abdel-Megeed A (2012) Impact of rock materials and biofertilizations on P and K availability for maize (*Zea maize*) under calcareous soil conditions. Saudi J Biol Sci 19:55–63

Aleksandrov VG, Blagodyr RN, Live IP (1967) Liberation of phosphoric acid from apatite by silicate bacteria. Microchem J 29:111–114

Altamare C, Norvell WA, Bjorkman T, Harman GE (1999) Solubilization of phosphates and micronutrients by the plant growth promoting and bacterial fungus *Trichodera harzianum* Rifai. Appl Environ Microbiol 65:2926–2933

Archana DS, Nandish MS, Savalagi VP, Alagawadi AR (2013) Characterization of potassium solubilizing bacteria (KSB) from rhizosphere soil. Bioinfolet 10: 248–257

Argelis DT, Gonzala DA, Vizcaino C, Gartia MT (1993) Biochemical mechanism of stone alteration carried out by filamentous fungi living in monuments. Biogeo Chem 19:129–147

Avakyan ZA, Karavaiko GI, Mel'nikova EO, Krutsko VS, Ostroushhko YI (1981) Role of microscopic fungi in weathering of rocks from a pegmatite deposit. Mikrobiologiya 50:115–120

Badr MA, Shafei AM, Sharaf El-Deen SH (2006) The dissolution of K and phosphorus bearing minerals by silicate dissolving bacteria and their effect on sorghum growth. Res J Agric Biol Sci 2:5–11

Basak BB, Biswas DR (2009) Influence of potassium solubilizing microorganism (*Bacillus mucilaginosus*) and waste mica on potassium uptake dynamics by Sudan grass (*Sorghum vulgare* Pers.) grown under two Alfisols. Plant Soil 317:235–255

Basak BB, Biswas DR (2010) Coinoculation of potassium solubilizing and nitrogen fixing bacteria on solubilization of waste mica and their effect on growth promotion and nutrient acquisition by a forage crop. Biol Fertil Soils 46:641–648

Belkanova NP, Karavajko GI, Avakyan ZA (1985) Cleavage of the siloxane bond in quartz by *Bacillus mucilaginosus*. Mikrobiologiya 54:27–30

Bennett PC, Choi WJ, Rogera JR (1998) Microbial destruction of feldspars. Mineral Manag 8(6):149–150

Berthelin J (1983) Microbial weathering processes. In: Krumbein WE (ed) Microbial geochemistry. Blackwell Scientific Publications, Boston, pp 223–262

Bertsch PM, Thomas GW (1985) Potassium status of temperature region soils. In: Munson RD (ed) Potassium in agriculture. ASA, CSSA and SSSP, Madison, pp 131–162

Brouder SM, Volenec JJ (2008) Impact of climate change on crop nutrient and water use efficiencies. Physiol Plant 133:705–724

Bunt JS, Rovira AD (1955) Microbiological studies of some sub Antarctic soils. J Soil Sci 6:119–128

Chandra K, Greep S, Ravindranath P, Sivathsa RSH (2005) Liquid biofertilizers. Regional Center for Organic Farming, Hebbal, Bangalore Chem Ind 44: 1376–1377

Chen YX, Lin Q, Lu F, He YF (2000) Study on detoxication of organic acid to radish under the stress of Pb and Cd. Acta Sci Circumstance 20:467–472

Christophe C, Turpault MP, Freyklett P (2006) Root associated bacteria contribute to mineral weathering and to mineral nutrition in trees and budgeting analysis. Appl Environ Microbiol 72:258–1266

Ciais P, Reichstein M, Viovy N, Granier A, Ogee J, Allard V et al (2005) Europe-wide reduction in primary productivity caused by the heat and drought in 2003. Nature 437:529–533

Ciobanu I (1961) Investigation on the efficiency of bacterial fertilizers applied to cotton. Cent Exp Ingras Bacteria Lucrari Stiint 3:203–214

Claassen N, Steingrobe B (1999) Mechanistic simulation models for a better understanding of nutrient uptake from soil. In: Rengel Z (ed) Mineral nutrition of crops. Fundamental mechanisms and implications. Haworth Press, New York, pp 327–367

Deng SB, Bai RB, Hu XM, Luo Q (2003) Characteristics of a bioflocculant produced by *Bacillus mucilaginosus* and its use in starch wastewater treatment. Appl Microbiol Biotechnol 60:588–593

Duff RB, Webley DM (1959) 2-ketoglutaric acid and natural chelator produced by soil bacteria. Chem Ind 44:1376–1377

Duff RB, Webley DM, Scott RO (1963) Solubilization of minerals and related materials by 2- ketogluconic acid producing bacteria. Soil Sci 5:105–114

El Dessougi H, Claassen N, Steingrobe B (2002) Potassium efficiency mechanisms of wheat, barley, and sugar beet grown on a K fixing soil under controlled conditions. J Plant Nutr Soil Sci 165:732–737

Friedrich S, Platonova NP, Karavaiko GI, Stichel E, Glombitza F (1991) Chemical and microbiological solubilization of silicates. Acta Biotechnol 11: 187–196

Glick BR (1995) The enhancement of plant growth by free-living bacteria. Can J Microbiol 41:109–117

Goldstein AH (1994) Involvement of the quinoprotein glucose dehydrogenase in the solubilization of exogenous mineral phosphates by Gram negative bacteria. In: Torriani-Gorni A, Yagil E, Silver E (eds) Phosphate in microorganisms: cellular and molecular biology. ASM, Washington, DC, pp 197–203

Gromov BV (1957) The microflora of rocks and primitive soil in some northern regions of the USSR. Microbiologia 26:52–54

Groudev SN (1987) Use of heterotrophic microorganisms in mineral biotechnology. Acta Biotechnol 7:299–306

Han HS, Lee KD (2005) Phosphate and potassium solubilizing bacteria effect on mineral uptake, soil availability and growth of eggplant. Res J Agric Biol Sci 1:176–180

Han HS, Supanjani E, Lee KD (2006) Effect of co-inoculation with phosphate and potassium solubilizing bacteria on mineral uptake and growth of pepper and cucumber. Plant Soil Environ 52:130–136

Hazen TC, Jimenez L, Victoria GL (1991) Comparison of bacteria from deep subsurface sediment and adjacent groundwater. Microbiol Ecol 22:293–304

Heinen W (1960) Silicon metabolism in microorganisms. Arch Microbiol 37:199–210

Herridge DF, Peoples MB, Boddey RM (2008) Global inputs of biological nitrogen fixation in agricultural systems. Plant Soil 311:1–18

Hinsinger P, Jaillard B (1993) Root-induced release of interlayer potassium and vermiculitization of phlogopite as related to potassium depletion in the rhizosphere of ryegrass. J Soil Sci 44:525–534

Holmes R (2011) Humphry Davy and the chemical moment. Clin Chem 57(11):1625–1631

Hopkins W (1995) Introduction to plant physiology. Wiley, INS, New York, pp 414–415

Hosseinpur AR, Motaghian HR, Salehi MH (2012) Potassium release kinetics and its correlation with pinto bean (*Phaseolus vulgaris*) plant indices. Plant Soil Environ 58:328–333

Hu X, Chen J, Guo J (2006) Two phosphate- and potassium-solubilizing bacteria isolated from Tianmu mountain, Zhejiang, China. World J Microbiol Biotechnol 22:983–990

Hutchen SE, Valsami JE, Eldowney MS (2003) The role of heterotrophic bacteria in feldspar dissolution. Mineral Manag 67:1151–1170

Johnston AE (2005) Understanding potassium and its use in agriculture. EFMA, Brussels

Kannan NM, Raj SA (1998) Occurrence of silicate solubilizing bacteria in rice ecosystem. Madras Agric J 85:47–50

Khudsen D, Peterson GA, Prov PF (1982) Lithium, sodium and potassium. In: Page AL (ed) Methods of soil analysis, part (2) agronomy monograph 9, 2nd edn. ASA and SSSA, Madison

Kraffczyk I, Trolldenier G, Beringer H (1984) Soluble root exudates of maize: influence of potassium supply and rhizosphere microorganisms. Soil Biol Biochem 16:315–322

Kumar A, Bahadur I, Maurya BR, Raghuwanshi R, Meena VS, Singh DK, Dixit J (2015) Does a plant growth-promoting rhizobacteria enhance agricultural sustainability? J Pure Appl Microbiol 9(1):715–724

Li YF (1994) The characteristics and function of silicate dissolving bacteria fertilizer. Soil Fertil 2:48–49

Li FC, Li S, Yang YZ, Cheng LJ (2006) Advances in the study of weathering products of primary silicate minerals, exemplified by mica and feldspar. Acta Petrol Mineral 25:440–448

Lian B, Fu PQ, Mo DM, Liu CQ (2002) A comprehensive review of the mechanism of potassium release by silicate bacteria. Acta Mineral Sinica 22:179

Lin QM, Rao ZH, Sun YX, Yao J, Xing LJ (2002) Identification and practical application of silicate-dissolving bacteria. Agric Sci China 1:81–85

Liu GY (2001) Screening of silicate bacteria with potassium releasing and antagonistic activity. Chin J Appl Environ Biol 7:66–68

Liu W, Xu X, Wu S, Yang Q, Luo Y, Christie P (2006) Decomposition of silicate minerals by *Bacillus mucilaginosus* in liquid culture. Environ Geochem Health 28:133–140

Liu D, Lian B, Dong H (2012) Isolation of *Paenibacillus* sp. and assessment of its potential for enhancing mineral weathering. Geomicrobiol J 29:413–421

Malinovskaya IM, Kosenko LV, Votselko SK, Podgorskii VS (1990) Role of *Bacillus mucilaginosus* polysaccharide in degradation of silicate minerals. Mikrobiologiya 59:49–55

Maurya BR, Meena VS, Meena OP (2014) Influence of Inceptisol and Alfisol's potassium solubilizing bacteria (KSB) isolates on release of K from waste mica. Vegetos 27(1):181–187

Meena OP, Maurya BR, Meena VS (2013) Influence of K-solubilizing bacteria on release of potassium from waste mica. Agric Sustain Dev 1(1):53–56

Meena VS, Maurya BR, Bahadur I (2014a) Potassium solubilization by bacterial strain in waste mica. Bangladesh J Bot 43(2):235–237

Meena VS, Maurya BR, Verma JP (2014b) Does a rhizospheric microorganism enhance K+ availability in agricultural soils? Microbiol Res 169:337–347

Meena RK, Singh RK, Singh NP, Meena SK, Meena VS (2015a) Isolation of low temperature surviving plant growth-promoting rhizobacteria (PGPR) from pea (*Pisum sativum* L.) and documentation of their plant growth promoting traits. Biocatal Agric Biotechnol. doi:10.1016/j.bcab.2015.08.006

Meena VS, Maurya BR, Verma JP, Aeron A, Kumar A, Kim K, Bajpai VK (2015b) Potassium solubilizing rhizobacteria (KSR): isolation, identification, and K-release dynamics from waste mica. Ecol Eng 81: 340–347

Mengel K, Kirkby EA (1987) Principles of plant nutrition. Internl Potash Inst Bern, Worblaufen-Bern, pp 200–210

Mengel K, Kirkby EA (2001) Principles of plant nutrition, 5th edn. Kluwer Acad. Publishers, Dordrecht, p 849

Mikhailouskaya N, Tcherhysh A (2005) K-mobilizing bacteria and their effect on wheat yield. Latvian J Agron 8:154–157

Moira EK, Henderson MEK, Duff RB (1963) The release of metallic and silicate ions from minerals, rocks and soils by fungal activity. J Soil Sci 14:237–245

Muentz A (1890) Surla decomposition desroches etla formation de la terre arable. C R Acad Sci 110: 1370–1372

Murali G, Gupta A, Nair RV (2005) Variations in hosting beneficial plant associated microorganisms by root (wilt) diseased and field tolerant coconut palms of west coast tall variety. Curr Sci 89:1922–1927

Muralikannan M (1996) Biodissolution of silicate, phosphate and potassium by silicate solubilizing bacteria in rice ecosystem. M.Sc. (Agric) thesis, Tamil Nadu Agricultural University, Coimbatore

Nayak B (2001) Uptake of potash by different plants with the use of potash mobilizing bacteria (*Frateuria aurantia*). M.Sc. (Agric) thesis, QUAT, Bhubaneswar

Norkina SP, Pumpyansakya LV (1956) Certain properties of silicate bacteria. Crop Sci Soc Jpn 28:35–40

Oborn I, Andrist-Rangel Y, Askekaard M, Grant CA, Watson CA, Edwards AC (2005) Critical aspects of potassium management in agricultural systems. Soil Use Manag 21:102–112

Palmer RJJ, Siebert J, Hirsch PR (1991) Biomass and organic acids in sandstone of a weathered building production by bacterial and fungal isolates. Microbiol Ecol 21:253–261

Park M, Singvilay O, Seok Y, Chung J, Ahn K, Sa T (2003) Effect of phosphate solubilizing fungi on P uptake and growth to tobacco in rock phosphate applied soil. Korean J Soil Sci Fertil 36:233–238

Parmar P (2010) Isolation of potassium solubilizing bacteria and their inoculation effect on growth of wheat (*Triticum aestivum* L. em. Thell.). M.Sc. thesis submitted to CCS Haryana Agricultural University, Hisar

Prajapati K, Sharma MC, Modi HA (2012) Isolation of two potassium solubilizing fungi from ceramic industry soils. Life Sci Leafl 5:71–75

Pretty KM, Stangel PJ (1985) Current and future use of world potassium. In: Munson RD (ed) Potassium in agriculture. American Society of Agronomy, Madison, pp 99–128

Purushothaman A, Chandramohan D, Natarajan R (1974) Distribution of silicate dissolving bacteria in velar estuary. Curr Sci 43:282–283

Qi Z, Spalding EP (2004) Protection of plasma membrane K^+ transport by the salt overly sensitive1 Na+-H+ antiporter during salinity stress. Plant Physiol 136: 2548–2555

Raj SA (2004) Solubilization on a silicate and concurrent release of phosphorus and potassium in rice ecosystem. In: Biofertilizer technology for rice based cropping system. Scientific Publishers, Jodhpur, pp 372–378

Rajan SSS, Watkinson JH, Sinclair AG (1996) Phosphate rock for direct application to soils. Adv Agron 57: 77–159

Ramarethinam S, Chandra K (2005) Studies on the effect of potash solubilizing/mobilizing bacteria *Frateuria aurantia* on brinjal growth and yield. Pestology 11: 35–39

Reitmeir RF (1951) Soil potassium. In: Norman AG (ed) Advances in agronomy, vol III, American Society of Agronomy. Academic, New York, pp 113–164

Rengel Z, Damon PM (2008) Crops and genotypes differ in efficiency of potassium uptake and use. Physiol Plant 133:624–636

Romheld V, Kirkby EA (2010) Research on potassium in agriculture: needs and prospects. Plant Soil 335: 155–180

Rus A, Lee B, Munoz-Mayor A, Sharkhuu A, Miura K, Zhu JK, Bressan RA, Hasegawa PM (2004) AtHKT1 facilitates Na^+ homeostasis and K^+ nutrition in planta. Plant Physiol 136:2500–2511

Schneider A, Tesileanu R, Charles R, Sinaj S (2013) Kinetics of soil potassium sorption desorption and fixation. Commun Soil Sci Plant Anal 44:837–849

Shehata MM, El-Khawas SA (2003) Effect of two biofertilizers on growth parameters, yield characters, nitrogenous components, nucleic acid contents, minerals, oil content, protein profiles and DNA banding pattern of sunflower (*Helianthus annus* L. cv. Vedock) yield. Pak J Biol Sci 6(14):1257–1268

Shen D (1997) Microbial diversity and application of microbial products for agricultural purposes in China. Agric Ecosyst Environ 62:237–245

Sheng XF (2005) Growth promotion and increased potassium uptake of cotton and rape by a potassium releasing strain of *Bacillus edaphicus*. Soil Biol Biochem 37:1918–1922

Sheng XF, He LY (2006) Solubilization of potassium bearing minerals by a wild type strain of *Bacillus edaphicus* and its mutants and increased potassium uptake by wheat. Can J Microbiol 52:66–72

Sheng XF, Huang WY (2002a) Study on the conditions of potassium release by strain NBT of silicate bacteria. Sci Agric Sin 35:673–677

Sheng XF, Huang WY (2002b) Mechanism of potassium release from feldspar affected by the strain NBT of silicate bacterium. Acta Pedol Sin 39:863–871

Sheng X, He L, Huang W (2001) The conditions of releasing potassium by a silicate dissolving bacterial strain NBT. Agric Sci China 1(6):662–666

Sheng XF, He LY, Huang WY (2002) The conditions for releasing potassium by a silicate dissolving bacterial strain NBT. Agric Sci China 1:662–666

Sheng XF, Xia JJ, Chen J (2003) Mutagenesis of the *Bacillus edaphicus* strain NBT and its effect on growth of chili and cotton. Agric Sci China 2:400–412

Sindhu SS, Gupta SK, Dadarwal KR (1999) Antagonistic effect of *Pseudomonas* spp. on pathogenic fungi and enhancement of plant growth in green gram (*Vigna radiata*). Biol Fertil Soils 29:62–68

Sindhu SS, Dua S, Verma MK, Khandelwal A (2010) Growth promotion of legumes by inoculation of rhizosphere bacteria. In: Khan MS, Zaidi A, Musarrat J (eds) Microbes for legume improvement. Springer-Wien, New York, pp 95–235

Singh G, Biswas DR, Marwah TS (2010) Mobilization of potassium from waste mica by plant growth promoting rhizobacteria and its assimilation by maize (*Zea mays*) and wheat (*Triticum aestivum* L.). J Plant Nutr 33: 1236–1251

Singh NP, Singh RK, Meena VS, Meena RK (2015) Can we use maize (*Zea mays*) rhizobacteria as plant growth promoter? Vegetos 28(1):86–99

Spalding EP, Hirsch RE, Lewis DR, Qi Z, Sussman MR, Lewis BD (1999) Potassium uptake supporting plant growth in the absence Of AKT1 channel activity: inhibition by ammonium and stimulation by sodium. J Gen Physiol 113:909–918

Sparks DL (1987) Potassium dynamics in soils. Adv Soil Sci 6:1–63

Sparks DL, Huang PM (1985a) Physical chemistry of soil potassium. In: Munson RD (ed) Potassium in agriculture. Soil Sci Soc. Am, Madison, pp 201–276

Sparks DL, Huang PM (1985b) Physical chemistry of soil potassium. In: Munson RD (ed) Potassium in agriculture. American Society of Agronomy, Madison, pp 201–276

Styriakova I, Styriak I, Galko I, Hradil D, Bezdicka P (2003) The release of iron bearing minerals and dissolution of feldspar by heterotrophic bacteria of Bacillus species. Ceram Silicaty 47:20–26

Sugumaran P, Janarthanam B (2007) Solubilization of potassium containing minerals by bacteria and their effect on plant growth. World J Agric Sci 3(3): 350–355

Supanjani HHS, Jung SJ, Lee KD (2006) Rock phosphate potassium and rock solubilizing bacteria as alternative sustainable fertilizers. Agron Sustain Dev 26:233–240

Surdam RC, MacGowan DB (1988) Oil field waters and sandstone diagenesis. Appl Geo Chem 2:613–620

Syers JK (1998) Soil and plant potassium in agriculture. The Fertiliser Society, York

Tandon HLS, Sekhon GS (1988) Potassium research and agricultural production in India. Fertilizer Development and Consultation Organization, New Delhi, p 144

Ullman WJ, Welch SA (2002) Organic ligands and feldspar dissolution. Geochem Soc 7:3–35

Ullman WJ, Kirchman DL, Welch SA (1996) Laboratory evidence for microbially mediated silicate mineral dissolution in nature. Chem Geol 132:11–17

Vainberg SN, Vlasov AS, Skripnik VP (1980) Enrichment of clay raw material using silicate bacteria. USSR Tr Mosk Khim Tekhnol Inst Im DI Mendeleeva 116: 34–37

Vandevivere P, Welch SA, Ullman WJ, Kirchman DJ (1994) Enhanced dissolution of silicate minerals by bacteria at near neutral pH. Microbiol Ecol 27: 241–251

Vessey KJ (2003) Plant growth promoting rhizobacteria as biofertilizers. Plant Soil 25:557–586

Wang HY, Shen QH, Zhou JM, Wang J, Du CW, Chen XQ (2011) Plants use alternative strategies to utilize non exchangeable potassium in minerals. Plant Soil 343:209–220

Webley DM, Henderson MEK, Taylor IF (1963) The microbiology of rocks and weathered stones. J Soil Sci 14:65–71

Welch SA, Ullman WJ (1993) The effect of organic acids on plagioclase dissolution rates and stoichiometry. Geochim Cosmochim Acta 57:2725–2736

Welch SA, Vandevivere P (2009) Effect of microbial and other naturally occurring polymers on mineral dissolution. Geomicrobiol J 12:227–238

Wu SC, Cao ZH, Li ZG, Cheung KC, Wong MH (2005) Effects of biofertilizer containing N fixer, P and K solubilizers and AM-fungi on maize growth: a greenhouse trial. Geoderma 125:155–166

Xiaoxi Z, Liu X, Tang J, Hu S, Jiang P, Li W, Xu L (2012) Characterization and potassium solubilizing ability of *Bacillus circulans* Z1-3. Adv Sci Lett 10:173–176

Xie JC (1998) Present situation and prospects for the world's fertilizer use. Plant Nutri Fertil Sci 4: 321–330

Yanai J, Linehan DJ, Robinson D, Young IM, Hackett CA, Kyuma K, Kosaki T (1996) Effects of inorganic nitrogen application on the dynamics of the soil solution composition in the root zone of maize. Plant and Soil 180:1–9

Zakaria AAB (2009) Growth optimization of potassium solubilizing bacteria isolated from biofertilizer. Malaysia Pahang: Bachelor of Chem. Eng. (Biotech.), Fac. of Chem., Natural Resources Eng. Univ pp. 40

Zhang C, Kong F (2014) Isolation and identification of potassium-solubilizing bacteria from tobacco rhizospheric soil and their effect on tobacco plants. Appl Soil Ecol 82:18–25

Zhang FS, Ma J, Cao YP (1997) Phosphorus deficiency enhances root exudation of low molecular weight organic acids and utilization of sparingly soluble inorganic phosphates by radish (*Raphanus sativus* L.) and rape (*Brassica napus* L.) plants. Plant Soil 196: 261–264

Zhao F, Sheng XF, Huang Z (2008) Isolation of mineral potassium solubilizing bacterial strains from agricultural soils in Shandong province. Biodivers Sci 16:593–600

Zhou H, Zeng X, Liu F, Qiu G, Hu Y (2006) Screening, identification and desilication of a silicate bacterium. J Cent South Univ Technol 13:337–341

Mycorrhizal Fungi: Role in the Solubilization of Potassium

6

Jose Alfonso Dominguez-Nuñez, Begoña Benito, Marta Berrocal-Lobo, and Ada Albanesi

Abstract

Mycorrhizae have been key living microorganisms in the plant colonization of terrestrial environments. The crucial roles of mycorrhizae are to provide plant nutrition and stress resistance for plant growth and development. Potassium (K^+) shows important functions in different metabolic processes and physiological functions and is the third crucial component of most crop fertilizers (after nitrogen and phosphorus). This chapter discusses the current state of knowledge about the role of mycorrhizal symbiosis in the release, solubilization, and transport of potassium from the soil to plants. Improved metagenomic analyses and the availability of transcriptome data provide promising tools for studying fungal weathering and the mycorrhizal contributions to K^+ solubilization and K^+ transport to plants. Additionally, some benefits of mycorrhizal potassium uptake in plants are described.

Keywords

Mycorrhizal fungi • K-solubilization • Symbiosis • Molecular biology

J.A. Dominguez-Nuñez (✉)
Department of Natural Systems and Resources, E.T.S.I. de Montes, Forestal y del Medio Natural, Polytechnic University of Madrid, Av/ Ciudad Universitaria s/n., 28040 Madrid, Spain
e-mail: josealfonso.dominguez@upm.es

B. Benito
Centre for Plant Biotechnology and Genomics (CBGP) U.P.M. – I.N.I.A., Parque Científico y Tecnológico de la U.P.M. Campus de Montegancedo, Pozuelo de Alarcón, 28223 Madrid, Spain

M. Berrocal-Lobo
Department of Natural Systems and Resources, E.T.S.I. de Montes, Forestal y del Medio Natural, Polytechnic University of Madrid, Av/ Ciudad Universitaria s/n., 28040 Madrid, Spain

Centre for Plant Biotechnology and Genomics (CBGP) U.P.M. – I.N.I.A., Parque Científico y Tecnológico de la U.P.M. Campus de Montegancedo, Pozuelo de Alarcón, 28223 Madrid, Spain

A. Albanesi
Department of Microbiology, Faculty of Agronomy & Agroindustries, National University of Santiago del Estero, Av/ Belgrano (S) 1912 (4200), Santiago del Estero, Argentina

V.S. Meena et al. (eds.), *Potassium Solubilizing Microorganisms for Sustainable Agriculture*,
DOI 10.1007/978-81-322-2776-2_6

6.1 Introduction

Potassium (K^+) is one of the most important macronutrients for all organisms. In plants, it represents 2–10 % of dry biomass, and approximately 100–200 mM is its optimal cytoplasmic concentration for enzyme activity (Leigh and Wyn Jones 1984). This cation (K^+) participates in different important processes, such as plasma membrane polarization, growth, modulation of the stomatal aperture, and adaptation to environmental changes (Broadley and White 2005; Wang and Wu 2013; Anschütz et al. 2014; Shabala and Pottosin 2014).

The proper maintenance of an elevated K^+ concentration in plant cells is vital for the smooth functioning of such physiological and metabolic processes. This cation is also the third critical component of most crop fertilizers (after nitrogen (N) and phosphorus (P); Benito and González-Guerrero 2014; Benito et al. 2014; Shin and Adams 2014). K^+ ions are most abundant in soil and their availability is low due to their strong mineral adsorption. Depending on the soil type, the K^+ concentration in soil solution is ~0.1–1 mM (Asher and Ozanne 1967). This low availability combined with the constitutive demand of plants lead to the formation of depleted areas around roots (Drew and Nye 1969). Consequently, plants must use efficient strategies to improve their K^+ acquisition from soil (Nieves-Cordones et al. 2014; Zörb et al. 2014), such as the uptake of high-affinity transport systems or the establishment of plant–microbe associations.

The most important components of plant-associated microbial populations are mycorrhizal fungi. Mycorrhizal symbioses are mutualistic interactions observed between the root systems of ~80–90 % of land plants and the mycelia of different fungi (Wang and Qiu 2006). Among the known mycorrhizal associations, mainly two forms are studied on the basis of their ecological importance: arbuscular mycorrhizae (AM) and ectomycorrhizae (ECM). Mycorrhizal fungi actively participate in plant development (Smith and Read 2008) via improvement of access to nutrients, particularly when resources become scarce, and in stress resistance. In turn, vegetable partners provide up to 20–25 % of photosynthetic carbohydrates to their symbionts (López et al. 2008).

Mycorrhizal fungi are able to absorb different macro- and micronutrients such as phosphorous (P), potassium (K), nitrogen (N), magnesium (Mg), calcium (Ca), iron (Fe), zinc (Zn), manganese (Mn), and copper (Cu) (Marschner and Dell 1994; Miransari 2013). It is estimated that more than 90 % of the phosphorus needs of the host plant are satisfied through the mycorrhizal fungus, in addition to ~50 % of the nitrogen demand and almost similar proportions of the other micronutrients. Interestingly, the mycorrhizae are also protective against the toxic concentrations of some of these elements, to the extent that at least some mycorrhizal fungi act as living buffers and provide important nutrients during availability of low concentrations and detoxify them during when they are present at higher levels. In spite of the wealth of knowledge available regarding N and P nutrition in mycorrhizal plants, very little is known about how K^+ homeostasis is affected through symbiosis (Smith and Read 2008). In a riparian area, one study identified a relationship between exchangeable K^+ and the AM fungal community composition but did not find a relationship between soil nitrogen or phosphorus concentrations and their richness, as well as the degree of colonization or community composition of AM fungal species (Beauchamp et al. 2006).

6.2 Plant Potassium Nutrition Associated with Mycorrhizal Symbiosis

6.2.1 Arbuscular Mycorrhizal Symbiosis

The available information on the uptake of macronutrient cations through AM plants is relatively inconsistent, in that increases or decreases, or the absence of effects, have been reported (Clark and Zeto 1996; Meena et al. 2014a). However, there are many studies demonstrating an improvement of K^+ nutrition in plants through AM symbiosis. Clark et al. (1999) observed that the acquisition of K^+ compared with that of calcium and magnesium was especially enhanced in AM switch grass grown in acid soil. K^+ enrichment of plants mycorrhized with the *Glomus* Br1 isolate has been observed in Zea mays root steles (Kaldorf et al. 1999). Additionally, when inocula of two AM fungus species, G. mosseae and G. intraradices, were applied in soil on a weight basis, increased K^+ uptake by a maize crop was recorded (Wu et al. 2005). Miransari et al. (2009a, b) investigated the effects of different species of mycorrhizal fungi, including G. etunicatum and G. mosseae, on nutrient uptake in corn and wheat in compacted soil. They indicated that mycorrhizal fungi were able to alleviate the unfavorable effects of compaction on corn and wheat growth by enhancing the uptake of various nutrients, such as K, Mg, and Ca. Cimen et al. (2010) investigated the effects of the mycorrhizal fungus G. intraradices on the uptake of different nutrients, including K, Mg, and Ca, by tomato (Lycopersicon esculentum L.). The fungus increased the uptake of K, Ca, and Mg by tomato (Meena et al. 2015b; Singh et al. 2015).

Zaefarian et al. (2011) investigated the effects of different fungal species, including G. intraradices, G. etunicatum, and G mosseae (as single treatments), and combined treatment with G. fasciculatum, Gigaspora hartiga, and G. mosseae on the uptake of N, K, Fe, P, zinc (Zn), and Cu. The fungal species were able to increase the uptake of all nutrients (with the exception of Fe) by alfalfa (Medicago sativa), and G. mosseae was the most efficient species in this regard. Similar results have been obtained in studies with Pelargonium peltatum (Perner et al. 2007) or Lactuca sativa (Baslam et al. 2013) inoculated with other AM fungi. Moreover, the effect of AM fungi combined with other microorganisms on the K nutrition of plants has also been studied. Wu et al. (2009) investigated the effects of Rhizobium and G. mosseae on Medicago sativa grown in three types of coal mine substrates. Single inoculations with R. meliloti showed it played a minimal role in M. sativa growth in the coal mine substrates. However, dual inoculation showed significantly greater effects on both growth of plants and nutrient acquisition as compared with single inoculation with either Rhizobium or G. mosseae, except in fly ash substrate. The greatest biomass, shoot length, and root, and also root N, K, and P contents, were recorded following the dual inoculation of the coal waste plus fly ash substrate. These results showed that G. mosseae increased plant K acquisition. However, these findings differed from those of Malcova et al. (2001), who observed that inoculation with G. claroideum BEG23 and BEG96 showed significantly decreased K acquisition by Calamagrostis epigejos grown in fly ash and pyrite smelter waste. The effect of AM symbiosis on the K^+ nutrition of plants was found using appropriate techniques. Assessment of the K^+ distribution in AM fungi (Rhizophagus irregularis) through particle-induced X-ray emission (PIXE) experiments (Johansson and Campbell 1988) showed high K^+ accumulation in spores (Pallon et al. 2007), vesicles (Olsson et al. 2011), and hyphae (Olsson et al. 2008).

PIXE analyses revealed higher K^+ concentrations in root sections of Aster tripolium mycorrhized by R. irregularis compared with control plants, suggesting a possible increase in K^+ uptake through AM colonization (Scheloske et al. 2004). K^+ transport is commonly visualized using rubidium (Rb^+) as an analog tracer. By measuring AM fungus-mediated Rb^+ acquisition by the soil, Hawkes and Casper (2002) elucidated putative competition mechanisms for four herbaceous species. Hence, the differences in the chemical characteristics of the tested substrates may contribute to the observed differences in K acquisition, as soil properties may affect the capacity of AM fungi to influence the growth and physiology of the host plant, along with other conditions (Clark and Zeto 2000; Kapoor et al. 2002; Wu et al. 2009; Meena et al. 2014a).

6.2.2 Ectomycorrhizal Symbiosis

As in AM fungi, the results of studies on ECM fungi addressing their role in K uptake by plants have not been consistent. Generally speaking, high heterogeneity of the acquisition rates of small mobile ions such as ammonium and K (but also of more mobile nitrate ions) has been investigated within the roots of a range of forest species, either ectomycorrhizal or not, with the root tips exhibiting much greater acquisition activity as compared with the more basal parts of roots (Plassard et al. 2002; Hawkins et al. 2008; Meena et al. 2013; Maurya et al. 2014). The role of ECM fungi in the acquisition of Mg and K is further supported through studies showing increased fungal exploitation of sites with more cation availability, presenting a striking parallel with proliferation of the intensive hyphae observed in patches with organic material like litter (Bending and Read 1995a, b) or when P or N was patchily supplied (Brandes et al. 1998; Jentschke et al. 2000). Additionally, it has been observed that ECM symbiosis improves K concentration in the tissue under non-sufficient conditions (Wallander 2000b), but this effect may be offset by plant growth promotion (Yuan et al. 2004). In a study performed in a stand of Douglas fir, Pseudotsuga menziesii, Fogel and Hunt (1983) reported that the mycorrhizae and roots contained larger reserves of potassium, nitrogen, phosphorus, and magnesium than the forest floor or soil fungi. Futai et al. (2008) reported that the ability of ECM fungal species to take up potassium or phosphorus may be significant in nitrogen-saturated forests.

Alves et al. (2010) reported that ECM fungi, particularly Pisolithus spp. isolates (UFSC-Pt22 and UFSC-Pt186), contributed to increasing the efficiency of alkaline breccia as a source of P and K during the growth of Eucalyptus dunnii seedlings. After 90 days, plant height, root length, shoot dry weight, potassium and phosphorus content, and mycorrhizal colonization were increased compared with controls. Forest trees deficient in K are commonly found in subtropical and tropical acid soils, and ectomycorrhizal inoculation may improve the growth and K nutrition of the trees in artificial plantations in these areas. Huang and Lapeyrie (1996) reported the utilization of hydrogen chloride (HCl) extractable K by Douglas fir inoculated with Laccaria bicolor S238N and the concomitant acceleration of the tree growth rate. Similarly, the growth of Pinus sylvestris seedlings was accelerated through Paxillus involutus when microcline was used as the K source (Wallander and Wickman 1999). Data obtained in multiple Rhizopogon spp. isolates obtained from the field showed that they caused significant K^+ sequestration in rhizomorphs, which could be crucial for forests subjected to very long periods of K^+ deprivation (Wallander et al. 2002; Wallander and Pallon 2005; Meena et al. 2014b).

Wallander et al. (2003) reported that another fungus, Suillus granulatus, can be considered an important K^+ accumulator. The greater mineral degradation efficiencies of these last two ECM fungi were suggested by the calcium-rich crystals originating through K^+-rich mineral apatite on rhizomorph surfaces. Thus, S. granulatus and Rhizopogon spp. could be considered as the key facilitators of K^+ fluxes between trees and soil in temperate forest ecosystems. Recently, an increase in the content of K^+ up to 35 % was found in Pinus pinaster mycorrhized through Hebeloma cylindrosporum after 2 months of culture under K^+ deficiency, suggesting that this fungus plays an important role in the adaptation of the pine species to limiting conditions (Garcia et al. 2014). K^+ assimilation was efficiently improved in mycorrhized E. globules and shoots of Acacia spirorbis through Pisolithus albus, by almost ~ 38 % (Jourand et al. 2014). However, the analysis of K concentrations in plant tissues failed to unequivocally demonstrate the involvement of ectomycorrhizae in soil K uptake and mobilization. Following ECM inoculation, promotion of K uptake by host trees from soils has been observed, but in some cases, K has been observed to be present at a lower concentration in the tissues of plants inoculated with mycorrhizal fungi compared with uninoculated controls (Bowen 1973; Smith and Read 1997; Kumar et al. 2015; Meena et al. 2015a). This may be due to a dilution effect resulting from mycorrhizal stimulation of host plant growth and suggests that K uptake is a better indicator than the K concentration regarding mycorrhizal effects on plant K nutrition (Yuan et al. 2004).

In several studies, different effects of ECM on K uptake by plants have been observed by the authors. Quercus faginea and Pinus halepensis colonized by Tuber melanosporum displayed a significant reduction of K^+ concentrations at an afforested site (Domínguez et al. 2006) and in a nursery (Dominguez-Nunez et al. 2015). However, in other experiments conducted in a nursery, the K^+ concentrations in seedlings were found to be either no different (Dominguez Núñez et al. 2008; 2009) or increased (Dominguez-Nuñez et al. 2012) following T. melanosporum inoculation. In another nursery experiment, the K^+ concentrations in P. halepensis seedlings were decreased by Pisolithus tinctorius inoculation (Dominguez-Nunez et al. 2013a).

The K^+ concentrations were not modified by Rhizopogon roseolus inoculation (Dominguez-Nunez et al. 2013b). These contradictory data again highlight that K^+ allocation from soil to plants via ECM fungi requires complementary functional investigations. Additionally, this effect has been analyzed and verified through appropriate techniques. K^+ fluxes from ECM fungi to host plants were first assessed by the determination of Rb^+ content (Rygiewicz and Bledsoe 1984; Jongbloed et al. 1991). Quantification of K^+ concentrations in Picea abies cultivated in medium through 230 μM K^+ suggested that approximately 5–6 % of the total K^+ came through the ECM fungus Paxillus involutus (Jentschke et al. 2001). Regarding the cellular distribution of K^+ within fungal hyphae, X-ray micro-analysis showed that K^+ was mainly localized in vacuoles of the ECM fungus Pisolithus tinctorius (Orlovich and Ashford 1993; Ashford et al. 1999). PIXE experiments on P. sylvestris Suillusluteus mycorrhizae reported that high K^+ concentrations were observed in the ECM vascular tissues (Turnau et al. 2001).

6.3 Mechanisms of K-Solubilization: Weathering

The weathering of minerals containing base cations by mycorrhizal fungi could have a significant effect on the plant acquisition and ecosystem cycling of base cations.

Laboratory studies on the dissolution of nepheline, feldspars, and other minerals, with or without the presence of soil mycorrhizae, have shown these minerals have the ability to release K in the appropriate concentration to meet the requirements of growing plants. When experiments were performed, the results were variable and also were not always substantiated via use of statistically rigorous design or analysis. Nevertheless, the most important positive results were shown for soils with very little cation exchange capacity and typical of oxisols encountered in tropical and subtropical regions and soils with very poor nutrient retention characteristics. However, use of crushed rock materials that might be collected through local sources has to be seriously considered as an alternative to conventional chemical fertilizers, which might be poorly accessible on the grounds of availability or cost (Manning 2010). Fungal leaching is largely associated with four main mechanisms: redoxolysis/reduction, acidolysis, complexolysis, and metal accumulation (Burgstaller and Schinner 1993; Adeleke et al. 2010). Most of these mechanisms are either directly or indirectly related to the capacity of fungi to produce various organic acids and ligands (Burgstaller and Schinner 1993; Jain and Sharma 2004).

There are few studies that have examined the effect of AM fungi on the weathering of K-containing minerals. AM can increase the solubility of mineral forms of K^+ by releasing protons, H^+, or CO_2 and organic acid anions such as oxalate, malate, and citrate. This process also increases N, K, Ca, and Fe in the plant leaves and fruits (Jones et al. 2009; Veresoglou et al. 2011; Yousefi et al. 2011). Moreover, the potential of ECM fungi to dissolve and take up K under different conditions has been widely studied, mainly through mica minerals (i.e., phlogopite, muscovite, and biotite), but also by using K-saturated silicate, vermiculite minerals (i.e., microcline and hornblende), and natural E-horizon soil (Paris et al. 1996; Wallander and Wickman 1999; Yuan et al. 2004; van Schöll et al. 2006b; Balogh-Brunstad et al. 2008; Calvaruso et al. 2009; Rosling 2009).

In several studies, it has been observed that ECM fungi did not grow well in various mineral substrates, with better ECM growth being observed in K-containing substrates. Rosling et al. (2004a) found that eight different ECM fungal cultures responded differently to culturing in the presence of marble, quartz, potassium feldspar, and apatite minerals. Increased carbon allocation through ECM tree seedlings showed more hyphal proliferation in patches with K-containing feldspar than in different patches with surrounding soil or in quartz (podzol E horizon) as demonstrated in a microcosm experiment (Rosling et al. 2004b). Wallander and Wickman (1999) reported that the availability of soluble potassium (K^+) from feldspars and biotite was correlated with the fungus biomass in a community composed of Suillus variegates, Paxillus involutus, and seedlings of Pinus sylvestris. Similarly, Wallander and Wickman (1999) compared ECM (colonized with Suillus variegatus) and non-mycorrhizal cultivation of Pinus sylvestris seedlings under pot experiments where the feldspar (microcline) or biotite was the source of potassium (K). The release of K from both minerals was enhanced (as measured by biomass K content) in the presence of mycorrhizal fungi. Pots with access to biotite exhibited soil solutions that were more rich in citric and oxalic acids, and the concentration of citric acid was positively correlated with the fungal biomass in the soil as well as foliar K^+ concentrations. However, in a K-limited forest, no increase in the hyphal colonization of buried biotite bags compared with controls was observed (Hagerberg et al. 2003). However, the K-solubilizing capacity of ECM fungi was questioned in previous studies.

Leyval and Berthelin (1989) reported that beech ECM–bacteria interactions influenced the release of Fe^{3+}, P, Mg^{2+}, and K^+, but later suggested (Leyval and Berthelin 1991) that rhizosphere bacteria were significantly responsible for the mineral nutrient solubilization and that the capacity of mycorrhizal fungi to increase the nutrient acquisition might be solely due to the greater absorptive surface area of mycorrhizal roots. Studies on the effects of mycorrhizae regarding the weathering of biotite (Wallander 2000b) and apatite (Wallander 2000a) suggest the exudation of oxalic acid through mycorrhizal fungi can influence the dissolution of apatite and increase P acquisition through P. sylvestris, but evidence of a direct effect of mycorrhizae on K^+ acquisition was less clear in the latter experiment.

Wallander and Hagerberg (2004) hypothesized that ECM fungi are unlikely to play an important role in the weathering of Mg- and K-containing minerals, as Mg and K deficiency decreases the belowground carbon allocation to roots, whereas P and N deficiency increases allocation to roots and the associated mycorrhizal fungi (Ericsson 1995). However, there are many studies examining the ability of ECM fungi to release K from minerals. Yuan et al. (2004) further investigated the effect of four fungal strains, Pisolithus XC1, Pisolithus spp., P. microcarpus, and Cenococcum geophilum SIV, collected from eucalyptus roots, on the degradation of vermiculite and phlogopite. The results were that all four strains were able to play important roles in weathering of the mineral phases and as a result release elemental potassium. Based on the weathering budgets, planted treatments involving P. sylvestris seedlings with or without mycorrhizal colonization showed importantly greater K mobilization compared with non-planted controls when K was supplied as microcline or biotite (Wallander and Wickman 1999), muscovite (van Schöll et al. 2006b), or E-horizon soil (van Hees et al. 2006b). Glowa et al. (2003) compared the ability of fungal Piloderma spp. to extract K^+ from chlorite, microcline, and biotite, and found that these species had the ability to acquire K^+ from all three minerals, with biotite being the most biodegradable. A Hebeloma (or - Inocybe-like) ECM morph type was shown to increase the release of sodium (Na) and K in the mycorrhizosphere in a spruce forest. Additionally, Yuan et al. (2000) reported that ectomycorrhizae could mobilize K^+ from clay minerals and, thus, increase the progress of K^+ acquisition by plants. Trees and their mycorrhizal fungi play important roles in replenishing the exchangeable pool of cations from the minerals (Paris et al. 1995b; Markewitz and Richter 2000; van Hees et al. 2006b). Furthermore, it has been shown that ECM has the capacity to mobilize K nutrients from organic substrates.

The proliferation of ECM hyphae in different patches of organic material collected through the FH horizon (humic layer) of the forest soil and inserted into peat-containing microcosms (Read 1991) showed that the mycelium has the ability to forage for nutrients contained in organic residues, which is further in accordance with the studies by Bending and Read (1995a) on the depletion of K, P, and N from similar organic patches. Although it was not clear if the saprotrophs play an important role in the observed nutrient mobilization, Bending and Read (1995b) measured elevated efficiencies of nutrient-mobilizing enzymes such as proteases, phosphomonoesterase, and polyphenol oxidase in substrates intensively colonized by ECM fungi. In most studies, the weathering activities and the K solubilization effects of ECM fungi have been associated with acidolysis and reduction of the soil pH because of LMWOA (low molecular weight organic acid) exudation by the ECM fungal hyphae. Studies by Ochs (1996) under in vitro conditions suggested that the exudation of simple organic ligands through fungi and mycorrhizal roots increased the dissolution of minerals, while exudates from humic substances and non-mycorrhizal roots did not significantly affect weathering. This interpretation was confirmed by Arocena et al. (1999), who investigated the chemistry of soil solutions from non-ectomycorrhizosphere and ectomycorrhizosphere soils under the subalpine fir Abies lasiocarpa.

In soil associated with fungi and mycorrhizal roots, the concentrations of the base cations Mg^{2+}, K^+, and Ca^{2+} were found to be higher as compared with soil devoid of hyphal rhizomorphs and roots. This was able to reduce the pH due to the exudation of different organic acids, as observed by Cromack et al. (1979), who suggested that the pH of soil colonized by mats of the ECM fungus Hysterangium crassum could be reduced to 4.9, compared with uncolonized soil, which exhibited a pH of 6.1. Among the LMWOAs, oxalate is likely one of the most important weathering agents exuded by ECM fungal hyphae and plant roots, according to most studies. Hyphae of Piloderma croceum have an important characteristic wall incrustation of calcium oxalate, as observed on mineral grain seedlings (van Schöll et al. 2006a). This species has been shown to enhance the exudation of oxalate compared with nonmycorrhizal-released K and Mg (Glowa et al. 2003; Arocena et al. 2004), but it did not show increased weathering of hornblende and muscovite in a study by van Schöll et al. (2006b). The processes of K liberation and transport from mineral sources appear to be associated with the displacement of interlayer K^+ from interlattice spaces and replacement by protons (from organic acids exuded by ECM fungi). Experimental evidence of the occurrence of mineral weathering via mycorrhizal hyphae has come through X-ray diffraction studies by Paris et al. (1995a, b, 1996) showing that fungi such as Pisolithus tinctorius 441 and Paxillus involutus (now considered a putative strain of P. microcarpus, according to Martin et al. 2002) are able to weather vermiculite and phlogopite mica, displacing interlayer Mg^{2+}, aluminum (Al^{3+}), and/or K^+ from interlattice spaces. Based on the observed exudation of protons and oxalate through examined Pisolithus isolates, the detected mineral weathering was assigned to the formation of Fe and Al oxalate and the ability of protons to travel within layer minerals and replace interlayer (Yuan et al. 2004). Oxalate chelation of these cations reveals the structural imbalance and may therefore participate in the dissolution and release of other elements such as K and P which are embedded in minerals (Yuan et al. 2004; Delvasto et al. 2009).

Second, both K (from minerals such as muscovite) and protons (from organic acids) are monovalent, but the size of protons is much smaller than that of K (Lapeyrie et al. 1987; Yuan et al. 2004). Thus, the size of protons gives them an advantage in replacing interlayer K contained in layer minerals, such as muscovite, which have been associated with iron ores, as investigated by Adeleke et al. (2010). Additionally, the slime produced through microorganisms largely consists of polysaccharides that can bind metals from solution (Banfield et al. 1999). In the presence of organic acid exudation, this process may create a concentration gradient that can enhance the movement of ions from minerals into solutions in response to low levels of K in the solutions (Delvasto et al. 2009; Adeleke et al. 2010). Complex-forming agents other than LMWOAs have been supported as a possible means of fungal weathering. The chemical weathering pathways induced may be significantly important for ECM fungi grown under combined cation limitation conditions (Paris et al. 1995b, 1996; van Schöll et al. 2006b).

The siderophore-producing fungi have the ability to induce weathering of crystalline and amorphous silicates at a higher rate than that expected from organic acids (Callot et al. 1987). Synergistic interaction effects of goethite dissolution have been shown in two ligand systems with oxalate and siderophores, which suggests that low concentrations of both ligands have the ability to provide an efficient weathering mechanism (Cheah et al. 2003). Van Hees et al. (2006c) reported that oxalate and ferricrocin (an intracellular siderophore) were consistently detected in exudations from the mycorrhizal fungus Hebeloma velutipes growing in symbiosis with P. sylvestris. ECM fungi exude different organic acids in the rhizosphere to start/initiate a chain of chemical reactions which lead to the breakdown of hard mineral materials. A shortage of base cations and the presence of certain minerals in the growth medium can enhance the production of various organic acids by ECM fungi (van Schöll et al. 2006a). The responses of ECM fungi to nutrient-limiting conditions have only been analyzed in depth in a few studies. In these studies, it was found that tunneled minerals emerged only when the very easily weatherable base cation-containing minerals had been strongly weathered (hornblende) or had disappeared (biotite), suggesting that ECM fungal weathering is driven through the (reduced) bioavailability of magnesium (Mg), Ca, and K (Hoffland et al. 2002; Smits 2005).

Rosling (2009) reviewed the weathering activity in response to K limitation described in several studies (Paris et al. 1995b, 1996; Wallander and Wickman 1999; Glowa et al. 2003; van Schöll et al. 2006a, b; van Hees et al. 2006a, b, c), four of which also examined Mg-limiting conditions. Paris et al. (1996) showed that two ECM fungi, P. tinctorius and P. involutus, have the ability to change their rate of oxalate exudation per unit of mycelial density during growth under combined Mg and K limitation. For P. tinctorius, the response was maintained on both nitrate and ammonium. Oxalate exudation by P. involutus was significantly enhanced in the Mg-free medium and was further enhanced in medium with neither K nor Mg, as compared with the complete and K-free medium. However, this was true possibly only when nitrogen (N) was supplied as nitrate; on the other hand, no oxalate was exuded during P. involutus growth on ammonium (Paris et al. 1996). The growth and activity of P. tinctorius was shown to affect the structure and composition of phlogopite with an enhanced cation exchange capacity (CEC) as a result of growth on Mg- and K-free medium. This treatment also showed important leaching of silicon (Si) and Al from the mineral (Paris et al. 1995b). Similar patterns were not shown with P. involutus in the same study. Structural alteration of phlogopite occurred only through nutrient treatment but not the influence of the N source on P. tinctorius. Colonization by P. involutus resulted in changes in the mineral only when the fungus was grown on ammonium.

These changes were more important and irreversible to a greater extent than those induced by P. tinctorius (Paris et al. 1995b). In a pot experiment (van Schöll et al. 2006b), three different ECM fungi, Suillus bovinus, Piloderma croceum, and P. involutus, were grown in symbiosis with P. sylvestris.

The systems were reacted with a complete nutrient solution, with or without muscovite as the main source of K and with or without Mg from the hornblende source. After 27 weeks of treatment the growth of seedlings colonized by P. involutus more significantly increased the weathering of muscovite, which was determined on the basis of the size of the non-mineral K pools (i.e., in the roots, in the soil, and in solution), than pine seedlings colonized by two other mycorrhizal fungi and non-mycorrhizal controls. P. involutus had the most efficient ability as a root colonizer tested in the experimental trial. However, the efficiency of the observed muscovite weathering have not been appear and as the result only the differences in soil exploration which measured as hyphal dry weight. There has been a positive correlation observed between the pH of the soil solution and total non-mineral K, suggesting that the muscovite weathering is caused by an acidification mechanism (van Schöll et al. 2006b). Three experiments applied during different growth conditions for the fungi—that is, symbiosis in pots, symbiosis in liquid and pure culture in liquid—were performed to analyze the exudation of LMWOAs in relation to Mg- and K-limiting growth conditions. These experiments demonstrated some important correlations (only seedlings colonized by P. involutus caused significantly enhanced oxalate exudation in K deficiency), but there was no regulated exudation pattern that could be demonstrated under the response of these nutrients (van Schöll et al. 2006a).

In a sterile multi-compartment system designed to examine the time of colonization and exudation rate per hyphal tip (van Hees et al. 2006c), the mycorrhizal fungus Hebeloma velutipes (UP184) (referred to as Hebeloma crustuliniforme in the paper) was grown under symbiosis with P. sylvestris, and the exudation of hydroxymate-type siderophores and LMWOAs was analyzed with and without available K in treatments. Only ferricrocin (an intracellular siderophore) and oxalate were consistently detected. The exudation rate of ferricrocin was significantly enhanced under K-limiting conditions. The effect of K limitation was studied earlier by application of the same isolate of H. velutipes that colonized Scots pine seedlings in a long-term column experiment (van Hees et al. 2006a, b). However, no treatment effects of mycorrhizal versus non-mycorrhizal growth or K-deficient versus K-sufficient growth were observed for total LMOWAs or dissolved organic carbon (DOC) (van Hees et al. 2006a). The treatments showed no effect on the total weathering budget. However, during K-limiting growth conditions, mycorrhizal colonization significantly enhanced the potassium (K) status of roots (van Hees et al. 2006b). H. velutipes may increase the K mobilization through increased exudation of siderophores or effective soil exploration. Because LMWOAs were the only possible weathering agents assessed in this study, no other information was obtained regarding other possible weathering mechanisms involved.

This ability of ECM fungi concerning the weathering of cation-containing minerals is being studied with a view toward future applications in biohydrometallurgy. LMWOAs are compounds that show great potential in the solubilization of complex or hard mineral materials (Paris et al. 1995b; Gadd 1999; van Breemen et al. 2000). The ECM fungi play a significant role in the mobilization of nutrients from some low-grade iron mineral ores (Adeleke et al. 2010). The reduction of particle size enhances the total particle surface area of minerals and influences the leaching process, as well as being expected to enhance the solubilization rate (Bosecker 1997; Vasan et al. 2001; Jain and Sharma 2004).

Adeleke et al. (2010) investigated the potential participation of ECM fungi in the mobilization of nutrients such as potassium (K) and phosphorus (P) from a typical insoluble iron ore. The results indicated the important potential of the four fungi to mobilize K and P from two types of iron ore, from various levels. The particle size, ore type, attachment of the fungi to the iron ore, and organic acid production were all found to play significant roles in the mobilization of nutrients from these ores. However, Adeleke et al. (2012) observed that the quantity of organic acid produced from Pisolithus tinctorius was not directly correlated with the mobilization of either K or P from the iron mineral ore, i.e., mineral weathering was influenced by a combination of factors and not only in response to organic acid.

6.4 Molecular Biology and Transport of Potassium in Mycorrhizal Interactions

Potassium is a macronutrient that is absolutely essential for the survival of plants and soil microorganisms. The availability of K^+ in natural environments is highly variable, and a large number of K^+ transport systems therefore exist in these organisms. Over the past few years, molecular biological approaches and functional characterization analyses have greatly advanced our understanding of K^+ transport in fungi and plants, and a large number of genes encoding K^+ transporters and channels have been identified (Mäser et al. 2001, 2002b; Véry and Sentenac 2002, 2003).

Considering that the K^+ concentrations and pH values of fungal and plant cells are considerably variable (Leigh and Wyn Jones 1984), it is not surprising that large K^+ transport systems and diversity in terms of ionic permeability and the affinity for and energetic coupling with other ions (H^+ and/or Na^+) exist. Some of the K^+ transporters characterized to date show a specific capacity to take up K^+ from very low K^+ concentrations (referred to as high-affinity K^+ transporters) and to maintain very high concentration ratios across the plasma membrane. Other K^+ transport systems are functional when the K^+ concentration is high (referred to as low-affinity K^+ transporters), or their activity is voltage membrane dependent (this is the case for the K^+ Shaker-like channels). Among the various K^+ transport system families identified thus far in nature, some are shared in fungi and plants, and others have been exclusively identified in one of these groups.

6.4.1 K^+ Transport Through Mycorrhizal Fungi

The K^+ transporters found in fungi belong to four families: Trk/Ktr/HKT, HAK transporters, ACU, and PAT ATPases (Souza and Gomes 1998; Rodríguez-Navarro 2000; Benito et al. 2004; Corratgé-Faillie et al. 2010). Excluding the PAT ATPases, which have been examined in very little studies (Souza and Gomes 1998; Fietto et al. 2002), the energization and ability of K^+ depletion have been different in the other three systems. A recent study identified 62 K^+ acquisition systems in fungal species for which complete genome sequences were available (most are non-mycorrhizal species). This work revealed that three types of K^+ acquisition systems (Trk/Ktr/HKT and ACU ATPases and HAK transporters) are widely found in different combinations across fungal species. PAT ATPases are present less frequently and are exceptional in Ascomycota (Benito et al. 2011). Regarding the fungal K^+ channels that may mediate K^+ movement, in contrast to the findings reported for animal or plant cells, very little is known about their function. These channels are less numerous and diverse than in plants and are basically grouped into two families: TOK channels, formed by the outward-rectifying channel (Ketchum et al. 1995); and an SKC family, similar to animal Shaker-like voltage-dependent K^+-selective channels (Lambilliotte et al. 2004).

6.4.1.1 ECM Symbiosis

Information recently obtained from the genomes of the ECM fungi Laccaria bicolor (Martin et al. 2008), Tuber melanosporum (Martin et al. 2010), and many others (www.genome.jgi.doe.gov) through BLAST search analyses against the protein sequences of K^+ transporters has allowed identification of homologous putative K^+ transporters operating in ECM fungi that are similar to the TRK transporters and channels described previously in fungi (Ko and Gaber 1991) and to TOK channels (Ketchum et al. 1995), HAK transporters (Bañuelos et al. 1995), and animal Shaker-like channels (SKC) (Papazian et al. 1987; Jan and Jan 1997).

Prior to the availability of these genomes, a Shaker-like channel and a Trk-type transporter were identified in an EST library from Hebeloma cylindrosporum (Lambilliotte et al. 2004). The member of the Trk/Ktr/HKT family was functionally characterized, indicating that HcTrk1 might transport not only K^+ but also Na^+. More recently, the application of H. cylindrosporum transgenic lines allowed localization of this transporter exclusively to the external hyphae of P. pinaster mycorrhizae (Garcia et al. 2014), supporting a specialized form of K^+ uptake involved in exploring soil regions in symbiotic conditions. The other authors identified the H. cylindrosporum EST library that belongs to the SKC family which represents voltage-dependent K^+-selective channels (Lambilliotte et al. 2004).

Interestingly, SKC channels have been found exclusively in Basidiomycota fungi and some other members of the basal fungi, but they are absent in sequenced Ascomycota. Additionally, analysis of the H. cylindrosporum genome has indicated the existence of a second Trk transporter and three HcTOK channels, two additional types of transport systems, and an HcHAK transporter. Based on the predicted Hebeloma inventory of K^+ transport systems, it could be hypothesized that during symbiosis, K^+ absorption from the soil takes place in the soil-exploring hyphae and surrounding root via HcTrk1 but also through other transporters, such as HcTrk2 and/or HcHAK. These transporters could operate in the extraradical mycelium, functioning as effective K^+ transporters, even in potassium (K^+) limiting zones, and acting as high-affinity K^+ transporters in a similar manner to other fungal TRK- and HAK-homologous proteins (Bañuelos et al. 2000). The K^+ supply of the host plant is expected to depend on the translocation rate of these K^+ ions through the fungal plasma membrane into the apoplastic interface and within the fungal symplast of the Hartig net. This would involve potassium (K^+) replenishment of the interfacial apoplast through fungal potassium (K^+) efflux systems, potentially mediated by a K^+-specific efflux ENA ATPase or by TOK channels (Benito et al. 2002), which may also operate in Hebeloma (Lambilliotte et al. 2004; Benito and González-Guerrero 2014). However, a precaution must be applied to these predictions due to the possible bidirectional efficiency of some transport systems under specific conditions and to their unknown subcellular localization. The functional and molecular characterization of these new candidates will determine their specific involvement in potassium (K^+) acquisition from the soil or in potassium (K^+) release to plant cells.

6.4.1.2 AM Symbiosis

The transporters and channels present in AM fungi are still a missing component of the process of K^+ transport from the soil to the host. Four sequences from Rhizophagus irregularis from an EST library (www.mycor.nancy.inra.fr) have been identified as K^+ transport systems (Casieri et al. 2013). Interestingly, three of these sequences encode SKC-type channels, a type of protein that has not been characterized in fungi to date, and one encodes a HAK transporter. However, no TRK or TOK members were identified from this library or from the sequence of the nuclear genome (www.genome.jgi.doe.gov). The functional characterization and analysis of these new candidates will provide more accurate information on the fungal molecular players involved in AM plant potassium (K^+) nutrition.

6.4.2 K^+ Transport Through Plants

In the past decade, there has been important progress in the identification of genes encoding potassium (K^+) transporters and channels in plants, as demonstrated in the numerous reviews on the subject (Véry and Sentenac 2003; Rodríguez-Navarro and Rubio 2006; Gierth and

Mässer 2007; Dreyer and Uozumi 2011). Most of the present knowledge on potassium (K^+) transport in plant systems has come from studies with Arabidopsis thaliana acting as a model (Alemán et al. 2011; Dreyer and Uozumi 2011; Coskun et al. 2014; Nieves-Cordones et al. 2014). It appears that the gene families involved in potassium (K^+) transport have been strongly conserved in higher plant species, in terms of both family gene numbers and structure.

The molecular families involved in K^+ transport systems that are active at the plasma membrane belong to the same fungal K^+ transport families, namely the Trk/Ktr/HKT transporter families and the HAK and Shaker K^+ channel family. However, the roles they play within plants, such as in K^+ acquisition from the soil via the roots, long-distance potassium (K^+) transport in the xylem and phloem vasculature, and potassium (K^+) accumulation and turgor-driven processes, such as guard cell movement and regulation of transpiration water loss, remain to be determined in detail. K^+ uptake in woody plants has been poorly studied, but BLAST search analyses conducted in the genomes of some woody plants that are able to form ECM associations have allowed the identification of two HKT, 22 HAK, and 11 voltage-gated Shaker-like K^+ channels in Populus trichocarpa (Gomez-Porras et al. 2012).

Two channels expressed in the root systems of Populus euphratica (PeKC1 and PeKC2) have been characterized as K^+ channels and may play a key role in potassium (K^+) nutrition in poplar trees (Zhang et al. 2010). However, their role in the context of mycorrhizal associations has not been elucidated. A similar situation occurs in AM plants, in which numerous K^+ transport systems can be identified through genome analysis, but there is no evidence regarding the involvement of any of these systems in K^+ nutrition during symbiosis. In an exceptional case, a HAK transporter has been identified that was observed to be upregulated 44-fold in Lotus japonicus mycorrhized through AM fungus Gigaspora margarita (Guether et al. 2009), which would suggest a putative role in mycorrhizal association. More recently, a SKOR channel from Z. mays was reported to be upregulated by AM colonization in response to salinity (Estrada et al. 2013). Future analyses of these first plant candidates are required to elucidate the molecular bases of potassium (K^+) acquisition through the plant–fungus interface (Lian et al. 2008; van Schöll et al. 2008; Zhang et al. 2010).

6.5 Benefits of Mycorrhizal Potassium Uptake for Plants

6.5.1 Improvement of Plant Stress

AM fungi have the ability to alleviate the unfavorable effects of stress under various stresses such as salinity, drought (Augé 2001), and soil compaction. The fungi efficiently alleviate stress by adjusting the uptake and, hence, the ratio of different nutrients in the plant, including K, Mg, Ca, Na, and Cl (Boomsma and Vyn 2008; Daei et al. 2009). Miransari et al. (2009a, b) evaluated the effect of soil compaction on the uptake of nutrients, including macro- and micronutrients, by mycorrhizal corn and wheat. Among the most important mechanisms through which the fungi alleviate stress is enhanced nutrient uptake of various nutrients, including N, P, K, Fe, Zn, Mn, and Cu. Interestingly, mycorrhizal corn plants were shown to be more sensitive to stress relative to mycorrhizal wheat plants, which may be due to their different root architecture. Additionally, the enhancement of plant salinity tolerance through AM symbiosis has been postulated for several decades (Hirrel and Gerdemann 1980; Ojala et al. 1983).

AM symbiosis alleviates the adverse effects of salinity by significantly increasing plant growth and nutrient uptake, including that of P, K, Mg, and Zn, and by decreasing the absorption of Na^+ and Cl^- (Miransari 2013). High uptake of K via mycorrhizal symbiosis under salinity is an important parameter alleviating the unfavorable effects of salinity by adjusting the acquisition of Na^+ and Cl^- by the host plant. Researchers have also indicated that mycorrhizal associations increase plant growth under salinity by enhancing the acquisition of P, K, Fe, Cu, and Zn (Al-Karaki et al. 2001; Al-Karaki 2006). Mardukhi et al. (2011) investigated the effects of G. intraradices mycorrhizal associations on nutrient uptake in wheat genotypes in saline soil under greenhouse and field conditions. Both single and combined use of G. mosseae and G. etunicatum were tested. Under greenhouse conditions, either single or co-inoculation of the mycorrhizal fungi significantly increased the uptake of various nutrients by the host plant, including Ca, N, K, P, Mg, Zn, Cu, Mn, and Fe. In addition, higher concentrations of K^+ and Ca^{2+} were shown to result from mycorrhizal inoculation under salinity stress.

The ratios of Ca^{2+} to Mg^{2+}, Ca^{2+} to Na^+, and K^+ to Na^+ are also modified by fungal treatment under salinity. The micronutrient concentrations of Mn^{2+}, Zn^{2+}, Cu, and Fe were demonstrated to be increased in plant tissues by fungi relative to the control treatment. Accordingly, mycorrhizal fungi alleviate salinity stress by increasing the integrity of cellular membranes (decreasing electrolyte leakage) and increasing the uptake of nutrients under salinity (Evelin et al. 2012). These nutrients are able to adjust plant water behavior by modifying the cell water potential and nutrient entry into the cell through the cellular membrane (Marschner 1995; Miransari 2012a, b).

AM colonization was found to enhance plant potassium (K^+) acquisition, whereas the sodium (Na^+) content was maintained at a low level when salt stress conditions were applied to Glycine max (Sharifi et al. 2007), Vicia faba (Rabie and Almadini 2005), Ocimum basilicum (Zuccarini and Okurowska 2008), Z. mays (Estrada et al. 2013), Acacia nilotica (Giri et al. 2007), or Olea europaea (Porras-Soriano et al. 2009). The data indicated that AM symbiosis improves the salt tolerance of the host plant by changing the K^+/Na^+ balance. This finding was corroborated by the detection of high internal potassium (K^+) concentrations in many AM fungi collected from natural saline sites (Hammer et al. 2011a). However, very little information is known about the physiological and molecular mechanisms that enhance this adaptation (Garcia and Zimmermann 2014). Recently, Estrada et al. (2013) reported differential expression levels during AM colonization for three potassium (K^+) transport systems of Z. mays that are putatively involved in Na^+/K^+ homeostasis (ZmSOS1), phloem loading/unloading (ZmAKT2), and xylem release (ZmSKOR). These results pave the way for the elucidation of plant proteins involved in the transport of potassium (K^+) arising through mycorrhizal fungi, especially during salt stress conditions.

Moreover, the impact of AM symbioses on drought stress resistance in different plants has been well studied (Harley and Smith 1983; Al-Karaki 1998; Porcel and Ruiz-Lozano 2004). The improvement of hydric stress tolerance has been observed to be accompanied by an elevation of potassium (K^+) concentrations, for example in Citrus tangerine (Wu and Xia 2006), involving a role of AM symbiosis through potassium (K^+) acquisition that is required for osmotic adjustment. El-Mesbahi et al. (2012) found that the hydraulic conductivity of AM-colonized Z. mays grown in hydric stress was increased through the supply of potassium (K^+) in the external medium. Moreover, the expression level of the plant aquaporin ZmPIP2 was found to be modulated by the potassium (K^+) supply under hydric stress, suggesting a tight link between the adaptation of AM plants to drought stress and potassium (K^+) resource availability.

In ECM plants, drought and salt stress tolerance linked to potassium (K^+) nutrition has been less well studied. Recently, Danielsen and Polle (2014) showed that there was an increase in K^+ in ECM poplar during drought conditions, suggesting a important role of mycorrhizal potassium (K^+) for environmental stress adaptation in this symbiotic interaction. However, under harsher stress, the transcript levels of stress marker genes (the ammonium transporter AMT3.1, the aquaporin PIP2.5, and ABA-responsive RD26) were found to be increased, regardless of the mycorrhizal population status of the trees (Bañuelos et al. 2000; Hinsinger et al. 2011).

The important role of ECM fungi in weathering and the subsequent acquisition of base cations may be increased significantly in situations where acid rain and intensive biomass harvesting increase losses and decrease the storage of Ca, K, and Mg. During the present stress conditions, the forest growth in a few stands may become limited through Mg and/or K, rather than

N (Landman et al. 1997; Übel and Heinsdorf 1997; Thelin 2000). Interestingly, a strong correlation observed between the distribution of P and K^+ during ECM and AM symbiosis has been reported (Smith et al. 1981; Cumming 1993; Wallander and Pallon 2005; Olsson et al. 2011). Moreover, K^+ and P have been found in the same fungal compartments in Paxillus involutus (Orlovich and Ashford 1993; Ashford et al. 1999) and Scleroderma verrucosum (Jung and Tamai 2013), such as vacuoles. K^+ is assumed to be one of the important counter-ions of polyphosphate (polyP) granules, mainly soluble polyP short chains, especially located in the fungal vacuoles (Bücking and Heyser 1999).

Garcia et al. (2014) recently revealed that overexpression of a potassium (K^+) transporter of H. cylindrosporum led to a change in K^+ and P translocation from the roots to the shoots of mycorrhized P. pinaster during K^+ deprivation, providing new evidence of the interaction of P and K^+ during their transport in ECM symbioses. All of these data appear to demonstrate that K^+ is not only a direct trophic element but also an "indirect trophic" component required for homeostasis and the correct transfer of various nutrients such as inorganic phosphate to the host plant (Garcia and Zimmermann 2014).

6.5.2 Protection Against Radiocesium Pollution

Radiocesium isotopes (134Cs, 137Cs) are important soil contaminants that can enter the food chain via intermediate plant acquisition (Delvaux et al. 2001). Because of the long half-lives of these two isotopes (2.1 years for 134Cs and 30.2 years for 137Cs) and the fact that they show biogeochemical behavior similar to that of potassium, which is a major nutrient, radiocesium is considered a threatening radiopollutant present in the environment. In the terrestrial ecosystems, the similarity between the biogeochemical routes of K and radiocesium implies a crucial hazard concerning radiocesium contamination of aboveground vegetation, as the mechanisms of root acquisition appear to be very closely related for the two elements (White and Broadley 2000; Tamponnet et al. 2008; Dupré de Boulois et al. 2008). ECM fungi have been identified as greater contributors to radiocesium immobilization in the soils. These fungi take up, translocate, and accumulate significant quantities of radiocesium in their mycelia and carpophores (Dupré de Boulois et al. 2008).

It has also been pointed out from clay minerals that ECM fungi can be involved in the mobilization of radiocesium. ECM fungi effuse protons that could replace interlayer potassium (K) in phlogopite, oxalate, and vermiculite minerals, which play important roles in biological weathering (Yuan et al. 2004). This mobilization, resulting in the acquisition and accumulation of potassium (K) through ECM fungi, likely also occurs for radiocesium, as potassium (K) and radiocesium occupy the same adsorption site on clay minerals (Delvaux et al. 2001; Dupré de Boulois et al. 2008). However, the relationship between radiocesium accumulation by ECM fungi and K levels in soil is unclear. Terada et al. (1998) reported that the acquisition and accumulation of radiocesium by Pleurotus ostreatus decreased with enhancing concentrations of potassium (K) in the growth media. However, dissimilar results were reported by Yoshida and Muramatsu (1998) through a series of mushrooms isolated from forest soil, where radiocesium and potassium (K^+) concentrations were not correlated, contrary to what was observed with the isolated plant samples; these authors found that the mechanism of radiocesium acquisition was different from that of potassium (K). It is nonetheless also probable that the efficiency of mushrooms in accumulating radiocesium and potassium (K) could be ascribed to other mechanisms following acquisition. The acquisition and accumulation of radiocesium by Hebeloma cylindrosporum is positively correlated with potassium (K) concentrations (Baeza et al. 2005).

Brunner et al. (1996) evaluated the cesium (Cs^+) to K^+ ratio in P. abies inoculated with Hebeloma crustuliniforme and revealed that lower acquisition of 134Cs by plants might be due to its retention in the outer hyphae for more important transfer of potassium (K^+) to the plant. By contrast, P. pinaster mycorrhized by Rizopogon roseolus displayed more increased concentrations of 134Cs than in non-mycorrhizal plants, whereas the potassium (K^+) content remained stable (Ladeyn et al. 2008), highlighting the greater importance of the considered host–symbiont pair (Garcia and Zimmermann 2014).

The influence of potassium (K^+) on 134Cs accumulation has also been reported in AM symbiosis, for Medicago truncatula colonized through R. irregularis. Radiocesium accumulation in plants appears to be inversely correlated with the potassium (K^+) content in the external medium (Gyuricza et al. 2010a). Based on these results, AM symbiosis, associated with a high potassium (K^+) concentration in the external medium, is important to avoid 134Cs accumulation in plants growing under contaminated soil. Interestingly, external P elicits the same effects as potassium (K^+) on 134Cs acquisition under AM colonization of M. truncatula (Gyuricza et al. 2010b), reinforcing the notion of a close relationship between P and K^+ through polyP synthesis, storage, and transport (Garcia and Zimmermann 2014). In contrast to these studies, Joner et al. (2004) conducted pot experiments using various external concentrations of potassium (K^+) ions, two AM fungal species, and three host plants, which indicated that the effects of AM on plant 134Cs and 137Cs accumulation could be negligible. Further investigations will be necessary to reach a conclusion regarding whether mycorrhizal associations play a important direct role through transport processes or are an indirect mechanism in plant protection against radiocesium-contaminated soils (Garcia and Zimmermann 2014).

6.6 Concluding Remarks

There are many examples of studies demonstrating an improvement of K^+ nutrition in plants through mycorrhizal symbiosis, especially under K^+-limiting conditions. Mycorrhizal fungi (ECM especially) play an important role in base cation nutrition in plants through mineral weathering. Some species of mycorrhizal fungi may also increase weathering rates by their chemical activity. The extent to which chemical weathering activity is increased during nutrient-limiting growth conditions is currently unclear, and only a little studies were reported by using realistic growth conditions for fungi. Mg- and K-limiting growth conditions result in enhanced siderophores and exudation of LMWOAs in some species of ectomycorrhizal fungi. To characterize the pathways involved in induced weathering in mycorrhizal fungi, it will be necessary to study the regulation of other weathering agents. Knowledge of these weathering processes may be of interest for applications such as biohydrometallurgy. The analysis of genomic and transcriptomic data provides a promising tool for studying fungal weathering and the mycorrhizal contribution to K^+ solubilization and K^+ transport to plants. Transporters from the Ktr/Trk/HKT and KUP/KT/HAK families could potentially take up potassium (K^+) from the soil. Additionally, TOK and Shaker-like K^+ channels likely participate in the transfer of K^+ from the arbuscule or Hartig net to the host plant cortical cells. K uptake by mycorrhizal fungi is also important under stress conditions, where the fungi can significantly increase plant nutrient uptake and play an important role in abiotic stress tolerance, the maintenance of P homeostasis, or the exclusion of soil contaminants such as radiocesium.

References

Adeleke RA, Cloete TE, Bertrand A, Khasa DP (2010) Mobilisation of potassium and phosphorus from iron ore by ectomycorrhizal fungi. World J Microbiol Biotechnol 26:1901–1913

Adeleke RA, Cloete TE, Bertrand A, Khasa DP (2012) Iron ore weathering potentials of ectomycorrhizal plants. Mycorrhiza 22:535–544

Alemán F, Nieves-Cordones M, Martínez V, Rubio F (2011) Root K^+ acquisition in plants: the Arabidopsis thaliana model. Plant Cell Physiol 52:1603–1612

Al-Karaki GN (1998) Benefit, cost and water-use efficiency of arbuscular mycorrhizal durum wheat grown under drought stress. Mycorrhiza 8:41–45

Al-Karaki GN (2006) Nursery inoculation of tomato with arbuscular mycorrhizal fungi and subsequent performance under irrigation with saline water. Sci Hortic 109:1–7

Al-Karaki GN, Hammad R, Rusan M (2001) Response of two tomato cultivars differing in salt tolerance to inoculation with mycorrhizal fungi under salt stress. Mycorrhiza 11:43–47

Alves L, Oliveira VL, Filho GNS (2010) Utilization of rocks and ectomycorrhizal fungi to promote growth of eucalypt. Braz J Microbiol 41:676–684

Anschütz U, Becker D, Shabala S (2014) Going beyond nutrition: regulation of potassium homoeostasis as a common denominator of plant adaptive responses to environment. J Plant Physiol 171:670–687

Arocena JM, Glowa KR, Massicotte HB, Lavkulich L (1999) Chemical and mineral composition of ectomycorrhizosphere soils of subalpine fir (Abies lasiocarpa (Hook.) Nutt.) in the Ae horizon of a luvisol. Can J Soil Sci 79:25–35

Arocena JM, Göttlein A, Raidl S (2004) Spatial changes of soil solution and mineral composition in the rhizosphere of Norway-spruce seedlings colonized by Piloderma croceum. J Plant Nutr Soil Sci 167:479–486

Asher CJ, Ozanne PG (1967) Growth and potassium content of plants in solution cultures maintained at constant potassium concentrations. Soil Sci 103:155–161

Ashford AE, Vesk PA, Orlovich DA, Markovina AL, Allaway WG (1999) Dispersed polyphosphate in fungal vacuoles in Eucalyptus pilularis/Pisolithus tinctorius ectomycorrhizas. Fungal Genet Biol 28:21–33

Augé RM (2001) Water relations, drought and vesicular–arbuscular mycorrhizal symbiosis. Mycorrhiza 11:3–42

Baeza A, Guillen J, Hernandez S, Salas A, Bernedo M, Manjon JL, Moreno G (2005) Influence of the nutritional mechanism of fungi (mycorrhize/saprophyte) on the uptake of radionuclides by mycelium. Radiochim Acta 93:233–238

Balogh-Brunstad Z, Kent Keller C, Thomas Dickinson J, Stevens F, Li CY, Bormann BT (2008) Biotite weathering and nutrient uptake by ectomycorrhizal fungus, Suillus tomentosus, in liquid culture experiments. Geochim Cosmochim Acta 72:2601–2618

Banfield JF, Barker WW, Welch SA, Taunton A (1999) Biological impact on mineral dissolution: application of the lichen model to understanding mineral weathering in the rhizosphere. Proc Natl Acad Sci U S A 96:3404–3411

Bañuelos MA, Klein RD, Alexander-Bowman SJ, Rodríguez-Navarro A (1995) A potassium transporter of the yeast Schwanniomyces occidentalis homologous to the Kup system of Escherichia coli has a high concentrative capacity. EMBO J 14:3021–3027

Bañuelos MA, Madrid R, Rodríguez-Navarro A (2000) Individual functions of the Hak and Trk potassium transporters of Schwanniomyces occidentalis. Mol Microbiol 37:671–679

Baslam M, Garmendia I, Goicoechea N (2013) The arbuscular mycorrhizal symbiosis can overcome reductions in yield and nutritional quality in greenhouse-lettuces cultivated at inappropriate growing seasons. Sci Hortic 164:145–154

Beauchamp VB, Stromberg JC, Stutz JC (2006) Arbuscular mycorrhizal fungi associated with Populus–Salix stands in a semiarid riparian ecosystem. New Phytol 170(2):369–380

Bending GD, Read DJ (1995a) The structure and function of the vegetative mycelium of ectomycorrhizal plants. V. Foraging behaviour and translocation of nutrients from exploited litter. New Phytol 130:401–409

Bending GD, Read DJ (1995b) The structure and function of the vegetative mycelium of ectomycorrhizal plants. VI. Activities of nutrient mobilizing enzymes in birch litter colonized by Paxillus involutus (Fr.) Fr. New Phytol 130:411–417

Benito B, González-Guerrero M (2014) Unravelling potassium nutrition in ectomycorrhizal associations. New Phytol 201:707–709

Benito B, Garciadeblas B, Rodríguez-Navarro A (2002) Potassium- or sodium-efflux ATPase, a key enzyme in the evolution of fungi. Microbiology 148:933–941

Benito B, Garciadeblas B, Schreier P, Rodríguez-Navarro A (2004) Novel P-type ATPases mediate high-affinity potassium or sodium uptake in fungi. Eukaryot Cell 3 (2):359–368

Benito B, Garciadeblas B, Fraile-Escanciano A, Rodríguez-Navarro A (2011) Potassium and sodium uptake systems in fungi. The transporter diversity of Magnaporthe oryzae. Fungal Genet Biol 48:812–822

Benito B, Haro R, Amtmann A, Cuin TA, Dreyer I (2014) The twins K^+ and Na^+ in Plants. J Plant Physiol 171:723–732

Boomsma C, Vyn T (2008) Maize drought tolerance: potential improvements through arbuscular mycorrhizal symbiosis? Field Crop Res 108:14–31

Bosecker K (1997) Bioleaching: metal solubilization by microorganisms. FEMS Microbiol Rev 20:591–604

Bowen GD (1973) Mineral nutrition of ectomycorrhizae. In: Marks GC, Kozlowski TT (eds) Ectomycorrhizae, their ecology and physiology. Academic, New York, pp 151–205

Brandes B, Godbold DL, Kuhn AJ, Jentschke G (1998) Nitrogen and phosphorus acquisition by the mycelium of the ectomycorrhizal fungus Paxillus involutus and its effect on host nutrition. New Phytol 140:735–743

Broadley MR, White PJ (2005) Plant nutritional genomics. Blackwell Publishing, CRC Press, Oxford

Brunner I, Frey B, Riesen TK (1996) Influence of ectomycorrhization and cesium/potassium ratio on uptake and localization of cesium in Norway spruce seedlings. Tree Physiol 16:705–711

Bücking H, Heyser W (1999) Elemental composition and function of polyphosphates in ectomycorrhizal fungi–an X-ray microanalytical study. Mycol Res 103:31–39

Burgstaller W, Schinner F (1993) Leaching of metals with fungi. J Biotechnol 27:91–116

Callot G, Maurette M, Pottier L, Dubois A (1987) Biogenic etching of microfractures in amorphous and crystalline silicates. Nature 328:147–149

Calvaruso C, Turpault M, Uroz S, Leclerc E, Kies A, Frey-Klett P (2009) Laccaria bicolor S238N improves Scots pine mineral nutrition by increasing root nutrient uptake from soil minerals but does not increase mineral weathering. Plant Soil 328:1–521

Casieri L, Lahmidi NA, Doidy J, Veneault-Fourrey C, Migeon A, Bonneau L et al (2013) Biotrophic transportome in mutualistic plant fungal interactions. Mycorrhiza 23:597–625

Cheah S-F, Kraemer SM, Cervini-Silva J, Sposito G (2003) Steady-state dissolution kinetics of goethite in the presence of desferrioxamine B and oxalate ligands: implications for the microbial acquisition of iron. Chem Geol 198:63–75

Cimen I, Pirinc V, Doran I, Turgay B (2010) Effect of soil solarization and arbuscular mycorrhizal fungus (Glomus intraradices) on yield and blossom-end rot of tomato. Int J Agric Biol 12:551–555

Clark RB, Zeto SK (1996) Growth and root colonization of mycorrhizal maize grown on acid and alkaline soil. Soil Biol Biochem 28:1505–1511

Clark RB, Zeto SK (2000) Mineral acquisition by arbuscular mycorrhizal plants. J Plant Nutr 23:867–902

Clark RB, Zobel RW, Zeto SK (1999) Effects of mycorrhizal fungus isolate on mineral acquisition by Panicum virgatum in acidic soil. Mycorrhiza 9:167–176

Corratgé-Faillie C, Jabnoune M, Zimmermann S, Véry AA, Fizames C, Sentenac H (2010) Potassium and sodium transport in non-animal cells: the Trk/Ktr/HKT transporter family. Cell Mol Life Sci 67:2511–2532

Coskun D, Britto DT, Kronzucker HJ (2014) The physiology of channel- mediated K^+ acquisition in roots of higher plants. Physiol Plant 151:305–312

Cromack K Jr, Sollins P, Graustein WC, Speidel K, Todd AW, Spycher G, Li CY, Todd RL (1979) Calcium oxalate accumulations and soil weathering in mats of the hypogeous fungus Hysterangium crassum. Soil Biol Biochem 11:463–468

Cumming JR (1993) Growth and nutrition of non mycorrhizal and mycorrhizal pitch pine (Pinus rigida) seedlings under phosphorus limitation. Tree Physiol 13:173–187

Daei G, Ardakani M, Rejali F, Teimuri S, Miransari M (2009) Alleviation of salinity stress on wheat yield, yield components, and nutrient uptake using arbuscular mycorrhizal fungi under field conditions. J Plant Physiol 166:617–625

Danielsen L, Polle A (2014) Poplar nutrition under drought as affected by ectomycorrhizal colonization. Environ Exp Bot 108:89–98

de Souza FSJ, Gomes SL (1998) A P-type ATPase from the aquatic fungus Blastocladiella emersonii similar to animal Na, K-ATPases. Biochim Biophys Acta 1383:183–187

Delvasto P, Ballester A, Muñóz JA, González F, Blazquez ML, Igual JM, Valverde A, García-Balboa C (2009) Mobilization of phosphorus from iron ore by the bacterium Burkholderia caribensis FeGL03. Miner Eng 22:1–9

Delvaux B, Kruyts N, Maes E, Smolders E (2001) Fate of radiocesium in soil and rhizosphere. In: Gobran GR, Wenzel WW, Lombi E (eds) Trace elements in the rhizosphere. CRC Press, BocaRaton, pp 61–91

Domínguez JA, Selva J, Rodríguez Barreal JA, Saiz de Omeñaca JA (2006) The influence of mycorrhization with Tuber melanosporum in the afforestation of a Mediterranean site with Quercus ilex and Quercus faginea. For Ecol Manag 231:226–233

Domínguez JA, Planelles R, Rodríguez Barreal JA, Saiz de Omeñaca (2008) The effect of Tuber melanosporum Vitt. mycorrhization on growth, nutrition, and water relations of Quercus petraea Liebl., Quercus faginea Lamk., and Pinus halepensis Mill. seedlings. New For 35:159–171

Domínguez Núñez JA, Planelles González R, Rodríguez Barreal JA, Saiz de Omeñaca González JA (2009) Influence of water-stress acclimation and Tuber melanosporum mycorrhization on Quercus ilex seedlings. Agrofor Syst 75(3):251–259

Dominguez JA, Martin A, Anriquez A, Albanesi A (2012) The combined effects of Pseudomonas fluorescens and Tuber melanosporum on the quality of Pinus halepensis seedlings. Mycorrhiza 22(6):429–436

Dominguez-Nuñez JA, Saiz M, Calderon C, Saiz de Omeñaca JA (2013) Short communication. Physiological effects of Rhizopogon Roseolus on Pinus halepensis seedlings. For Syst 22(3):568–572

Domínguez-Núñez JA, Muñóz D, de la Cruz A, Saiz de Omeñaca JA (2013) Effects of Pseudomonas fluorescens on the water parameters of mycorrhizal and non-mycorrhizal seedlings of Pinus halepensis. Agronomy 3:571–582

Dominguez-Nuñez JA, Medina M, Berrocal-Lobo M, Anriquez A, Albanesi A (2015) The combined effects of Pseudomonas fluorescens CECT 844 and the black truffle co inoculation on Pinus nigra seedlings. iForest – Biogeosciences and Forestry. (On-line. doi:10.3832/ifor1334-007)

Drew MC, Nye PH (1969) The supply of nutrient ions by diffusion to plant roots in soil. II. The effect of root hairs on the uptake of potassium by roots of ryegrass (Lolium multiflorum). Plant Soil 31:407–424

Dreyer I, Uozumi N (2011) Potassium channels in plant cells. FEBS J 278:4293–4303

Dupré de Boulois H, Joner EJ, Leyval C, Jakobsen I, Chen BD, Roos P, Thiry Y, Rufyikiri G, Delvaux B, Declerck S (2008) Role and influence of mycorrhizal fungi on radiocesium accumulation by plants. J Environ Radioact 99:785–800

El-Mesbahi MN, Azcón R, Ruiz-Lozano JM, Aroca R (2012) Plant potassium content modifies the effects of arbuscular mycorrhizal symbiosis on root hydraulic properties in maize plants. Mycorrhiza 22:555–564

Ericsson T (1995) Growth and shoot–root ratio of seedlings in relation to nutrient availability. Plant Soil 168:205–214
Estrada B, Aroca R, Maathuis FJM, Barea JM, Ruiz-Lozano JM (2013) Arbuscular mycorrhizal fungi native from a Mediterranean saline area enhance maize tolerance to salinity through improved ion homeostasis. Plant Cell Environ 36:1771–1782
Evelin H, Giri B, Kapoor R (2012) Contribution of Glomus intraradices inoculation to nutrient and mitigation of ionic imbalance in NaCl-stressed Trigonella foenum-graecum. Mycorrhiza 22:203–217
Fietto LG, Pugliese L, Gomes SL (2002) Characterization and expression of two genes encoding isoforms of a putative Na, K-ATPase in the chytridiomycete Blastocladiella emersonii. Biochim Biophys Acta 1576:59–69
Fogel R, Hunt G (1983) Contribution of mycorrhizae and soil fungi to nutrient cycling in a Douglas-fir ecosystem. Can J For Res 13:219–232
Futai K, Taniguchi T, Kataoka R (2008) Ectomycorrhizae and their importance in forest ecosystems. In: Siddiqui A, Futai (eds) Mycorrhizae: sustainable agriculture and forestry. Springer Science + Business Media B.V, Netherlands. pp 241–285
Gadd GM (1999) Fungal production of citric and oxalic acid: importance in metal speciation, physiology and biogeochemical processes. Adv Microb Physiol 41:47–92
Garcia K, Zimmermann SD (2014) The role of mycorrhizal associations in plant potassium nutrition. Front Plant Sci 5:337
Garcia K, Delteil A, Conéjéro G, Becquer A, Plassard C, Sentenac H, Zimmermann S (2014) Potassium nutrition of ectomycorrhizal Pinus pinaster: overexpression of the Hebeloma cylindrosporum HcTrk1 transporter affects the translocation of both K^+ and phosphorus in the host plant. New Phytol 201:951–960
Gierth M, Mäser P (2007) Potassium transporters in plants – involvement in K1 acquisition, redistribution and homeostasis. FEBS Lett 581:2348–2356
Giri B, Kapoor R, Mukerji KG (2007) Improved tolerance of Acacia nilotica to salt stress by arbuscular mycorrhiza, Glomus fasciculatum, maybe partly related to elevated K^+/Na^+ ratios in root and shoot tissues. Microb Ecol 54:753–760
Glowa KR, Arocen JM, Massicotte HB (2003) Extraction of potassium and/or magnesium from selected soil minerals by Piloderma. Geomicrobiol J 20:99–111
Gomez-Porras JL, Riaño-Pachón DM, Benito B, Haro R, Sklodowski K, Rodríguez-Navarro A, Dreyer I (2012) Phylogenetic analysis of K^+ transporters in bryophytes, lycophytes, and flowering plants indicates a specialization of vascular plants. Front Plant Sci 3:167
Guether M, Balestrini R, Hannah M, He J, Udvardi MK, Bonfante P (2009) Genome-wide reprogramming of regulatory networks, transport, cell wall and membrane biogenesis during arbuscular mycorrhizal symbiosis in Lotus japonicus. New Phytol 182:200–212
Gyuricza V, Declerck S, Dupré de Boulois H (2010a) Arbuscular mycorrhizal fungi decrease radiocesium accumulation in Medicago truncatula. J Environ Radioact 101:591–596
Gyuricza V, Dupré de Boulois H, Declerck S (2010b) Effect of potassium and phosphorus on the transport of radiocesium by arbuscular mycorrhizal fungi. J Environ Radioact 101:482–487
Hagerberg D, Thelin G, Wallander H (2003) The production of ectomycorrhizal mycelium in forests: relation between forest nutrient status and local mineral sources. Plant Soil 252:279–290
Hammer EC, Nasr H, Pallon J, Olsson PA, Wallander H (2011) Elemental composition of arbuscular mycorrhizal fungi at high salinity. Mycorrhiza 21:117–129
Harley JL, Smith SE (1983) Mycorrhizal symbiosis. Academic, London
Hawkes CV, Casper BB (2002) Lateral root function and root overlap among mycorrhizal and nonmycorrhizal herbs in a Florida shrub land, measured using rubidium as a nutrient analog. Am J Bot 89:1289–1294
Hawkins B, Boukcim H, Plassard C (2008) A comparison of ammonium, nitrate and proton net fluxes along seedling roots of Douglas-fir and lodgepole pine grown and measured with different inorganic nitrogen sources. Plant Cell Environ 31:278–287
Hinsinger P, Brauman A, Devau N, Gérard F, Jourdan C, Laclau CP, Le Cadre E, Jaillard B, Plassard C (2011) Acquisition of phosphorus and other poorly mobile nutrients by roots. Where do plant nutrition models fail? Plant Soil 348:29–61
Hirrel MC, Gerdemann JW (1980) Improved growth of onion and bell pepper in saline soils by two vesicular-arbuscular mycorrhizal fungi. Soil Sci Soc Am J 44:654–655
Hoffland E, Giesler R, Jongmans T, Van Breemen N (2002) Increasing feldspar tunneling by fungi across a North Sweden podzol chronosequence. Ecosystems 5:11–22
Huang JG, Lapeyrie F (1996) Ability of ectomycorrhizal fungus Laccaria bicolor S238N to increase the growth of Douglas fir seedlings and their phosphorus and potassium uptake. Pedosphere 6:217–224
Jain N, Sharma D (2004) Biohydrometallurgy for nonsulfidic minerals-a review. Geomicrobiol J 21:135–144
Jan LY, Jan YN (1997) Cloned potassium channels from eukaryotes and prokaryotes. Annu Rev Neurosci 20:91–123
Jentschke G, Brandes B, Kuhn AJ, Schröder WH, Becker JS, Godbold DL (2000) The mycorrhizal fungus Paxillus involutus transports magnesium to Norway spruce seedlings. Evidence from stable isotope labeling. Plant Soil 220:243–246
Jentschke G, Brandes B, Kuhn AJ, Schröder WH, Godbold DL (2001) Interdependence of phosphorus, nitrogen, potassium and magnesium translocation by the ectomycorrhizal fungus Paxillus involutus. New Phytol 149:327–337

Johansson SAE, Campbell JL (1988) PIXE, a novel technique for elemental analysis. Wiley, Chichester

Joner EJ, Roos P, Jansa J, Frossard E, Leyval C, Jakobsen I (2004) No significant contribution of arbuscular mycorrhizal fungi to transfer of radiocesium from soil to plants. Appl Environ Microbiol 70:6512–6517

Jones DL, Nguyen C, Finlay RD (2009) Carbon flow in the rhizosphere: carbon trading at the soil-root interface. Plant Soil 321:5–33

Jongbloed RH, Clement JMAM, Borst-Pauwels GWFH (1991) Kinetics of NH^{+}_{4} and K^{+} uptake by ectomycorrhizal fungi. Effect of NH^{+}_{4} on K^{+} uptake. Physiol Plant 83:427–432

Jourand P, Hannibal L, Majorel C, Mengant S, Ducousso M, Lebrun M (2014) Ectomycorrhizal Pisolithus albus inoculation of Acacia spirorbis and Eucalyptus globulus grown in ultramafic topsoil enhances plant growth and mineral nutrition while limits metal uptake. J Plant Physiol 171:164–172

Jung NC, Tamai Y (2013) Polyphosphate (phytate) formation in Quercus acutissima-Scleroderma verrucosum ectomycorrhizae supplied with phosphate. J Plant Interact 8:291–303

Kaldorf M, Kuhn AJ, Schröder WH, Hildebrandt U, Bothe H (1999) Selective element deposits in maize colonized by a heavy metal tolerance conferring arbuscular mycorrhizal fungus. J Plant Physiol 154:718–728

Kapoor R, Giri B, Mukerji KG (2002) Soil factors in relation to distribution and occurrence of vesicular arbuscular mycorrhiza. In: Mukerji KG, Manoharachary C, Chamola BP (eds) Techniques in mycorrhizal studies. Kluwer Academic Publishers, Boston, pp 51–85

Ketchum KA, Joiner WJ, Sellers AJ, Kaczmarek LK, Goldstein SAN (1995) A new family of outwardly rectifying potassium channel proteins with 2 pore domains in tandem. Nature 376:690–695

Ko CH, Gaber RF (1991) TRK1 and TRK2 encode structurally related K^{+} transporters in Saccharomyces cerevisiae. Mol Cell Biol 11:4266–4273

Kumar A, Bahadur I, Maurya BR, Raghuwanshi R, Meena VS, Singh DK, Dixit J (2015) Does a plant growth-promoting rhizobacteria enhance agricultural sustainability? J Pure Appl Microbiol 9(1):715–724

Ladeyn I, Plassard C, Staunton S (2008) Mycorrhizal association of maritime pine, Pinus pinaster, with Rhizopogon roseolus has contrasting effects on the uptake from soil and root-to-shoot transfer of ^{137}Cs, ^{85}Sr and ^{95}mTc. J Environ Radioact 99:853–863

Lambilliotte R, Cooke R, Samson D, Fizames C, Gaymard F, Plassard C et al (2004) Large-scale identification of genes in the fungus Hebeloma cylindrosporum paves the way to molecular analyses of ectomycorrhizal symbiosis. New Phytol 164:505–513

Landmann G, Hunter IR, Hendershot W (1997) Temporal and spatial development of magnesium deficiency in forest stands in Europe, North America and New Zealand. In: Hüttl RF, Schaaf W (eds) Magnesium deficiency in forest ecosystems. Kluwer, Dordrecht

Lapeyrie F, Chilvers GA, Bhem CA (1987) Oxalic acid synthesis by the mycorrhizal fungus Paxillus involutus (Batsch. Ex Fr.) Fr. New Phytol 106:139–146

Leigh RA, Wyn Jones RG (1984) A hypothesis relating critical potassium concentrations for growth to the distribution and functions of this ion in the plant cell. New Phytol 97:1–13

Leyval C, Berthelin J (1989) Interactions between Laccaria laccata-Agrobacterium radiobacter and beech roots influence on phosphorus, potassium, magnesium and iron mobilisation from minerals and plant growth. Plant Soil 117:103–110

Leyval C, Berthelin J (1991) Weathering of a mica by roots and rhizospheric microorganisms of pine. Soil Sci Soc Am J 55:1009–1016

Lian B, Wang B, Pan M, Liu C, Teng HH (2008) Microbial release of potassium from K-bearing minerals by thermophilic fungus Aspergillus fumigatus. Geochim Cosmochim Acta 72:87–98

López MF, Dietz S, Grunze N, Bloschies J, Weiss M, Nehls U (2008) The sugar porter gene family of Laccaria bicolor: function in ectomycorrhizal symbiosis and soil-growing hyphae. New Phytol 180:365–378

Malcova R, Albrechtova J, Vosatka M (2001) The role of the extraradical mycelium network of arbuscular mycorrhizal fungi on the establishment and growth of Calamagrostis epigejos in industrial waste substrates. Appl Soil Ecol 18:129–142

Manning D (2010) Mineral sources of potassium for plant nutrition. A review. Agron Sustain Dev 30:281–294

Mardukhi B, Rejali F, Daei G, Ardakani MR, Malakouti MJ, Miransari M (2011) Arbuscular mycorrhizas enhance nutrient uptake in different wheat genotypes at high salinity levels under field and greenhouse conditions. C R Biol 334:564–571

Markewitz D, Richter DD (2000) Long-term soil potassium availability from a Kanhapludult to an aggrading loblolly pine ecosystem. For Ecol Manag 130:109–129

Marschner H (1995) Mineral nutrition of higher plants, 2nd edn. Academic, London

Marschner H, Dell B (1994) Nutrient uptake in mycorrhizal symbiosis. Plant Soil 159:89–102

Martin F, Díez J, Dell B, Delaruelle C (2002) Phylogeography of the ectomycorrhizal Pisolithus species as inferred from nuclear ribosomal DNA ITS sequences. New Phytol 153:345–357

Martin F, Aerts A, Ahren D, Brun A, Danchin EGJ, Duchaussoy F et al (2008) The genome of Laccaria bicolor provides insights into mycorrhizal symbiosis. Nature 452:88–92

Martin F, Kohler A, Murat C, Balestrini R, Coutinho PM, Jaillon O et al (2010) Périgord black truffle genome uncovers evolutionary origins and mechanisms of symbiosis. Nature 464:1033–1038

Mäser P, Thomine S, Schroeder JI et al (2001) Phylogenetic relationships within cation transporter families of Arabidopsis. Plant Physiol 126:1646–1667

Mäser P, Gierth M, Schroeder JI (2002) Molecular mechanisms of potassium and sodium uptake in plants. Plant Soil 247:43–54

Maurya BR, Meena VS, Meena OP (2014) Influence of Inceptisol and Alfisol's potassium solubilizing bacteria (KSB) isolates on release of K from waste mica. Vegetos 27(1):181–187

Meena OP, Maurya BR, Meena VS (2013) Influence of K-solubilizing bacteria on release of potassium from waste mica. Agric Sustain Dev 1(1):53–56

Meena VS, Maurya BR, Bahadur I (2014a) Potassium solubilization by bacterial strain in waste mica. Bangladesh J Bot 43(2):235–237

Meena VS, Maurya BR, Verma JP (2014b) Does a rhizospheric microorganism enhance K+ availability inagricultural soils? Microbiol Res 169:337–347

Meena RK, Singh RK, Singh NP, Meena SK, Meena VS (2015a) Isolation of low temperature surviving plant growth-promoting rhizobacteria (PGPR) from pea (*Pisum sativum* L.) and documentation of their plant growth promoting traits. Biocatal Agric Biotechnol. doi:10.1016/j.bcab.2015.08.006

Meena VS, Maurya BR, Verma JP, Aeron A, Kumar A, Kim K, Bajpai VK (2015b) Potassium solubilizing rhizobacteria (KSR): isolation, identification, and K-release dynamics from waste mica. Ecol Eng 81:340–347

Miransari M (2012a) Soil microbes and environmental health. Nova Publishers, Hauppauge

Miransari M (2012b) Soil nutrients. Nova Publishers, Hauppauge

Miransari M (2013) Arbuscular mycorrhizal fungi and uptake of nutrients. In: Aroca R (ed) Symbiotic endophytes, vol 37. Springer, Berlin, pp 253–270

Miransari M, Rejali F, Bahrami HA, Malakouti MJ (2009a) Effects of soil compaction and arbuscular mycorrhiza on corn (*Zea mays* L.) nutrient uptake. Soil Tillage Res 103:282–290

Miransari M, Rejali F, Bahrami HA, Malakouti MJ (2009b) Effects of arbuscular mycorrhiza, soil sterilization, and soil compaction on wheat (*Triticum aestivum* L.) nutrients uptake. Soil Tillage Res 104:48–55

Nieves-Cordones M, Alemán F, Martínez V, Rubio F (2014) K^+ uptake in plant roots. The systems involved, their regulation and paralleles in other organisms. J Plant Physiol 171:688–695

Ochs M (1996) Influence of humified and non-humified natural organic compounds on mineral dissolution. Chem Geol 132:119–124

Ojala JC, Jarrell WM, Menge JA (1983) Influence of mycorrhizal fungi on the mineral nutrition and yield of onion in saline soil. Agron J 75:255–259

Olsson PA, Hammer EC, Wallander H, Pallon J (2008) Phosphorus availability influences elemental uptake in the mycorrhizal fungus *Glomus intraradices*, as revealed by particle-induced X-ray emission analysis. Appl Environ Microbiol 74:4144–4148

Olsson PA, Hammer EC, Pallon J, van Aarle IM, Wallander H (2011) Elemental composition in vesicles of an arbuscular mycorrhizal fungus, as revealed by PIXE analysis. Fungal Biol 115:643–648

Orlovich DA, Ashford AE (1993) Polyphosphate granules are an artefact of specimen preparation in the ectomycorrhizal fungus *Pisolithus tinctorius*. Protoplasma 173:91–102

Pallon J, Wallander H, Hammer E, Arteaga Marrero N, Auzelyte V, Elfman M et al (2007) Symbiotic fungi that are essential for plant nutrient uptake investigated with NMP. Nucl Inst Methods Phys Res B 260:149–152

Papazian DM, Schwarz TL, Tempel BL, Jan YN, Jan LY (1987) Cloning of the genomic and complementary DNA from Shaker, a putative potassium channel gene from Drosophila. Science 237:749–753

Paris F, Bonnaud P, Ranger J, Lapeyrie F (1995a) In vitro weathering of phlogopite by ectomycorrhizal fungi, I. Effects of K^+ and Mg^+ deficiency on phyllosilicate evolution. Plant Soil 177:191–201

Paris F, Bonnaud P, Ranger J, Robert M, Lapeyrie F (1995b) Weathering of ammonium- or calcium-saturated 2:1 phyllosilicates by ectomycorrhizal fungi in vitro. Soil Biol Biochem 27:1237–1244

Paris F, Botton B, Lapeyrie F (1996) In vitro weathering of phlogopite by ectomycorrhizal fungi.2. Effect of K^+ and Mg^{2+} deficiency and N sources on accumulation of oxalate and H^+. Plant Soil 179:141–150

Perner H, Schwarz D, Bruns C, Mäder P, George E (2007) Effect of arbuscular mycorrhizal colonization and two levels of compost supply on nutrient uptake and flowering of pelargonium plants. Mycorrhiza 17:469–474

Plassard C, Guerin Laguette A, Véry AA, Casarin V, Thibaud JB (2002) Local measurements of nitrate and potassium fluxes along roots of maritime pine. Effects of ectomycorrhizal symbiosis. Plant Cell Environ 25:75–84

Porcel R, Ruiz-Lozano JM (2004) Arbuscular mycorrhizal influence on leaf water potential, solute accumulation, andoxidative stress in soybean plants subjected to drought stress. J Exp Bot 55:1743–1750

Porras-Soriano A, Soriano-Martín ML, Porras-Piedra A, Azcón R (2009) Arbuscular mycorrhizal fungi increased growth, nutrient uptake and tolerance to salinity in olive trees under nursery conditions. J Plant Physiol 66:1350–1359

Rabie GG, Almadini AM (2005) Role of bioinoculants in development of salt-tolerance of Vicia faba plants under salinity stress. Afr J Biotechnol 4:210–222

Read DJ (1991) Mycorrhizas in ecosystems. Experientia 47:376–391

Rodríguez-Navarro A (2000) Potassium transport in fungi and plants. Biochim Biophys Acta 1469:1–30

Rodríguez-Navarro A, Rubio F (2006) High-affinity potassium and sodium transport systems in plants. J Exp Bot 57:1149–1160

Rosling A (2009) Trees, mycorrhiza and minerals field relevance of In vitro experiments. Geomicrobiol J 26:389–401

Rosling A, Lindahl BD, Finlay RD (2004a) Carbon allocation to ectomycorrhizal roots and mycelium colonizing different mineral substrates. New Phytol 162:795–802

Rosling A, Lindahl BD, Taylor AFS, Finlay RD (2004b) Mycelial growth and substrate acidification of ectomycorrhizal fungi in response to different minerals. FEMS Microbiol Ecol 47:31–37

Rygiewicz PT, Bledsoe CS (1984) Mycorrhizal effects on potassium fluxes by northwest coniferous seedlings. Plant Physiol 76:918–923

Scheloske S, Maetz M, Schneider T, Hildebrandt U, Bothe H, Povh B (2004) Element distribution in mycorrhizal and nonmycorrhizal roots of the halophyte Aster tripolium determined by proton induced X-ray emission. Protoplasma 223:183–189

Shabala S, Pottosin I (2014) Regulation of potassium transport in plants under hostile conditions: implications for abiotic and biotic stress tolerance. Physiol Plant 151:257–279

Sharifi M, Ghorbanli M, Ebrahimzadeh H (2007) Improved growth of salinity-stressed soy bean after inoculation with pre-treated mycorrhizal fungi. J. Plant Physiol 164:1144–1151

Shin R, Adams E (2014) Transport, signaling and homeostasis of potassium and sodium in plants. J Integr Plant Biol 56:231–249

Singh NP, Singh RK, Meena VS, Meena RK (2015) Can we use maize (*Zea mays*) rhizobacteria as plant growth promoter? Vegetos 28(1):86–99

Smith SE, Read DJ (1997) Mycorrhizal symbiosis, 2nd edn. Academic, London

Smith SE, Read DJ (2008) Mycorrhizal symbiosis, 3rd edn. Academic, New York

Smith SE, Smith FA, Nicholas DJD (1981) Effects of endomcorrhizal infection on phosphate and cation uptake by Trifolium subterraneum. Plant Soil 63:57–64

Smits MM (2005) Ectomycorrhizal fungi and biogeochemical cycles of boreal forests. PhD thesis, Wageningen University. www.library.wur.nl/wda/dissertations

Tamponnet C, Martin-Garin A, Gonze MA, Parekh N, Vallejo R, Sauras-Year T et al (2008) An overview of BORIS: bioavailability of radionuclides in soils. J Environ Radioact 99:820–830

Terada H, Shibata H, Kato F, Sugiyama H (1998) Influence of alkali elements on the accumulation of radiocesium by mushrooms. J Radioanal Nucl Chem 235:195–200

Thelin G (2000) Nutrient imbalance in Norway spruce. Ph.D. thesis, Lund University

Turnau K, Berger A, Loewe A, Einig W, Hampp R, Chalot M et al (2001) Carbon dioxide concentration and nitrogen input affect the C and N storage pools in Amanita muscaria-Picea abies mycorrhizae. Tree Physiol 21:93–99

Übel E, Heinsdorf D (1997) Results of long-term K and Mg fertilizer experiments in afforestation. For Ecol Manag 91:47–52

van Breemen N, Lundström US, Jongmans AG (2000) Do plants drive podzolization via rock-eating mycorrhizal fungi? Geoderma 94:163–171

van Hees PAW, Rosling A, Essén S, Godbold DL, Jones DL, Finlay RD (2006a) Oxalate and ferricrocin exudation by the extramatrical mycelium of an ectomycorrhizal fungus in symbiosis with Pinus sylvestris. New Phytol 169:367–378

van Hees PAW, Rosling A, Finlay RD (2006b) The impact of trees, ectomycorrhiza and potassium availability on simple organic compounds and dissolved organic carbon in soil. Soil Biol Biochem 38:1912–1923

van Hees PAW, Rosling A, Lundström US, Finlay RD (2006c) The biogeochemical impact of ectomycorrhizal conifers on major soil elements (Al, Fe, K and Si). Geoderma 136:364–377

van Schöll L, Hoffland E, van Breemen N (2006a) Organic anion exudation by ectomycorrhizal fungi and Pinus sylvestris in response to nutrient deficiencies. New Phytol 170:153–163

van Schöll L, Smits MM, Hoffland E (2006b) Ectomycorrhizal weathering of the soil minerals muscovite and hornblende. New Phytol 171:805–814

van Schöll L, Kuyper TW, Smits MM, Landeweert R, Hoffland E, van Breemen R (2008) Rock-eating mycorrhizas: their role in plant nutrition and biogeochemical cycles. Plant Soil 303:35–47

Vasan SS, Modak JM, Natarajan KA (2001) Some recent advances in the bioprocessing of bauxite. Int J Miner Process 62:173–186

Veresoglou SD, Mamolos AP, Thornton B, Voulgari OK, Sen R, Veresoglou S (2011) Medium-term fertilization of grassland plant communities masks plant species-linked effects on soil microbial community structure. Plant Soil 344:187–196

Véry A-A, Sentenac H (2002) Cation channels in the Arabidopsis plasma membrane. Trends Plant Sci 7:168–175

Véry A-A, Sentenac H (2003) Molecular mechanisms and regulation of K^+ transport in higher plants. Annu Rev Plant Biol 54:575–603

Wallander H (2000a) Uptake of P from apatite by *Pinus sylvestris* seedlings colonized by different ectomycorrhizal fungi. Plant Soil 218:249–256

Wallander H (2000b) Use of strontium isotopes and foliar K content to estimate weathering of biotite induced by pine seedlings colonised by ectomycorrhizal fungi from two different soils. Plant Soil 222:215–229

Wallander H, Hagerberg D (2004) Do ectomycorrhizal fungi have a significant role in weathering of minerals in forest soil? Symbiosis 27:249–257

Wallander H, Pallon J (2005) Temporal changes in the elemental composition of *Rhizopogon* rhizomorphs during colonization of patches with fresh organic matter or acid-washed sand. Mycologia 97:295–303

Wallander H, Wickman T (1999) Biotite and microcline as potassium sources in ectomycorrhizal and non-mycorrhizal *Pinus sylvestris* seedlings. Mycorrhiza 9:25–32

Wallander H, Johansson L, Pallon J (2002) PIXE analysis to estimate the elemental composition of ectomycorrhizal rhizomorphs grown in contact with different minerals in forest soil. FEMS Microbiol Ecol 39:147–156

Wallander H, Mahmood S, Hagerberg D, Johansson L, Pallon J (2003) Elemental composition of ectomycorrhizal mycelia identified by PCR-RFLP analysis and grown in contact with apatite or wood ash in forest soil. FEMS Microbiol Ecol 44:57–65

Wang B, Qiu YL (2006) Phylogenetic distribution and evolution of mycorrhizas in land plants. Mycorrhiza 16:299–363

Wang Y, Wu WH (2013) Potassium transport and signaling in higher plants. Annu Rev Plant Biol 64:4.1–4.26

White PJ, Broadley MR (2000) Mechanisms of caesium uptake by plants. New Phytol 147:241–256

Wu QS, Xia RX (2006) Arbuscular mycorrhizal fungi influence growth, osmotic adjustment and photosynthesis of *citrus* under well watered and water stress conditions. J Plant Physiol 163:417–425

Wu SC, Cao ZH, Li ZG, Cheung KC, Wong MH (2005) Effects of biofertilizer containing N-fixer, P and K solubilizers and AM fungi on maize growth: a greenhouse trial. Geoderma 125:155–166

Wu FY, Bi YL, Wong MH (2009) Dual inoculation with an arbuscular mycorrhizal fungus and *Rhizobium* to facilitate the growth of alfalfa on coal mine substrates. J Plant Nutr 32:755–771

Yoshida S, Muramatsu Y (1998) Concentrations of alkali and alkaline earth elements in mushrooms and plants collected in a Japanese pine forest, and their relationship with 137Cs. J Environ Radioact 41:183–205

Yousefi AA, Khavazi K, Moezi AA, Rejali F, Nadian NH (2011) Phosphate solubilizing bacteria and arbuscular mycorrhizal fungi impacts on inorganic phosphorus fractions and wheat growth. World Appl Sci J 15 (9):1310–1318

Yuan L, Fang DH, Wang ZH, Shun H, Huang JG (2000) Bio-mobilization of potassium from clay minerals: I. By ectomycorrhizas. Pedosphere 10:339–346

Yuan L, Huang J, Li X, Christie P (2004) Biological mobilization of potassium from clay minerals by ectomycorrhizal fungi and eucalypt seedling roots. Plant Soil 262:351–361

Zaefarian F, Rezvani M, Rejali F, Ardakani MR, Noormohammadi G (2011) Effect of heavy metals and arbuscular mycorrhizal fungal on growth and nutrients (N, P, K, Zn, Cu and Fe) accumulation of alfalfa (*Medicago sativa* L.). Am Eurasian J Agric Environ 11:346–352

Zhang H, Yin W, Xia X (2010) Shaker-like potassium channels in *Populus*, regulated by the CBL-CIPK signal transduction pathway, increase tolerance to low-K $^+$ stress. Plant Cell Rep 29:1007–1012

Zörb C, Senbayram M, Peiter E (2014) Potassium in agriculture-status and perspectives. J Plant Physiol 171:656–669

Zuccarini P, Okurowska P (2008) Effects of mycorrhizal colonization and fertilization on growth and photosynthesis of sweet basil under salt stress. J Plant Nutr 31:497–513

Diversity of Potassium-Solubilizing Microorganisms and Their Interactions with Plants

7

Encarna Velázquez, Luis R. Silva, Martha-Helena Ramírez-Bahena, and Alvaro Peix

Abstract

Potassium is, together with nitrogen and phosphorus, an essential element for plant nutrition, being the third element in the classical chemical fertilizers NPK. Deficiencies in K results in plants with poorly developed roots, small seeds and lower yields, and therefore the availability of this element is crucial for plant growth and development. The plants take this element from soil, but high amounts of K present in soils correspond to insoluble forms from rocks and silicate minerals. The solubilization of K by microorganisms is then a reliable alternative to make available this element for plants. In this chapter, we review the microorganisms reported as K solubilizers and their phylogenetic diversity, including filamentous fungi such as *Penicillium* or *Aspergillus*, yeasts such as *Torulaspora* and bacteria of many different genera from phyla *Proteobacteria*, *Firmicutes*, *Actinobacteria* and *Bacteroidetes*. Some of these potassium-solubilizing microorganisms (KSMs) play a positive role in the growth of different plants, which is relevant to sustainable agriculture schemes, and therefore this is a promising field of study in the 'greening' era.

Keywords

Solubilization • Microorganisms • K solubilizers • *Penicillium* • *Aspergillus*

E. Velázquez
Departamento de Microbiología y Genética. Laboratorio 209. Edificio Departamental de Biología. Doctores de la Reina s/n, Universidad de Salamanca, Salamanca 37007, Spain

Unidad Asociada Grupo de Interacción Planta-Microorganismo Universidad de Salamanca-CSIC, Salamanca, Spain

L.R. Silva
REQUIMTE/Laboratório de Farmacognosia, Departamento de Química, Faculdade de Farmácia, Universidade do Porto, Rua de Jorge Viterbo Ferreira, n.° 228, 4050-313 Porto, Portugal

M.-H. Ramírez-Bahena • A. Peix (✉)
Unidad Asociada Grupo de Interacción Planta-Microorganismo Universidad de Salamanca-CSIC, Salamanca, Spain

Instituto de Recursos Naturales y Agrobiología, IRNASA-CSIC, Cordel de Merinas, 40-52, 37008 Salamanca, Spain
e-mail: alvaro.peix@csic.es

V.S. Meena et al. (eds.), *Potassium Solubilizing Microorganisms for Sustainable Agriculture*,
DOI 10.1007/978-81-322-2776-2_7

7.1 Introduction

Potassium (K) is the third essential element together with nitrogen (N) and phosphorus (P) for plant development, and they form the NPK chemical fertilizers used in both intensive and extensive agriculture. The negative effects of an inadequate supply of K include poorly developed roots, slow growth, small seeds and lower yields of crops as well as higher sensitivity to diseases and pests (Meena et al. 2014b, 2015a). Although K reserves in soil are generally large, most of the soil K (~98 %) is not directly available for plant uptake (Zorb et al. 2014). Therefore soil KSMs play an essential role in plant development and some of them are included in commercial biofertilizers (Ramarethinam and Chandra 2006). As occurs in the case of P solubilization, the major mechanism of K mineral solubilization is the action of organic acids synthesized by rhizospheric microorganisms, and also the production of capsular polysaccharides is involved in dissolution of minerals to release potassium (Sheng and He 2006; Liu et al. 2006; Meena et al. 2015b; Singh et al. 2015). It has been shown that potassium-solubilizing bacteria (KSB) can convert the insoluble or mineral structural K compounds into soluble forms in soil making them available to the plants (Zeng et al. 2012). Therefore, it has been proposed that KSMs can be used to improve the productivity of agricultural lands and to restore the productivity of degraded, marginally productive and unproductive agricultural soils (Meena et al. 2014b).

Although the ability to produce indole-3-acetic acid or siderophores as well as to solubilize potassium is reported for some strains isolated from rhizospheric soil (Zhao et al. 2008), K solubilization is not commonly included among the in vitro tests for searching mechanisms of plant growth promotion by microorganisms, in contrast with the case of P solubilization (Peix et al. 2001). For this reason, knowledge on the ability of strains for K solubilization is scarce, and it is only known in a few species for the moment. Further studies about the ability to solubilize potassium of bacteria and fungi isolated from plant rhizosphere and soil will likely considerably increase significantly the number of species known to possess this ability. In this article, we review the microbial species in which the K solubilization ability has been reported for enhancement of sustainable agriculture (Meena et al. 2013, 2015a).

7.2 Sources of Potassium-Solubilizing Microorganisms (KSMs)

The sources from which KSMs have been isolated since the first evidences of microbial involvement in K solubilization in the twenty-first century have been revised by some authors (Meena et al. 2014b; Shanware et al. 2014; Sindhu et al. 2014). In the last decade, KSMs were isolated mainly in Asian countries from different sources, and their ability to solubilize K has been checked using different substrates. Although mica mines are a source of KSB (Gundala et al. 2013), most of KSM strains isolated in India came from soil and rhizosphere of different plants. KSB strains able to release K from microcline, orthoclase, muscovite, mica or potassium muriate have been isolated from soils cultivated with different plants (Ramarethinam and Chandra 2006; Sugumaran and Janarthanam 2007; Bagyalakshmi et al. 2012a; Maurya et al. 2014; Saiyad et al. 2015) and from soil samples of ceramic industries, where bacterial and fungal strains presented K solubilization from potassium aluminium silicate minerals (Prajapati and Modi 2012; Prajapati et al. 2012). The rhizospheric soils have been shown as good sources of KSB (Archana et al. 2013; Padma and Sukumar 2015) that are able to release K from different K insoluble sources, as occurs with strains isolated from black pepper rhizosphere using wood ash (Sangeeth et al. 2012) and with those isolated from wheat and maize rhizospheres using media supplemented with different types of mica (Parmar and Sindhu 2013; Meena et al. 2014a).

Several KSB have been also isolated from soils and plant rhizospheres in China. KSB able to release K from montmorillonite, kaolinite and K feldspar (Hu et al. 2006), silicate materials (Liu et al. 2006), potassium-bearing rock (Liu et al. 2012) and feldspar (Zeng et al. 2012) have been isolated from different soils in this country. Also, KSB have been isolated from cotton rhizosphere able to release K from illite (Sheng and He 2006) and from rhizospheric soil of different plants able to release K from K feldspar in solid or liquid media (Zhao et al. 2008; Zhang and Kong 2014; Meena et al. 2014a, b; Kumar et al. 2015).

Few works have been developed in other Asian countries such as those by Diep and Hieu (2013) who isolated from soils/weathered rocks in Vietnam more than 20 KSB strains able to solubilize K after the addition of apatite to culture media. Zarjani et al. (2013) isolated six KSB strains from different Iranian soils able to solubilize K after adding acid-leached soil, biotite and muscovite to the media, biotite being the best substrate to solubilize K for all the strains tested. Leaungvutiviroj et al. (2010) found several KSB strains in rhizospheric soil in Thailand able to release K from K feldspar. Outside Asia, the type strain of *Paenibacillus mucilaginosus* KCTC 3870^T was isolated from a rock in Russia and was reported as KSB (Han and Lee 2005). In Brazil, two fungal strains able to solubilize K from alkaline ultramafic rock powder were isolated from sugar cane rhizosphere by Rosa-Magri et al. (2012) and from sugar cane vinasses by Lopes-Assad et al. (2010).

7.3 Diversity of Potassium-Solubilizing Microorganisms (KSMs)

The available data showed that most KSMs belong to domain *Bacteria* (KSB) whose classification is mainly based in the analysis of the 16S rRNA gene (Woese 2000; Sheng and Huang 2002). Most of KSB strains reported to date belong to the phylum *Firmicutes* that comprises gram-positive bacteria with low G + C percentage, followed by the phylum *Proteobacteria* that includes gram-negative bacteria and by *Actinobacteria* comprising gram-positive bacteria with high G + C percentage. Only two species from phylum *Bacteroidetes* comprising gram-negative bacteria have been reported as potassium solubilizers (Table 7.1). Although the KSB strains reported in some papers were not identified (Parmar and Sindhu 2013; Meena et al. 2014a; Padma and Sukumar 2015), identified on the basis of phenotypic characteristics (Zeng et al. 2012; Sugumaran and Janarthanam 2007) or without references to the methodology used for identification (Bagyalakshmi et al. 2012b; Saiyad et al. 2015), in most studies the strains were identified by 16S rRNA gene sequencing. Even in some specific cases, a precise identification was carried out with the analysis of housekeeping genes such as *gyrB* that allowed a more accurate identification at species level than 16S rRNA gene (Hu et al. 2010).

Only few KSM strains reported to date belong to the domain *Eukarya* and specifically to the kingdom *Fungi* (KSF) and phylum *Ascomycota* which include filamentous fungi and yeasts (Table 7.1). Although macroscopic and microscopic features for fungi identification are still in use, sequencing of 5.8-ITS fragment located between the 18S and 28S rRNA genes is the current criterion for this purpose, which allows an accurate identification at genus, species and strain levels (Schoch et al. 2012). Since gene sequencing is currently an affordable technique for most laboratories, it is desirable that all the future studies on the diversity of KSMs are based on molecular approaches in order to have a correct identification of isolates at species level for both bacteria and fungi which is essential to the knowledge of biodiversity of KSMs as well as for the selection of safe species for the man and the animals to formulate biofertilizers.

7.3.1 Gram-Negative K-Solubilizing Bacteria (KSB)

Most of gram-negative KSB strains isolated to date belong to the phylum *Proteobacteria*. The KSB strains from class *Alphaproteobacteria* belong to the family *Rhizobiaceae* and have been identified by 16S rRNA gene sequencing.

Table 7.1 Classification of potassium-solubilizing microorganisms (KSM)

Strain	Isolation source	Closest related species (16S rRNA gene)	Reference
Domain *Bacteria*, **phylum** *Proteobacteria*, **class** *Alphaproteobacteria*, **order** *Rhizobiales*, **family** *Rhizobiaceae*			
CH9E	Weathered materials of denatured rock mountain in Vietnam	*A. tumefaciens*[a]	Diep and Hieu (2013)
GL11	Tobacco rhizosphere in China	*A. tumefaciens*[a]	Zhang and Kong (2014)
AFM2	Rhizospheric soil in China	*A. tumefaciens*[a]	Zhao et al. (2008)
Class *Betaproteobacteria*, **order** *Burkholderiales*, **family** *Burkholderiaceae*			
GL13	Tobacco rhizosphere in China	*B. cepacia*	Zhang and Kong (2014)
No name	Soil in India	*B. metallica*	Saiyad et al. (2015)
Class *Betaproteobacteria*, **order** *Burkholderiales*, **family** *Alcaligenaceae*			
IARI-HHS1-31	Wheat in India	*A. piechaudii*	Verma et al. (2015)
Class *Gammaproteobacteria*, **order** *Enterobacteriales*, **family** *Enterobacteriaceae*			
KSB 8	Ceramic industry soil in India	*E. hormaechei*	Prajapati and Modi (2012)
XF11, GL7, XF2, XF5, GL10	Tobacco rhizosphere in China	*K. variicola*, *E. cloacae*, *E. asburiae*	Zhang and Kong (2014)
		E. aerogenes, *Pantoea agglomerans*	
No name	Soil in India	*E. asburiae*	Saiyad et al. (2015)
IARI-HHS2-58	Wheat in India	*Klebsiella* sp	Verma et al. (2015)
Order *Pseudomonadales*, **family** *Pseudomonadaceae*			
No name	Tea soil in India	*P. putida*	Bagyalakshmi et al. (2012a, b)
Order *Xanthomonadales*, **family** *Xanthomonadaceae*			
No name	Agricultural soils in India	*F. aurantia*	Ramarethinam and Chandra (2006)
IARI-HHS1-20, IARI-HHS2-27	Wheat in India	*S. maltophilia*, *Stenotrophomonas* sp.	Verma et al. (2015)
Phylum *Bacteroidetes*, **order** *Cytophagales*, **family** *Cytophagaceae*			
TC1D	Weathered materials of denatured rock mountain in Vietnam	*Flectobacillus* sp	Diep and Hieu (2013)
Order *Flavobacteriales*, **family** *Flavobacteriaceae*			
JM19	Tobacco rhizosphere in China	*M. odoratimimus*	Zhang and Kong (2014)
Phylum *Firmicutes*, **order** *Bacillales*, **family** *Bacillaceae*			
TC1A, CH7A	Weathered materials of denatured rock mountain in Vietnam	*B. cereus*, *B. coagulans*	Diep and Hieu (2013)
TD6B, CH7D		*B. subtilis*, *B. megaterium*	
Z 1–3	Soil in China	*B. circulans*	Zeng et al. (2012)
JK3, JK4, JK5, JK6, JK7	Iranian soils	*B. megaterium*	Zarjani et al. (2013)
KT7/2	Rhizospheric soil in Thailand	*B. subtilis*	Leaungvutiviroj et al. (2010)
SVNM9	Mica mine in India	*B. amyloliquefaciens*	Gundala et al. (2013)
PB16	Soil in India	*B. coagulans*	Saiyad et al. (2015)
No name	Soil in India	*B. megaterium*	Saiyad et al. (2015)

(continued)

Table 7.1 (continued)

Strain	Isolation source	Closest related species (16S rRNA gene)	Reference
IARI-HHS1-2, IARI-HHS2-13, IARI-HHS2-30, IARI-HHS2-45	Wheat in India	*B. megaterium*, *B. horikoshii*, *B. amyloliquefaciens*, *Bacillus* sp.	Verma et al. (2015)
IARI-HHS2-48	Wheat in India	*Exiguobacterium antarcticum*	Verma et al. (2015)
Oder *Bacillales*, **family** *Paenibacillaceae*			
HKK-2	Agronomic soil in Hong Kong	*B. mucilaginosus*[b]	Wu et al. (2005) and Hu et al. (2010)
KCTC 3870^T	Rock in Russia	*B. mucilaginosus*[b]	Han and Lee (2005) and Hu et al. (2010)
MCRCp1	Soil in India	*B. mucilaginosus*[b]	Sugumaran and Janarthanam (2007)
KNP 413, KNP 414	Mountain soil in China	*B. mucilaginosus*[a]	Hu et al. (2006) and Hu et al. (2010)
No name	Unknown	*B. mucilaginosus*[b]	Liu et al. (2006) and Hu et al. (2010)
AC2, AHZ1	Rhizospheric soil in China	*B. mucilaginosus*[b]	Zhao et al. (2008) and Hu et al. (2010)
NBT	Rhizospheric soil of cotton in China	*B. edaphicus*	Hu et al. (2010)
KT	Soil in China	*P. harenae*, *Paenibacillus agarexedens*, *Paenibacillus alkaliterrae*	Liu et al. (2012)
IISRBK2	Rhizosphere of black pepper in India	*P. glucanolyticus*	Sangeeth et al. (2012)
Phylum *Actinobacteria*, **order** *Actinomycetales*, **family** *Micrococcaceae*			
JK2	Iranian soil	*Arthrobacter* sp.	Zarjani et al. (2013)
Order *Actinomycetales*, **family** *Microbacteriaceae*			
DNV16	Weathered materials of denatured rock mountain in Vietnam	*M. hominis*	Diep and Hieu (2013)
JM17	Tobacco rhizosphere in China	*M. foliorum*	Zhang and Kong (2014)
Domain *Fungi*, **phylum** *Ascomycota*, **class** *Saccharomycetes*, **order** *Saccharomycetales*, **family** *Saccharomycetaceae*			
1S112	Sugar cane rhizosphere in Brazil	*T. globosa*	Rosa-Magri et al. (2012)
Class *Eurotiomycetes*, **order** *Eurotiales*, **family** *Aspergillaceae*			
TH003	Compost in China	*A. fumigatus*	Lian et al. (2008)
CCT4355	Sugar cane vinasse in Brazil	*A. niger*	Lopes-Assad et al. (2010)
KF1, KF2	Ceramic industry soil in India	*A. niger*, *A. terreus*	Prajapati et al. (2012)

[a]This species is currently named *R. radiobacter* (Young et al. 2001)
[b]*B. mucilaginosus* and *B. edaphicus* were reclassified as *P. mucilaginosus* and *P. edaphicus*, respectively, by Hu et al. (2010)

The strain GL11 isolated from tobacco rhizosphere in China showed 100 % sequence identity to the species *Agrobacterium tumefaciens* (Zhang and Kong 2014), and the strains AFM2, isolated from rhizospheric soil in China (Zhao et al. 2008), and CH9E, isolated from weathered materials of a denatured rock mountain in Vietnam (Diep and Hieu 2013), presented in both cases 99 % similarity in their 16S rRNA sequences with respect to *A. tumefaciens*. This species is currently named *Rhizobium radiobacter* after the reclassification of the species *A. tumefaciens* into *A. radiobacter* (Sawada et al. 1993) and the later reclassification of the whole genus *Agrobacterium* into genus *Rhizobium* (Young et al. 2001).

Within the class *Betaproteobacteria* and the order *Burkholderiales*, a strain isolated from tobacco rhizosphere soil has been reported to date as K solubilizer by Zhang and Kong (2014). This strain, namely, GL13, was identified as *Burkholderia cepacia* from family *Burkholderiaceae* based on 16S rRNA gene sequencing. A second strain of genus *Burkholderia*, *B. metallica*, was recently isolated from soil in India although no data were reported about the methodology used for identification (Saiyad et al. 2015). Within the same order, but from family *Alcaligenaceae*, a KSB strain isolated from wheat has been also recently identified on the basis of its 16S rRNA gene sequence as *Achromobacter piechaudii* (Verma et al. 2015).

Most of gram-negative KSB strains isolated to date belong to class *Gammaproteobacteria* and to family *Enterobacteriaceae*. Species from genera *Klebsiella*, *Enterobacter* and *Pantoea* were isolated from tobacco rhizosphere in China (Zhang and Kong 2014) and identified by 16S rRNA gene sequencing. Several strains were closely related to *K. variicola*, *P. agglomerans*, *E. asburiae*, *E. aerogenes* and *E. cloacae* with similarities ranging from 99 % to 100 %. Two KSB isolated from different soils in India were identified as *E. hormaechei* and *E. asburiae* (Prajapati and Modi 2012; Saiyad et al. 2015), and a KSB strain isolated from wheat in the same country was classified into genus *Klebsiella* through 16S rRNA gene sequencing (Verma et al. 2015). Within class *Gammaproteobacteria*, strains from other three genera isolated in India have been reported as K solubilizers, *Pseudomonas putida* from family *Pseudomonadaceae* isolated from a tea soil (Bagyalakshmi et al. 2012b), *Frateuria aurantia* from family *Xanthomonadaceae* isolated from an agricultural soil (Ramarethinam and Chandra 2006) and *Stenotrophomonas maltophilia* and *Stenotrophomonas* spp. isolated from wheat in India (Verma et al. 2015).

The other phylum of gram-negative bacteria from which KSB have been reported is the phylum *Bacteroidetes* which is represented by two species isolated in Vietnam and China whose 16S rRNA gene was also analysed. The strain TC1D isolated from weathered materials of a denatured rock mountain in Vietnam was classified into genus *Flectobacillus* from family *Cytophagaceae* without assigning it to a species (Diep and Hieu 2013), and the strain JM19 was identified as *Myroides odoratimimus*, which belongs to the family *Flavobacteriaceae*, and was isolated from tobacco rhizosphere soil (Zhang and Kong 2014).

7.3.2 Gram-Positive K-Solubilizing Bacteria

Most KSB isolated to date belong to the phylum *Firmicutes* which comprises gram-positive bacteria of low G + C content, and all of them are sporulating rods that belong to the class *Bacilli*, with families *Bacillaceae* and *Paenibacillaceae* similarly represented among KSB. Within family *Bacillaceae*, several KSB strains have been classified into genus *Bacillus*, from which the strain Z 1–3, isolated from soil in China, was identified as *B. circulans* only on the base of phenotypic approaches (Zeng et al. 2012). Recently two strains from *B. coagulans* and *B. megaterium* have been isolated from soil in India whose identification method was not reported (Saiyad et al. 2015). The remaining strains were identified on the basis of their 16S rRNA gene, such as the strain KT7/2 isolated from rhizospheric soil in Thailand (Leaungvutiviroj et al. 2010) and the strain TD6B isolated from weathered materials of denatured rock mountain in Vietnam, that were closely related to *B. subtilis* (Diep and Hieu 2013). These authors also isolated several strains whose 16S rRNA gene was equidistant to different species, as occurs in the case of strains CH7A, closely related to *B. subtilis* and *B. cereus*, and TC1A, closely related to *B. cereus* and *B. coagulans*. KSB strains presenting 99 % similarity in their 16S rRNA gene sequence with respect to *B. megaterium* were reported in Vietnam and Iran (Diep and Hieu 2013; Zarjani et al. 2013). The strain SVNM9 isolated in a mica mine in India also presented 99 % similarity with respect to its closest relative species *B. amyloliquefaciens* (Gundala et al. 2013), and Verma et al. (2015) identified several KSB strains isolated from wheat in the same country as *B. megaterium*, *B. horikoshii*, *B. amyloliquefaciens*, *Bacillus* sp. and *Exiguobacterium antarcticum*.

One of the most widely known species of KSB is *B. mucilaginosus* that together with *B. edaphicus* has a high K-solubilizing activity (Han and Lee 2005; Hu et al. 2006; Liu et al. 2006; Sheng and He 2006; Wu et al. 2006). These two species have been reclassified into genus *Paenibacillus* as *P. mucilaginosus* and *P. edaphicus*, respectively (Hu et al. 2010), and then they currently belong to family *Paenibacillaceae* instead to family *Bacillaceae*. The reclassification was based on the 16S rRNA and *gyrB* analyses showing that the KSB strains KNP 413 and KNP 414 isolated in a mountain soil in China are phylogenetically close to the type strain of *P. mucilaginosus*. The complete genome of the silicate strain KNP 414 was recently sequenced (Lu et al. 2013). Several KSB strains from these two species have been isolated in different countries, such as the strains AC2 and AHZ1 isolated from rhizospheric soil in China identified as *P. mucilaginosus* by 16S rRNA gene sequencing (Zhao et al. 2008) and the strain MCRCp1 isolated from soil in India that was identified as *P. mucilaginosus* on the basis of its phenotypic characteristics (Sugumaran and Janarthanam 2007).

The KSB strain NBT isolated from cotton rhizosphere in China was identified as *P. edaphicus* through 16S rRNA gene sequencing as reported Sheng and He (2006). Several mutants of this strain were obtained and one of them was able to release higher amounts of potassium from both feldspar and illite substrates (Sheng and He 2006). Other KSB strains have been classified into genus *Paenibacillus* on the basis of their 16S rRNA gene sequences, such as the strain IISRBK2 that was isolated from rhizosphere of black pepper and identified as *P. glucanolyticus* (Sangeeth et al. 2012). The strain KT isolated from soil in China with similarities lower than 97 % with respect to several *Paenibacillus* species, *P. harenae*, *P. agarexedens* and *P. alkaliterrae* (Liu et al. 2012), probably belongs to a currently undescribed species of this genus. The other phylum of gram-positive bacteria in which KSB have been reported is *Actinobacteria*, which comprises bacteria with high G + C content, with all KSB strains being reported to date to belong to families *Microbacteriaceae* and *Micrococcaceae*. Two KSB strains from family *Microbacteriaceae* were isolated from tobacco rhizosphere in China and from weathered materials of denatured rock mountain in Vietnam, respectively. The strain JM17 was identified as *Microbacterium foliorum* with 100 % similarity in its 16S rRNA gene (Zhang and Kong 2014), and the strain DNV16 was closely related to *M. hominis* with 99 % similarity in the same gene (Diep and Hieu 2013). A strain named JK2 isolated from soil in Iran was classified into the genus *Arthrobacter* from family *Micrococcaceae* by 16S rRNA gene sequencing (Zarjani et al. 2013).

7.3.3 K-Solubilizing Fungi (KSF)

There are very few studies about KSF, even less than those for bacteria; nevertheless, some strains of filamentous fungi and yeasts have been reported as KSF in the last years. All of them belong to the phylum *Ascomycota* and were isolated in Asia and South America from different sources. The only yeast reported to date as KSF, strain 1S112, was isolated from sugarcane rhizosphere in Brazil and releases potassium from alkaline ultramafic rock powder (Rosa-Magri et al. 2012). It was identified by sequencing of the 5.8S rRNA fragment as *Torulaspora globosa*, classified within family *Saccharomycetaceae*. The remaining fungal KSMs known to date belong to the family *Aspergillaceae* and to the genus *Aspergillus*. A KSF strain of *A. fumigatus*, TH003, was isolated from compost in China by Lian et al. (2008) and identified by 5.8S-ITS sequence.

In the work of Lopes-Assad et al. (2010), a strain of *A. niger* isolated from sugar cane vinasse in Brazil, CCT4355, and a reference strain from the same species, ATCC9142, were found to be good K solubilizers releasing potassium from alkaline ultramafic rock powder. There are no data about the identification method used for the identification of strain CCT4355. Also, a strain from *A. niger*, KF1, was isolated from a ceramic industry soil in India together with a strain of *A. terreus*, KF2; they were identified on the basis of their macroscopic and microscopic morphology (Prajapati et al. 2012). According to the results of this work, the strain of *A. niger* showed the highest K solubilization and acid production by utilizing feldspar and potassium aluminium silicate as insoluble sources of potassium.

7.4 KSMs as Plant Growth Promoters

The effects of KSMs on plant growth and yield were studied several decades ago by Aleksandrov (1985) who showed that the application of organo-minerals combined with silicate bacteria enhanced plant growth and yield of maize and wheat. From this date ahead, several studies have been carried out using KSR as inoculants of different plants that have been revised by some authors (Meena et al. 2015a; Sindhu et al. 2014). In the last decade, several studies carried out in hydroponic, microcosm and field conditions have been performed about the effect on plant growth and K content of plants cultivated with or without addition of soluble or insoluble potassium sources. In experiments in hydroponic conditions, a KSB strain from *P. mucilaginosus* was able to increase the dry biomass and the K uptake by wheat and maize in the presence of mica as insoluble K source with respect to uninoculated controls (Singh et al. 2010). Some microcosm experiments have been performed without addition of potassium sources to the soil showing that, for example, the inoculation of groundnuts with the KSB strain *P. mucilaginosus* MCRCp1 isolated from soil in India increased the K content in the soil, plant yield and oil content of the seeds (Sugumaran and Janarthanam 2007).

Nevertheless, most of microcosms experiments have been carried out combining KSB and K insoluble sources, as those performed with the type strain of *P. mucilaginosus* KCTC 3870^T in combination with illite powder that increases K content in the soil as well as growth and K uptake of several plants such as eggplant (Han and Lee 2005), pepper and cucumber (Han et al. 2006). Sheng (2005) also performed a microcosm assay including treatments of cotton and rape plants treated with the KSB strain *P. edaphicus* NBT isolated from rhizospheric soil of cotton in China and an insoluble source of K. The results showed that plant species treated with this combination presented similar shoot and root weights as those treated with soluble K and significantly higher than those from uninoculated controls. This treatment also increased the N and P content in both plants and soil. Increases ranging from 26 % to 30 % were found in both plants when illite was added to soil as source of insoluble K. The strain NBT and two of its mutant strains were also able to increase the shoot and root growth of wheat and its content in N, P and K when it was cultivated in a soil with low content of available K (Sheng et al. 2008). Also the inoculation of corn (*Zea mays*) and Chinese kale (*Brassica alboglabra*) with *B. subtilis* KT7/2 and K_2O as potassium source led to an increase in dry weight of shoots and roots as well as an increase in K content in soil (Leaungvutiviroj et al. 2010). Recently, it has been reported that the inoculation of the KSB strains GL7, JM3, XF4 and XF11 isolated from tobacco rhizosphere in China in combination with K feldspar powder showed a beneficial effect on plant growth and nutrient uptake by tobacco seedlings. The plant dry weight and the uptake of both K and N by tobacco seedlings increased significantly with respect to uninoculated controls mainly when the strain inoculated was *Klebsiella variicola* XF11 (Zhang and Kong 2014).

Finally, some field experiments have been performed, and the results showed that the inoculation with a KSB and phytohormone-producing strain of *P. putida* isolated from a tea soil in India in combination with a soluble NPK chemical fertilizer increases the yield of tobacco green leaf and K content both in crop shoots and soil (Bagyalakshmi et al. 2012b). Also the inoculation with *Frateuria aurantia* improve the crop yield of eggplants (Ramarethinam and Chandra 2006) and also increase biomass, nutrient content, potassium uptake and leaf quality of tobacco (Subhashini 2015). All these studies showed a positive effect of inoculation of KSB in plant growth for different crops with or without addition of K insoluble sources and then they can be considered as good biofertilizers, and concretely a biofertilizer based on the KSB *Frateuria aurantia* is already commercialized in India under the name Symbion-K (Ramarethinam and Chandra 2006).

7.5 Future Perspectives for KSMs

To date KSMs, and particularly KSB, have been mostly isolated from Asian soils; there are few studies about KSMs in other continents. Undoubtedly after studying KSMs in more diverse geographical locations, the biodiversity of these microorganisms will increase. These studies will allow the selection of the best K solubilizers that when combined with other plant growth-promoting microorganisms (PGPMs) can substitute or decrease the need of using chemical fertilizers in crops. In this sense, some works performed dual co-inoculations with KSB and N-fixing bacteria, KSB and PSB that increased the availability of N, K and P in the soil as well as plant growth and nutrient uptake. For example, the co-inoculation of the KSB *P. mucilaginosus* and the N-fixing bacteria *Azotobacter chroococcum* promotes the plant growth of *Sorghum vulgare* and the K and N uptake with respect to the uninoculated controls and also with respect to the plants treated with the two bacterial species separately (Basak and Biswas 2009; Basak and Biswas 2010). Also, the co-inoculation of the KSB *P. mucilaginosus* with the PSB *B. megaterium* in conjunction with amendment of rock P or K materials increased both P and K uptake and promoted the growth of eggplant (Han and Lee 2005), pepper and cucumber (Han et al. 2006). Other studies reported that the co-inoculation of different plants with several bacteria has different plant growth promotion abilities and these bacteria with mycorrhizal fungi. In the study of Leaungvutiviroj et al. (2010), KSB (*B. subtilis*), PSB (*Burkholderia unamae*), N-fixing bacteria (*Azotobacter tropicalis*) and an auxin-producing strain (KJB9/2) were inoculated in maize and Chinese kale showing a growth enhancement ranging from four to seven times when rock phosphate and K feldspar were combined with the biofertilizers with respect to untreated controls. Recently, field experiments carried out in Egypt showed that inoculation of combined N-fixing and P- and K-solubilizing bacteria together with humus increases the productivity of *Sorghum bicolor* and significant increase of dry weight was also found after the single inoculation of a KSB of *B. circulans* (Afifi et al. 2014).

Several studies also showed the positive effects of combination of KSB, PSB and N-fixing bacteria with mycorrhiza, such as that of Wu et al. (2005) who showed that the inoculation of maize with a combination of the N-fixer *A. chroococcum*, the P solubilizer *B. megaterium*, the K solubilizer *B. mucilaginosus* and the AM fungus *G. mosseae* resulted in the highest biomass and seedling height. The microbial inocula increased the N, P and K uptake as well as improved soil properties, such as organic matter content and total N in soil. Very recently, Padma and Sukumar (2015) also showed a positive effect on the plant growth of mulberry plants inoculated with KSB combined with other bioinoculants including N-fixers, P solubilizers and mycorrhiza. The inoculation with biofertilizers able to solubilize potassium can also influence the plant content in fatty acids, as shown by Ahmed and El-Araby (2012) using potash biofertilizer in sunflowers. In field experiments, they found that the application of these biofertilizers increase the oil content in linoleic and linolenic acids. These results showed that KSMs not only mobilize K to plants improving its growth but they can increase the bioactive compounds in plants. On the other hand, Wu et al. (2006) inoculated a combination of the N-fixer *A. chroococcum*, the P solubilizers *B. megaterium* and the K solubilizers *B. mucilaginous* in *Brassica juncea* plants to evaluate the effect on the plant uptake of heavy metals from Pb-Zn mine tailings. The presence of these beneficial bacteria stimulated plant growth and protected the plant from metal toxicity obtaining a higher efficiency of phytoextraction compared with control treatments.

7.6 Conclusions

The currently known KSMs have been isolated from different sources, mainly soil and plant rhizosphere, and they are able to release K from different insoluble K sources. The studies performed to date on KSMs showed their high diversity since they are represented by fungi and bacteria from very different phylogenetic groups. Although few studies have been still performed with K solubilizers worldwide, the available data suggest that they may be used in phytoremediation and that biofertilizers including these bacteria alone or in combination with other PGPR could permit the substitution or decrease of chemical fertilizer use within sustainable agriculture schemes, and therefore this is a promising field of study in the 'greening' era.

Acknowledgements The authors would like to thank our numerous collaborators and students involved in our research on plant-microbe interactions over the years. Funding was provided by Spanish national and regional (Junta de Castilla y León) governments.

References

Afifi MMI, El-Sayed GAM, Manal A, El-Gamal H, Massoud ON (2014) Synergistic effect of biofertilizers containing N-fixer, P and K solubilizers and humic substances on *Sorghum bicolor* productivity. Middle East J Appl Sci 4:1065–1074

Ahmed HFS, El-Araby MMI (2012) Evaluation of the influence of nitrogen fixing, phosphate solubilizing and potash mobilizing biofertilizers on growth, yield, and fatty acid constituents of oil in peanut and sunflower. Afr J Biotechnol 11:10079–10088

Aleksandrov VG (1985) Organo-mineral fertilizers and silicate bacteria. Dokl Akad Nauk 7:43–48

Archana DS, Nandish MS, Savalagi VP, Alagawadi AR (2013) Characterization of potassium solubilizing bacteria (KSB) from rhizosphere soil. Bioinfolet 10: 248–257

Bagyalakshmi B, Ponmurugan P, Balamurugan A (2012a) Impact of different temperature, carbon and nitrogen sources on solubilization efficiency of native potassium solubilizing bacteria from tea (*Camellia sinensis*). J Biol Res 3:36–42

Bagyalakshmi B, Ponmurugan P, Marimuthu S (2012b) Influence of potassium solubilizing bacteria on crop productivity and quality of tea (*Camellia sinensis*). Afr J Agric Res 7:4250–4259

Basak BB, Biswas DR (2009) Influence of potassium solubilizing microorganism (*Bacillus mucilaginosus*) and waste mica on potassium uptake dynamics by sudangrass (*Sorghum vulgare* Pers.) grown under two alfisols. Plant Soil 317:235–255

Basak BB, Biswas DR (2010) Co-inoculation of potassium solubilizing and nitrogen fixing bacteria on solubilization of waste mica and their effect on growth promotion and nutrient acquisition by a forage crop. Biol Fertil Soils 46:641–648

Diep CN, Hieu TN (2013) Phosphate and potassium solubilizing bacteria from weathered materials of denatured rock mountain, Ha Tien, Kiên Giang province, Vietnam. Am J Life Sci 1:88–92

Gundala PB, Chinthala P, Sreenivasulu B (2013) A new facultative alkaliphilic, potassium solubilizing, *Bacillus* sp. SVUNM9 isolated from mica cores of Nellore district, Andhra Pradesh, India. Res Rev J Microbiol Biotechnol 2:1–7

Han HS, Lee KD (2005) Phosphate and potassium solubilizing bacteria effect on mineral uptake, soil availability and growth of eggplant. Res J Agric Biol Sci 1: 176–180

Han HS, Supanjani, Lee KD (2006) Effect of co-inoculation with phosphate and potassium solubilizing bacteria on mineral uptake and growth of pepper and cucumber. Plant Soil Environ 52: 130–136

Hu X, Chen J, Guo J (2006) Two phosphate- and potassium-solubilizing bacteria isolated from Tianmu Mountain, Zhejiang, China. World J Microbiol Biotechnol 22:983–990

Hu XF, Li SX, Wu JG, Wang JF, Fang QL, Chen JS (2010) Transfer of *Bacillus mucilaginosus* and *Bacillus edaphicus* to the genus *Paenibacillus* as *Paenibacillus mucilaginosus* comb. nov. and *Paenibacillus edaphicus* comb. nov. Int J Syst Evol Microbiol 60: 8–14

Kumar A, Bahadur I, Maurya BR, Raghuwanshi R, Meena VS, Singh DK, Dixit J (2015) Does a plant growth-promoting rhizobacteria enhance agricultural sustainability? J Pure Appl Microbiol 9(1):715–724

Leaungvutiviroj C, Ruangphisarn P, Hansanimitkul P, Shinkawa H, Sasaki K (2010) Development of a new biofertilizer with a high capacity for N2 fixation, phosphate and potassium solubilization and auxin production. Biosci Biotechnol Biochem 74: 1098–1101

Lian B, Wang B, Pan M, Liu C, Teng HH (2008) Microbial release of potassium from K-bearing minerals by thermophilic fungus *Aspergillus fumigatus*. Geochim Cosmochim Acta 72:87–98

Liu W, Xu X, Wu X, Yang Q, Luo Y, Christie P (2006) Decomposition of silicate minerals by *Bacillus mucilaginosus* in liquid culture. Environ Geochem Health 28:133–140

Liu D, Lian B, Dong H (2012) Isolation of *Paenibacillus* sp. and assessment of its potential for enhancing mineral weathering. Geomicrobiol J 29:413–421

Lopes-Assad ML, Avansini SH, Rosa MM, de Carvalho JRP, Ceccato-Antonini SR (2010) The solubilization of potassium-bearing rock powder by *Aspergillus niger* in small-scale batch fermentations. Can J Microbiol 56:598–605

Lu JJ, Wang JF, Hu XF (2013) Genome sequence of growth-improving *Paenibacillus mucilaginosus* strain KNP414. Genome Announc 1:e00881–13

Maurya BR, Meena VS, Meena OP (2014) Influence of inceptisol and alfisol's potassium solubilizing bacteria (KSB) isolates on release of K from waste mica. Vegetos 27:181–187

Meena OP, Maurya BR, Meena VS (2013) Influence of K-solubilizing bacteria on release of potassium from waste mica. Agric Sustain Dev 1(1):53–56

Meena VS, Maurya BR, Bahadur I (2014a) Potassium solubilization by bacterial strain in waste mica. Bangladesh J Bot 43(2):235–237

Meena VS, Maurya BR, Verma JP (2014b) Does a rhizospheric microorganism enhance K+ availability in agricultural soils? Microbiol Res 169:337–347

Meena RK, Singh RK, Singh NP, Meena SK, Meena VS (2015a) Isolation of low temperature surviving plant growth-promoting rhizobacteria (PGPR) from pea (*Pisum sativum* L.) and documentation of their plant growth promoting traits. Biocatal Agric Biotechnol. doi:10.1016/j.bcab.2015.08.006

Meena VS, Maurya BR, Verma JP, Aeron A, Kumar A, Kim K, Bajpai VK (2015b) Potassium solubilizing rhizobacteria (KSR): isolation, identification, and K-release dynamics from waste mica. Ecol Eng 81: 340–347

Padma SD, Sukumar J (2015) Response of mulberry to inoculation of potash mobilizing bacterial isolate and other bio-inoculants. Glob J Biosci Biotechnol 4: 50–53

Parmar P, Sindhu SS (2013) Potassium solubilization by rhizosphere bacteria: influence of nutritional and environmental conditions. J Microbiol Res 3:25–31

Peix A, Rivas-Boyero AA, Mateos PF, Rodríguez-Barrueco C, Martínez-Molina E, Velázquez E (2001) Growth promotion of chickpea and barley by a phosphate solubilizing strain of *Mesorhizobium mediterraneum* under growth chamber conditions. Soil Biol Biochem 33:103–110

Prajapati KB, Modi HA (2012) Isolation and characterization of potassium solubilizing bacteria from ceramic industry soil. CIBTech J Microbiol 1:8–14

Prajapati KB, Sharma MC, Modi HA (2012) Isolation of two potassium solubilizing fungi from ceramic industry soils. Life Sci Leafl 5:71–75

Ramarethinam S, Chandra K (2006) Studies on the effect of potash solubilizing bacteria *Frateuria aurantia* (Symbion-K- liquid formulation) on Brinjal (*Solanum melongena* L.) growth and yield. Pestology 11:35–39

Rosa-Magri MM, Avansini SH, Lopes-Assad ML, Tauk-Tornisielo SM, Ceccato-Antonini SR (2012) Release of potassium from rock powder by the yeast *Torulaspora globosa*. Braz Arch Biol Technol 55:577–582

Saiyad SA, Jhala YK, Vyas RV (2015) Comparative efficiency of five potash and phosphate solubilizing bacteria and their key enzymes useful for enhancing and improvement of soil fertility. Int J Sci Res Pub 5:1–6

Sangeeth KP, Bhai RS, Srinivasan V (2012) *Paenibacillus glucanolyticus*, a promising potassium solubilizing bacterium isolated from black pepper (*Piper nigrum* L.) rhizosphere. J Spices Aromat Crop 21: 118–112

Sawada H, Ieki H, Oyaizu H, Matsumoto S (1993) Proposal for rejection of *Agrobacterium tumefaciens* and revised descriptions for the genus *Agrobacterium* and for *Agrobacterium radiobacter* and *Agrobacterium rhizogenes*. Int J Syst Bacteriol 43:694–702

Schoch CL, Seifert KA, Huhndorf A, Robert V, Spouge JL, Levesque CA, Chen W, Fungal Barcoding Consortium (2012) Nuclear ribosomal internal transcribed spacer (ITS) region as a universal DNA barcode marker for fungi. PNAS USA 109:6241–6246

Shanware AS, Kalkar SA, Trivedi MM (2014) Potassium solublisers: occurrence, mechanism and their role as competent biofertilizers. Int J Curr Microbiol App Sci 3:622–629

Sheng XF (2005) Growth promotion and increased potassium uptake of cotton and rape by a potassium releasing strain of *Bacillus edaphicus*. Soil Biol Biochem 37:1918–1922

Sheng XF, Huang WY (2002) Mechanism of potassium release from feldspar affected by the strain NBT of silicate bacterium. Acta Pedol Sin 39(6):863–871

Sheng XF, He LY (2006) Solubilization of potassium-bearing minerals by a wild-type strain of *Bacillus edaphicus* and its mutants and increased potassium uptake by wheat. Can J Microbiol 52:66–72

Sheng XF, Zhao F, He H, Qiu G, Chen L (2008) Isolation, characterization of silicate mineral solubilizing Bacillus globisporus Q12 from the surface of weathered feldspar. Can J Microbiol 54:1064–1068

Sindhu SS, Parmar P, Phour M (2014) Nutrient cycling: potassium solubilization by microorganisms and improvement of crop growth. In: Parmar N, Shing A (eds) Geomicrobiology and biogeochemistry soil biology. Springer, Heidelberg, pp 175–198

Singh G, Biswas DR, Marwaha TS (2010) Mobilization of potassium from waste mica by plant growth promoting rhizobacteria and its assimilation by maize (*Zea mays*) and wheat (*Triticum aestivum* L.): a hydroponics study under phytotron growth chamber. J Plant Nutr 33: 1236–1251

Singh NP, Singh RK, Meena VS, Meena RK (2015) Can we use maize (*Zea mays*) rhizobacteria as plant growth promoter? Vegetos 28(1):86–99

Subhashini DV (2015) Growth promotion and increased potassium uptake of tobacco by potassium mobilizing bacterium *Frateuria aurantia* grown at different potassium levels in vertisols. Commun Soil Sci Plant Anal 46:210–220

Sugumaran P, Janarthanam P (2007) Solubilization of potassium containing minerals by bacteria and their effect on plant growth. World J Agr Sci 3:350–355

Verma P, Yadav AN, Khannam KS, Panjiar N, Kumar S, Saxena AK, Suman A (2015) Assessment of genetic diversity and plant growth promoting attributes of psychrotolerant bacteria allied with wheat (*Triticum aestivum*) from the northern hills zone of India. Ann Microbiol. doi:10.1007/s13213-014-1027-4

Woese CR (2000) Interpreting the universal phylogenetic tree. Proc Natl Acad Sci U S A 97:8392–8396

Wu SC, Cao ZH, Li ZG, Cheung KC, Wong MH (2005) Effects of biofertilizer containing N-fixer, P and K solubilizers and AM fungi on maize growth: a greenhouse trial. Geoderma 125:155–166

Wu SC, Cheung KC, Luo YM, Wong MH (2006) Effects of inoculation of plant growth-promoting rhizobacteria on metal uptake by *Brassica juncea*. Environ Pollut 140:124e135

Young JM, Kuykendall LD, Martínez-Romero E, Kerr A, Sawada H (2001) A revision of *Rhizobium* Frank 1889, with an emended description of the genus, and the inclusion of all species of *Agrobacterium* Conn 1942 and *Allorhizobium undicola* de Lajudie et al. 1998 as new combinations:*Rhizobium radiobacter*, *R. rhizogenes*, *R. rubi*, *R. undicola* and *R. vitis*. Int J Syst Evol Microbiol 51:89–103

Zarjani JK, Aliasgharzad N, Oustan S, Emadi M, Ahmadi A (2013) Isolation and characterization of potassium solubilizing bacteria in some Iranian soils. Arch Agron Soil Sci 59:1713–1723

Zeng X, Liu X, Tang J, Hu S, Jiang P, Li W, Xu L (2012) Characterization and potassium-solubilizing ability of *Bacillus circulans* Z1-3. Adv Sci Lett 10:173–176

Zhang C, Kong F (2014) Isolation and identification of potassium-solubilizing bacteria from tobacco rhizospheric soil and their effect on tobacco plants. Appl Soil Ecol 82:18–25

Zhao F, Sheng XF, Huang Z (2008) Isolation of mineral potassium solubilizing bacterial strains from agricultural soils in Shandong province. Biodivers Sci 16:593–600

Zörb C, Senbayram M, Peiter E (2014) Potassium in agriculture – status and perspectives. J Plant Physiol 171:656–669

Interactions Between Arbuscular Mycorrhizal Fungi and Potassium-Solubilizing Microorganisms on Agricultural Productivity

8

Perumalsamy Priyadharsini and Thangavelu Muthukumar

Abstract

Microbial interaction is a key determinant of soil fertility, plant health, and crop productivity. Arbuscular mycorrhizal fungi (AMF) belonging to the phylum *Glomeromycota* are a ubiquitous component of most natural and agroecosystems. These fungi associate with most of the plant species and provide several benefits including better nutrition and increased tolerance to various biotic and abiotic stresses. Mycorrhizal symbiosis can affect the microbial population and their activity in the rhizosphere both qualitatively and quantitatively including mineral solubilization by microorganisms. These changes are mediated through the so-called mycorrhizosphere effect resulting from direct or indirect changes in root exudation (composition and quantity) patterns or through fungal exudates. In most instances, the interaction between AMF and nutrient-solubilizing microorganisms is synergistic resulting in stimulation of plant growth through improved nutrient acquisition and inhibition of plant pathogens. The ecological impact of AMF interactions with microorganisms involved in potassium solubilization is not well resolved compared to those involved in phosphate solubilization. Although direct studies on the interactions between AMF and potassium-solubilizing microorganisms (KSMs) on plant growth are limited, studies on plant growth-promoting microorganisms (PGPMs) and AMF do involve organisms with K-solubilizing capabilities. Evidence does exist on the influence of KSMs on mycorrhizal formation and function. Interactions between AMF and KSMs are vital in sustainable low-input crop production systems that rely on biological processes to achieve improved plant growth and yield in addition to maintaining soil fertility. This article examines the interactions between AMF and KSMs on plant growth, development, and crop productivity.

P. Priyadharsini • T. Muthukumar (✉)
Root and Soil Biology Laboratory, Department of Botany, Bharathiar University, Coimbatore 641 046, Tamil Nadu, India
e-mail: tmkum@yahoo.com

V.S. Meena et al. (eds.), *Potassium Solubilizing Microorganisms for Sustainable Agriculture*,
DOI 10.1007/978-81-322-2776-2_8

Keywords
Mycorrhiza • Rhizosphere • Organic acids • Mycorrhizosphere

8.1 Introduction

The terrestrial plants depend on soil for their mineral nutrition, and microorganisms in the soil play a pivotal role in the modification and availability of nutrients in the soil. The carbon essential to drive the microbial process in the soil is derived from plants either directly or indirectly. Different estimates suggest that around 5–20 % of the photosynthetic carbon is released by plants into the rhizosphere. The root exudates released simple hexose sugars, mucilages, and organic anions, along with more complex carbon derived from sloughed cells and root turnover (Jones et al. 2009). The carbon released by plant roots is available to soil microorganisms and results in a significant increase in microbial biomass and their activity in the rhizosphere (Chen et al. 2002). In addition to organic compounds, plant roots also release proton, oxygen, and water into the rhizosphere (Nannipieri et al. 2007). Nutrients are one of the major environmental factors that limit plant growth and productivity in most terrestrial ecosystems (Lambers et al. 2009). Plant species vary in their capacity to acquire nutrients from soil (Lambers et al. 2009). Some plant species are capable of acquiring nutrients from deficient soils, whereas others are unable to do so and, therefore, cannot persist on such a soil (Lambers et al. 2008; Meena et al. 2013; Maurya et al. 2014).

Like nitrogen (N) and phosphorus (P), potassium (K) is also one of the important and essential macronutrients for all organisms, including plants. It plays a major role in the growth, development, and metabolism of plants. Potassium deficiency in plants results in poor development of roots and reduced plant growth and yield. The function of K in plants includes regulation of turgor, charge balance, movement, and protein synthesis (Leigh and Wyn-Jones 1984). Terrestrial plants grow on a wide range of soil K concentrations ranging from 10 μm to 10 mM (Marschner 1995) and represents 2–10 % of the dry mass (Leigh and Wyn-Jones 1984). In spite of their abundance, K availability is extremely low in many soils, especially those under intensive agriculture. This low availability of K is due to their strong adsorption with soil minerals. Around 90–98 % of the soil K is held firmly in crystal lattice structure of minerals and, therefore, is unavailable to plants. Only 0.1–2.0 % of the total soil K is available to plants in water-soluble or exchangeable form (Zörb et al. 2014). The low availability of K in the soils combined with its high demand by plants results in the formation of the depletion zones around the roots (Drew and Nye 1969). Subsequently, plant needs to develop efficient strategies to enhance K uptake from soil (Zörb et al. 2014), which includes association with microorganisms in the soil (Garcia and Zimmermann 2014). Solubilization and release of K from insoluble minerals by K-solubilizing bacteria (KSB) is one of the natural ways by which K is made available to plants. Plants have adopted this natural process to make available K in the rhizosphere through maintaining the microbial populations and their activities under deficient conditions. In addition, K acquisition from soils with low K concentrations in soil solution as well as plant growth can also be enhanced by mycorrhizal symbioses (Meena et al. 2014a; Singh et al. 2015).

The aim of this article is to discuss the interaction between arbuscular mycorrhizal fungi (AMF) and KSMs and its subsequent influence on plant growth. Plant growth-promoting rhizobacteria (PGPR) other than those branded as KSB also influence AMF symbiosis and their activity resulting in improved K acquisition by plants (Table 8.1). Such types of organisms and their activities are also included in this article. Since it is impossible to prepare an

Table 8.1 Response of interactions between plant growth-promoting rhizobacteria (PGPR) and arbuscular mycorrhizal fungi (AMF) on plant tissue potassium (K) concentrations

Host plant	PGPR	AMF	Tissue K	Response	Reference
Acacia senegal	*P. fluorescens*	Natural species	Shoot	Increase	Ndoye et al. (2012)
Allium cepa	*A. brasilense*, *A. chorococcum*, *Burkholderia cepacia*	*R. clarus*, *Rhizophagus fasciculatus*	Shoot	Increase	Pulido et al. (2003)
Calendula officinalis	*A. chroococcum*, *P. fluorescens*, *Azospirillum lipoferum*	Efficient indigenous	Leaf/ root	Increase	Hosseinzadah et al. (2011)
Capsicum chinense	*A. chroococcum*, *A. brasilense*	Commercial inoculums	Shoot	Increase	Constantino et al. (2008)
Carica papaya	*Bacillus consortium*	*F. mosseae* *G. manihotis*	Stem/ leaves	Increase/ decrease	Jaizme-Vega et al. (2006)
Casuarina equisetifolia	*Paenibacillus polymyxa*	*Funneliformis geosporum*	Shoot/ root	Increase	Muthukumar and Udaiyan (2010)
Coleus forskohlii	*P. fluorescens*	*Scutellospora* spp.	Shoot	Increase	Srimathi Priya and Kumutha (2009)
Dendrocalamus strictus	*P. polymyxa*, *A. brasilense*	*Glomus aggregatum*	Shoot/ rhizome/ root	Increase	Muthukumar and Udaiyan (2006)
Glycyrrhiza glabra	*B. coagulans*	*G. aggregatum*	Root	Increase	Selvaraj and Sumithra (2011)
Helianthus annuus	*P. fluorescens* strains	*C. etunicatum*, *R. intraradices*	Shoot	Increase	Shirmardi et al. (2010)
Lycopersicon esculentum	*A. brasilense*, *A. chorococcum*, *Burkholderia cepacia*	*R. clarus* *R. fasciculatus*	Shoot	Increase	Pulido et al. (2003)
Medicago sativa L. var. Valenciana	*P. aeruginosa*, *B. cepacia*, *Hafnia alvei*, *Enterobacter cloacae*	Consortium of *Glomus* spp.	Shoot	Increase	Chamizo et al. (2009)
Musa acuminata Colla AAA cv. "Grande Naine"	*Bacillus* strains	*G. manihotis*	Shoot	Increase	Rodríguez-Romero et al. (2005)
Ocimum basilicum	*B. subtilis*	*R. intraradices*	Leaves	Increase	Abdel-Rahman et al. (2011)
Solanum viarum	*B. coagulans*, *Trichoderma harzianum*	*G. aggregatum*	Leaves/ root	Increase	Hemashenpagam and Selvaraj (2011)
Sphaeranthus amaranthoides	*B. subtilis*, *Trichoderma viride*	*Glomus walkeri*	Leaf	Increase	Sumithra and Selvaraj (2011)
Triticum aestivum	*Pseudomonas* spp.	Indigenous AMF consortium	Grains	Increase/ decrease	Roesti et al. (2006)
Vetiveria zizanioides	*Stenotrophomonas maltophilia*, *Agrobacterium tumefaciens*, *Azospirillum* spp., *B. subtilis*	Commercial inoculums	Shoot	Increase	Bhromsiri and Bhromsiri (2010)

exhaustive review involving all the studies and molecular mechanisms involved in these interactions and as it exceeds the limits of this short contribution, this article summarizes the main aspects with limited discussion of the underlying mechanisms.

8.2 Do K and P Solubilization Share a Common Mechanism?

Soil microorganisms, including bacteria and fungi, release K by excreting organic acids like

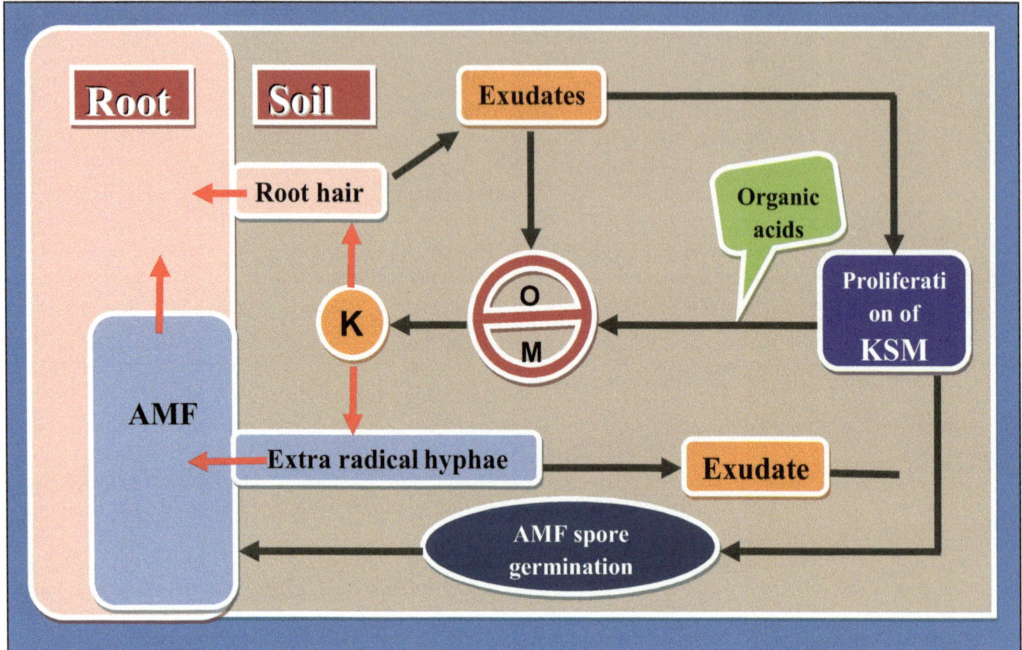

Fig. 8.1 Schematic representation of the interaction between potassium-solubilizing microorganisms (KSMs) and arbuscular mycorrhizal fungi (AMF) on potassium (K^+) availability in agricultural soil and acquisition by plant roots (For abbreviations: *OK* organic potassium, *MK* mineral potassium)

citric, oxalic, tartaric, succinic, and α-ketogluconic acids (Fig. 8.1). These acids dissolve rock K or chelate the primary mineral ions (e.g., silicon) to release K in the soil solution (Bennett et al. 1998; Basak and Biswas 2010). Similarly, P solubilization also occurs through microbial processes like organic acid production and proton extrusion (Govindasamy et al. 2010). The low molecular weight organic acids produced by microorganisms mainly include gluconic and ketogluconic acids (Deubel et al. 2000). In addition, other organic acids of microbial origin include mixtures of lactic, isovaleric, isobutyric, and acetic acids. This indicates that P-solubilizing ability can also enhance the availability of K in the soil. For example, inoculation of the phosphate-solubilizing bacteria (PSB) *Bacillus megaterium* var. *phosphaticum* increased the availability of K in the rhizosphere of pepper (*Capsicum annum*) and cucumber (*Cucumis sativus*) by 10.44 % and 8.09 %, respectively. Similarly, in the same study, inoculation of K-solubilizing *Paenibacillus mucilaginosus* (*Bacillus mucilaginosus*) increased the P availability by 6.25 and 8.97 % suggesting that these microorganisms can affect solubilization of other mineral elements (Han et al. 2006). This increased availability was also reflected in the K content of plants where pepper plants inoculated with PSB had 9.96 % more K in their shoots and 9.46 % more K in their roots. Similarly, cucumber plants inoculated with PSB had 5.26 % and 4.38 % higher K in their shoots and roots, respectively (Han et al. 2006). A similar observation was also reported for PSB *B. megaterium* var. *phosphaticum* and KSB *B. mucilaginosus* inoculation in maize under calcareous soil conditions (Abou-el-Seoud and Abdel-Megeed 2012). Inoculation of P-solubilizing *Pseudomonas putida* and N-fixer and P solubilizer *Bacillus polymyxa* increased K concentration in *Stevia rebaudiana* by 40.61 % and 44.55 %, respectively, compared to uninoculated plants. Hu et al. (2006) isolated two strains of *P. mucilaginosus* (KNP413 and KNP414) that were capable of solubilizing both P and K from soils of Tianmu Mountain, Zhejiang, China. These bacterial strains dissolved the phosphate source, phosphorite, as well as K mineral sources like montmorillonite, kaolinite, and potassium feldspar (Meena et al. 2014b; Kumar et al. 2015). However, the mechanisms by which these bacteria solubilize the nutrients are yet to be ascertained. These studies clearly suggest that PSB, in addition to improving P availability in the soil or P concentrations in plant tissues, also increases the availability and concentrations of K in soil and plant tissues.

8.3 Arbuscular Mycorrhizal Fungi (AMF)

The obligate symbiotic fungi that form arbuscular mycorrhiza belong to a small fungal phylum *Glomeromycota* (Schüβler et al. 2001). This group of fungi is named after the characteristic structure they produce within plant roots called "arbuscules." These structures act as an interface in transferring the fungal-acquired mineral nutrients to plant root cells. AMF also produce vesicles, intra- and intercellular hyphae and/or hyphal coils within roots, and extraradical mycelium and auxiliary cells in the soil. Majority of AMF species produce spores in soil, while some produce spores both within the host roots and in the surrounding soil. Fossil evidence suggests that AMF have coevolved with the emergence of first land plants. This symbiosis is observed in all terrestrial ecosystems and in 80 % of all terrestrial plant species (Smith and Read 2008). AMF act as a bridge, connecting soil and plant roots. The AMF help their plant hosts by taking up and transferring soil mineral nutrients, especially phosphate (Smith and Read 2008). Other benefits of the symbiosis to plants include improved water relations, interplant transfer of nutrients, enhanced tolerance to various biotic and abiotic stresses, and positive changes in soil structure. In turn, plants provide 4–20 % of the photosynthets allocated to the roots for AMF (Smith and Read 2008). In addition, AMF influence soil microbial communities through their external hyphal networks in the soil (Andrade et al. 1997; Bonfante and Anca 2009). The term "mycorrhizosphere" was introduced to refer to the soil region that was influenced by mycorrhiza. The mycorrhizosphere encompasses the rhizosphere, the region of the soil influenced by plant roots, and the hyphosphere, the region of the soil influenced by the fungal hyphae. The extent of extraradical mycelial network formed in the soil tends to vary with AMF species. Taxa belonging to genera like *Gigaspora* and *Scutellospora* form extensive extraradical hyphae per unit length of colonized roots, while others like *Acaulospora*, *Funneliformis*, *Glomus*, and *Rhizophagus* produce few extensive extraradical hyphal networks (Treseder 2013). This clearly shows that the extent of plants mycorrhizosphere could vary with associating AMF taxa. The influence of AMF hyphae on soil microorganisms is due to the presence of 4–10 m of fungal hyphae per gram soil (Schreiner et al. 1997). The presence of AMF affects both the number and diversity of microorganisms in the soil (Miransari 2011). AMF interact with a wide range of soil organisms like nodulating bacteria, PGPR, saprophytic soil fungi, nematodes, as well as deleterious plant pathogens. Some of the interactions between AMF and soil microorganisms are important as they substantially modify plant growth and fitness.

8.4 Role of AMF in K Uptake

The role of AMF in plant K nutrition has been extensively reviewed recently by Garcia and Zimmermann (2014). Although not extensively studied as for P, available evidences indicate that AMF could play a substantial role in the K nutrition of plants. The uptake of K by AMF and its subsequent translocation to plants are well documented (Perner et al. 2007; Baslam et al. 2013). Nevertheless, species of AMF differ in their ability to affect K nutrition of plants. Veresoglou et al. (2011) conducted microcosm experiments to assess the contributions of *Rhizophagus intraradices* (*Glomus intraradices*) (B.B/E) and *Gigaspora margarita* (BEG 34) symbioses to mineral nutrition of buckhorn plantain (*Plantago lanceolata*) in a native, nutrient-limited coastal dune soil. The results of the experiments suggested that *R. intraradices*-inoculated plants had 23.3–55.8 % more total K in their tissues when compared to *G. margarita*-inoculated plants. Further, the overall contribution of the two isolates to plant nutrition was also not similar as evidenced by the differences in plant tissue N/P/K ratios. *Rhizophagus intraradices*-inoculated plants generally exhibited nutrient ratios indicative of P limitation, whereas for *G. margarita*-mycorrhized plants, corresponding ratios strongly implied either N or K limitation (Veresoglou et al. 2011). Particle-induced X-ray emission experiments involving the AMF *R. intraradices* suggest that the K taken up by AMF were accumulated in various fungal structures like hyphae, vesicles, and spores (Johansson and Campbell 1988; Pallon et al. 2007; Olsson et al. 2008, 2011). Although transporters and channels of AMF involved in P transport are characterized, these are missing for K (Garcia and Zimmermann 2014). Four sequences (three coding for Shaker-like channels and one for KT/KUP/HAK transporter) have been identified to be involved in K transport from an ETS library of *Rhizophagus irregularis* (Casieri et al. 2013). Nevertheless, no Trk or Tok members were identified from this library as for the ectomycorrhizal symbiosis.

8.5 Soil Microorganisms and AMF

Several biotic and abiotic factors in soil are known to affect the production and viability of AMF spores. Among biotic factors, soil microorganisms like bacteria and fungi influence the viability and function of AMF spores in soil (Walley and Germida 1996). These microorganisms associated with AMF spores may colonize the spores superficially or can exist as intracellular entities commonly referred to as bacteria-like organelles (Bianciotto et al. 1996; Budi et al. 1999). Both gram-positive (*Bacillus* spp.) and gram-negative (*Pseudomonas*, *Burkholderia*, *Flavobacterium* spp.) bacteria tend to associate with AMF spores (Xavier and Germida 2003). Although bacteria-like organelles found within intact AMF spores do not affect their viability (Bianciotto et al. 1996), bacteria or fungi colonizing the spore surface may positively or negatively alter spore function (Mayo et al. 1986; Lee and Koske 1994). Germination of AMF spores under in vitro condition occurs only in the presence of certain bacteria. For example, germination of *Glomus versiforme* spores was stimulated by *Pseudomonas* and *Corynebacterium* spp., under in vitro conditions (Mayo et al. 1986). An in vitro assessment of the effect of *Rhizophagus clarus* (*Glomus clarum*) spore-associated bacteria on spore germination revealed that most of the bacteria isolated from the spore surface generally did not significantly alter spore function (Xavier and Germida 2003). Some bacteria isolated from clean, decontaminated spores either inhibited or stimulated germination of *R. clarus* spores. Contrarily, *P. putida* isolated from AMF spores inhibited germination of *R. clarus* spores (Walley and Germida 1997). As direct fungal contact was essential to stimulate AMF spore germination, it was not necessary to inhibit spore germination which occurred due to the production of volatile metabolites by the bacteria. In addition, inoculation of a stimulatory bacterial isolate (*Bacillus pabuli*) significantly enhanced mycorrhization of pea (*Pisum sativum*) plants by *R. clarus* over those plants co-inoculated with an inhibitory bacterial isolate (*Bacillus chitinosporus*) (Xavier and Germida 2003). The bacteria isolated from spores of AMF (*R. intraradices*, *Funneliformis mosseae*, *Glomus mosseae*) with the ability to inhibit the growth of plant pathogens were tested for their efficacy to influence AMF colonization and plant growth in potato (Bharadwaj et al. 2008). The colonization of potato roots by *F. mosseae* was sevenfold higher in the presence of the *Pseudomonas* FWC70 isolate in a greenhouse and was six- to ninefold higher in the presence of the three bacterial isolates, namely, *Pseudomonas* FWC70, *Stenotrophomonas* FWC94, and *Arthrobacter* FWC110, in an outdoor pot experiment. In addition, *Pseudomonas* isolates stimulated several growth traits of potato as well as antagonistic activity against pathogens like *Erwinia carotovora*, *Phytophthora infestans*, and *Verticillium dahliae*. These results showed that some of the spore-associated bacteria are multifunctional and may impart multiple benefits like enhanced mycorrhizal colonization and plant growth as well as antagonistic activity against pathogens (Bharadwaj et al. 2008). Three probable endobacteria (*Bacillus*, *Paenibacillus* spp.) isolated from spores of *G. margarita* were shown to stimulate the hyphal growth of *G. margarita* by three- to fivefold (Cruz and Ishii 2012). A scanning electron microscopic examination of the spore and hyphal surface revealed bacterial aggregation resembling biofilms on the fungal surfaces (Meena et al. 2015a, b). The presence of possible biofilm on the fungal surface leads to the speculation that the bacteria may stimulate AMF hyphal growth to colonize plant roots under normal as well as flooded soil conditions (Cruz and Ishii 2012).

8.6 Influence of AMF on Rhizosphere Bacteria

Differences in microbial populations occur in the rhizosphere of mycorrhizal and non-mycorrhizal plants (Secilia and Bagyaraj 1987). Nevertheless, results of studies examining the AMF interactions with rhizosphere bacteria are variable. AMF have positive, negative, or neutral effects on rhizosphere bacteria. Rapid colonization by *R. intraradices* reduced populations of fluorescent pseudomonads (*P. putida*, *P. fluorescence*) in the rhizosphere of cucumber (*Cucumis sativus*) (Paulitz and Linderman 1989). This reduction in the bacterial populations was more pronounced during the early stages of plant growth than during the later stages. The negative influence of *R. intraradices* on fluorescent pseudomonads during early stages of plant growth was attributed to the changes in the quality and quantity of root exudates released into the rhizosphere by mycorrhizal roots. Generally, AMF utilize large quantities of host carbon during the establishment of the symbiosis resulting in reduced availability of carbon to the rhizosphere microorganisms. This is evidenced by the lack of reduction in the populations of pseudomonads in the rhizosphere of cucumber colonized by *Claroideoglomus etunicatum* (*Glomus etunicatum*) which was a slow colonizer and produced less amount of fungal biomass both in the soil and roots (Paulitz and Linderman 1989). A reduction in the populations of fluorescent pseudomonads in the rhizosphere of mycorrhizal compared to non-mycorrhizal subterranean clover has also been reported (Meyer and Linderman 1986). In contrast to these, Mansfeld-Giese et al. (2002) did not find any significant difference in the total bacterial number between *R. intraradices*-colonized and *R. intraradices*-non-colonized cucumber plants. However, major differences between mycorrhizal and non-mycorrhizal treatments were observed for a few bacterial species and certain bacteria like those of the genus *Paenibacillus* which were more frequent in rhizosphere of mycorrhizal plants, indicating that bacteria within this genus may live in close association with mycelium of these fungi (Mansfeld-Giese et al. 2002). Likewise, certain bacterial genera including *Arthrobacter* and *Bacillus* were more prevalent in the hyphosphere or the soil around specific AMF hyphae, while *Pseudomonas* spp. were found to be more abundant in the rhizosphere of sorghum (*Sorghum bicolor*) (Artursson et al. 2005).

Wamberg et al. (2003) examined the influence of the AMF *R. intraradices* on microbial activity in the rhizosphere of pea (cv. *Solara*) plants raised in γ-irradiated soil in pots at sufficient N and limiting P. The influence of the AMF on soil microbial activity in the rhizosphere varied depending on the growth phase of the plant. Although AMF did not affect total bacterial numbers, it affected the composition of the rhizosphere bacterial community with some bacteria being stimulated and others being repressed as revealed from DNA analysis (DGGE) (Wamberg et al. 2003). The presence of *F. mosseae* and *R. intraradices* positively influenced the populations like *B. megaterium* and *B. mucilaginous* as evidenced by the presence of a positive correlation between AMF colonization and PSB and KSB populations in the rhizosphere (Wu et al. 2005). Further, inoculation of *F. mosseae* or *R. intraradices* increased the populations of K solubilizers in the rhizosphere of maize by 3.07–15.38 % compared to uninoculated control (Wu et al. 2005). All these studies suggest that the association of some bacteria with AMF can be more specific and may be mediated by some kind of communication between the bacteria and AMF possibly in the form of fungal exudates (Artursson et al. 2005, 2006).

The chemotactic responses of the PGPR *Azotobacter chroococcum* and *P. fluorescens* to roots of *R. fasciculatus*-colonized tomato plants were examined by Gupta (2003). A greater number of bacterial cells of wild strains were generally attracted toward *R. fasciculatus*-colonized tomato roots than non-AMF tomato roots. The concentrations of amino acids, organic acids, and sugars of tomato plants were higher than in non-AMF plants (Gupta 2003). Citric and malic acids were the predominant constituents in tomato root exudates. These organic acids exuded by mycorrhizal roots served as a strong chemoattractant for bacteria, especially *P. fluorescens*. In contrast, sugars served as a stronger attractant for *A. chroococcum* than amino acids, and the response was weakest for organic acids suggesting that the quality of AMF root exudates can affect different bacteria variedly (Gupta 2003).

An in vitro experiment using a split Petri dish system to extract soluble substances released by the extraradical mycelium of *R. intraradices* revealed that AMF hyphal exudates could have antagonistic as well as stimulatory effects on fungi and bacteria (Filion et al. 1999). Toljander et al. (2007) also studied the potential of AMF mycelial exudates to stimulate an extracted community of soil bacteria as well as to modify its composition. The results of the study indicated that the bacterial community composition was significantly influenced by the presence of hyphal exudates. The frequency of occurrence of many *Gammaproteobacteria*, including a taxon within the *Enterobacteriaceae*, increased in response to the presence of AMF mycelial exudates. Characterization of the mycelial exudates of *Glomus* spp. MUCL 43205 (GINCO, Louvain-la-Neuve, Belgium) using proton nuclear magnetic resonance (HNMR) spectrometry revealed the presence of low molecular weight compounds like formiate; acetate; α- and β-glucose, a starch-like compound (possibly glycogen); and putative di- and oligosaccharides (Toljander et al. 2007). These clearly show that the mycelial products may play an important role in direct interactions between AMF and other soil microorganisms. In spite of these evidences, the relative importance of AMF mycelial exudates for the development of soil microbial communities in the field remains to be investigated.

8.7 Influence of KSM on AMF Colonization

8.7.1 Bacteria

The influence of mineral-solubilizing microorganisms (MSMs) on AMF formations can be varied. Results of studies involving KSB and AMF interactions revealed that the associations could be positive, neutral, or negative. Inoculation of microbial consortia involving *B. mucilaginous* has been shown to increase root colonization by *F. mosseae* and *R. intraradices* in maize (Wu et al. 2005). Root colonization of sweet potato (*Ipomoea batatas*) by *R. fasciculatus* was increased by ~6 % when the AMF were co-inoculated with *Pseudomonas aeruginosa* (Vogeti et al. 2011). In contrast, dual inoculation of *Glomus manihotis* and *Bacillus* spp. slightly decreased AMF colonization of the in vitro-raised banana plantlets when compared to plantlets inoculated with *G. manihotis* alone (Rodriguez-Romero et al. 2005). A marginal decrease in AMF colonization was also reported for *Begonia malabarica* co-inoculated with *Bacillus coagulans* and *F. mosseae* (Selvaraj et al. 2008; Meena et al. 2015a).

Inoculation of in vitro-raised *B. tulda* with the KSB *Frateuria aurantia* had no significant influence on root colonization by indigenous AMF. Nevertheless, AMF spore numbers were higher for KSB-inoculated soils compared to uninoculated soils (Ghori et al. 2014). Influence of PGPR on AMF colonization can also vary with AMF species. As co-inoculation of *Bacillus subtilis* increased average colonization by *R. fasciculatus* in *Pelargonium graveolens* than those inoculated with AMF alone over two growing seasons, it decreased average root colonization by *G. aggregatum* and *R. intraradices* (Alam et al. 2011; Meena et al. 2015b). In another study, Li et al. (2008) examined the influence of 17 strains of *Paenibacillus* isolated from the mycorrhizal and non-mycorrhizal cucumber rhizosphere on AMF formation and growth of cucumber by two AMF, *R. intraradices* and *F. mosseae*. Although mycorrhizal formation by *R. intraradices* and *F. mosseae* was not affected by strains of *Paenibacillus polymyxa*, one strain of *Pseudomonas macerans* reduced root colonization of *R. intraradices* by 25 % than those inoculated with AMF alone.

One of the major bottlenecks that prevent the widespread use of AMF in plant production systems is the inability of AMF to grow in axenic cultures. Bacteria belonging to the genus *Paenibacillus* that live intracellularly in the ectomycorrhizal fungus *Laccaria bicolor* have been shown to positively influence the symbiosis (Bertaux et al. 2003). Hildebrandt et al. (2002) reported that *Paenibacillus validus* isolated as a contaminant from plates containing germinating spores of *R. intraradices* promoted growth of the fungus until sporulation independently of any plant influence. However, spores formed were smaller in size than those originally used for inoculation. In a later study, Hildebrandt et al. (2006) showed that the in vitro spores of *R. intraradices* produced in the presence of two strains of *P. validus* (DSM ID617 and ID618) were similar in size to those that were inoculated, and spores formed in axenic cultures were infective.

Bacteria that stimulate mycorrhizal formation or function are termed as mycorrhizal helper bacteria (MHB). Most of the MHB include species of *Bacillus* and *Pseudomonas*. These bacteria may affect AMF formation and function through alterations in root cell permeability, penetration of the fungus into the host root, changes in root exudates, and phytohormone production, alleviating adverse effects of environmental parameters on mycelial development and stimulating the growth of plant root hairs (Miransari 2011).

8.7.2 Fungi

Filamentous fungi belonging to genera like *Aspergillus*, *Penicillium*, and *Fusarium* are known solubilizers of P and K (Gaur 1990; Lopes-Assad et al. 2010). Generally, fungi belonging to the genus *Aspergillus* play a multifaceted role like mineral solubilization, plant growth promotion, and synergistic interactions with AMF. Lian et al. (2007) suggested that *Aspergillus fumigatus* solubilizes K by at least three ways: (1) complexation of soluble organic ligands, (2) immobile biopolymers such as the insoluble components of secretion, and (3) mechanical forces in association with the direct physical contact between cells and mineral particles. Among the various *Aspergillus* species, *Aspergillus niger* is an exceptionally efficient producer of organic acids and is known to solubilize P and K (ultramafic) rocks (Silva Filho et al. 2002; Caravaca et al. 2004; Lopes-Assad et al. 2010) either by direct dissolution of the minerals or as a cation chelant, releasing soluble nutrients (Sperber 1958; Cerezine et al. 1988). Studies on the interaction between the K-solubilizing fungi and AMF are limited. *Aspergillus niger* significantly decreased percent germination and length of hyphae of germinated *F. mosseae* spores (McAllister et al. 1995). This negative effect of *A. niger* on *F. mosseae* spore germination and the hyphal length was due to the soluble and volatile compounds produced by *A. niger*. The negative effect of *A. niger* volatile compounds on *F. mosseae* spore germination was greater when *A. niger* was grown on malt extract agar. The percentage of root colonization in maize by *F. mosseae* was unaffected when the AMF were inoculated 2 weeks after *F. mosseae* inoculation. But, AMF colonization of maize decreased when *A. niger* was inoculated simultaneously or 2 weeks before *F. mosseae* inoculation (McAllister et al. 1995). Similarly, populations of *A. niger* decreased when inoculated simultaneously or 2 weeks after *F. mosseae* inoculation. But populations of *A. niger* were unaffected when it was inoculated 2 weeks before *F. mosseae* (McAllister et al. 1995). In a field experiment, Caravaca et al. (2004) evaluated the effect of *R. intraradices* or *Glomus deserticola* inoculation and the addition of sugar beet, rock phosphate (RP), and *A. niger* on the establishment of *Dorycnium pentaphyllum* seedlings. The results of the study showed that the highest levels of AMF colonization occurred in the seedlings inoculated with AMF when grown in the non-amended soil. However, sugar beet, RP, and *A. niger* application had a negative effect on the AMF colonization levels. *Aspergillus tubingensis* strain that increases the availability of K from different sources like bauxite, gypsum, top soil, and sludge was also shown to significantly increase colonization in Bermuda grass (*Cynodon dactylon*) by native AMF (Babu and Reddy 2011). Tomato plants (*Lycopersicon esculentum*) co-inoculated with *A. niger* and *F. mosseae* had 25 % more of their root length colonized by *F. mosseae* compared to plants inoculated with AMF alone (Velazquez et al. 2005).

8.8 Interactions Between KSMs and AMF on Plant Growth and K Content

The KSMs can improve the efficiency of the AMF symbiosis by increasing bioavailable K in the soil. In soils with low K availability, KSMs may release K ions from sparingly soluble inorganic and organic K compounds (Meena et al. 2014a). This increases the available K pool in the soil for the extraradical AMF hyphae to take and transfer it onto the plant roots (Garcia and Zimmermann 2014). This is clearly evidenced by a study (Rodriguez-Romero et al. 2005) where the influence of the AMF, *G. manihotis* and *Bacillus* spp., on growth and nutrient uptake of in vitro-raised banana plantlets (*Musa acuminata* Colla AAA cv. Grande Naine) was examined. Although the K content of *Bacillus* spp.-inoculated plantlets was lower than control, its co-inoculation with the AMF resulted in plant K content that was 47 % higher as compared to control plantlets.

Ghori et al. (2014) studied the influence of AMF and PGPR that included the KSB *F. aurantia* on growth of in vitro-propagated bamboo (*Bambusa tulda*). The results of this study indicated that inoculation of the microbial consortia (*F. mosseae*, *Azospirillum brasilense*, and *F. aurantia*) increased plant height by 157 % compared to control and 13–125 % than the individual inoculation of the microbes. Similarly, the number of branches and leaves produced was also 312–541 % higher compared to control and 268–119 % and 422–110 % compared to individual inoculations.

Inoculation of *B. coagulans* has been shown to increase the growth and K content of leaves in *B. malabarica* by 186 and 27 % than uninoculated control plants (Selvaraj et al. 2008). However, dual inoculation of *B. coagulans* along with *F. mosseae* increased plant growth by 256 and K content of leaves by 468 % (Selvaraj et al. 2008). In maize, inoculation of bacterial consortia comprising *Azospirillum chroococcum* (nitrogen fixing), *B. megaterium* (PSB), and *P. mucilaginous* (KSB) along with the AMF *F. mosseae* or *R. intraradices* significantly increased K uptake compared to uninoculated plants or those fertilized with urea, KH_2PO_4, and KCl (Wu et al. 2005). Available evidence does suggest that soil and environmental conditions have a greater impact on the functioning efficacy of PGPR and AMF. For example, Adesemoye et al. (2008) tested the hypothesis that microbial inoculants which enhance crop growth and productivity also remove more nutrients like K from the soil in a field study spanning over two growing seasons (2006 and 2007). The results of this study indicated the K uptake of the PGPR (*Bacillus* spp.)- and AMF (*R. intraradices*)-inoculated field corn (*Zea mays*) was more effective in season 2006 as compared to 2007 which was affected by drought.

Interactions among AMF and PGPR are not limited to increase the concentration of K in vegetative parts but also in reproductive parts of plants. For example, a study involving inoculation of PGPR including *P. putida* and AMF consortia consisting of *R. intraradices*, *F. mosseae*, and *C. etunicatum* on K content and antioxidant activity in tomato showed that inoculation of AMF along with *P. putida* increased the shoot and fruit K contents by 13.26 and 14.21 % compared to those inoculated with bacteria alone. In addition, bacterium and AMF dual-inoculated plants had higher lycopene content (78.03 vs. 58.55 mg/kg f.wt.) and antioxidant activity (46.45 vs. 42.95) as compared to those inoculated with bacteria alone (Ordookhani et al. 2010; Meena et al. 2014b).

The beneficial effect of interaction between *F. mosseae* and *P. fluorescens* on growth of common bean (*Phaseolus vulgaris*) under different levels of NaCl-induced salinity was examined by Younesi and Moradi (2014). Although the benefit of dual inoculation increased with increasing salinity levels when compared to control, there was a reduction in benefit when compared to *P. fluorescens*-inoculated plants (Younesi and Moradi 2014). In contrast, dual-inoculated plants had higher concentrations of K as compared to uninoculated plants or plants inoculated with the symbionts individually. Other studies have also indicated positive influence of interactions between AMF and PGPR on plant K content (Table 8.1).

8.9 Conclusions

It is evident from the available information that enhancing microbial-mediated K availability in agricultural soils is feasible. This can be achieved by two ways: (i) management of indigenous microorganisms to optimize their capacity to mobilize K and (ii) introduction of specific microbial inoculants. Nevertheless, to achieve maximum successes with the microbial-mediated processes, a thorough understanding of microbial interactions across various groups of organisms in soil is essential. In addition, the influence of other factors, especially the management practices in agroecosystems on the outcome of these interactions, is vital, and inorganic fertilizer application or cropping history has been shown to influence both formation and functioning of AMF symbiosis. The diversity and complicity in the interactions existing in soil microbial communities have been realized only with the advent of molecular tools and metagenomic approaches. Our current understanding of the interactions between KSM and AMF is mostly based on organisms that can be cultured. Unfortunately, this represents only a small percentage of the actual microbial diversity present in the soil as evidenced by the results of molecular-based studies. Further, most of the studies to understand the influence of tripartite interactions between KSM, AMF, and plants have been performed under controlled conditions involving a limited number of microbial and plant species. Future studies with more plant and microbial species under realistic field conditions are necessary to understand the better contribution of KSM and AMF interaction on the mobilization of soil K and its subsequent acquisition by plants. This could result in the development of more K-efficient and sustainable plant production systems as well as improve our understanding of microbial interactions

References

Abdel-Rahman SSA, Abdel-Kader AAS, Khalil SE (2011) Response of three sweet basil cultivars to inoculation with *Bacillus subtilis* and arbuscular mycorrhizal fungi under salt stress conditions. Nat Sci 9: 93–111

Abou-el-Seoud II, Abdel-Megeed A (2012) Impact of rock materials and biofertilizations on P and K availability for maize (*Zea maize*) under calcareous soil conditions. Saudi J Biol Sci 19:55–63

Adesemoye AO, Torbert HA, Kloepper JW (2008) Enhanced plant nutrient use efficiency with PGPR and AMF in an integrated nutrient management system. Can J Microbiol 54:876–886

Alam M, Khaliq A, Sattar A, Shukla RS, Anwar M, Dharni S (2011) Synergistic effect of arbuscular mycorrhizal fungi and *Bacillus subtilis* on the biomass and essential oil yield of rose-scented geranium (*Pelargonium graveolens*). Arch Agron Soil Sci 57: 889–898

Andrade G, Mihara KL, Linderman RG, Bethlenfalvay GJ (1997) Bacteria from rhizosphere and hyphosphere soils of different arbuscular-mycorrhizal fungi. Plant Soil 192:71–79

Artursson V, Finlay RD, Jansson JK (2005) Combined bromodeoxyuridine immunocapture and terminal restriction fragment length polymorphism analysis highlights differences in the active soil bacterial metagenome due to *Glomus mosseae* inoculation or plant species. Environ Microbiol 7:1952–1966

Artursson V, Finlay RD, Jansson JK (2006) Interactions between arbuscular mycorrhizal fungi and bacteria and their potential for stimulating plant growth. Environ Microbiol 8:1–10

Babu AG, Reddy AS (2011) *Aspergillus tubingensis* improves the growth and native mycorrhizal colonization of bermuda grass in Bauxite residue. Biorem J 15:157–164

Basak BB, Biswas DR (2010) Coinoculation of potassium solubilizing and nitrogen fixing bacteria on solubilization of waste mica and their effect on growth promotion and nutrient acquisition by a forage crop. Biol Fertil Soils 46:641–648

Baslam M, Garmendia I, Goicoechea N (2013) The arbuscular mycorrhizal symbiosis can overcome reductions in yield and nutritional quality in greenhouse-lettuces cultivated at inappropriate growing seasons. Sci Hortic 164:145–154

Bennett PC, Choi WJ, Rogera JR (1998) Microbial destruction of feldspars. Mineral Mag 8:149–150

Bertaux J, Schmid M, Chemidlin Prevost-Boure N, Chrin JL, Hartmann A, Garbaye J, Frey-Klatt P (2003) In situ identification of intracellular bacteria related to *Paenibacillus* ssp. In the mycelium of the ectomycorrhizal fungus *Laccaria bicolor* S238N. Appl Environ Microbiol 68:4243–4248

Bharadwaj DP, Lundquist P-O, Alströma S (2008) Arbuscular mycorrhizal fungal spore-associated bacteria affect mycorrhizal colonization, plant growth and potato pathogens. Soil Biol Biochem 40:2494–2501

Bhromsiri C, Bhromsiri A (2010) The effects of plant growth-promoting rhizobacteria and arbuscular mycorrhizal fungi on the growth, development and nutrient uptake of different vetiver ecotypes. Thai J Agric Sci 43:239–249

Bianciotto V, Minerdi D, Perotto S, Bonfante P (1996) Cellular interactions between arbuscular mycorrhizal fungi and rhizosphere bacteria. Protoplasma 193: 123–131

Bonfante P, Anca IA (2009) Plants, mycorrhizal fungi, and bacteria: a network of interactions. Annu Rev Microbiol 63:363–383

Budi SW, van Tuinen D, Martinotti G, Gianinazzi S (1999) Isolation from the *Sorghum bicolor* mycorrhizosphere of a bacterium compatible with arbuscular mycorrhiza development and antagonistic towards soil borne fungal pathogens. Appl Environ Microbiol 65: 5148–5150

Caravaca F, Alguacil MM, Azcónb R, Díaz G, Roldán A (2004) Comparing the effectiveness of mycorrhizal inoculation and amendment with sugar beet, rock phosphate and *Aspergillus niger* to enhance field performance of the leguminous shrub *Dorycnium pentaphyllum* L. Appl Soil Ecol 25:169–180

Casieri L, Lahmidi NA, Doidy J, Veneault-Fourrey C, Migeon A, Bonneau L, Courty PE, Garcia K, Charbonnier M, Delteil A, Brun A, Zimmermann S, Plassard C, Wipf D (2013) Biotrophic transportome in mutualistic plant fungal interactions. Mycorrhiza 23:597–625

Cerezine PC, Nahas E, Banzatto DA (1988) Phosphate accumulation by *Aspergillus niger* from fluorapatite. Appl Microbiol Biotechnol 29:501–505

Chamizo A, Ferrera-Cerrato R, González-Chávez MC, Ortiz-Solorio CA, Santizo-Rincón JA, Alarcón LVA (2009) Alfalfa inoculation with arbuscular mycorrhizal fungi and rhizobacteria in two soil types. Terra Latinoam 27:197–205

Chen CR, Condron LM, Davis MR, Sherlock RR (2002) Phosphorus dynamics in the rhizosphere of perennial ryegrass (*Lolium perenne* L.) and radiata pine (*Pinus radiata* D.Don). Soil Biol Biochem 34:487–499

Constantino M, Gómez-Álvarez R, Álvarez-Solís JD, Geissen V, Huerta E, Barba E (2008) Effect of inoculation with rhizobacteria and arbuscular mycorrhizal fungi on growth and yield of *Capsicum chinense* Jacquin. J Agric Rural Dev Trop Subtrop 109:169–180

Cruz AF, Ishii T (2012) Arbuscular mycorrhizal fungal spores host bacteria that affect nutrient biodynamics and biocontrol of soil borne pathogens. Biol Open 1: 52–57

Deubel A, Gransee A, Merbak W (2000) Transformation of organic rhizodeposits by rhizoplane bacteria and its influence on the availability of tertiary calcium phosphate. J Plant Nutr Soil Sci 163:387–392

Drew MC, Nye PH (1969) the supply of nutrient ions by diffusion to plant roots in soil. II. The effect of root hairs on the uptake of potassium by roots of ryegrass (*Lolium multiflorum*). Plant Soil 31:407–424

Filion M, St-Arnaud M, Fortin JA (1999) Direct interaction between the arbuscular mycorrhizal fungus *Glomus intraradices* and different rhizosphere microorganisms. New Phytol 141:525–533

Garcia K, Zimmermann SD (2014) The role of mycorrhizal associations in plant potassium nutrition. Front Plant Sci 5:1–9

Gaur AC (1990) Phosphate solubilizing microorganisms as biofertilizers. Omega Scientific Publishers, New Delhi

Ghori TK, Anusuya D, Geetha M (2014) Response of papaya to inoculation with *Frateuria aurentia* (potassium mobilizer) and plant growth promoting rhizomicroorganism (PGPR). Indian J Appl Res 4:86–87

Govindasamy V, Senthilkumar M, Magheshwaran V, Kumar U, Bose P, Sharma V, Annapurna K (2010) *Bacillus* and *Paenibacillus* spp.: potential PGPR for sustainable agriculture. In: Maheshwari DK (ed) Plant growth and health promoting bacteria, microbiology monographs, vol 18. Springer, Heidelberg, pp 333–364

Gupta SS (2003) Chemotactic response of plant-growth-promoting bacteria towards roots of vesicular-arbuscular mycorrhizal tomato plants. FEMS Microbiol Ecol 45:219–227

Han HS, Supanjani, Lee KD (2006) Effect of co-inoculation with phosphate and potassium solubilizing bacteria on mineral uptake and growth of pepper and cucumber. Plant Soil Environ 52:130–136

Hemashenpagam N, Selvaraj T (2011) Effect of arbuscular mycorrhizal (AM) fungus and plant growth promoting rhizomicroorganisms (PGPRs) on medicinal plant *Solanum viarum* seedlings. J Environ Biol 32: 579–583

Hildebrandt U, Janetta K, Bothe H (2002) Towards growth of arbuscular mycorrhizal fungi independent of a plant host. Appl Environ Microbiol 68:1919–1924

Hildebrandt U, Ouziad F, Marner F-J, Bothe H (2006) The bacterium *Paenibacillus validus* stimulates growth of the mycorrhizal fungus *Glomus intraradices* up to the formation of fertile spores. FEMS Microbiol Lett 254: 258–267

Hosseinzadah F, Satei A, Ramezanpou MR (2011) Effects of mycorrhiza and plant growth promoting rhizobacteria on growth, nutrients uptake and physiological characteristics in *Calendula officinalis* L. Middle East J Sci Res 8:947–953

Hu XF, Chen J, Guo JF (2006) Two phosphate and potassium solubilizing bacteria isolated from Tianmu mountain, Zhejiang, China. World J Microbiol Biotechnol 22:983–990

Jaizme-Vega MDC, Rodríguez-Romer AS, Núñez LAB (2006) Effect of the combined inoculation of arbuscular mycorrhizal fungi and plant growth-promoting rhizobacteria on papaya (*Carica papaya* L.) infected with the root-knot nematode *Meloidogyne incognita*. Fruits 61:151–162

Johansson SAE, Campbell JL (1988) PIXE, a novel technique for elemental analysis. Wiley, Chichester

Jones DL, Nguyen C, Finlay RD (2009) Carbon flow in the rhizosphere: carbon trading at the soil-root interface. Plant Soil 321:5–33

Kumar A, Bahadur I, Maurya BR, Raghuwanshi R, Meena VS, Singh DK, Dixit J (2015) Does a plant growth-promoting rhizobacteria enhance agricultural sustainability? J Pure Appl Microbiol 9(1): 715–724

Lambers H, Chapin FS III, Pons TL (2008) Plant physiological ecology, 2nd edn. Springer, New York

Lambers H, Mougel C, Jaillard B, Hinsinger P (2009) Plant-microbe-soil interactions in the rhizosphere: an evolutionary perspective. Plant Soil 321:83–115

Lee PJ, Koske RE (1994) *Gigaspora gigantea*: parasitism of spores by fungi and actinomycetes. Mycol Res 98: 458–466

Leigh RA, Wyn-Jones RG (1984) A hypothesis relating critical potassium concentrations for growth to the distribution and functions of this ion in the plant cell. New Phytol 97:1–13

Li B, Ravnskov S, Xie GL, Larsen J (2008) Differential effects of *Paenibacillus* spp. on cucumber mycorrhizas. Mycol Prog 7:277–284

Lian B, Wang B, Pan M, Liu C, Teng HH (2007) Microbial release of potassium from K-bearing minerals by thermophilic fungus *Aspergillus fumigatus*. Geochim Cosmochim Acta 72:87–98

Lopes-Assad ML, Avansini SM, Rosa MM, de Carvalho JRP, Ceccato-Antonini SR (2010) The solubilization of potassium-bearing rock powder by *Aspergillus niger* in small-scale batch fermentations. Can J Microbiol 56:598–605

Mansfeld-Giese K, Larsen J, Bødker L (2002) Bacterial populations associated with mycelium of the arbuscular mycorrhizal fungus *Glomus intraradices*. FEMS Microbiol Ecol 41:133–140

Marschner H (1995) Mineral nutrition of higher plants, 2nd edn. Academic, London

Maurya BR, Meena VS, Meena OP (2014) Influence of inceptisol and alfisol's potassium solubilizing bacteria (KSB) isolates on release of K from waste mica. Vegetos 27(1):181–187

Mayo K, Davis RE, Motta J (1986) Stimulation of germination of spores of *Glomus versiforme* by spore-associated bacteria. Mycologia 78:426–431

McAllister CB, Garcia-Romera I, Martin J, Godeas A, Ocampo JA (1995) Interaction between *Aspergillus niger* van Tiegh. and *Glomus mosseae*. (Nicol. & Gerd.) Gerd. & Trappe. New Phytol 129:309–316

Meena OP, Maurya BR, Meena VS (2013) Influence of K-solubilizing bacteria on release of potassium from waste mica. Agric Sustain Dev 1(1):53–56

Meena VS, Maurya BR, Bahadur I (2014a) Potassium solubilization by bacterial strain in waste mica. Bangladesh J Bot 43(2):235–237

Meena VS, Maurya BR, Verma JP (2014b) Does a rhizospheric microorganism enhance K+ availability in agricultural soils? Microbiol Res 169:337–347

Meena RK, Singh RK, Singh NP, Meena SK, Meena VS (2015a) Isolation of low temperature surviving plant growth-promoting rhizobacteria (PGPR) from pea (*Pisum sativum* L.) and documentation of their plant growth promoting traits. Biocatal Agric Biotechnol. doi:10.1016/j.bcab.2015.08.006

Meena VS, Maurya BR, Verma JP, Aeron A, Kumar A, Kim K, Bajpai VK (2015b) Potassium solubilizing rhizobacteria (KSR): isolation, identification, and K-release dynamics from waste mica. Ecol Eng 81: 340–347

Meyer JR, Linderman RG (1986) Selective influence on populations of rhizosphere or rhizoplane bacteria and actinomycetes by mycorrhizas formed by *Glomus fasciculatum*. Soil Biol Biochem 18:191–196

Miransari M (2011) Interactions between arbuscular mycorrhizal fungi and soil bacteria. Appl Microbiol Biotechnol 89:917–930

Muthukumar T, Udaiyan K (2006) Growth of nursery grown bamboo to arbuscular mycorrhizal fungi and plant growth promoting rhizobacteria in two soil types with and without fertilizer application. New For 81: 469–485

Muthukumar T, Udaiyan K (2010) Growth response of *Casuarina equisetifolia* to bioinoculants under tropical nursery conditions. New For 40:101–118

Nannipieri P, Ascher J, Ceccherini MT, Landi L, Pietramellara G, Renella G, Valori F (2007) Microbial diversity and microbial activity in the rhizosphere. Ci Suelo (Argentina) 25:89–97

Ndoye F, Kane A, Bakhoum N, Sanon A, Fall D, Diouf D, Sy MO, Noba K (2012) Response of *Acacia senegal* (L.) Willd. seedlings and soil bio-functioning to inoculation with arbuscular mycorrhizal fungi, rhizobia and *Pseudomonas fluorescens*. Afr J Microbiol Res 6:7176–7184

Olsson PA, Hammer EC, Wallander H, Pallon J (2008) Phosphorus availability influences elemental uptake in the mycorrhizal fungus *Glomus intraradices*, as revealed by particle-induced X-ray emission analysis. Appl Environ Microbiol 74:4144–4148

Olsson PA, Hammer EC, Pallon J, van Aarle IM, Wallander H (2011) Elemental composition in vesicles of an arbuscular mycorrhizal fungus, as revealed by PIXE analysis. Fungal Biol 115:643–648

Ordookhani K, Khavazi K, Moezzi A, Rejali F (2010) Influence of PGPR and AMF on antioxidant activity, lycopene and potassium contents in tomato. Afr J Agric Res 5:1108–1116

Pallon J, Wallander H, Hammer E, Arteaga Marrero N, Auzelyte V, Elfman M, Kristiansson P, Nilsson C, Olsson PA, Wegdén M (2007) Symbiotic fungi that are essential for plant nutrient uptake investigated with NMP. Nucl Instrum Methods Phys Res B 260: 149–152

Paulitz TC, Linderman RG (1989) Interactions between fluorescent pseudomonads and VA mycorrhizal fungi. New Phytol 113:37–45

Perner H, Schwarz D, Bruns C, Mader P, George E (2007) Effect of arbuscular mycorrhizal colonization and two levels of compost supply on nutrient uptake and flowering of pelargonium plants. Mycorrhiza 17: 469–474

Pulido LE, Cabero A, Medina N (2003) Biofertilization using rhizobacteria and AMF in the production of tomato (*Lycopersicon esculentum* Mill.) and onion (*Allium cepa* L.) seedlings. II. Root colonization and nutritional status. Cultivos Tropicales 24:5–13

Rodríguez-Romero AS, Guerra MSP, Jaizme-Vega MDC (2005) Effect of arbuscular mycorrhizal fungi and rhizobacteria on banana growth and nutrition. Agron Sustain Dev 25:395–399

Roesti D, Gaur R, Johri BN, Imfeld G, Sharma S, Kawaljeet K, Aragno M (2006) Plant growth stage, fertiliser management and bio-inoculation of arbuscular mycorrhizal fungi and plant growth promoting rhizobacteria affect the rhizobacterial community structure in rain-fed wheat fields. Soil Biol Biochem 38:1111–1120

Schreiner RP, Mihara KL, McDaniel H, Bethlenfalvay GJ (1997) Mycorrhizal fungi influence plant and soil functions and interactions. Plant Soil 188:199–210

Schüßler A, Scharzott D, Walker C (2001) A new fungal phylum, the glomeromycota: phylogeny and evolution. Mycol Res 105:1413–1421

Secilia J, Bagyaraj DJ (1987) Bacteria and actinomycetes associated with pot cultures of vesicular-arbuscular mycorrhizas. Can J Microbiol 33:1069–1073

Selvaraj T, Sumithra P (2011) Effect of *Glomus aggregatum* and plant growth promoting rhizomicroorganisms on growth, nutrition and content of secondary metabolites in *Glycyrrhiza glabra* L. Indian J Appl Pure Biol 26:283–290

Selvaraj T, Rajeskumar S, Nisha MC, Wondimo L, Tesso M (2008) Effect of *Glomus mosseae* and plant growth promoting rhizo-microorganisms (PGPRs) on growth, nutrients and content of secondary metabolites in *Begonia malabarica* Lam. Maejo Int J Sci Technol 2:516–525

Shirmardi M, Savaghebi GR, Khavazi K, Akbarzadeh A, Farahbakhsh M, Rejali F, Sadat A (2010) Effect of microbial inoculants on uptake of nutrient elements in two cultivars of sunflower (*Helianthus annuus* L.) in saline soils. Not Sci Biol 2:57–66

Silva Filho GN, Narloch C, Scharf R (2002) Solubilizac¸a˜o de fosfatos naturais por micr-organismos isolados de cultivos de *Pinus* e *Eucalyptus* de Santa Catarina. Pesq Agrop Bras 37:847–854 [In Portuguese]

Singh NP, Singh RK, Meena VS, Meena RK (2015) Can we use maize (*Zea mays*) rhizobacteria as plant growth promoter? Vegetos 28(1):86–99

Smith SE, Read DJ (2008) Mycorrhizal symbiosis. Academic, New York, p 787

Sperber JI (1958) The incidence of apatite-solubilizing organisms in the rhizosphere and soil. Aust J Agric Res 9:778–781

Srimathi Priya L, Kumutha K (2009) Effect of arbuscular mycorrhizal inoculum on enzyme activities and microbial population in the rhizosphere *of Colous forskohlii* Briq. Mycorrhiza News 20:14–21

Sumithra P, Selvaraj T (2011) Influence of *Glomus walkeri* Blaszk and Renker and plant growth promoting rhizomicroorganisms on growth, nutrition and content of secondary metabolites in *Sphaeranthes amaranthoides* (L.) Burm. J Agric Technol 7: 1685–1692

Toljander JF, Lindahl BD, Paul LR, Elfstrand M, Finlay RD (2007) Influence of arbuscular mycorrhizal mycelial exudates on soil bacterial growth and community structure. FEMS Microbiol Ecol 61:295–304

Treseder KK (2013) The extent of mycorrhizal colonization of roots and its influence on the plant growth and phosphorus content. Plant Soil 371:1–13

Velázquez MS, Elíades LA, Irrazabal GB, Saparrat CM, Cabello MN (2005) Mycobization with *Glomus mosseae* and *Aspergillus niger* in *Lycopersicon esculentum* plants. J Agric Technol 1:315–326

Veresoglou SD, Mamolos AP, Thornton B, Voulgari OK, Sen R, Veresoglou S (2011) Medium-term fertilization of grassland plant communities masks plant species-linked effects on soil microbial community structure. Plant Soil 344:187–196

Vogeti S, Brunda devi K, Tilak KVBR, Bhadraiah B (2011) Dual inoculation of *Glomus fasciculatum* (Taxter) Migula on nutrient levels in sweet potato (*Ipomoea batatas* L.) Lam. Proc Nat Acad Sci India Sect B 81:428–432

Walley FL, Germida JJ (1996) Failure to decontaminate *Glomus clarum* NT4 spores is due to spore wall-associated bacteria. Mycorrhiza 6:43–49

Walley FL, Germida JJ (1997) Response of spring wheat (*Triticum aestivum*) to interactions between *Pseudomonas* species and *Glomus clarum* NT4. Biol Fertil Soils 24:365–371

Wamberg C, Christensen S, Jakobsen I, Müller AK, Sørensen SJ (2003) The mycorrhizal fungus (*Glomus intraradices*) affects microbial activity in the rhizosphere of pea plants (*Pisum sativum*). Soil Biol Biochem 35:1349–1357

Wu SC, Cao ZH, Li ZG, Cheung KC, Wong MH (2005) Effects of biofertilizer containing N-fixer, P and K solubilizers and AM fungi on maize growth: a greenhouse trial. Geoderma 125:155–166

Xavier LJC, Germida JJ (2003) Bacteria associated with *Glomus clarum* spores influence mycorrhizal activity. Soil Biol Biochem 35:471–478

Younesi O, Moradi A (2014) Effects of plant growth-promoting rhizobacterium (PGPR) and arbuscular mycorrhizal fungus (AMF) on antioxidant enzyme activities in salt-stressed bean (*Phaseolus vulgaris* L.). Agriculture (Poľnohospodárstvo) 60:10–21

Zorb C, Senbayram M, Peiter E (2014) Potassium in agriculture – status and perspectives. J Plant Physiol 171:656–669

Can Potassium-Solubilising Bacteria Mitigate the Potassium Problems in India? 9

Madhumonti Saha, Bihari Ram Maurya, Indra Bahadur, Ashok Kumar, and Vijay Singh Meena

Abstract

The nutrients are one of the basic inputs for achieving increased crop productivity, and potassium is one of them. In a recently published nutrient budget, it has been shown that potassium became unbalanced through increased removal and increased system leakage, simplification (monoculture) and increased demand for rapid plant growth. Addressing this nutrient imbalance and shortages in developing countries, it would be required to enhance the world production of potassium fertiliser. India is not self-sufficient in producing potassium fertiliser. The entire quantities of potassium fertilisers are imported. Therefore efficient management in soil fertility is a prime concern to the scientist in view of the increasing gap between production and consumption of fertiliser with time. So, nowadays, crop nutrient stability is important to mitigate the exhaustion of soil nutrient stock for sustaining the crop yield and to reinstate the soil fertility. In these circumstances, it is suitable to study the present knowledge of substitutive sources of K. So this assessment imparts an opportunity to flourish native mineral sources of K as a substitute to impede costly remunerative fertilisers. So, bio-intrusion of potassium minerals with potassium-solubilising microorganism (KSM) would be a preferable and

M. Saha (✉)
Department of Soil Science and Agricultural Chemistry, Institute of Agricultural Sciences, Banaras Hindu University, Varanasi 221005, Uttar Pradesh, India

Department of Agricultural Chemistry and Soil Science, Faculty of Agriculture, Bidhan Chandra Krishi Viswavidyalaya, Mohanpur, West Bengal 741 252, India
e-mail: madhumonti2609@gmail.com

B.R. Maurya • I. Bahadur
Department of Soil Science and Agricultural Chemistry, Institute of Agricultural Sciences, Banaras Hindu University, Varanasi 221005, Uttar Pradesh, India

A. Kumar
Department of Botany, MMV, Banaras Hindu University, Varanasi 221005, Uttar Pradesh, India

V.S. Meena
Department of Soil Science and Agricultural Chemistry, Institute of Agricultural Sciences, Banaras Hindu University, Varanasi 221005, Uttar Pradesh, India

ICAR-Vivekananda Institute of Hill Agriculture, Almora 263 601, Uttarakhand, India

V.S. Meena et al. (eds.), *Potassium Solubilizing Microorganisms for Sustainable Agriculture*, DOI 10.1007/978-81-322-2776-2_9

feasible technology to solubilise unavailable form of K minerals into available pool for plant uptake and efficiently used as a source of K fertiliser for maintaining the ecological balance and sustaining agricultural production and environmental quality.

Keywords
Potassium-solubilising bacteria • K minerals • Mitigation • Nutrient budget • Sustainability

9.1 Introduction

Agriculture is the root of Indian economy, where the land-person ratio is rapidly narrowing; the only means of meeting needs of agricultural product is through increasing productivity. Consequently the productive resources of this country are adversely affected. Immediate attention is needed for management of soil fertility in both terms of quality and quantity. Potassium is known as the quality element since it has many roles in improving quality of horticultural crops, as well as others. In fruit and vegetable crops, it influences colour, shape, size, taste, shelf life and processing characteristics. It increases vitamin C content and improves utilisation of N and thereby influences protein formation in plants (Srinivasarao et al. 2007). Among the essential plant nutrients, potassium is required in comparatively enormous amount by plants to raise the yield as it has a greater consequence. These fertilisers assume a great significance and constitute one of the key inputs for achieving high productivity. The cracks between removal and application of K fertilisers to crops are widening. Therefore it is comparative to understand the dynamics of potassium in soil matching with potassium needed crops to provide balance nutrition and maintain potassium status in Indian soils (Khudsen et al. 1982; Krieg and Holt 1984; Herrera et al. 1993; Goldstein et al. 1994; Glick 1995; Liu 2001; Hutchens et al. 2003; Meena 2013). In this context, to introduce an alternative indigenous K source is indispensible. The alternative is exploiting other natural resources such as K-bearing minerals which include K feldspar, leucite, K mica (e.g. biotite, phlogopite and glauconite) and clay minerals such as illite. India is favoured to get hold of the world's largest bed of micas extended over about 3888 km^2 in Munger district of Bihar and Koderma and Giridih districts of J0068arkhand (Maurya et al. 2015). Bio-formulations are defined as living active artefacts which contain one or more functional microbial strains which are convenient and cost-effective material. So, most bio-formulations are meant for field application; it is essential that suitable carrier materials are used to maintain cell viability under adverse environmental conditions (Fig. 9.1).

9.1.1 Potassium Mining from Indian Soils

There is an increasing concern about the sustainability of Indian agriculture because of deterioration in soil fertility. According to Ghosh and Hasan (1980), soil test results for potassium (K) fertility status among Indian agricultural soils are categorised accordingly as ~21% low, 51% medium and 28 % high. The potassium content of Indian soils has commonly been regarded as enough even if K deficiency is potent in Indian soils owing to the accelerated advancement of agriculture. Since the use of potassium has covered a lot of plant activities, depletion potassium uptake can cause problem for plant growth. There are several factors that

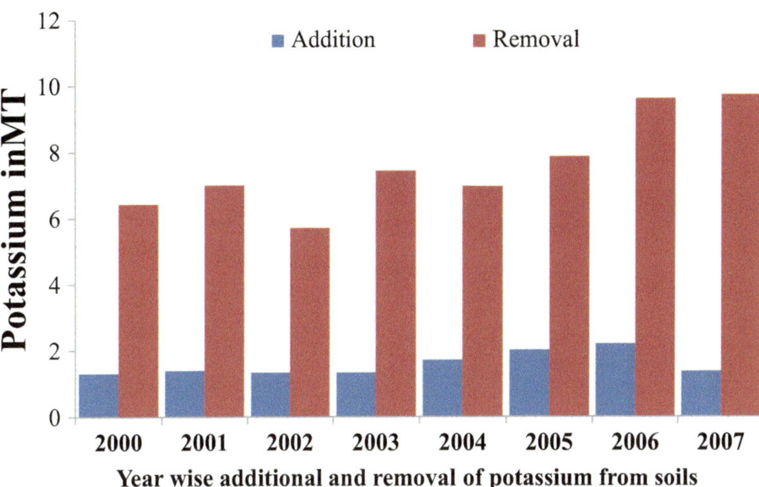

Fig. 9.1 Year-wise addition and removal of K from Indian soils (Adapted from Pathak et al. 2009)

Table 9.1 Critical limit of available K (NH_4OAC—extractable) in soils for different crops based on data from different regions of India

Crop	Soil and State	Critical limit (mg K kg^{-1} Soil)
Rice	Alluvial, Uttar Pradesh	117–120
Wheat	Alluvial, Uttar Pradesh	95–100
Maize	Calcareous, Bihar	81
Pearl millet	Alluvial, Andhra Pradesh	160
Potato	Submontane, Himachal Pradesh	120
Chickpea	Alluvial Karha, Uttar Pradesh	105–137

Adapted from Rao and Srivastava 2012

lead to this problem, for instance, low soil potassium-supplying capacity, insufficient application of mineral potassium fertiliser and biofertiliser and complete removal of plant straw. There are various approaches that help in the availability of potassium in the soil (Maurya et al. 2014; Meena et al. 2014a; Singh et al. 2015) (Table 9.1).

However, the concentration of potassium is disturbed by inadequate fertiliser use, cropping sequence, soil weathering and system removal (Sheng and Huang 2002). Thus, the amount present is inadequate to satisfy crop requirement. So, the uptake of K to the plants directly depends on the soil solution K and indirectly on K present in soil solid, which maintains the buffering capacity. Potassium status of soils varies with the clay mineralogy, CEC, soil texture, soil moisture, soil temperature and liming practices. In general, exchangeable potassium is more in black cotton and vertic type of soil than in alluvial, red and lateritic soil. Illite-dominated alluvial soils contain largest amount of nonexchangeable and total potassium. The kaolinite-dominant red and lateritic soils contain lowest amount of all forms of potassium (Rao et al. 2011).

9.1.2 Potassium in Soil (Forms of K)

Potassium is available in four forms in the soil—(1) K in primary mineral crystal structure, (2) K in nonexchangeable form in secondary minerals, (3) K in exchangeable form on soil colloidal surfaces and (4) K^+ ion present in soil solution. Potassium constitutes about 2.5 % of the earth's crust and actual concentrations in the soil of this nutrient vary widely ranging from 0.04 % to 3.0 % (Mengel and Kirkby 1987) (Fig. 9.2).

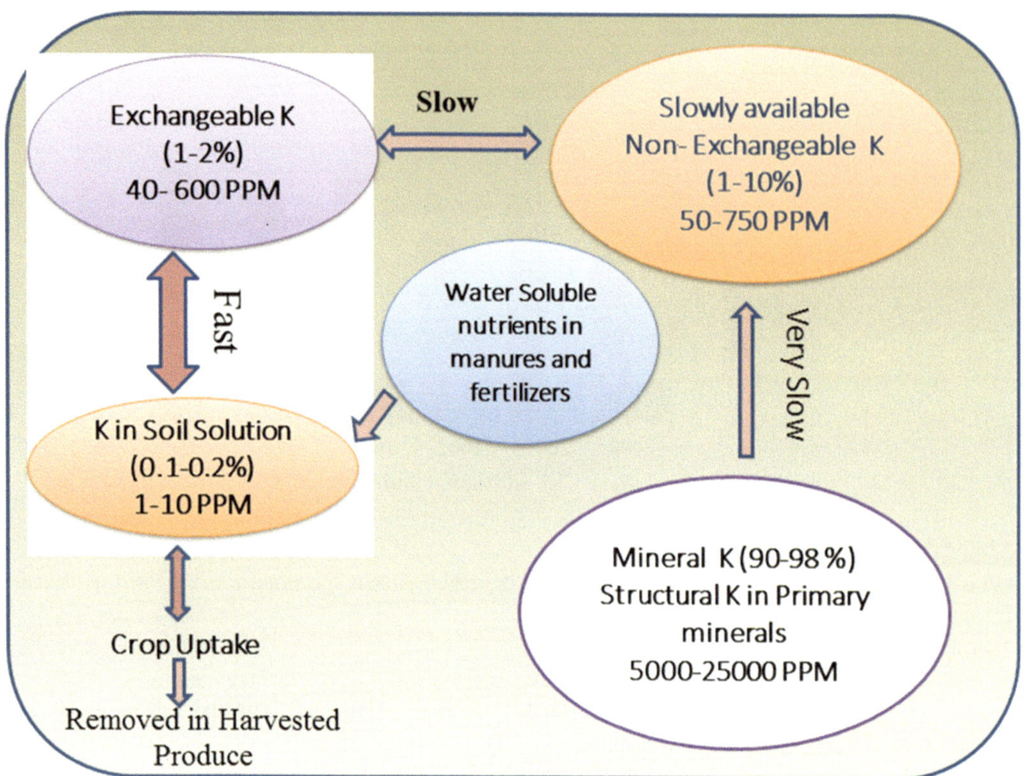

Fig. 9.2 The potassium dynamics in soils and its different forms in soil system

Table 9.2 Potassium requirement of different crops

Crop	Total K uptake (kg ha^{-1})
Wheat	137
Rice	180
Pigeon pea	16
Mustard	133
Sugarcane	270
Tobacco	65
Alfalfa	669
Potato	119
Banana	1053
Pineapple	440

Out of the four forms of potassium found in the soil, some 90–98 % of all soil K in a mineral is in relatively unavailable form, mostly in crystal structure of feldspar and micas. Only 1–2 % of the total soil K is readily available and exists in two forms: (1) solution K and (2) exchangeable K—which are in dynamic equilibrium. The mineral and nonexchangeable K can be seen as an important buffer mechanism for soil solution K. Depletion of K in soil solution and exchangeable form due to plant uptake and removal, luxury consumption, erosion and leaching losses causes the release of nonexchangeable K to the exchangeable form.

Potassium is the second most abundant plant nutrient and also a nonrenewable resource. Potassium has many functions in plant growth such as the smooth progress of cell division and growth, increase of disease resistance and drought tolerance and regulation of the opening and closing of the stomata required for osmotic regulation (Table 9.2). Besides, potassium is essential for photosynthesis process and acts as key to activate enzymes to metabolise carbohydrates for the manufacture of amino acids and proteins. Furthermore, potassium assimilates transport during plant ontogeny, and one of the most important influences is the improvement in oil content in plants (Meena et al. 2014b; Kumar et al. 2015).

When potassium in the rock minerals is solubilised by microorganisms by the production and release of organic acids, these are called plant growth-promoting rhizobacteria (Han and Lee 2006). Potassium-solubilising plant growth-promoting rhizobacteria such as *Acidithiobacillus ferrooxidans*, *B. edaphicus*, *B. mucilaginosus*, *Burkholderia*, *Paenibacillus* spp. and *Pseudomonas* has been reported to release potassium in accessible form from potassium-bearing minerals in soils (Liu et al. 2012; Meena et al. 2015a). Thus, to shorten the practice of agrochemicals, potassium-solubilising plant growth-stimulating rhizobacteria is used as biofertiliser for agricultural enhancement which is a platform for eco-friendly crop production. Moreover, due to imbalanced fertiliser application, inadequacy of potassium status in soil is one of the crucial restrictions in crop production. The aim is to explore an alternative indigenous source of potassium for plant availability and to sustain potassium level in soils for viable crop production (Kumar and Dubey 2012).

9.2 Potassium-Solubilising Bacteria

The use of potassium materials directly may be more advantageous agronomically and more convenient environmentally than soluble forms of potassium. Besides, natural sources of K materials are more economical; however, most of them are not immediately accessible to a plant because of their releasing diagram of K in soil which is very slow, and their treatment as fertiliser usually accounts for insignificant yield increase of current crop. Thus, some microbial strains are identified which are efficient of solubilising potassium minerals rapidly and can sustain our prevailing reservoirs and shrink environmental threat caused by excessive use of chemical fertilisers.

Microorganisms are generally known as potassium-solubilising or potassium-dissolving bacteria which solubilise K-bearing minerals to available K for plants. Potassium-solubilising microorganisms (KSM), as biofertilisers, have become of interest in many agricultural countries as significant area of cultivated soils which are insufficient in soil available potassium (Xie 1998). The first sign of microbial engrossment in solubilisation of rock potassium was described by Muentz (1890). Bio-formulation studies of plant growth-promoting rhizobacteria that may be limited to the entire replacement of chemical fertilisation, pesticides, provide a sustainable background and access to raise crop production and plant health status. Several microorganisms such as *Bacillus extorquens*, *Aspergillus niger* and *Clostridium pasteurianum* were found to be able to multiply on orthoclase, microcline, muscovite, micas and biotite under laboratory conditions (Reitmeir 1951; Meena et al. 2015a).

9.2.1 Importance of Potassium Solubilisation

Potassium-solubilising bacteria (KSB) were derived to diffuse potassium, silicon and aluminium from insoluble K-bearing aluminosilicate minerals such as orthoclase, micas, illite and biotite by secreting different organic acids which present a potent mechanism for using these indigenous resources, in a pragmatic and prudent way. Management of natural resources would lessen the dependence on inorganic fertilisers which is environmentally safe and economically stable. The distributions of silicate-dissolving microbes in seawater were studied by Purushothaman et al. (1974). They suggested that these microbes play an important role in turning of excreting metabolic outgrowth that combine with the mineral surface in marine environment by dissolving silicon from silicate clays.

Barker et al. (1998) found that intact microbial respiration, deterioration of particulate organic carbon and their dissolution organic carbon raise carbonic acid concentration at surfaces of minerals in soils which can enhance mineral weathering rates by endorsement of proton dissolution mechanism. Consequently, fixed form of potassium by - mineral-dissolving bacteria was scanned in pure culture demonstration to diagnose the characterisation of these bacteria for releasing K and SiO_2.

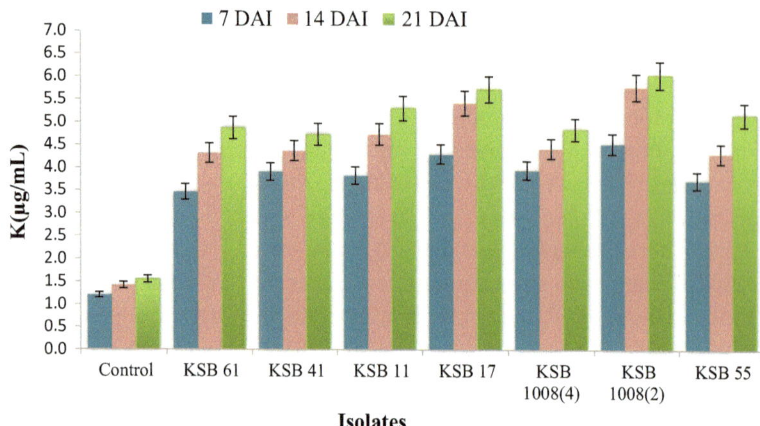

Fig. 9.3 Effect of KSB on K solubilisation in Aleksandrov broth media under different incubation periods

Various researches on mineral weathering reported that *Bacillus* species has the potentiality to solubilise K in the culture medium containing K minerals (Vandevivere et al. 1994). They introduced that *B. mucilaginosus* optimises the releases of K and SiO_2 from the crystal structure of primary minerals by secreting organic acids. In general, potassium-solubilising bacteria (KSB) play an important role in reducing nutrient deficiency in soil (Fig. 9.3).

9.2.2 Mechanisms of K Solubilisation

Hutchens reported that certain bacteria produces mucilage like exopolysaccharides which form an envelope around the bacterial cell and attack the silicate minerals chelating with silicon and consequently facilitates K from those structure. The switchover of crystalline biotite, mica, vermiculite and certain rocks to amorphous form occurred due to the activity of some organic syntheses produced by microbial metabolism (Weed et al. 1969). It was found that various organic ligands include exudates, extracellular enzymes, metabolic by-products and chelates, and both light and compound organic acids heighten the dissolution of aluminosilicate mineral both in field and laboratory experiments by Grandstaff (1986), Welch and Ullman (1993) and Argelis et al. (1993) who studied the in situ experiment on weathering of sand stone, granite and limestone and isolated three fungal species, i.e. *Cladosporium cladosporioides* and *Penicillium frequentans*, which have the ability to produce enormous amounts of organic acids like oxalic, citric and gluconic acids during their metabolism in broth media that regulate higher degradation of silicate clays, mica and feldspar from sandstone, granite and also limestone by decreasing the pH of the environment. According to Chen et al. (2007), some strains of *B. megaterium* and *Arthrobacter* spp. were able to produce organic acids and mono-hydroxamate siderophore. These organic acids and siderophores could perform a vital part in the solubilisation of elements such as K, Si and Fe from the liquid medium containing acid-leached soil, muscovite and biotite (Hutchens et al. 2003). There was a mutual relationship between soil microflora and minerals possessed in soil environment, and these interactions have been extensively studied to produce technologies such as biomineralization, bioremediation and biohydrometallurgy (Rawlings 2002; Reitmeir 1951; Ullaman et al. 1996; Requena et al. 1997; Srinivasarao and Takkar 1997; Venkateswarlu et al. 2012; Meena 2015b).

Jones et al. (2003) studied that organic acids have been considered to execute many roles in soil including root nutrient recovery, breaking down of mineral, microbial chemotaxis and biodegradation of metal. Further the production of capsular polysaccharides helps in weathering illite and feldspar minerals to release K (Sheng et al. 2006). Other potential hypotheses/mechanisms are to mobilise soil K resources by the production of biofilms on the mineral surface rhizospheric soil by some conclusive bacterial strains (Balogh-Brunstad et al. 2008). Essential nutrients have to be supplemented through various sources of fertiliser for higher productivity and maintenance of soil health. It was found that ectomycorrhizal hearting networks and arbuscular structure of non-ectomycorrhizal trees, implanted in biofilms, conducted nutrients to the host. These results indicated that biofilms assist to stimulate the weathering of minerals and thereby increase uptake of nutrients to the plant. Some rock-decaying fungi (ectomycorrhizae) have the ability to ooze out organic anions which have low molecular weight to a limit that forms microscopic tunnels in the vicinity of exudates at hyphal tips within minerals present in soils by that weathering rates are consequently increased (Van Scholl et al. 2008).

9.3 Plant Growth-Promoting Activates of KSB

Specifically, plant growth-promoting bacteria (PGPR) have been described to be a major element for plant growth initiation under negative balance of nutrient in soil. Efficient plant growth-promoting microbes (bacteria) must be able to explore the root system throughout the growing season (Defreitas and Germida 1992). PGPR can correct plant growth, advances plant nutrition, promotes root growth pattern, reforms plant competitiveness and calibrates to any stresses in plants by external factors. A diverse number of PGPR in conjunction with associative bacteria such as *Bacillus*, *Pseudomonas*, *Azospirillum* and *Enterobacter* have been adopted for their auspicious responses on plant growth (Kloepper et al. 1991; Ho flich et al. 1994). Certain investigation clearly exhibited the effect of PGPR on the growth of the same crop under various conditions (Boelens et al. 1993; Javed and Arshad 1997). Apart from the K-solubilising abilities, the KSMs have the ability to form plant growth hormones, ammonia and siderophore, and to solubilise phosphorus sufficient for explanation, etc.

9.4 Responses of Potassium-Solubilising Bacteria on Plant Growth and Yield

Confined KSB isolated from rock and mineral samples indicated higher activity to potassium dynamics in acid-leached soil and improved the growth of green gram seedling. In a field experiment, it was recorded that *Bacillus mucilaginosus* is able to mobilise potassium from K-containing minerals, and it has a beneficial effect on nutrient uptake and growth of cucumber and pepper on inceptisol of Korea (Han et al. 2006). Mikhailouskaya and Tcherhysh (2005) studied the responses of K-mobilising bacteria successively growing on eroded soil which are analogous with yields of wheat

Table 9.3 Evaluation of KSB on growth parameters of tea plants

Treatment details	Green leaf yield (kg made tea ha^{-1})	Productivity index (PI)	Banji content (%)
N_{100} P_{100} K_{100} + KSB	2624.3 ± 65.2e	1.2 ± 0.1e	40.0 ± 1.0b
N_{100} P_{100} K_{75} + KSB	2957.0 ± 75.7g	1.6 ± 0.1g	36.0 ± 2.0a
N_{100} P_{100} K_{50} + KSB	2524.7 ± 52.8a	1.3 ± 0.1f	41.7 ± 0.6b
N_{100} P_{100} K_{25} + KSB	2358.0 ± 43.1b	0.9 ± 0.1b	48.3 ± 1.5c
N_{100} P_{100} K_{100} + KSB + VC	2562.7 ± 38.9d	1.2 ± 0.2c	50.0 ± 1.0d
N_{50} P_{50} K_{50}+ KSB+VC	2451.3 ± 41.1c	1.2 ± 0.1c	55.3 ± 4.1e
N_{100} P_{100} K_{100} alone	2739.3 ± 59.7f	1.5 ± 0.1d	53.7 ± 5.5f
Control (N_0 P_0 K_0)	1904.3 ± 83.1a	0.7 ± 0.2a	63.0 ± 2.0g
SEm±	41.39	0.05	2.22
CD at $P = 0.05$	119.14	0.16	6.41

Values are mean of three replications with standard deviations. The values followed by the same letter are not significantly different ($P < 0.05$) as measured by Duncan's multiple range test, *VC* vermicompost, *KSB* potassium-solubilising bacteria. A higher alphabet indicates the improvements obtained due to treatment

Adapted from Bagyalakshmi et al. (2012)

(1.04 t ha^{-1}) on slightly eroded soil without bacterial inoculation (Table 9.3).

It has been shown that the production of brinjal, plant growth and availability of K significantly increased compared to control as a result of inoculation of potash-solubilising bacteria (*Frateuria aurantia*). *F. aurantia* belonging to the family *Pseudomonadaceae* obtained from the agricultural soils of Coimbatore region of Tamil Nadu, India, was solubilised with K considerably, and this promoted crop yield (Ramarethinam and Chandra 2005).

Han et al. (2006) found that the dry weight of maize plants inoculated with bacteria (potassium dissolving) is increased by ~23 %, compared to the noninoculated treatment. Maize and wheat plants inoculated with *Bacillus mucilaginosus*, *Azotobacter chroococcum* and *Rhizobium* resulted in significantly higher mobilisation of potassium from waste mica, which sequentially serves as an essence for the supply of potassium for plant growth (Singh et al. 2010). Basak and Biswas (2009) found that K uptake is enhanced significantly by increasing the amount of mica application up to 100 mg K kg^{-1} soil with suitable potassium-solubilising bacteria, which was 36.0 % and 13.1 % higher over control and application of mica at 50 K kg^{-1} soil, respectively, in Sudan grass. Microbes are really effective in inducing plant growth as they secrete plant growth stimulators and enhance seedling initiation and promote root growth. They also play a substantial role in decomposition of organic materials and compost enrichment (Bahadur et al. 2014).

9.5 Future Scope

However, most of these incubation experiments were carried out in the laboratory. Currently, little information is available on the field application of such methods, which is most likely due to difficulties in soil inoculation under field conditions. Concerning to assess the efficacy of such applications for agricultural production systems, we need more field studies that evaluate their effect on both soil properties and crop growth. A better understanding of the mechanisms behind the K release from soil minerals is a key to developing new approaches for sustainable agriculture. However, the origin of K absorbed by the crop and the contribution of nonexchangeable K are extremely difficult to estimate due to a lack of suitable methods under field conditions or even in microcosm experiments. Nevertheless, what we can conclude is that crops have differences in transforming nonexchangeable K to soluble forms. Therefore, in K-limited areas, the selection of certain species or varieties that are efficient in solubilising K via exudates should have a great potential to increase resource use efficiency.

9.6 Conclusions

The diversity of potassium-solubilising microorganisms plays a key role in agriculture which can be used as bio-inoculants to sustain the release of potassium and make it available for the plant. However, very little information regarding the role of potassium-solubilising bacteria (KSB) in solubilisation of K mineral is available. It is necessary to look for KSB biofertilisers integrated with reduced level of K fertilisers which in turn provide high crop yield with low cost. Therefore, in this perspective, the current chapter is outlined to highlight the contributions of K-solubilising microorganisms from rhizospheric soil to develop efficient indigenous microbial consortia which are required for enhancing plant growth and yield of different crops further improving the soil fertility. The negative balance of potassium may be refined by considering maximum area under conservation agriculture with nonconventional sources of potassium in combination with biological methods which may further reduce the long-term negative adjustment of K in Indian agriculture for evergreen revolution.

Acknowledgments The authors are sincerely thankful to the head of the Department of Soil Science and Agricultural Chemistry, Institute of Agricultural Sciences, Banaras Hindu University, Varanasi, UP, India, and they gratefully acknowledged their valuable suggestion, cooperation and research facilities during the course of investigation.

References

Argelis DT, Gonzala DA, Vizcaino C, Gartia MT (1993) Biochemical mechanism of stone alteration carried out by filamentous fungi living in monuments. Biogeo Chem 19:129–147

Bagyalakshmi B, Ponmurugan P, Balamurugan A (2012) Impact of different temperature, carbon and nitrogen sources on solubilization efficiency of native potassium solubilizing bacteria from tea (Camellia sinensis). J Biol Res 3(2):36–42

Bahadur I, Meena VS, Kumar S (2014) Importance and application of potassic biofertilizer in Indian agriculture. Int Res J Biol Sci 12:80–85

Balogh-Brunstad Z, Keller CK, Gill RA, Bormann BT, Li CY (2008) The effect of bacteria and fungi on chemical weathering and chemical denudation fluxes in pine growth experiments. Biogeochem 88:153–167

Barker WW, Welch SA, Chu S, Banfield F (1998) Experimental observations of the effects of bacteria on aluminosilicates weathering. Am Mineral 83: 1551–1563

Basak BB, Biswas DR (2009) Influence of potassium solubilizing microorganism (*Bacillus mucilaginosus*) and waste mica on potassium uptake dynamics by sudan grass (*Sorghum vulgare* Pers.) grown under two Alfisols. Plant Soils 317:235–255

Boelens J, Zoutmann D, Cambell J, Verstraete W (1993) The use of bioluminescence as a reporter to study the adherence of the plant growth promoting *Rhizopseudomonas* 7NSK2 and ANP15 to canola roots. Can J Microbiol 39:329–334

Chen Z, Cuin TA, Zhou M, Twomey A, Naidu BP, Shabala S (2007) Compatible solute accumulation and stress-mitigating effects in barley genotypes contrasting in their salt tolerance. J Exp Bot 58: 4245–4255

Defreitas JR, Germida JJ (1992) Growth promotion of winter wheat by fluorescent *Pseudomonas* under field conditions. Soil Biol Biochem 24:1137–1146

Ghosh AB, Hasan R (1980) Soil fertility map of India. Indian Agricultural Research Institute, New Delhi

Glick BR (1995) The enhancement of plant growth by free-living bacteria. Can J Microbiol 41:109–117

Goldstein SAN, Pheasant DJ, Miller C (1994) Neuron 12: 1377–1388

Grandstaff DE (1986) The dissolution rate of Forsteritic Olivine from Hawaiian Beach Sand. In: Colman SM, Dethier DP (eds) Rates of chemical weathering of rocks and minerals. Academic Press, New York, pp 41–59

Han HS, Lee KD (2006) Effect of co-inoculation with phosphate and potassium solubilizing bacteria on mineral uptake and growth of pepper and cucumber. Plant Soil Environ 52:130–136

Han HS, Supanjani P, Lee KD (2006) Effect of co-inoculation with phosphate and potassium solubilizing bacteria an mineral uptake and growth of pepper and cucumber. Plant Soil Environ 52(3): 130–136

Herrera MA, Salamanka CP, Barea JM (1993) Inoculation of woody legumes with selected arbuscular mycorrhizal fungi and rhizobia to recover desertified Mediterranean ecosystems. Appl Environ Microbiol 59: 129–133

Ho flich G, Wiehe W, Khn G (1994) Plant growth stimulation with symbiotic and associative rhizosphere microorganisms. Experientia 50:897–905

Hutchens SE, Valsami JE, Eldowney MS (2003) The role of heterotrophic bacteria in feldspar dissolution. Minerol Mag 67:1151–1170

Javed NP, Arshad M (1997) Growth promotion of two wheat cultivars by plant growth promoting rhizobacteria. Pak J Bot 29:243–248

Jones DL, Shannon DV, Murphy D, Farrar J (2003) Role of dissolved organic nitrogen (DON) in soil N cycling in grassland soils. Soil Biol Biochem 36:749–756

Khudsen D, Peterson GA, Prov PF (1982) Lithium, sodium and potassium. In: Page AL (ed) Methods of soil analysis part (2) agronomy monograph 9, 2nd edn. ASA and SSSA, Madison

Kloepper JW, Zablowicz RM, Tipping B, Lifshitz R (1991) Plant growth mediated by bacterial rhizosphere colonizers. In: Keister DL, Gregan B (eds) The rhizosphere and plant growth, vol 14. BARC symposium, pp 315–326

Krieg NR, Holt JG (1984) Bergey's manual of systematic bacteriology. 8th edn. Williams & Wilkins, Baltimore

Kumar P, Dubey RC (2012) Plant growth promoting rhizobacteria for biocontrol of phytopathogens and yield enhancement of Phaseolus vulgaris. J Curr Pers Appl Microbiol 1:6–38

Kumar A, Bahadur I, Maurya BR, Raghuwanshi R, Meena VS, Singh DK, Dixit J (2015) Does a plant growth promoting rhizobacteria enhance agricultural sustainability? J Pure Appl Microbiol 9(1): 715–724

Liu GY (2001) Screening of silicate bacteria with potassium releasing and antagonistic activity. Chin J Appl Environ Biol 7:66–68

Liu D, Lian B, Dong H (2012) Isolation of Paenibacillus sp. and assessment of its potential for enhancing mineral weathering. Geophys J Roy Astron Soc 29: 413–421

Maurya BR, Meena VS, Meena OP (2014) Influence of inceptisol and alfisol's potassium solubilizing bacteria (KSB) isolates on release of K from waste mica. Vegetos 27(1):181–187

Maurya BR, Kumar A, Raghuwanshi R, Bahadur I, Meena VS (2015) Effect of phosphate solubilizing isolates on growth, yield and phosphate acquisition by rice and wheat crops. Afr J Microbiol Res 9(12): 1367–1375

Meena OP, Maurya BR, Meena VS (2013) Influence of K-solubilizing bacteria on release of potassium from waste mica. Agric Sustain Dev 1(1):53–56

Meena VS, Maurya BR, Bahadur I (2014a) Potassium solubilization by bacterial strain in waste mica. Bangladesh J Bot 43(2):235–237

Meena VS, Maurya BR, Verma JP (2014b) Does a rhizospheric microorganism enhance K+ availability in agricultural soils? Microbiol Res 169:337–347

Meena RK, Singh RK, Singh NP, Meena SK, Meena VS (2015a) Isolation of low temperature surviving plant growth-promoting rhizobacteria (PGPR) from pea (*Pisum sativum* L.) and documentation of their plant growth promoting traits. Biocatal Agric Biotechnol. doi:10.1016/j.bcab.2015.08.006

Meena VS, Maurya BR, Verma JP, Aeron A, Kumar A, Kim K, Bajpai VK (2015b) Potassium solubilizing rhizobacteria (KSR): isolation, identification, and K-release dynamics from waste mica. Ecol Eng 81: 340–347

Mengel K, Kirkby EA (1987) Principles of plant nutrition, 4th edn. International Potash Institute, Bern

Mikhailouskaya N, Tcherhysh A (2005) K-mobilizing bacteria and their effect on wheat yield. Latnian J Agron 8:154–157

Muentz A (1890) Surla decomposition desroches etla formation de la terre arable. C R Acad Sci 110: 1370–1372

Pathak H, Aggarwal PK, Singh SD (2009) Climate change impact, adaptation and mitigation in agriculture: methodology for assessment and applications. Indian Agricultural Research Institute, New Delhi, p 302

Purushothaman DA, Chandramohan S, Natarajan R (1974) Distribution of silicate dissolving bacteria in vellar estuary. Curr Sci 43:282–283

Ramarethinam S, Chandra K (2005) Studies on the effect of potash solubilizing/mobilizing bacteria *Frateuria aurantia* on brinjal growth and yield. Pestol 11: 35–39

Rao AS, Srivastava S (2012) Assessment of potassium availability in vertisols and its implication on fertilizer K recommendations theme: refinement of K recommendations in vertisols. Indian Institute of Soil Science Nabibagh, Berasia Road, BhopalEvent: 7th IPI-FAI round table in collaboration with IPNI New Delhi

Rao S, Srinivasarao C, Srivastava S (2011) Potassium status and crop response to potassium on the soils of agro-ecological regions of India. International Potash Institute, Basel, p 185

Rawlings DE (2002) Heavy metal mining using microbes. Annu Rev Microbiol 56:65–91

Reitmeir RF (1951) Soil potassium. In: Norman AG (ed) Advances in agronomy II. Academic Press, New York, pp 113–164

Requena BN, Jimenez I, Toro M, Barea JM (1997) Interactions between plant-growth-promoting rhizobacteria (PGPR), arbuscular mycorrhizal fungi and Rhizobium spp. In the rhizosphere of Anthyllis cytisoides, a model legume for revegetation in Mediterranean semiarid ecosystems. New Phytol 136:667–677

Sheng XF, Huang WY (2002) Mechanism of potassium release from feldspar affected by the strain NBT of silicate bacterium. Acta Pedol Sin 39:863–871

Sheng XF, Xia JJ, Cheng J (2006) Mutagenesis of the *Bacillus edaphicus* strain NBT and its effect on growth of chilli and cotton. Agric Sci Chin 37: 342–349

Singh G, Biswas DR, Marwah TS (2010) Mobilization of potassium from waste mica by plant growth promoting rhizobacteria and its assimilation by maize (*Zea mays*) and wheat (*Triticum aestivum* L.). J Plant Nutri 33: 1236–1251

Singh NP, Singh RK, Meena VS, Meena RK (2015) Can we use maize (*Zea mays*) rhizobacteria as plant growth promoter? Vegetos 28(1):86–99

Srinivasarao CH, Takkar PN (1997) Evaluation of different extractants for measuring the soil potassium and determination of critical levels for plant available K in smectitic soils for sorghum. J Indian Soc Soil Sci 45: 113–119

Srinivasarao C, Vittal KPR, Tiwari KN, Gajbhiye PN, Kundu S (2007) Categorization of soils based on potassium reserves and production systems: implications in K management. Aust J Soil Res 45:438–447

Ullaman WJ, Kirchman DL, Welch WA (1996) Laboratory evidence by microbially mediated silicate mineral dissolution in nature. Chem Geol 132:11–17

Vandevivere P, Welch SA, Ullman WJ, Kirchman DJ (1994) Enhanced dissolution of silicate minerals by bacteria at near neutral pH. Microb Ecol 27:241–251

Van Schöll L, Kuyper TW, Smits MM, Landeweert R, Hoffland E, van Breemen N (2008) Rock-eating mycorrhizas: their role in plant nutrition and biogeochemical cycles. Plant and Soil 303:35–40

Venkateswarlu B, Singh AK, R Srinivasa Rao CG, Kumar KA, Virmani SM (2012) Natural resource management for accelerating agricultural productivity. Studium Press (India) Pvt. Ltd, New Delhi, p 234

Weed SB, Davey CB, Cook MG (1969) Weathering of mica by fungi. Soil Sci Soc Am 33:702–706

Welch SA, Ullman WJ (1993) The effect of organic acids on plagioclase dissolution rates and stoichiometry. Geochim Cosmochim Acta 57:2725–2736

Xie JC (1998) Present situation and prospects for the world's fertilizer use. Plant Nutri Fertil Sci 4:321–330

10 Mechanism of Potassium Solubilization in the Agricultural Soils by the Help of Soil Microorganisms

Sajid Masood and Asghari Bano

Abstract

Potassium (K) is an essential macronutrient, largely required for normal functioning of plants; however, its solubilization is an important detriment for limiting agricultural productivity. This happens mainly because of insoluble forms of K, i.e. silicate minerals are present in the earth's crust. These minerals gradually dissolute K through different weathering processes. However, the use of certain microbes can assist the solubilization of K in addition to physical and chemical weathering of K minerals. These microorganisms, particularly bacteria, convert insoluble form of K to soluble forms, viz. acidification, chelation and exchange reactions. In this chapter, efforts are made to describe the possible ways of K release by the microbes. In addition, potential of different microbial species is summarized. Moreover, suggestions are provided for the sustainable plant production. In this way, the substantial decrease in the cost of commercial fertilizers could be expected.

Keywords

Dissolution • Minerals • Microbes • Potassium • Solubilization • Weathering

10.1 Introduction

K is the third major nutrient after nitrogen and phosphorus, required for plant growth and development (Hu et al. 2006; Parmar and Sindhu 2013). This is involved in several plant physiological and biochemical functions like cell osmotic regulation and enzyme activation when present in the soil solution (Valmorbida and Boaro 2007). However, soil solution or soluble K fraction in the soils is quite low. This is mainly due to the reason that a large fraction of the K is present as insoluble rocks, minerals and other deposits (Goldstein 1994). K in various rocks like igneous, metamorphic and sedimentary rocks can be used as fertilizers under suitable

S. Masood • A. Bano (✉)
Department of Plant Sciences, Quaid-i-Azam University, Islamabad 45320, Pakistan
e-mail: bano.asghari@gmail.com

V.S. Meena et al. (eds.), *Potassium Solubilizing Microorganisms for Sustainable Agriculture*,
DOI 10.1007/978-81-322-2776-2_10

management practices (Rosa-Magri et al. 2012). Therefore, the effective incorporation of rocks into the soils may reduce the use of fertilizers and chemicals for sustainable agriculture. Nevertheless, the long time is required to release the minerals from the rocks and to enhance the solubility of major elements of plant nutrition, making this alternative unfeasible.

In soils, the most essential K-containing minerals are feldspar, biotite, illite, muscovite and alkali, which discharge K through weathering (Marschner 2011), dissolution reactions and exchange processes, although the solubility of these minerals is a major issue, but the enhanced K dissolution can be achieved with the help of potassium-solubilizing microorganisms (KSMs), including bacteria, moulds and actinomycetes, which are capable of releasing the K from silicate minerals through the process of decomposition (Weed et al. 1969). The mechanism that lies behind the K release is an acid production by the microorganisms that favours mineral dissolution by absorbing soluble K. This mechanism of K release by potassium-solubilizing bacteria (KSB) such as *B. megaterium* and *B. mucilaginosus* has been reported previously (Han and Lee 2005). Several species of moulds like *Aspergillus*, *Penicillium* and *Fusarium* and bacteria such as *Bacillus*, *Pseudomonas* and *Micrococcus* have been reported to solubilize K from the insoluble K minerals (Gaur 1990). After 21 days of incubation (DAI), the mould, *A. niger* (CCT4355), produced medium acidification that caused the K release from the ultramafic alkaline and phlogopite (Lopes-Assad et al. 2006; Meena et al. 2013; Maurya et al. 2014). Similarly, Aleksandrov et al. (1967) found an increase in K, silicon and aluminium solubilization through the action of silicate bacteria on insoluble minerals.

10.2 Occurrence of K and Its Sources

K belongs to the group of alkali metals and carries the 1^+ oxidation state. It is widely distributed in the earth's crust and ranked seventh most abundant element. It constitutes ~2.6 % of the earth's crust and present in a variety of compounds; the most common compounds are KCl and KOH. Most common K-containing minerals in the earth's crust are feldspars and clays. However, the K contents in the soils may vary depending on the nature of parent material and the rate of weathering. Due to the slow rate of weathering, K release is slow to provide large amounts of K to the crops. On the other hand, weathering of primary minerals results in the formation of secondary minerals which may serve as a K source in the soil. These minerals include micaceous minerals like illite and vermiculite leading to the soil development.

In general, large quantities of K are usually bound in the primary minerals of the sand- and silt-sized fractions than the secondary minerals (clay fraction of the soil). According to Brady and Weil (2007), ~90–98 % of the total soil K remains in the structures of the minerals, whereas 2–10 % of K is either fixed or exchangeable and water-soluble ion. Scheffer and Schachtschabel (1989) reported that total K contents in the soils have been ranged 0.2–3.3 %; of the total K, less than 100 ppm is exchangeable or dissolved in soil water. This is mainly due to the reason that K occurs in three different forms in the soils. The first is "unavailable K" which is present in the crystalline structures of mica, feldspar and clays. Plants cannot take this insoluble form of K. The second form of K is "slowly available K" that is trapped in interlayer spaces of certain clay minerals. The third important form of the K is "readily available K" which is dissolved K in soil, water or on the surface of certain clay minerals. Apart from this, K also occurs in water bodies, for example, K concentrations up to 390 mg/L in seawater have been reported previously (Sparks 1989; Meena et al. 2015b; Singh et al. 2015).

Numerous sources like rocks, minerals and organic deposits are adding the K in water and soil systems. Among these sources, minerals and rocks are considered as the major sources of K in the environment. K is added in the soils through chemical fertilizers which are also prepared from mined potash salts, but are unable to meet the crop demands (Manning 2010). Hence, the

Table 10.1 K and K_2O concentrations in K-containing silicate minerals along with their dissolution rates

Mineral family	K mineral ore	Contains hydroxyl	Chemical formulae	% K_2O	% K	Relative dissolution rate
Feldspar/ Feldspathoid	Nepheline	No	(Na, K) $AlSiO_4$	15.7	13.0	40–100
	Feldspar	No	$KAlSi_3O_8$	16.9	14.0	1–2
	Leucite	No	$KAlSi_2O_6$	21.6	17.9	ND
	Kalsilite	No	$KAlSiO_4$	29.8	24.7	ND
Mica	Muscovite	Yes	$KAl_3Si_3O_{10}(OH)_2$	10.9	9.0	N/A
	Biotite	Yes	$K_2Fe_6Si_6Al_2O_{20}(OH)_4$	7.6	9.2	N/A
	Phlogopite	Yes	$K_2Mg_6Si_6Al_2O_{20}(OH)_4$	11.3	9.4	N/A

Adapted and modified from Manning (2010)
ND not detected

possible use of K silicate minerals as an alternative approach has been investigated in the last years (Harley and Gilkes 2000). However, a number of factors are responsible for K release from the silicate minerals like soil pH, temperature, moisture, microorganisms, the reactive surface area and vegetation type. As well known, these minerals are soil-forming materials widely distributed in the earth's crust, thereby providing much of the original K content of a soil. Therefore, silicate minerals or rocks have been considered as a major source of plant nutrients by many authors (Harley and Gilkes 2000; Gillman et al. 2002; van-Straaten 2007; Meena et al. 2014a, 2015a). Silicate minerals containing K as a source have been summarized in Table 10.1 by Manning (2010). The relative percentages of K and K_2O in silicate minerals have been given against the different formulae of silicate minerals. Other than these sources, organic matter and plant residues also release K into the soils after decomposition. However, their decomposition is dependent on the organic matter percentage in the soils, type of soil, moisture, soil temperature and crop species. In general, the average K contents in living plant tissues are ~0.3 %, but increased in a dry plant matter by 1.5–4 %. This assimilated K in plant material becomes readily available when the plants die or shed their leaves and branches at maturation stage.

10.3 Functions of K in Plants

K is considered as the third most essential macronutrient after nitrogen (N) and phosphorus (P) required by all plants. It plays a crucial role in plant cells, however, dependent on K-specific membrane-bound proteins. These functions include the maintenance of electrochemical gradients at membrane sites (Cheeseman and Hanson 1979), cell expansion, which are turgor-related processes (Dolan and Davies 2004), formation of pollen tube (Mouline et al. 2002) and stomatal regulation (Dietrich et al. 2001). Apart from these, K has a role in enzyme activation (Suelter 1970) and xylem transport of nitrate (Ben-Zioni et al. 1971) and other ions, especially monovalent cations. It is evident that phloem sugar translocation (Cakmak et al. 1994a, b) and the charge stabilization of anions within the cell (Clarkson and Hanson 1980) also take place with the help of K (Meena et al. 2014b; Kumar et al. 2015).

As well known, K is directly involved in plant physiological processes like electrochemical gradient of the plasma membrane, plant growth and yield and opening and closing of the stomata and provides resistance against the environmental changes (Broadley and White 2005; Anschütz et al. 2014; Shabala and Pottosin 2014). Therefore, high amounts of K are required for the normal functioning of the plants (Benito et al. 2014; Shin and Adams 2014). Despite of the higher abundance of K in the soil, it is less available due to strong mineral adsorption that ranges from 0.1 to 1 mm depending on soil type (Asher and Ozanne 1967). Similarly, plant demands are often coupled with the low availability of K. Hence, the higher acquisition of K can be achieved through the role of soil biota in K solubilization. In this chapter, we have described the mechanism of action of soil biota on K-containing minerals indicating new directions in K research.

10.4 Role of Soil Microorganisms in K Solubilization

10.4.1 General Importance

Different rocks can be used as fertilizers after dissolution reactions and, therefore, have been shown to increase the agronomic potential (Song and Huang 1988; Kalinowski and Schweda 1996). On the other hand, microbial species such as actinomycetes, other bacteria and moulds are capable of solubilizing the potassium-containing silicate minerals through the process of decomposition (Weed et al. 1969). This relationship between microorganisms and minerals may potentially affect any biochemical, biological and geological weathering taking place at or near the surface of the earth. Weathering of minerals and rocks is the deterioration or disintegration of rocks into fragments arbitrated through different processes, acting synergistically to ascertain the configuration of the biosphere, hydrosphere, atmosphere and lithosphere (Vaughan et al. 2002). In this regard, filamentous fungi, animals and plants have also been shown to augment the desegregation of rock substrates by physical forces, such as the pressure of growing hyphae and root penetration (Burford et al. 2003). Bacteria, algae, fungi and plants contribute to the biochemical weathering of rocks through the organic acid production, other metabolites and, especially, CO_2 that leads to carbonic acid production, resulting in soil acidification (Sterflinger 2000; Gadd and Sayer 2000) as shown in Fig. 10.1. Now, the question has arisen as to how K is solubilized by soil organisms or what kinds of processes are involved in the K solubilization from K-containing minerals in closed association with soil microorganisms. The mechanism lies behind the K solubilization, and processes are explained in following subheadings.

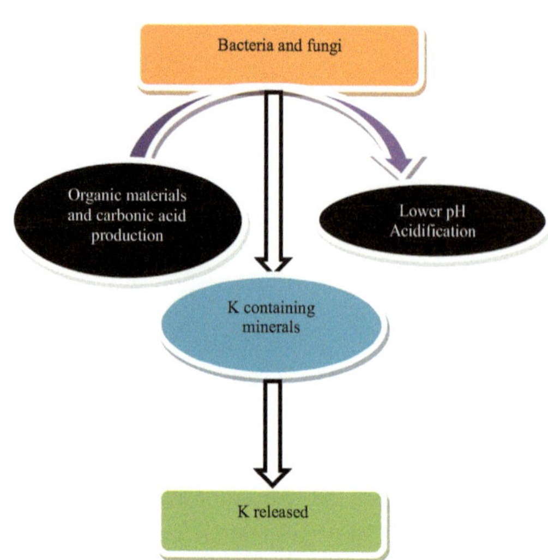

Fig. 10.1 General processes involved in K release from K-containing minerals by microorganisms

10.4.2 K Solubilization by Microbial Species

Soil hosts a variety of soil microorganisms including bacteria and fungi. These microorganisms enhance the solubility of K minerals through different processes like enzymolysis, capsule absorption, complexation through extracellular polysaccharides and acidolysis (Avakyan 1984; Friedrich et al. 1991; Welch et al. 1999; Meena et al. 2014b). Therefore, KSMs are considered most important microorganisms for solubilizing the fixed form of K in the soil through ion exchange processes. While reviewing the K release from the feldspar and illite minerals through silicate bacteria, Lian (1998) and Lian et al. (2002) suggested a stage model (Fig. 10.2) and proposed the development of bacteria-mineral complexes supported by extracellular polysaccharides. Consequently, K solubilization and mineral dissolution in the microenvironments were enhanced through the soil acidification in the presence of organic ligands that stimulated surface reactions. Furthermore, they reported that, for phyllosilicates like illite, minor organic ligands may impetus into the interlayer spaces to push the K out (Lian et al. 2002). These dynamics of K in soils

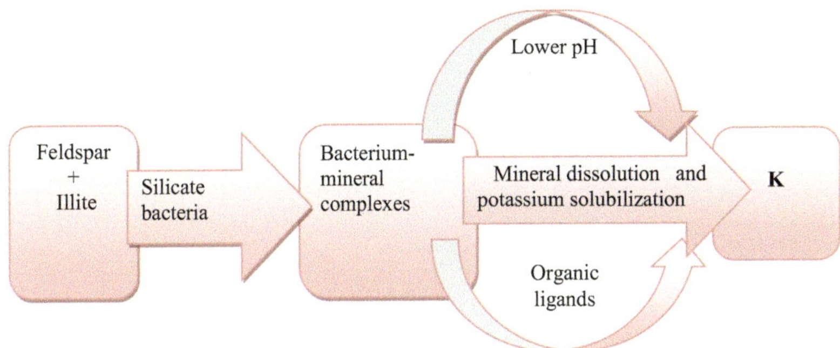

Fig. 10.2 Schematic presentation of K solubilization by the bacteria after forming the mineral-soluble complexes

were pointed out that K is liberated from the mica to water-soluble and transferable pool of K due to inoculation of mica using *B. mucilaginosus* in the soils (Basak and Biswas 2009). The solubilization of K from the rock and minerals attained by the microbes was first reported by Muentz (1890). Evidence also exists that silicate bacteria were effective for the release of K and other elements of silicate minerals (Aleksandrov et al. 1967). This K release was resulted by the strains of bacteria in the presence of oxygen and acidic pH. However, their efficiency of releasing K from the minerals is dependent upon the nature of K-containing rocks and minerals and aerobic conditions (Sheng and Huang 2002). Other factors also have a great impact on the K solubilization when the bacteria are used, for example, pH. Parmar and Sindhu (2013) observed the maximum K solubilization at pH 7.0 of the broth media. It is possible that different bacteria may perform differently at varying pH levels, but significantly solubilize the K in the soils.

This is supported by numerous studies where a wide variety of bacteria like *B. mucilaginosus*, *B. edaphicus*, *B. circulans*, *Burkholderia*, *Acidithiobacillus ferrooxidans*, *Paenibacillus* spp. and *Pseudomonas* spp. have been proven to solubilize the K from K-containing minerals in the soils (Liu et al. 2012; Meena et al. 2014a, b, c). Accordingly, Archana et al. (2013) reported that in in vitro conditions, KSMs like *B. extorquens*, *Clostridium pasteurianum* and *A. niger* enhanced the mineral weathering of mica, biotite, orthoclase, microclase and muscovite minerals. Zhang and Kong (2014) isolated and identified the KSB through the 16 rDNA sequences. Further, they compared the performance of KSB both in solid and liquid media for the solubilization of K-feldspar. They observed that strains *Pantoea agglomerans*, *Agrobacterium tumefaciens*, *Microbacterium foliorum*, *Myroides odoratimimus*, *Burkholderia cepacia*, *Enterobacter aerogenes*, *E. cloacae* and *E. asburiae* remain effective in K solubilization in both media. Several other studies also demonstrated the significant role of plant growth-promoting rhizobacteria (PGPR) in K solubilization and its mobilization in the plant root systems (Kumar and Singh 2001; Kukreja et al. 2004; El-Fattah et al. 2013).

Fungi are also considered as a biological weathering mediator of rocks, minerals and building materials due to the secretion of protons, organic acids and other metabolites along with their filamentous growth (Gu et al. 1998; Gadd 1999). It has also been reported that fungi play an important role in the solubilization of K from aluminosilicate minerals (Yuan et al. 2000, 2004). For example, *A. fumigatus* releases potassium in three possible ways as shown in Fig. 10.3a. First, it forms the complexes with soluble organic ligands. Thereafter, the formation of immobile biopolymers takes place, whereas the third possible way is the mechanical forces along with the direct contact among fungal filaments and K-containing minerals (Lian et al. 2008). As fungal hyphae deeply penetrate in already formed cavities, the cracks and fissures develop as a result of biochemical weathering of minerals (Sterflinger 2000). On the other hand, K release through the fungi is controlled by four major processes which are given in Fig. 10.3b. These

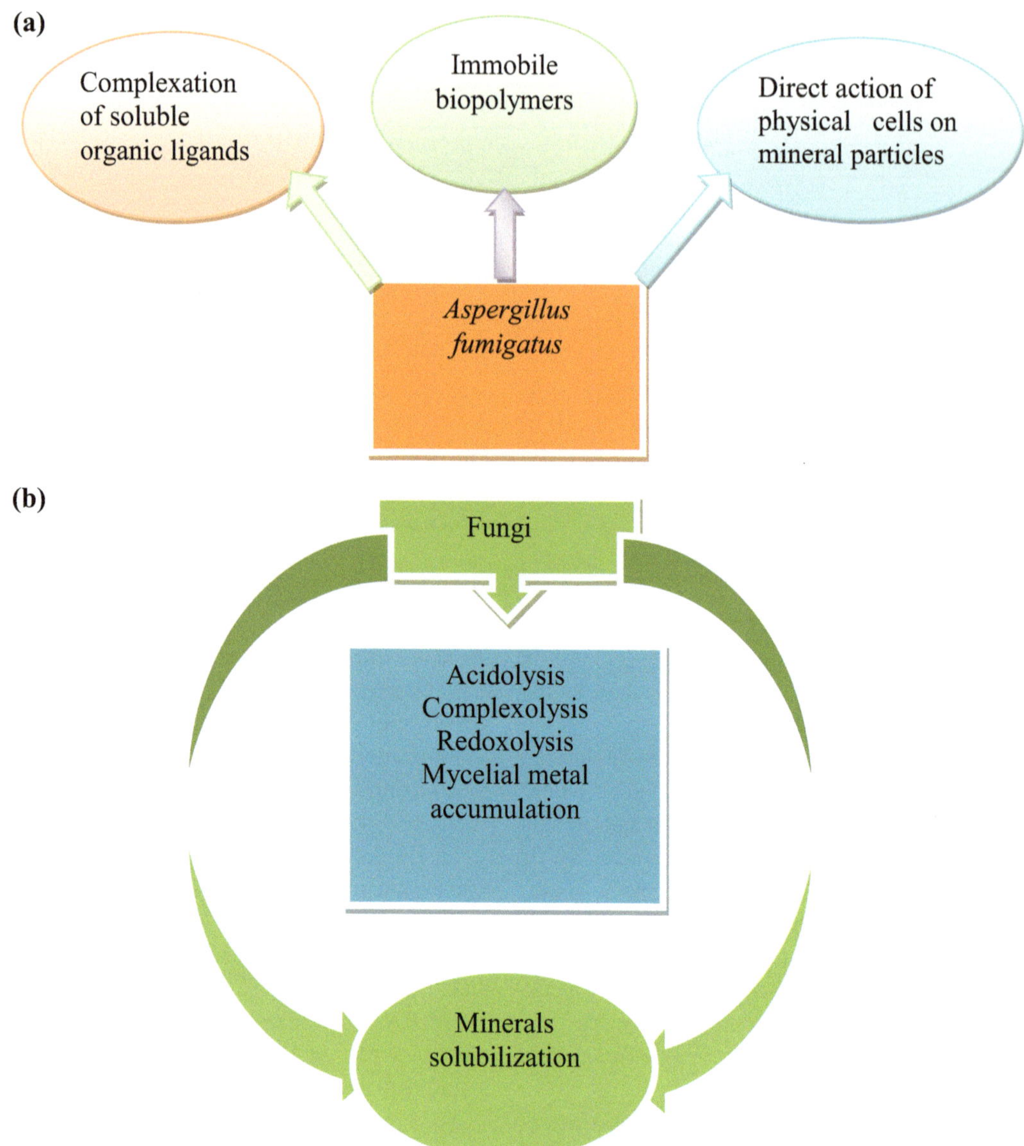

Fig. 10.3 Potassium solubilization by (**a**) *Aspergillus fumigatus* and (**b**) processes of mineral solubilization by fungi

processes include acidolysis and complexolysis, redoxolysis and metal accumulation by mycelium, respectively (Kumar and Kumar 1999; Gadd 2001). Further, they reported that the lower molecular weight organic acids are responsible for the biochemical dissolution of minerals.

Various reports are available describing the K solubilization from K-containing minerals by different fungal strains. For example, Yuan et al. (2004) collected fungal strains *Pisolithus* XC1, *P. microcarpus* and *Cenococcum geophilum* SIV from the roots of eucalyptus and observed significant weathering of minerals like phlogopite and vermiculite. Likewise, *Piloderma* has also been reported to enhance the K release from biotite, chlorite and microcline, thereby reducing the nutritional requirements of the crops (Glowa et al. 2003). In addition to abovementioned fungal strains, the role of ectomycorrhizae (ECM) in K solubilization cannot be overruled, because they have shown to enhance the K uptake after releasing K from the clay minerals (Yuan et al. 2000). According to Uroz et al. (2007), ECM mediate biochemical and physical weathering of minerals, particularly in forest ecosystems. This results in increased nutrient concentrations, including K.

10.4.3 Major Mechanistic Processes Responsible for K Solubilization

Numerous processes are involved in K solubilization as given below.

10.4.3.1 Weathering

Breakdown of rocks and minerals is accelerated by soil microorganisms. This is also called as biotic weathering. Soil microorganisms aid the phenomenon of physical and chemical weathering. For example, symbiotic association of fungi and algae (lichen) enhances the physical weathering by transferring the loose material and rock fragments; these are attached to the lower side and detached when the lichens contract during dry spells. On the other hand, soil microorganisms, especially rhizobacteria, produce 2-ketogluconic acid that binds Ca and, hence, acts as strong weathering agent in basic rocks. Alternatively, soil microorganisms respire and produce CO_2, which reacts with water and forms carbonic acid. This carbonic acid production contributes in chemical weathering by dissolving $CaCO_3$ and K minerals through dissolution reactions.

10.4.3.2 Ion Exchange Processes

Plants always require balance and adequate levels of nutrients during the entire growth period. These requirements are replenished through the ion exchange processes; these include fixation, solubilization, adsorption and desorption in the soils. Plants only take the nutrients or ions which are coming in the soil solution after the equilibrium between the soil adsorption sites and soil solution is maintained. Chemical equilibrium between the soil solution phase and solid phase is described by Lindsay (1979). As the soil is negatively charged, therefore, it adsorbs maximum of the nutrients or ions, especially K, which is fixed on the soil surface rather coming in the soil solution. However, the production of H^+ and acids by the soil microorganisms contributes in ion exchange processes. In this way, K is released into the soil solution and becomes readily available to the plants. Similar charge species also contribute in ion exchange processes when their occurrence is higher in the soil solution. For example, when Ca^{2+} is present in excess in the soil solution, it can exchange two K^+ ions from the soil adsorption sites and contributes in desorption and K solubilization. These ions also exchange the K which is trapped in the interlayer spaces of the minerals to some extent.

10.4.3.3 Decomposition of Organic Residues

The third important process of K solubilization or K release in the soils is decomposition. Decomposition of organic material adds the K and other minerals in the soils. Plant debris, leaves, twigs and surface litter are cycled due to the action of soil microorganisms on them. These microorganisms get their food from the material and convert it into an inorganic form. This results in large amounts of nutrient release in the soils including K.

10.5 Factors Responsible for K Solubilization

There are a number of factors responsible for K solubilization in association with soil microorganisms. These factors include the type and nature of rocks and minerals, temperature, moisture, rate of reaction, pH and organic acids, etc. Here, we have discussed the major factors which are responsible in K solubilization. Contribution from the other factors is negligible, whereas the nature and type of rocks have been discussed above.

10.5.1 Soil pH

Soil pH plays a crucial role in K solubilization because it can be easily released by decreasing the pH of extracting solution. In 1953a, b, York and co-workers concluded that extracting solution of ammonium acetate having a pH of 4.8 easily solubilizes K than the same solution having a pH of 7.0. Rich and Black (1964) have also reported that in acid systems, small hydronium ions are more easily exchanged with K as compared to large hydronium basic cations. Normally, fixation of K is higher at high pH and lower at decreasing pH levels, which is explained by two theories. The first theory demonstrates that in an acidic environment, hydronium ions compete with K ions for fixation sites. The second theory exhibits that acidic soils enhance the oxygen protonation by changing the geometry and/or reducing the charge layer which results in weakening of the electrostatic bonds of interlayer K ions (De-Villiers and Jackson 1967; Thompson and Ukrainczyk 2002). When the pH is higher, fixation of K is greater in the soils, and the K forms like carbonate and bicarbonates are more fixed due to alkaline environment as compared to the chloride form of K. Acidic pH has a strong influence on K solubilization in the soils. The study by Wood and de-Turk (1941) also explains that more K is transitioned from acid-soluble pool to acid-insoluble pool over time, by splitting the fixed potassium pool into acid-soluble and acid-insoluble pool (Fig. 10.4).

10.5.2 Production of Organic Acids and Other Organic Materials

Soil microorganisms have good weathering ability due to the production of proton, organic acid, siderophores and organic ligands. Organic acids like citric acid, formic acid, malic acid, oxalic acid, etc. are produced after the decomposition of organic matter or from their own secretions. These organic acids enhance the K solubilization by providing proton for the soil adsorption sites and complexation or chelation with Ca^{2+} ions; this process dissolves the minerals and rocks containing K. In confirmation to this, Sheng et al. (2003) reported that the enhanced soil mineral dissolution by organic compounds like acetate, citrate and oxalate is secreted by microorganisms. Another study by Styriakova et al. (2003) observed that K can be solubilized by complex formation between organic acids and metal ions like Fe^{2+}, Al^{3+} and Ca^{2+}. It was also reported that the rocks and K-containing minerals such as illite, micas and orthoclases are solubilized by *B. mucilaginous*, a KSB through organic acid exudation (Ullaman et al. 1996). Solubilization of feldspar and illite (K minerals) was enhanced by PGPR through the production of capsular polysaccharides and organic acids such as tartaric and oxalic acids (Sheng and He 2006). Further, they reported that oxalic acid was effective in solubilization of Nanjing feldspar, whereas oxalic and tartaric acids enhanced the K solubilization of Suzhou

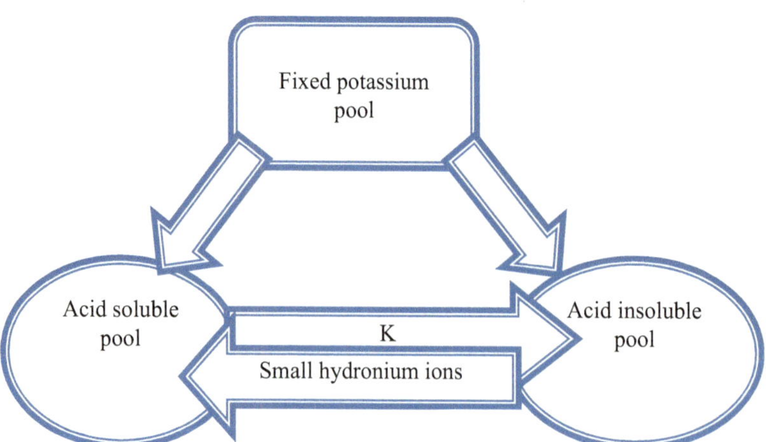

Fig. 10.4 Exchange of K with small hydronium ions in soil acid systems for its solubilization

illite mineral. Similarly, Tan et al. (2014) also reported that PGPR produce acids, alkalis or chelating agents which enhance the release of K from K-bearing minerals such as muscovite mica. The principle mechanism for the solubilization of elements by PGPR involves lowering of soil pH, mineralization through chelation-mediated complexes and secretion of organic and inorganic acids (Khan et al. 2009).

10.6 Concluding Remark

It is summarized that the main source of K in the soils is minerals containing K. These minerals don't release K easily. However, the action of soil microorganisms on minerals releases K in the soils and makes it available for the plants. KSMs contribute in physical and chemical weathering through dissolution, decomposition, chelation-complexation and ion exchange processes. Decreased pH corresponded to increased production of organic acids which shows the higher activity of soil microorganisms. Consequently, KSMs solubilize the K from its unavailable K pool in the soil. Certain bacteria and fungi species indicated in this chapter have shown a considerable potential to solubilize the K that is closely associated with the major factors which control the K solubilization. It is further suggested that the use of a particular species of bacteria and fungi in the soils can minimize the maximum cost of chemical fertilizers; otherwise, K fertilizers are costly as compared to other fertilizers.

References

Aleksandrov VG, Blagodyr RN, Ilev IP (1967) Liberation of phosphoric acid from apatite by silicate bacteria. Mikrobiol Z (Kiev) 29:111–114

Anschütz U, Becker D, Shabala S (2014) Going beyond nutrition: regulation of potassium homoeostasis as a common denominator of plant adaptive responses to environment. J Plant Physiol 171:670–687

Archana DS, Nandish MS, Savalagi VP, Alagawadi AR (2013) Characterization of potassium solubilizing bacteria (KSB) from rhizosphere soil. BIOINFOLET-A Q J Life Sci 10:248–257

Asher CJ, Ozanne PG (1967) Growth and potassium content of plants in solution cultures maintained at constant potassium concentrations. Soil Sci 103:155–161

Avakyan ZA (1984) Silicon compounds in solution bacteria quartz degradation. Microbiology 54:301–307

Basak BB, Biswas DR (2009) Influence of potassium solubilizing microorganism (*Bacillus mucilaginosus*) and waste mica on potassium uptake dynamics by Sudan grass (*Sorghum vulgare* Pers.) grown under two Alfalfa soils. Plant Soil 317:235–255

Benito B, Haro R, Amtmann A, Cuin TA, Dreyer I (2014) The twins K^+ and Na^+ in plants. J Plant Physiol 171:723–732

Ben-Zioni A, Vaadia Y, Lips SH (1971) Nitrate uptake by roots as regulated by nitrate reduction products of the shoot. Physiol Plant 24:288–290

Brady NC, Weil RR (2007) The nature and properties of soils, 14th edn. Prentice Hall, Upper Saddle River

Broadley MR, White PJ (2005) Plant nutritional genomics. Blackwell Publishing CRC Press, Oxford

Burford EP, Fomina M, Gadd GM (2003) Fungal involvement in bio-weathering and biotransformation of rocks and minerals. Miner Mag 67:1127–1155

Cakmak I, Hengeler C, Marschner H (1994a) Changes in phloem export of sucrose in leaves in response to phosphorus, potassium and magnesium deficiency in bean plants. J Exp Bot 45:1251–1257

Cakmak I, Hengeler C, Marschner H (1994b) Partitioning of shoot and root dry matter and carbohydrates in bean plants suffering from phosphorus, potassium and magnesium deficiency. J Exp Bot 45:1245–1250

Cheeseman JM, Hanson JB (1979) Mathematical analysis of the dependence of cell potential on external potassium in corn roots. Plant Physiol 63:1–4

Clarkson DT, Hanson JB (1980) The mineral nutrition of higher-plants. Annu Rev Plant Physiol Plant Mol Biol 31:239–298

De-Villiers JM, Jackson ML (1967) CEC variations with pH in soil clays. Soil Sci Soc Am Proc 31:473–476

Dietrich P, Sanders D, Hedrich R (2001) The role of ion channels in light-dependent stomatal opening. J Exp Bot 52:1959–1967

Dolan L, Davies J (2004) Cell expansion in roots. Curr Opin Plant Biol 7:33–39

El-Fattah DAA, Wedad EE, Mona SZ, Mosaad KH (2013) Effect of carrier materials, sterilization method, and storage temperature on survival and biological activities of *Azotobacter chroococcum* inoculant. Ann Agric Sci 58:111–118

Friedrich S, Platonova NP, Karavaiko GI (1991) Chemical and microbiological solubilization of silicates. Acta Biotechnol 3:187–196

Gadd GM (1999) Fungal production of citric acid and oxalic acid: importance in metal speciation, physiology and biogeochemical processes. Adv Microbiol Physiol 41:47–92

Gadd GM (2001) Metal transformations. In: Gadd GM (ed) Fungi in bioremediation. Cambridge University Press, Cambridge, UK, pp 359–382

Gadd GM, Sayer JA (2000) Fungal transformation of metals and metalloids. In: Lovley DR (ed) Environmental microbe-metal interactions. American Society for Microbiology, Washington, DC, pp 237–256

Gaur AC (1990) Phosphate solubilizing microorganisms as biofertilizers. Omega Scientific Publisher, New Delhi

Gillman GP, Burkett DC, Coventry RJ (2002) Amending highly weathered soils with finely ground basalt rock. Geochemistry 17:987–1001

Glowa KR, Arocena JM, Massicotte HB (2003) Extraction of potassium and/or magnesium from selected soil minerals by *Piloderma*. Geomicrobiol J 20:99–111

Goldstein AH (1994) Involvement of the quinoprotein glucose dehydrogenase in the solubilization of exogenous phosphates by gram-negative bacteria. Phosphate in microorganisms: cellular and molecular biology. ASM Press, Washington, DC, pp 197–203

Gu JD, Ford TE, Berke NS, Mitchell R (1998) Biodeterioration of concrete by the fungus *Fusarium*. Int Biodet Biodeg 41:101–109

Han HS, Lee DK (2005) Phosphate and potassium solubilizing bacteria effect on mineral uptake, soil availability and growth of eggplant. Res J Agr Biol Sci 1: 176–180

Harley AD, Gilkes RJ (2000) Factors influencing the release of plant nutrients from silicate rock powders: a geochemical overview. Nutr Cycl Agroecosys 56: 11–36

Hu X, Chen J, Guo J (2006) Two phosphate and potassium-solubilizing bacteria isolated from Tianmu mountain, Zhejiang, China. World J Microbiol Biotechnol 22:983–990

Kalinowski BE, Schweda P (1996) Kinetics of muscovite, phlogopite, and biotite dissolution and alteration at pH 1–4, room temperature. Geochim Cosmochim Acta 60:367–385

Khan AA, Jilani G, Akhtar MS, Naqvi SMS, Rasheed M (2009) Phosphorus solubilizing bacteria: occurrence, mechanisms and their role in crop production. J Agric Biol Sci 1:48–58

Kukreja K, Suneja S, Goyal S, Narula N (2004) Phytohormone production by *Azotobacter*: a review. Agric Rev 25:70–75

Kumar R, Kumar AW (1999) Biodeterioration of stone in tropical environments: an overview. The J. Paul Getty Trust, Los Angeles

Kumar V, Singh KP (2001) Enriching vermicompost by nitrogen fixing and phosphate solubilizing bacteria. Bioresour Tech 76:173–175

Kumar A, Bahadur I, Maurya BR, Raghuwanshi R, Meena VS, Singh DK, Dixit J (2015) Does a plant growth-promoting rhizobacteria enhance agricultural sustainability? J Pure Appl Microbiol 9(1):715–724

Lian B (1998) A study on how silicate bacteria GY92 dissolve potassium from illite. Acta Mineral Sin 18: 234–238

Lian B, Fu PQ, Mo DM, Liu CQ (2002) A comprehensive review of the mechanism of potassium releasing by silicate bacteria. Acta Mineral Sin 22:179–183

Lian B, Wang B, Pan M, Liu C, Teng HH (2008) Microbial release of potassium from K-bearing minerals by thermophilic fungus *Aspergillus fumigatus*. Geochim Cosmochim Acta 72:87–98

Lindsay WL (1979) Soil chemical equilibria. Wiley, New York

Liu D, Lian B, Dong H (2012) Isolation of *Paenibacillus* sp. and assessment of its potential for enhancing mineral weathering. Geomicrobiol J 29:413–421

Lopes-Assad ML, Rosa MM, Erler G, Ceccato-Antonini SR (2006) Solubilização de pó-de-rocha por *Aspergillus niger*. Espaço e Geografia 9:1–16

Manning DAC (2010) Mineral sources of potassium for plant nutrition: a review. Agron Sust Develop 30: 282–294

Marschner P (2011) Marschner's mineral nutrition of higher plants, 3rd edn. Academic, London

Maurya BR, Meena VS, Meena OP (2014) Influence of Inceptisol and Alfisol's potassium solubilizing bacteria (KSB) isolates on release of K from waste mica. Vegetos 27(1):181–187

Meena OP, Maurya BR, Meena VS (2013) Influence of K-solubilizing bacteria on release of potassium from waste mica. Agric Sustain Dev 1(1):53–56

Meena VS, Maurya BR, Bahadur I (2014a) Potassium solubilization by bacterial strain in waste mica. Bangladesh J Bot 43(2):235–237

Meena VS, Maurya BR, Verma JP (2014b) Does rhizospheric microorganism enhance K^+ availability in agricultural soils? Can J Microbiol 52:66–72

Meena VS, Maurya BR, Verma JP (2014c) Does a rhizospheric microorganism enhance K+ availability in agricultural soils? Microbiol Res 169:337–347

Meena RK, Singh RK, Singh NP, Meena SK, Meena VS (2015a) Isolation of low temperature surviving plant growth-promoting rhizobacteria (PGPR) from pea (*Pisum sativum* L.) and documentation of their plant growth promoting traits. Biocatal Agric Biotechnol 4 (4):806–811. doi:10.1016/j.bcab.2015.08.006

Meena VS, Maurya BR, Verma JP, Aeron A, Kumar A, Kim K, Bajpai VK (2015b) Potassium solubilizing rhizobacteria (KSR): isolation, identification, and K-release dynamics from waste mica. Ecol Eng 81: 340–347

Mouline K, Very AA, Gaymard F, Boucherez J, Pilot G, Devic M, Bouchez D, Thibaud JB, Sentenac H (2002) Pollen tube development and competitive ability are impaired by disruption of a shaker K(+) channel in Arabidopsis. Genes Dev 16:339–350

Muentz A (1890) Sur la decomposition des roches etla formation de la terre arable. CR Acad Sci 110: 1370–1372

Parmar P, Sindhu SS (2013) Potassium solubilization by rhizosphere bacteria: influence of nutritional and environmental conditions. J Microbiol Res 3:25–31

Rich CI, Black WR (1964) Potassium exchange as affected by cation size, pH, and mineral structure. Soil Sci 97:382–390

Rosa-Magri MM, Avansini SH, Lopes-Assad ML, Tauk-Tornisielo SM, Sandra Regina Ceccato-Antonini SR (2012) Release of potassium from rock powder by the yeast *Torulaspora globosa*. Braz Arch Biol Technol 55:577–582

Scheffer F, Schachtschabel P (1989) Lehrbuch der Bodenkunde. 12 Ferdinand Enke Verl Stuttgart, Germany

Shabala S, Pottosin I (2014) Regulation of potassium transport in plants under hostile conditions: implications for abiotic and biotic stress tolerance. Physiol Plant 15:257–279

Sheng XF, He LY (2006) Solubilization of potassium bearing minerals by a wild type strain of *Bacillus edaphicus* and its mutants and increased potassium uptake by wheat. Can J Microbiol 52:66–72

Sheng XF, Huang WY (2002) Study on the conditions of potassium release by strain NBT of silicate bacteria. Sci Agric Sinica 35:673–677

Sheng XF, Xia JJ, Chen J (2003) Mutagenesis of the *Bacillus edphicaus* strain NBT and its effect on growth of chili and cotton. Agric Sci China 2:40–41

Shin R, Adams E (2014) Transport, signalling and homeostasis of potassium and sodium in plants. J Integ Plant Biol 56:231–249

Singh NP, Singh RK, Meena VS, Meena RK (2015) Can we use maize (*Zea mays*) rhizobacteria as plant growth promoter? Vegetos 28(1):86–99

Song SK, Huang PM (1988) Dynamics of potassium release from potassium-bearing minerals as influenced by oxalic and citric acids. Soil Sci Soc Am J 52: 383–390

Sparks DL (1989) Kinetics of soil chemical processes. Academic, San Diego, CA

Sterflinger K (2000) Fungi as geologic agents. Geomicrobiol J 17:97–124

Styriakova I, Styriak I, Hradil D, Bezdicka P (2003) The release of iron bearing minerals and dissolution of feldspar by heterotrophic bacteria of *Bacillus* species. Ceram-Silikaty 47:20–26

Suelter CH (1970) Enzymes activated by monovalent cations. Science 168:789–795

Tan KZ, Radziah O, Halimi MS, Khairuddin AR, Habib SH, Shamsuddin ZH (2014) Isolation and characterization of rhizobia and plant growth promoting rhizobacteria and their effects on the growth of rice seedlings. Am J Agric Biol Sci 9:342–360

Thompson ML, Ukrainczyk L (2002) Micas. In: Dixon JB, Schulze DG (eds) Soil mineralogy with environmental applications, Book Series No. 7. Soil Science Society of America, Madison, pp. 431–466

Ullaman WJ, Kirchman DL, Welch WA, Vandevivere P (1996) Laboratory evidence by microbially mediated silicate mineral dissolution in nature. Chem Geol 132:11–17

Uroz S, Calvaruso C, Turpault MP, Pierrat JC, Mustin C, Frey-Klett P (2007) Effect of the mycorrhizosphere on the genotypic and metabolic diversity of the bacterial communities involved in mineral weathering in a forest soil. Appl Environ Microbiol 73:3019–3027

Valmorbida J, Boaro CSF (2007) Growth and development *of Mentha piperita* L. in nutrient solution as affected by rates of potassium. Braz Arch Biol Technol 50:379–384

van Straaten P (2007) Agrogeology, the use of rocks for crops. Enviroquest Ltd, Cambridge/Ontario

Vaughan DJ, Pattrick RAD, Wogelius RA (2002) Minerals metals and molecules: ore and environmental mineralogy in the new millennium. Miner Mag 66:653–676

Weed SB, Davey CB, Cook MG (1969) Weathering of mica by fungi. Soil Sci Soc Am J 33:702–706

Welch SA, Barker WW, Banfield JF (1999) Microbial extracellular polysaccharides and plagioclase dissolution. Geochim Cosmochim Acta 63:1405–1419

Wood LK, de Turk EE (1941) Absorption of potassium in soils in non-replaceable forms. Soil Sci Soc Am Proc 5:152–161

York ET, Bradfield R, Peech M (1953a) Calcium- potassium interactions in soils and plants. I: lime-induced potassium fixation in Mardin silt loam. Soil Sci 76: 379–387

York ET, Bradfield R, Peech M (1953b) Calcium- potassium interactions in soils and plants. II: reciprocal relationship between calcium and potassium in plants. Soil Sci 76:481–491

Yuan L, Fang DH, Wang ZH, Shun H, Huang JG (2000) Bio-mobilization of potassium from clay minerals: I by ectomycorrhizas. Pedosphere 10:339–346

Yuan L, Huang JG, Li XL, Christie P (2004) Biological mobilization of potassium from clay minerals by ectomycorrhizal fungi and eucalypt seedling roots. Plant Soil 262:351–361

Zhang C, Kong F (2014) Isolation and identification of potassium-solubilizing bacteria from tobacco rhizospheric soil and their effect on tobacco plants. Appl Soil Ecol 82:18–25

Regulation of Plant Physiology and Antioxidant Enzymes for Alleviating Salinity Stress by Potassium-Mobilizing Bacteria

11

Yachana Jha and R.B. Subramanian

Abstract

Potassium (K) is among the most important essential macronutrients for plant growth. About 98 % of the potassium in the earth's crust exists in insoluble forms as rocks and silicate minerals, resulting in very low concentrations of soluble potassium in the soil for plant growth and development. Rhizosphere bacteria are a group of metal-mobilizing, plant growth–promoting bacteria having the ability to solubilize potassium from insoluble potassium rocks. KSM (a potassium-solubilizing microorganism) is a metal-mobilizing, plant growth–promoting bacterium living symbiotically in/on the root surface and helps directly or indirectly in promoting plant growth via solubilization of insoluble minerals (K and P), assisting in resource acquisition (macro- and micronutrients), production of phytohormones and secretion of different regulatory chemicals in the purlieu of the rhizosphere of the plant root. KSM such as *Bacillus* spp. and *Pseudomonas* spp. are the most dominant plant growth–promoting bacteria (PGPB) of rhizospheric soils. Inoculation of KSM is found to be promising to induce growth of plants under 2.3–3.5 dSm^{-1} salinity levels and low availability of P and K, protecting the plants from salinity injury by enhancing their growth-related physiology and lipid peroxidation. These KSM help in the decrement of lipid peroxidation and enhance the stability of the plant cell membrane for the survival of the plant under salt stress. Inoculation of plants with such beneficial root-associated bacteria could provide salt tolerance to plants as these isolates also reside within the root, which is the plant part first directly in contact with saline soil. Hence it serves as a useful tool for alleviating salinity stress as well as in uptake of important mineral nutrients. The diversity of potassium-solubilizing microbes (KSM) and ability to mobilize important macronutrients from insoluble to soluble forms through biological

Y. Jha (✉)
NV Patel College of Pure and Applied Sciences, SP University, VV Nagar, Anand 388120, India
e-mail: yachanajha@ymail.com

R.B. Subramanian
BRD School of Biosciences, Sardar Patel University, PostBox-39VV Nagar, Anand 388120, India
e-mail: subramanianrb@gmail.com

V.S. Meena et al. (eds.), *Potassium Solubilizing Microorganisms for Sustainable Agriculture*,
DOI 10.1007/978-81-322-2776-2_11

conversion make them a good choice for agricultural use. Application of such efficient KSM aims to develop future prospects to provide a sustainable environmental system in different crop fields under both normal and stress conditions.

Keywords
KSM • Potassium solubilisation • Photosynthesis • Root morphology • Lipid peroxidation • SDS profile • Salinity

11.1 Introduction

Saline soil distribution has increased throughout the ecosphere, especially in dry and semi-dry areas where agriculture performs under irrigation (Pessarakli and Szabolcs 1999). Potassium deficiency frequently compounds the problems of saline soil (Booker 1991). High salinity affects plant growth through many destructive effects such as an osmotic effect and harmfulness of salt ions, as well as variations in the biological and chemical properties of soil (Keren 2000). It also suppresses potassium uptake by plant roots and reduces the available potassium by absorption processes and low solubility of the K mineral. Since potassium is a critical nutrient influencing plant growth, this adversely affects plant growth under stress conditions. Nowadays application of inorganic fertilizers is the most common approach to improve soil fertility. However, the application of potassium supplements in chemical fertilizers is quickly fixed to the insolubilized forms, especially in saline soil, and this is the reason for low availability of potassium. Recent studies have proved that the use of biofertilizer combined with 25 % of chemical fertilizer gives good results for plant growth in the long term (Kramany et al. 2007). Nowadays, biofertilizer is an excellent substitute for inorganic fertilizer to upsurge soil fertility and enhance crop production in eco-friendly farming. Furthermore, the use of biofertilizer has gained momentum in recent years since chemical fertilizers are costly and have a hazardous effect on the ecosystem (Aseri et al. 2008).

Potassium solubilisation by KSM facilitates production of organic acids (Han and Lee 2005; Meena et al. 2013; Maurya et al. 2014). Fundamentally, potassium is a macronutrient for plants and can increase the plant height and fresh weight (Hossein et al. 2011). However, in recent decades, the issue of sustainable soil potassium management has somewhat been overlooked due to the possible environmental impact of nitrogen and phosphorus, which has been considered a more prominent problem. Furthermore, the use of biofertilizer somehow has not had continuous and constant effects compared with chemical fertilizer, as its mobilization is very slow. The mechanisms and interfaces among these microbes and plants are still not well understood, particularly for actual applications (Barea et al.1997; Meena et al. 2015b; Singh et al. 2015).

In plants, potassium also acts as a regulator, as it is a constituent of more than 65 different enzyme systems of drought tolerance and water-use efficiency. All crops require potassium, especially high-carbohydrate plants such as rice and potato. In addition, it has been shown that for optimum growth, crops need more potassium than usual (Simonsson et al. 2007). Although potassium is not an essential component of cell structure, it regulates many biochemical processes essential for growth, development and seed production. Potassium is present in four forms in the soil: as an interchangeable cation tightly bound to the soil mineral surfaces or organic matter; as an anion (K^+) in water extract of soil; tightly fixed in weathered micaceous resources; and in the matrix of K-containing primary crystals. The potassium-solubilizing plant growth–promoting bacteria may prove a useful tool in developing approaches to enable plant growth in saline soils.

Therefore, the application of potassium-solubilizing, plant growth–promoting microbes in agronomic practice would not only counterbalance the high production cost of potassium fertilizers, but also transform the insoluble potassium in the soil or in fertilizers applied for crop growth and yield (Meena et al. 2014a, 2015b).

Salt stress adversely affects plant nutrient acquisition, especially in the root, resulting in a significant decrease in shoot dry biomass. Inoculation of plants with KSM efficiently encourages the indigenous KSM population in the rhizosphere of the plant roots and enhances plant dry biomass, the root/shoot dry weight ratio and nutrient uptake by plants irrespective of salinity. KSM inoculation reduces the lethal effects of salt ions on plant growth by enabling better nutrient (NPK) absorption and greater accumulation of ions in different root tissues, when salinity is within acceptable limits. More specifically, soil-borne *Pseudomonas* and *Bacillus* have received specific consideration because of their outstanding root-colonizing ability, catabolic flexibility and ability to produce a wide range of metabolites that help the plant to survive in diverse biotic and abiotic stress conditions (Jha and Subramanian 2011b). The *Bacillus* strain could relieve the effects of salinity stress, and treatment with the bacterium *B. cereus* in soybean has been shown to alleviate the adverse effect of salinity in terms of the decline in growth, shoot and root dry mass, and number of leaves (Chakraborty et al. 2011).

The collaboration of plant growth–promoting bacteria, especially KSM, and their effect on the biological growth response of plants under soil salinity are complex. Inoculation of KSM in plants, alone or in groups, can confer tolerance against adverse environmental conditions and also improve potassium availability. This is correlated with variations in the antioxidant enzyme activity, level of oxidative damage by lipid peroxidation activity, photosynthesis rate, leaf greenness and other growth promotion parameters.

11.2 Isolation and Identification of KSM and Its Quantitative Potassium Solubilization

Salt ions limit cellular progression and affect plant physiology, thus for plant growth promotion, dealing with the consequences of stress is very important. Potassium accessibility for crop plants in soil is generally very low, as about 95–97 % of the entire potassium in the soil is in inaccessible mineral forms. Furthermore, the obsession of supplementary nutrients/fertilizers in soil diminishes the efficacy of applied P and K fertilizers, and thus a bulk amount of supplementary fertilizers becomes inaccessible to plants (Lian et al. 2002; Kumar et al. 2015b).

This results in frequent application of potassium fertilizers in the crop field, which not only increase the cost but are also environmentally objectionable (Fig. 11.1). Therefore, it is the need of the hour to find an environmental friendly and economically realistic choice for enhanced crop production in low-K soils. The metal-mobilizing, plant growth–promoting bacteria contribute significantly in the mobilization of insoluble forms of soil minerals (Supanjani et al. 2006). In this situation, microorganisms with potassium-solubilizing ability, known as KSM, may help the plant to acquire K for plant growth and hence are a feasible supplement to chemical fertilizers (Khan et al. 2006). The plant rhizosphere is the most active environmental place where frequent inter- and intra-species communications of bacteria, fungi and protozoa take place, due to the availability of rich and varied microbial nutrition secreted by the plant root (Bais et al. 2006). For the maintenance of plant health, the rhizospheric bacterial population plays a significant role in nutrient uptake, and development of tolerance against different environmental stress is now well known. So, biofertilization by potassium-solubilizing, plant growth–promoting bacteria enhances the nutrient

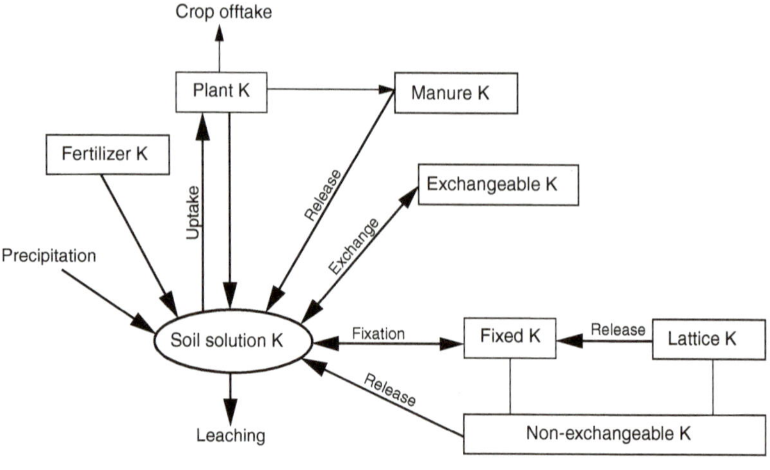

Fig. 11.1 The potassium cycle in the soil–plant–animal system (Syers 1998)

Fig. 11.2 Aleksandrov agar plates showing the potassium-solubilizing activity of *B. pumulis* and *P. pseudoalcaligenes*

conditions of the plant by integrating N fixation and NPK solubilisation. Therefore, the numerous KSM residing in the rhizosphere of the plant can provide K to the plant from the insoluble form from the sources otherwise poorly accessible by numerous mechanisms and are considered as a capable biofertilizer (Zaidi et al. 2009). KSM genera like *Azotobacter*, *Bacillus*, *Beijerinckia*, *Burkholderia*, *Enterobacter*, *Erwinia*, *Flavobacterium*, *Microbacterium*, *Pseudomonas*, *Rhizobium* and *Serratia* have been reported as the utmost substantial potassium solubilisation bacteria (Bhattacharyya and Jha 2012; Meena et al. 2015a). KSM was isolated by culturing of serially diluted (up to 10^6) enriched samples on Aleksandrov agar medium constituted with 1 g glucose, 0.05 M $MgSO_4.7H_2O$, 0.0005 M $FeCl_3$, 0.01 M $CaCO_3$, 0.2 M $CaPO_4$ and 0.5 M potassium aluminium silicate, agar 3 % pH 6.5 (Sugumaran and Janartham 2007) and incubated at 37 °C for 1 week. Colonies exhibiting clear zones on the 10^4, 10^5 and 10^6 dilution plates (Fig. 11.2) were selected as potential potassium solubilizers. On the basis of the zone activity of the different isolates, further screening was carried by using Khandeparkar's selection ratio:

$$\begin{aligned}\text{Ratio} &= \text{D/d} \\ &= \text{Diameter of zone of clearance/} \\ &\quad \text{Diameter of growth}\end{aligned}$$

Normally, the solubilisation of inorganic potassium by the soil bacteria occurs with the production of low weight organic acids (Zaidi et al. 2009). Mobilization of insoluble potassium from the aluminium silicate minerals by the KSM is due to the action of organic acids like citric, oxalic, malic, succinic and tartaric acid (Jha and Subramanian 2013b). Importantly, K solubilisation and mineralization can coincide in the same bacterial strain (Tao et al. 2008). Bacterial isolates like *Pseudomonas* and *Bacillus* have been found to be effective with reference to their K-solubilising competence (Table 11.1). The growth promotion efficiency of the isolates is analysed by their ability to solubilize potassium. K is one of the major macronutrients in the prerequisites for plants, and most potassium in soil is present in an insoluble form and cannot be utilized by the plants (Pradhan and Sukla 2005; Meena et al. 2013; Maurya et al. 2014).

11.3 Titratable Acidity and Organic Acid Production

Inorganic K solubilisation by KSM occurs principally by low molecular weight organic acid

Table 11.1 Potassium released, titratable acidity, organic acid concentration and pH during solubilization of tri-calcium phosphate over an incubation period of 12 days by the KSB

Days	pH		K release (μg K ml^{-1})		Titratable acidity ($\times 10^{-2}$)		Organic acid ($\times 10^{-2}$)	
	P. pseudoalcaligenes	*B. pumilus*	*P. pseudoalcaligenes*	*B. pumilus*	*P. pseudoalcaligenes*	*B. pumilus*	*P. pseudoalcaligenes*	*B. pumilus*
0	7.00 ± 0.01	7.00 ± 0.05	73.50 ± 2.12	74.5 ± 06.14	0	0	0	0
3	5.03 ± 0.03	5.00 ± 0.03	215.00 ± 4.63	191.5 ± 09.12	7.49 ± 0.11	13.3 ± 0.12	5.42 ± 1.19	1120 ± 1.66
7	3.51 ± 0.01	5.03 ± 0.01	535.50 ± 11.57	212.5 ± 15.09	23.40 ± 0.22	15.3 ± 0.15	4.18 ± 1.39	12.20 ± 2.12
9	3.80 ± 0.05	5.55 ± 0.03	626.50 ± 17.20	355.0 ± 08.10	27.70 ± 0.13	19.0 ± 0.15	3.42 ± 1.80	7.92 ± 1.15
12	3.90 ± 0.05	5.83 ± 0.02	874 ± 20.13.10	192.5 ± 19.15	29.50 ± 0.16	10.1 ± 0.10	1.98 ± 2.14	5.72 ± 1.19

production, by: (i) dropping the pH; (ii) improving chelation of the cations bound to K; (iii) contending with K for adsorption sites in the soil; and (iv) forming soluble complexes with metal ions associated with insoluble K (Ca, Al, Fe); thus K is released. The drop in the pH of the culture medium showed the production of organic acids by the K-solubilizing microorganisms (Jha and Subramanian 2014b; Maliha et al. 2004). Such an oxidation pathway takes place on the external face of the cytoplasmic membrane (Zaidi et al. 2009). Titratable acidity (TA) is determined by titrating 1 ml of culture filtrate against 0 10 mM NaOH in the presence of phenolphthalein (Whitelaw et al.1999). To estimate the production of organic acid by isolates, 1 ml of culture supernatant is used and the volume is increased to 5 ml with distilled water. To this, 0.05 ml ammonium chloride, 0.05 ml magnesium sulphate (0.5 M) and a pinch of Eriochrome T dye are added. This solution is titrated with 0.05 M ethylene diaminetetraacetic acid (EDTA). The end point of the reaction is given by the appearance of a blue colour. The result is expressed in mmol L^{-1} (Welcher 1958). After the ninth day of inoculation, the K produced and TA by *Bacillus* improve, while potassium production and TA are enhanced by *Pseudomonas* after 12 days of inoculation in the medium (Table 11.1). Such mineral K mobilization due to secretion of organic acids, such as gluconic and 2-ketogluconic acids, was reported by Zaidi et al. (2009). For the mobilization of potassium, production of organic acid is a well-recognized mechanism and is the cause of the decrease in the pH of the media by the isolates (Han et al. 2006; Jha et al. 2014b; Meena et al. 2015b).

An opposite relationship exists between the pH of the medium and the soluble K content, and a positive relationship exists between the soluble K content and titratable acid production, suggesting that acidification of the medium can assist in K solubilisation. A plant inoculated with more than one KSM has enhanced potassium assimilation ability compared with a plant inoculated with a single KSM, suggesting that strains in combination with others proceed synergistically in K solubilisation, and mixed application of such KSMs improves the K requirement of the crops.

However, the insufficiencies of potassium become a problem for crop yield because in crop fields, potassium is simply not available due to use by the crop, runoff, leaching and soil erosion (Sheng and Huang 2002). So the direct use of rock potassium (K) and phosphate (P) can be agronomically more beneficial and ecologically more practicable than use of soluble P and K (Rajan et al. 1996). Also, rock P and K supplies are an economical source of P and K, but both of these minerals are not easily available to a plant because minerals from rock are released very slowly and this is a constant reason for inadequate crop yield (Zapata and Roy 2004).

11.4 Rice Cultivation, Inoculation and Effect on Growth Parameters Under Saline Stress

KSMs are characterized by the following intrinsic uniqueness: (i) they must have ability to colonize the rhizosphere; (ii) they must persist, reproduce and compete with other surrounding microorganisms, for their own survival and the benefit of the host plant; and (iii) they must help the host (plant) grow and survive in adverse conditions (Jha and Subramanian 2014a). When the plant has been inoculated with metal-mobilizing, plant growth–promoting bacteria and planted in a crop field, this exerts a positive effect on plant growth promotion and survival. Seven-day-old KSM-inoculated plants were carefully removed from different test tubes inoculated with a strain of KSM bacterium, and planted in a pot. All seedlings were grown for 4 weeks in the absence of any nutrient supplements and the pot was kept in a greenhouse at 20–25 °C temperatures with relative humidity of 75–85 %. Salinity badly affected the growth of the plant, irrespective of the association with beneficial KSM action, as a decrease in plant growth is the consequence of the modification of many physiological events in the plant, such as reduction in chlorophyll content in the leaves, photosynthetic activity, nutrient uptake

Table 11.2 Effect of KSM on dry weight, plant height, leaf greenness, photosynthetic rate and phosphorus concentration under saline conditions

Salinity %	Treatment	DW (g plant^{-1})	PH (cm)	SPAD value	Photosynthetic rate (μmol CO_2 m^{-2} s^{-1}) (Li-Cor)	K conc. (mg kg^{-1})
0.3 dSm^{-1} control	No inoculation	2.43^{d}	62.1^{d}	43.3$^{c, d}$	23.4^{d}	58,710^{d}
	B. pumulis	2.68^{c}	69.3^{c}	44.1^{bc}	25.1^{c}	60,131^{c}
	P. pseudoalcaligenes	2.87^{b}	74. 2$^{a, b}$	44.9$^{a, b}$	27.7$^{a, b}$	64,223^{b}
	Co-inoculation	2.93^{a}	78.4^{a}	45.2^{a}	29.3^{a}	71,142^{a}
2.1 dSm^{-1}	No inoculation	2.03^{d}	54.2^{d}	39.8$^{c, d}$	19.2$^{c, d}$	54,123^{d}
	B. pumulis	2.23^{c}	61.4^{c}	40.3$^{b, c}$	20.8$^{b, c}$	61,341^{c}
	P. pseudoalcaligenes	2.41^{b}	68.5$^{a, b}$	41.2$^{a, b}$	21.7$^{a, b}$	68,250^{b}
	Co-inoculation	2.58^{a}	71.2^{a}	43.1^{a}	22.5^{a}	71,322^{a}
3.9 dSm^{-1}	No inoculation	1.98^{d}	42.3^{d}	36.1$^{c, d}$	18.6$^{c, d}$	45,132^{d}
	B. pumulis	2.21^{c}	48.4^{c}	37.2$^{b, c}$	19.7$^{b, c}$	47,242$^{b, c}$
	P. pseudoalcaligenes	2.34^{b}	55.2$^{a, b}$	38.3$^{a, b}$	20.5$^{a, b}$	48,121$^{a, b}$
	Co-inoculation	2.46^{a}	58.6^{a}	39.2^{a}	21.3^{a}	49,650
5.4 dSm^{-1}	No inoculation	1.73^{d}	34.1^{d}	32.2$^{c, d}$	16.2$^{c, d}$	39,415^{d}
	B. pumulis	1.86^{c}	37.5^{c}	33.3$^{b, c}$	17.2$^{b, c}$	41,153$^{b, c}$
	P. pseudoalcaligenes	1.98^{b}	43.7^{b}	34.1$^{a, b}$	18.4$^{a, b}$	43,972^{b}
	Co-inoculation	2.12^{a}	47.1^{a}	34.9^{a}	19.3^{a}	44,716^{a}
6.9 dSm^{-1}	No inoculation	1.52^{d}	28.6$^{c, d}$	29.3$^{c, d}$	13.4$^{c, d}$	35,162^{d}
	B. pumulis	1.74^{c}	31.4^{c}	30.1$^{b, c}$	14.2$^{b, c}$	36,343$^{b, c}$
	P. pseudoalcaligenes	1.86^{b}	37.2$^{a, b}$	31.3$^{a, b}$	15.3$^{a, b}$	37,320$^{a, b}$
	Co-inoculation	1.92^{a}	41.3^{a}	32.2^{a}	16.5^{a}	38,835
9.8 dSm^{-1}	No inoculation	1.21^{d}	21.7^{d}	25.4$^{c, d}$	10.4$^{c, d}$	32,174^{d}
	B. pumulis	1.42^{c}	26.3^{c}	26.3^{bc}	11.2$^{b, c}$	33,232^{c}
	P. pseudoalcaligenes	1.56^{b}	32.4^{b}	27.2$^{a, b}$	12.3$^{a, b}$	34,241$^{a, b}$
	Co-inoculation	1.75^{a}	37.1^{a}	27.9^{a}	13.4^{a}	35,716^{a}

Values are mean of three replications. Means within columns sharing the same letters are not significantly different ($P \leq 0.05$; LSD test)

and induction of antioxidant enzyme activity. However, the plants inoculated with KSM had a low degree of growth suppression, and the inoculated plants showed greater plant height by 12–26 % in the control condition and 13–70 % under salinity, and greater dry weight by 8–20 % in the control condition and 2–41 % under salinity, than the non-inoculated plants, as shown in Table 11.2. The techniques used by the plants against the damaging effects of salinity stress are very complex, so the growth and development mechanisms of the plants require a sensible balance with the target of saline stress effectors and supplements. The existence of *Azotobacter*, *Azospirillum*, *Bacillus*, *Bradyrhizobium*, *Pseudomonas* and *Rhizobium* has been reported from dry ecologies, acid soils, saline soils, alkaline zones and extremely tough hill slopes (Selvakumar et al. 2009; Jha and Subramanian 2013a) and may serve as a useful tool to enhance plant productivity under adverse environmental conditions.

11.5 Effect of KSM on Photosynthesis Under Saline Stress in Plants

Plant organelles like chloroplasts, mitochondria and peroxisomes, having a large number of oxidoreductase enzymes, are the site for intense oxidizing metabolic activity causing a high rate of electron flow and are the main source of reactive oxygen species (ROS) generation in plants. In plants, singlet oxygen molecules are mostly formed in the chlorophyll (Chl) and its tetrapyrrole metabolic intermediates in the presence of light. Chlorophyll is the site of photosynthesis and is the most abundant pigment in land plants. It acts as the central light energy–harvesting unit and is a component of not only the light reaction but also the dark reaction of photosynthesis in plants. The light harvested by the photosystems is used to excite the electrons, which permits their conversion into an electrochemical potential for production of chemical energy (ATP) via photophosphorylation. But incomplete oxidation results in the inefficient transfer of energy and is the cause of the formation of triplet-state Chl which counters with triplet oxygen to produce the highly reactive singlet oxygen molecules. In the plant, the reactive oxygen molecule is captured by the carotenoids to protect the plant from oxidative stress (Skovsen et al. 2005).

In the plant, photosynthesis is the chief process that produces highly reactive singlet oxygen molecules known as ROS in the chloroplasts, and can be a source for photoinhibitory and photooxidative damage (Edreva 2005). Besides the electron transport in photosystems, nicotinamide adenine dinucleotide phosphate (NADPH)-dependent oxidase stimulation and various types of biotic and abiotic stress are the other main causes of the ROS generation in the plant (Vranova et al. 2002). In the plant, the light-mediated conversion of reactive oxygen to superoxide takes place by a reaction linked to the photorespiratory cycle in the peroxisome by the help of antioxidant enzymes. Salinity causing the destruction of chlorophyll (chlorosis) is very common, resulting in inhibition of photosynthesis. Besides, potassium is essential for the photosynthesis process and acts as a key for photosynthesis enzymes to initiate the formation of carbohydrate, which is metabolized into amino acids and proteins in the plant. Plants inoculated with *Pseudomonas* and *Bacillus* (KSM) in non-saline conditions, as well as at different level of salinity, showed significantly high leaf greenness and photosynthetic rates compared with non-inoculated control plants (Jha and Subramanian 2013c). The plants inoculated with KSM showed 3–5 % higher leaf greenness in the control condition and 2–10 % under salinity. Similarly, the photosynthetic rate was also higher by 9–26 % in the control condition and 5–20 % under salinity. This may be because these isolates help the plants with water absorption and retention, and these findings were similar to those of Han and Lee (2005). Potassium has many functions in plant growth, such as the progress of cell division and growth, development of resistance to disease and drought tolerance. Potassium regulates the water status of the plant by controlling the opening and closing of the stomata and is required for osmotic regulation.

11.6 Effect of KSM on Nutrient Acquisition by Plant Roots Under Saline Stress

Water tension in plants, caused by high salinity, is one of the severe aspects that limits crop yield.

Salinity is one of the chief abiotic factors that restricts crop production by affecting almost plant functions, due to osmotic stress. Fundamentally, K^+ is readily water soluble and highly mobile and is transported in the plant xylem (Lack and Evans 2005). This transport of potassium can be facilitated either by potassium channels against the electrochemical gradient by the help of membrane potential or by secondary transporters (Bashan et al. 2004). Kaymak et al. (2008) reported that *Bacillus* spp. are very efficient in modifying various root parameters such root length and the dry matter content of the root as well as root functions in mint. The presence of denser root hairs increases the surface area of the root, which enhances water as well as mineral uptake (Raven and Edwards 2001). Improved root growth was suggested as a probable mechanism by which KSM enhances plant growth (Fallik et al. 1994). Inoculation of plants with both *Pseudomonas* and *Bacillus* facilitated enhanced root length as well as root hair development in general, but no anatomical change was observed in the xylem vessels under salinity (Jha and Subramanian 2013a). The root is the plant part that helps in procurement of nutrients from soils for the plant and is also the part that first interfaces with the abiotic and biotic components of soil. The root interfaces with physical, chemical and biological components of the soil in its surroundings (rhizosphere) (Hinsinger et al. 2009). Salinity dominated by Na^+ and Cl^- not only reduces Ca^{2+} and K^+ availability, but also reduces Ca^{2+} and K^+ mobility and transport to the growing parts of plants, affecting the quality of both vegetative and reproductive organs. The KSM strains vary greatly in their effect on the concentration of inorganic nutrients in plant leaves under soil salinity conditions. But inoculation with KSM increases the foliar K concentration in both normal and saline states. Plants inoculated with KSM showed 3–6 % enhanced foliar K concentrations in the normal state, 6–14 % in the low-salinity state and 4–8 % enhanced foliar K concentration in the high-salinity state. By better understanding of rhizosphere interfaces and association with soil bacteria, there is a chance to enhance the competence of nutrient uptake by plants (Meena et al. 2014a, b).

11.7 Effect of KSM on Accumulation of Osmoprotectant Under Saline Stress in Plants

The growth, development and yield of various crops are undesirably affected by numerous environmental factors like nutrient imbalances (including mineral toxicities and deficiencies), drought, salinity and extremes of temperature. Salinity is among the chief environmental limitations for crop production worldwide. Salinity is nowadays a major problem due to increases in sea water levels, limiting crop cultivation over large areas of land all over the world (Mandhania et al. 2006). This is more prominent in coastal and desert regions of the world. Salinity causes various types of physiological and biochemical changes in the plant. The major effect of salinity is accumulation of osmoprotectants, such as glycine-betaine derivatives and proline (Yancy et al. 1982). Osmoprotectants are low molecular weight, non-toxic, neutral solutes that help the plant with water retention under salinity. Though K is the major osmoticum, accumulation of compatible solutes is an important component of the adaptive mechanism under drought stress conditions (McCue and Hanson 1990). In water deficit, interaction of K and applied water, resulting in an increase in the sugar concentration in different cultivars, has been reported in several studies. Fixation of assimilated carbon into a particular plant metabolite is determined by the capacity of the plant for photoassimilate production, which decreases under drought stress conditions. Akatska and Nelson (1966) indicated that K increased starch synthetase activity and protected the enzyme from thermal inactivation. To alleviate the damaging effects of salt ions, plant cells accumulate osmoprotectant, which decreases the cell water potential. Due to the soil salinity, movement of water is restricted through the root and causes osmosis; for defending against the instant effect of water shortage, accumulation of the osmoprotectant takes place for osmoregulation. Such an osmoprotectant permits supplementary water to be taken up from the environment to counteract the effect of osmosis. Salt induces osmotic stress by limiting the absorption of water from the soil and ionic stress resulting from high concentrations of potentially toxic salt ions within plant cells. Salt stress is also linked to oxidative stress as a consequence of the generation of ROS, such as superoxide ions, hydrogen peroxide and hydroxyl radicals, which are detrimental to plant survival under salt stress. High salt ion concentrations cause a disturbance in membrane permeability expressed by an increase in solute leakage, which causes an imbalance of cellular ions resulting in ion toxicity (Cheesman 1988). This study demonstrated that the bacterial treatment influenced the extent of water stress, and *Pseudomonas* efficiently protected the host plant against the detrimental effects of salt. Greater hydration induced by the KSM might be attributed to increased water-use efficiency. Under salinity, many plants naturally accumulate osmolytes like proline and glycine-betaine. High levels of proline and glycine-betaine enable the plants to maintain an osmotic balance when growing under low water potential. It has been reported that proline and glycine-betaine protect plants against osmotic stresses, not only by adjusting osmotic pressure, but also by stabilizing many functional units such as complex II electron transport, membranes and proteins and enzymes such as RUBISCO (Ashraf and Foolad 2007). Glycine-betaine (N, N′, N0″-trimethyl-glycine) and proline are naturally synthesized and accumulated in bacteria, cyanobacteria, algae, fungi, animals and many plants. But important crop plants like arabidopsis, mustard and rice do not have the ability for proline and glycine-betaine production and accumulation under salinity stress (Rhodes and Hanson 1993). The study showed that plants like rice, when inoculated with selected KSM, acquire the ability to accumulate osmolytes like proline and glycine-betaine, which help the plant to survive under salinity stress (Jha et al. 2011a). Thus, to increase the salt stress tolerance in plants, diverse methods have been surveyed to intensify the accumulation of such osmoprotectants in plants grown under salinity stress. To overcome water deficit, plants have developed mechanisms for physiological adaptation, such as improvement of water-use efficiency by regulation of stomatal closure, development of root systems to acquire water, accumulation of osmoprotectants and control of water permeability by aquaporins (Kumar et al. 2003).

11.8 Effect of KSM on Lipid Peroxidation Activity Under Saline Stress in Plants

Plants growing on land and being sessile experience a multitude of environmental stresses. These environmental stresses are salinity, drought, radiation, extremes of temperature, water logging and mineral deficiency, etc. Such stress is nowadays one of the main challenges in the world's mission to produce food for all, as it affect the overall crop production worldwide. Salt stress causes diverse adverse effects on every organ of the plant, and the most affected parts are cell membranes (Shannon 1984). Previous studies suggested that accumulation of K^+ ions in the plant cell under salinity could help the plant survive in an adverse environment, as cationic solutes like K^+ ions are accountable for stomatal movements in the plant under low leaf water status. Salinity reduces the photosynthetic capacity of the plant due to low leaf water status and limited stomatal function. In such conditions, plants face partial nutrient imbalance and membrane disruption (Jha and Subramanian 2014a). In a plant growing in aerobic conditions, many oxidative reactions take place in subcellular organelles such as the chloroplast, mitochondrion and peroxisome. These subcellular organelles are common sites for the production of ROS. Although ROS are normally formed in low concentrations in plants under regular growth conditions, under environmental stress their production is increased during mitochondrial respiration, photorespiration and photosynthesis. The ROS are very reactive in nature and have the tendency to interact with any cell components such as the cell membrane, pigments, proteins, lipids, DNA and other important metabolites, which results in cell destruction (Casano et al. 1997). Plants have a natural capacity to detoxify ROS by induction of different types of antioxidants. Antioxidants are usually divided into two different types, i.e. enzymatic and non-enzymatic. Plant cell membranes are constituted by two types of lipids—the dominant storage lipid, triacylglycerols (TAGs), and a minor component, steryl esters. In the plant, TAGs and other neutral lipids are the hydrophobic barrier for protection from biotic and abiotic stress. Malondialdehyde (MDA) content is used as an indicator for the lipid peroxidation level in the plant leaf (Rao and Sresty 2000). MDA content is determined by using the thiobarbituric acid reaction, as an end product of lipid peroxidation. Free radical–induced oxidative damage in the cell is reflected by the peroxidation of lipids in plant cell membranes under salinity (Sudhakar et al. 2001; Jha et al. 2014a). Enhanced lipid peroxidation has been demonstrated in plants to increase salinity, while plants inoculated with selected KSM showed decreased MDA activity (Jha and Subramanian 2015). A reduction in potassium levels eventually inhibits plant growth as it hampers protein and lipid metabolism. Reduced availability of water will change the mineral distribution, increase membrane instability, increase the respiration rate and fail to maintain the turgor pressure and some of the events that prevail during this osmotic and ionic stress episode (Kumar et al. 2015a; Meena et al. 2015a; Jha and Subramanian 2014b).

11.9 Study of the Effect of KSM on Protein Expression by Protein Profiling

The induction of stress proteins in KSM-inoculated plants was analysed on 10 % SDS-PAGE by using the method of Laemmli (1970) in the seedlings of plants exposed to two different concentration of NaCl. Fresh plant leaves (2 g) were homogenized by mortar-grounded material in liquid nitrogen with an extraction buffer (20 mM Tris-HCl, pH 8.0, 5 mM EDTA, 50 mM leupeptin, 1 mM pepstatin A, 10 mM 3,4-dichloroisocumarine, 1 mM phenylmethylsulfonyl fluoride and 0.05 % SDS, from an experimental set). The samples bands were compared with the standard protein molecular weight marker bands (protein ladder) in the electrophorogram to determine the molecular weight of the protein. A few new bands of low molecular weight proteins were observed in the SDS gel, which may have some function in helping the plants in adverse condition. Polymorphic bands expressed in KSM-treated plants under control conditions were not seen in plant under stress, while a few more bands expressed under stress were not seen under control conditions (Fig. 11.3). A set of bands was present in all treatments, showing that the KSM induced different small protein molecules in plant under stress as well as control conditions to establish itself in the host plant and to protect the plant under salinity. Hettema et al. 1998 reported that low molecular weight polypeptides may act as molecular chaperones. The molecular chaperones bind with denatured proteins to maintain them in a state that allows further refolding (Lee and Vierling 2000).

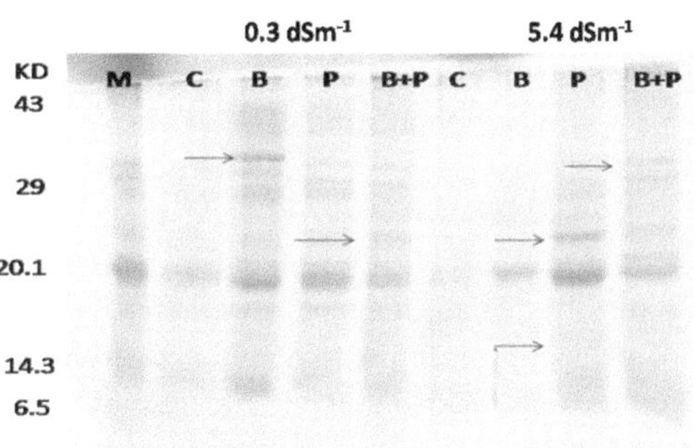

Fig. 11.3 Polyacrylamide gel of total soluble protein showing the differential expression of low molecular weight proteins due to inoculation of KSM and salinity. Loaded samples were adjusted to a constant amount of protein (15 μg). *M* marker (KD), *C* Control, *B* Plant inoculated with *B. pumulis*, *P P. pseudoalcaligenes* and *B* + *P* inoculated with both the isolates

11.10 Concluding Remarks and Future Perspectives

Mineral fertilizers first became familiar in an earlier era in the agriculture field, to reduce the difficulties faced by farmers and for sustainable production of crops. However, mineral fertilizers gradually started to express their side effects on human health and ecosystems. To overcome the drawback of chemical fertilizers and to increase plant nutrient status and yield in a sustainable way, the biofertilizers then began to be used. Moreover, KSM species specifically are well known for their capability to solubilize rock to enhance K availability in agricultural soils, besides increasing mineral contents in plants. This kind of bacteria is a very useful tool for solubilisation of potassium, which is the main nutrient, after nitrogen and phosphorus, for plants. Defence and antioxidant enzymes could be induced simply by inoculation with KSM in the plant prior to biotic and abiotic stress.

Plants inoculated with such KSM have enhanced plant growth and acquire a better capacity for salt tolerance, correlated with regulation of ion concentrations. To grow food for all, the development of genetically modified crops has raised hopes, but there has been only limited progress in realizing these hopes. So the use of biofertilizers, especially KSM, may be a beneficial means for enhanced plant growth and yield for a growing population. The extensive use of chemical fertilizers in agriculture may reduce with the wide-scale use of KSM biofertilizers. Furthermore, application of KSM biofertilizers is readily available to poor as well as rich farmers of the world. In future, KSM biofertilizers are a probable alternative to the chemical supplements used in the agriculture field that have several harmful effects on human health and the environment. Further investigation and understanding of mechanisms of KSM-mediated phytostimulation would help us to find more capable rhizobacterial strains having the ability to function efficiently under different agro-ecological conditions for sustainable agriculture.

References

Akatsuka T, Nelson OE (1966) Starch granule-bound adenosine diphosphate glucose starch glucosyltransferases of maize seeds. J Biol Chem 241: 2280–2286

Aseri GK, Jain N, Panwar J, Rao AV, Meghwal PR (2008) Biofertilizers improve plant growth, fruit yield, nutrition, metabolism and rhizosphere enzyme activities of pomegranate (*Punica granatum* L.) in Indian Thar desert. Sci Horticult 117:130–135

Ashraf M, Foolad MR (2007) Role of glycine betaine and proline in improving plant abiotic stress resistance. Environ Exp Bot 59:206–216

Bais HP, Weir TL, Perry LG, Gilroy S, Vivanco JM (2006) The role of root exudates in rhizosphere interactions with plants and other organisms. Annu Rev Plant Biol 57:233–266

Barea JM, Azcon-Aguilar C, Azcon R (1997) Interactions between mycorrhizal fungi and rhizosphere microorganisms within the context of sustainable soil–plant systems. In: Gange AC, Brown VK (eds) Multitrophic interactions in terrestrial systems. Blackwell Science, Cambridge, pp 65–77

Bashan Y, Holguin G, de-Bashan LE (2004) Azospirillum-plant relationships: physiological, molecular, agricultural, and environmental advances (1997–2003). Can J Microbiol 50:521–577

Bhattacharyya PN, Jha D (2012) Plant growth–promoting rhizobacteria (PGPR): emergence in agriculture. World J Microbiol Biotechnol 28:1327–1350

Casano LM, Gomez LD, Lascano HR, Gonzalez C, Trippi VS (1997) Inactivation and degradation of CuZn – SOD by active oxygen species in wheat chloroplasts exposed to photo-oxidative stress. Plant Cell Physiol 38:433–440

Chakraborty U, Swarnendu R, Chakraborty AP, Dey P, Chakraborty B (2011) Plant growth promotion and amelioration of salinity stress in crop plants by a salt-tolerant bacterium. Recent Res Sci Technol 3(11):61–70

Cheesman JM (1988) Mechanisms of salinity tolerance in plant. Plant Physiol 87:547–550

Edreva A (2005) Generation and scavenging of reactive oxygen species in chloroplasts: a submolecular approach. Agric Ecosyst Environ 106:119–133

El Kramany MF, Bahr AA, Mohamed MF, Kabesh MO (2007) Utilization of bio-fertilisers in field crops production 16-groundnut yield, its components and seeds content as affected by partial replacement of chemical fertilizers by bio-organic fertilizers. J Appl Sci Res 3(1):25–29

Fallik E, Sarig S, Okon Y (1994) Morphology and physiology of plant roots associated with *Azospirillum*. In: Okon Y (ed) Azospirillum/plant associations. CRC Press, Florida, pp 77–85

Hossein H, Daliri MS, Mobaser HR, Moosavi AA (2011) Effect of different nitrogen and potassium fertilizer levels on quality and quantity yield of flue-cured tobacco (Coker 347). World Appl Sci J 15(7):941–946

Han HS, Lee KD (2005) Physiological responses of soybean – inoculation of *Bradyrhizobiumjaponicum* with PGPB in saline soil conditions. Res J Agric Biol Sci 1(3):216–221

Han HS, Supanjani E, Lee KD (2006) Effect of co-inoculation with phosphate and potassium solubilizing bacteria on mineral uptake and growth of pepper and cucumber. Plant Soil Environ 52(3): 130–136

Hettema EH, Ruigrok CC, Koerkamp MG, van den Berg M, Tabak HF, Distel B, Braakman I (1998) The cytosolic DnaJ-like protein djp1p is involved specifically in peroxisomal protein import. J Cell Biol 142(2):421–434

Hinsinger P, Bengough AG, Vetterlein D, Young IM (2009) Rhizosphere: biophysics, biogeochemistry and ecological relevance. Plant Soil. doi:10.1007/S11104-008-9885-9

Jha Y, Subramanian RB (2011) Endophytic Pseudomonas pseudoalcaligenes shows better response against the Magnaporthegrisea than a rhizospheric Bacillus pumilus in Oryzasativa (Rice). Arch Phytopathol Plant Protect 44(6):592–604

Jha Y, Subramanian RB (2013a) Rhizobacteria regulates physiology and enzyme levels in paddy under salinity. J Appl Bot Food Qual 85:168–173

Jha Y, Subramanian RB (2013b) Root associated bacteria from the rice antagonizes the growth of Magnaporthe grisea. J Plant Pathol Microb 4:164. doi:10.4172/2157-7471.1000164

Jha Y, Subramanian RB (2013c) Paddy inoculated with PGPR show better growth physiology and nutrient content under salinity. Chil J Agric Res 73(1):213–219

Jha Y, Subramanian RB (2014a) PGPR regulate caspase-like activity, programmed cell death, and antioxidant enzyme activity in paddy under salinity. Physiol Mol Biol Plants 20(2):201–207

Jha Y, Subramanian RB (2014b) Characterization of root-associated bacteria from paddy and its growth-promotion efficacy.3. Biotech 4(3):325–330

Jha Y, Subramanian RB (2015) Reduced apoptosis like cell death and improved cell membrane integrity in paddy under salinity by root associate bacteria. Theor Exp Plant Physiol 27:227–235

Jha Y, Subramanian RB, Patel S (2011) Combination of endophytic and rhizospheric plant growth promoting rhizobacteria in *Oryzasativa* shows higher accumulation of osmoprotectant against saline stress. Acta Physiol Plant 33:797–802

Jha Y, Sablok G, Naidu Subbarao N, Sudhakar R, TurabeFazil MHU, Subramanian RB, Squartini, Kumar S (2014a) Bacterial-induced expression of RAB18 protein in *Orzyasativa* salinity stress and insights into molecular interaction with GTP ligand. J Mol Recognit 27:521–527

Jha Y, Subramanian RB, Jethwa R, Patel N (2014b) Isolation and identification of PGPR from rhizobacteria from *Suaedanudiflora* plant and its effect on maize (pioneer 30 v92). Indian J Plant Prot 42(4):422–429

Kaymak HC, Yarali F, Guvenc I, Donmez MF (2008) The effect of inoculation cuttings with plant growth Rhizobacteria (PGPR) on root formation of mint (*MenthapiperitaL*). Afr J Biotechnol 7:4479–4483

Keren R (2000) Salinity. In: Sumner ME (ed) Handbook of soil science. CRC Press, Boca Raton, pp G3–G25

Khan MS, Zaidi A, Wani PA (2006) Role of phosphate-solubilizing microorganisms in sustainable agriculture – a review. Agron Sustain Dev 27:29–43

Kumar SG, Reddy AM, Sudhakar C (2003) Nacl effects on proline metabolism in two high yielding genotypes of mulberry (*Morusalba* L.) with contrasting salt tolerance. Plant Sci 165:1245–1251

Kumar A, Bahadur I, Maurya BR, Raghuwanshi R, Meena VS, Singh DK, Dixit J (2015a) Dose a plant growth promoting rhizobacteria enhance agricultural sustainability. J Pure Appl Microbiol 9(1):715–724

Kumar A, Maurya BR, Raghuwanshi R (2015b) Characterization of bacterial strains and their impact on plant growth promotion and yield of wheat and microbial populations of soil. Afr J Agric Res 10(12): 1367–1375. doi:10.5897/AJAR2014.8894

Lack AJ, Evans DE (2005) Instant notes in plant biology, 1st edn, vol 7. Bios Scientific Publishers, Oxford, pp 68–71

Laemmli UK (1970) Cleavage of structural proteins during the assembly of the head of bacteriophage T4. Nature 227(259):680–685

Landon JR (1991) Booker tropical soil manual. A handbook for soil survey and agricultural land evaluation in the tropics and subtropics. Booker Tate Essex: Longman Scientific & Technical Publishers, Harlow

Lee GJ, Vierling E (2000) A small heat shock protein cooperates with heat shock protein 70 systems to reactivate a heat- denatured protein. Plant Physiol 1: 189–198

Lian B, Fu P, Mo DM, Liu CQ (2002) A comprehensive review of the mechanism of potassium release by silicate bacteria. Acta Mineral Sin 22:179–183

Maliha R, Samina K, Najma A, Sadia A, Farooq L (2004) Organic acids production and phosphate solubilization by phosphate solubilizing microorganisms under in vitro conditions. Pak J Biol Sci 7:187–196

Mandhania S, Madan S, Sawhney V (2006) Antioxidant defense mechanism under salt stress in wheat seedling. Biol Plant 50:227–231

Maurya BR, Meena VS, Meena OP (2014) Influence of Inceptisol and Alfisol's potassium solubilizing bacteria (KSB) isolates on release of K from waste mica. Vegetos 27(1):181–187

McCue KF, Hanson AD (1990) Drought and salt tolerance: towards understanding and application. Trends Biotechnol 8:358–362

Meena OP, Maurya BR, Meena VS (2013) Influence of K-solubilizing bacteria isolates on release of K from waste mica. Agric Sustain Dev 1(1):53–56

Meena VS, Maurya BR, Bahadur I (2014a) Potassium solubilisation by bacteria strain in waste mica. Bangladesh J Bot 43(2):235–237

Meena VS, Maurya BR, Verma JP (2014b) Does a rhizospheric microorganism enhance K^+ availability in agricultural soil? Microbiol Res 169:337–347

Meena RK, Singh RK, Singh NP, Meena SK, Meena VS (2015a) Isolation of low temperature surviving plant growth promoting bacteria (PGPR) from pea (*PisumSativum* L) and demonstration of their plant growth promoting trait. Biocatal Agric Biotechnol. doi:10.1016/j.bcab.2015.08.006

Meena VS, Maurya BR, Verma JP, Arora A, Kumar A, Kim K, Bajpai VK (2015b) Potassium solubilizing rhizobactera (KSR): isolation, identification and K-release dynamics from waste mica. Ecol Eng 81: 340–347

Pessarakli M, Szabolcs I (1999) Soil salinity and sodicity as particular plant/crop stress factors. In: Pessarakli M (ed) Handbook of plant and crop stress, 2nd edn. Marcel Dekker, New York, pp 1–15

Pradhan N, Sukla LB (2005) Solubilization of inorganic phosphate by fungi isolated from agriculture soil. Afr J Biotechnol 5:850–854

Rajan SSS, Watkinson JH, Sinclair AG (1996) Phosphate rock of for direct application to soils. Adv Agron 57: 77–159

Rao MKV, Sresty TVS (2000) Antioxidative parameters in the seedlings of pigeonpea (*Cajanuscajan* L. *Millspaugh*) in response to Zn and Ni stresses. Plant Sci 157:113–128

Raven JA, Edwards D (2001) Roots: evolutionary origins and biogeochemical significance. J Exp Bot 52: 381–401

Rhodes D, Hanson AD (1993) Quaternary ammonium and tertiary sulfonium compounds in higher-plants. Annu Rev Plant Phys 44:357–384

Selvakumar G, Joshi P, Nazim S, Mishra PK, Bisht JK, Gupta HS (2009) Phosphate solubilization and growth promotion by *Pseudomonas fragi* CS11RH1 (MTCC 8984) a psychrotolerant bacterium isolated from a high altitude Himalayan rhizosphere. Biologia 64:239–245

Shannon MC (1984) Breeding, selection, and the genetics of salt tolerance. In: Wiley J (ed) Salinity tolerance in plants strategies for crop improvement. Wiley, New York, pp 231–254

Sheng XF, Huang WY (2002) Mechanism of potassium release from feldspar affected by the strain NBT of silicate bacterium. Acta Pedol Sin 39:863–871

Simonsson M, Andersson S, Andrist-angel Y, Hillier S, Mattsson l, Oborn I (2007) Potassium release and fixation as a function of fertilizer application rate and soil parent material. Geoderma 140:188–198

Singh NP, Singh RK, Meena VS, Meena RK (2015) Can we use maize (*Zea mays*) rhizobacteria as plant growth promoter? Vegetos 28(1):86–99

Skovsen E, Snyder JW, Lambert JD, Ogilby PR (2005) Lifetime and diffusion of singlet oxygen in a cell. J Phys Chem B 109:8570–8573

Sudhakar C, Lakshmi A, Giridarakumar S (2001) Changes in the antioxidant enzyme efficacy in two high yielding genotypes of mulberry (*Morusalba* L.) under NaCl salinity. Plant Sci 161:613–619

Sugumaran P, Janartham B (2007) Solubilization of potassium containing minerals by bacteria and their effect on plant growth. World J Agric Sci 3(3): 350–355

Supanjani Han HS, Jung SJ, Lee KD (2006) Rock phosphate potassium and rock solubilizing bacteria as alternative sustainable fertilizers. Agron Sustain Dev 26:233–240

Syers JK (1998) Soil and plant potassium in agriculture. Proceedings no. 411, The International Fertiliser Society York, UK: 32 pp

Tao GC, Tian SJ, Cai MY, Xie GH (2008) Phosphate solubilizing and -mineralizing abilities of bacteria isolated from. Pedosphere 18:515–523

Vranova E, Inze D, Van Breusegem F (2002) Signal transduction during oxidative stress. J Exp Bot 53: 1227–1236

Welcher FJ (1958) The analytical uses of ethylene diamine tetraacetic acid (EDTA). D. Van Nostrand-company, Princeton

Whitelaw MA, Harden TJ, Helyar KR (1999) Phosphate solubilisation in solution culture by the soil fungus *Penicilliumradicum*. Soil Biol Biochem 31:655–665

Yancy PH, Clark ME, Hand SC, Bowlus RD, Somero GN (1982) Living with water stress, evolution of osmolytes systems. Science 217:1214–1223

Zaidi A, Khan MS, Ahemad M, Oves M (2009) Plant growth promotion by phosphate solubilizing bacteria. Acta Microbiol Immunol Hung 56:263–284

Zapata F, Roy RN (2004) Use of phosphate rock for sustainable agriculture. FAO and IAEA, Rome

Can *Bacillus* spp. Enhance K^+ Uptake in Crop Species

12

Muhammad Yasin, Iqra Munir, and Muhammad Faisal

Abstract

For all organisms potassium (K) acts as an essential cationic nutrient. In plants, K regulates growth and movement of various plant organs and plays a crucial role in biotic and abiotic stress tolerance mechanisms. The rhizospheric microorganisms play an important role in K biogeochemical cycling and may influence K^+ uptake in crop species by direct and indirect mechanisms. The present article's major emphasis is on the characterization of K-solubilizing bacteria (KSB) and their effect on K uptake, plant development, and growth. Moreover, under various environmental conditions, the importance of K in crop yield improvement and the concentration of K in various plant parts were also described, and the possible molecular mechanism behind this phenomenon will be elaborated.

Keywords

Potassium • Rhizosphere • Microorganisms • Biogeochemical cycling • K uptake

12.1 Introduction

Potassium (K) is an essential nutrient element for sustainable crop production and productivity. K is the seventh most common element in earth's crust and lithosphere contains ~2.5 % K. In soil K is present in various chemical forms such as soil minerals and exchangeable and non-exchangeable forms. Among these K forms in soil, the concentration of soluble K is usually very less (~2 %), whereas ~ 98 % of K exists in the form of insoluble rocks and minerals (Goldstein 1994). The silicate minerals of K include orthoclase, muscovite, feldspar, mica, biotite, smectite, vermiculite, etc. (Ullman et al. 1996). K is the second most abundant element in plants which comprises 2 to 10 % of total plant dry biomass.

M. Yasin (✉) • I. Munir • M. Faisal (✉)
Department of Microbiology and Molecular Genetics, University of the Punjab, Quaid-e-Azam Campus, Lahore 54590, Pakistan
e-mail: yasin_mmg@yahoo.com; mohdfaysal@yahoo.com

V.S. Meena et al. (eds.), *Potassium Solubilizing Microorganisms for Sustainable Agriculture*,
DOI 10.1007/978-81-322-2776-2_12

In plants, K regulates plant growth, phloem transport, tropisms, and regulation of stomatal aperture and plays an important role in biotic and abiotic stress tolerance (Anschütz et al. 2014). At the cell level, K plays various functions such as protein synthesis, electrical neutralization of anionic charges, regulation of the osmotic potential, and long- and short-term control of membrane polarization (Cherel et al. 2014). Activity of ~70 enzymes was shown to be sensitive to K^+ concentration (Marschner 1995; Sheng et al. 2002). The importance of K fertilization has been recognized a long time ago. In the twentieth century, green revolution in agriculture resulted in dramatically enhanced crop yield, and global demand for K-based fertilizers is increasing day by day. In 2012 over 50 MT of potash were produced. A judicious use of NPK-containing fertilization is necessary for optimal plant growth and development. However, in field the efficiency of the NPK fertilizers is lower because of chemical fixation of these elements. Although K occurs naturally in soil, its availability to crops is generally low because ~90 to 98 % of K is present as minerals in soil and it is not available for nutrient uptake by plant roots. However, through various mechanisms, the rhizosphere microorganisms, both bacteria and fungi can enhance the solubilization of such essential chemical elements, and thus they become more available for uptake by plant roots (Parmar and Sindhu 2013; Meena et al. 2013; Maurya et al. 2014).

12.2 Potassium-Solubilizing Bacteria (KSB)

There is a wide range of soil microorganisms that are involved in solubilization of K and convert the fixed and insoluble forms of K into soluble and available forms which can be easily taken up by plants roots (Li et al. 2006; Keshavarz et al. 2013; Gundala et al. 2013; Meena et al. 2014a; Singh et al. 2015). In soil, K-mobilizing microbial communities may enhance the plant growth and crop yield and are economically viable and environment friendly. The solubilization of rock K by microorganisms was first reported by Muntz (1890). Different types of K-solubilizing bacteria have been isolated from different locations of agricultural lands and found to solubilize K and silicates from insoluble forms of these minerals and thus play an important role in nutrient cycle (Aleksandrov et al. 1967; Meena et al. 2014b).

Some of the previously reported K-solubilizing bacterial species include *Bacillus edaphicus* (Sheng 2005), *B. circulans* (Lian et al. 2002), *Acidithiobacillus ferrooxidans*, *Burkholderia* spp. (Sheng and He 2006; Sheng and Huang 2002), *Enterobacter hormaechei* (Prajapati et al. 2013), *B. mucilaginosus* (Keshavarz et al. 2013), *Paenibacillus mucilaginosus* (Liu et al. 2012; Hu et al. 2006), *Sphingomonas* spp. (Uroz et al. 2007), *Arthrobacter* spp. (Keshavarz et al. 2013), and *Paenibacillus glucanolyticus* (Sangeeth et al. 2012); however, among these *B. edaphicus* and *B. mucilaginosus* (Zhao et al. 2008; Sheng 2005; Lian et al. 2002; Li et al. 2006; Rajawat et al. 2012; Kumar et al. 2015; Meena et al. 2015a) exhibited high K-solubilizing and K-mobilizing activity (Table 12.1 and Fig. 12.1).

12.3 K Solubilization Mechanisms

KSB solubilize K by decreasing the pH, chelating the cations bounded to K, and acidolysis of surrounded area by K-solubilizing microorganisms. Decreases in pH suggest the release of protons and organic acids by KSB (Uroz et al. 2009; Keshavarz et al. 2013; Parmar and Sindhu 2013). The KSMs synthesize and discharge organic acids in the environment which acidify the microbial cell and its surrounding environment which eventually allows the release of K^+ from minerals by acidification and protonation (Goldstein 1994). Among these different organic acids which are involved in K solubilization include gluconic acid, succinic acid, α-ketogluconic acid, oxalic acid, citric acid, malic acid, acetic acid, and lactic acid which are the major ones produced by *Bacillus* spp. (Taha et al. 1969; Bajpai and Sundara 1971; Sheng and He 2006; Gundala

Table 12.1 Isolation sources of KSMs from K-bearing minerals and rhizospheric soil of various plants

Isolation source	Reference
Wheat	Parmar and Sindhu (2013)
Valencia orange	Shaaban et al. (2012)
Rice	Muralikannan and Anthomiraj (1998)
Tea	Bagyalakshmi et al. (2012)
Potato	Abdel-Salam and Shams (2012)
Black pepper	Sangeeth et al. (2012)
Chili, sorghum, bajra, and maize	Archana et al. (2013)
Common bean	Kumar et al. (2012)
Feldspar	Sheng et al. (2008)
Potato-soybean cropping sequence	Biswas (2011)
Iranian soils	Keshavarz et al. (2013)
Ceramic industry soil	Prajapati and Modi (2012)
Soil, rock, and mineral	Sugumaran and Janarthanam (2007)
Soil of Tianmu Mountain, Zhejiang province (China)	Hu et al. (2006)
Ha Tien Mountain, Vietnam	Diep and Hieu (2013)
Tobacco rhizosphere	Zhang and Kong (2014)
Mica core of Andhra Pradesh	Gundala et al. (2013)
Bio-fertilizers	Zakaria (2009)
Tomato, banana, groundnut, cotton, and soybean	Archana et al. (2012)

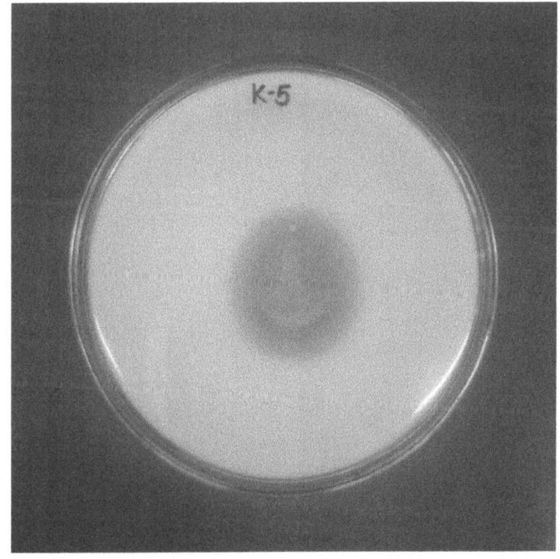

Fig. 12.1 Formation of solubilization zone by KSB strains K5 in Aleksandrov medium

et al. 2013; Meena et al. 2014a). The production of organic acid can be detected through enzymatic methods and high-performance liquid chromatography (HPLC) (Archana et al. 2013). However, acidification is not the only mechanism for K solubilization, because in some circumstances the decrease in pH did not affect the K solubilization (Rosa-Magri et al. 2012; Keshavarz et al. 2013; Meena et al. 2015b).

12.4 The Factor Influences the K Solubilization?

As we know, most of the K reserves of soil are not available to plants. Hence, it is necessary to supply the soluble form of K for plant availability. In soil, K mobilization is affected by many biotic and abiotic environmental factors like soil properties (physicochemical characteristics, aeration, pH), presence of mycorrhizal fungi and rhizosphere microorganisms especially bacteria, and the composition of plant root exudates. These parameters may also influence the K mobilization ability of KSB. Gahoonia et al. (1997) explained three hypothetical processes: (a) mineral fragmentation caused by the activity of root by increasing the bacterial effect on mineral mobilization due to increased surface area for reactivity, (b) root exudates helping indirectly by providing the substrates for the production of weathering metabolites by bacteria, and (c) besides producing weathering agents, bacteria

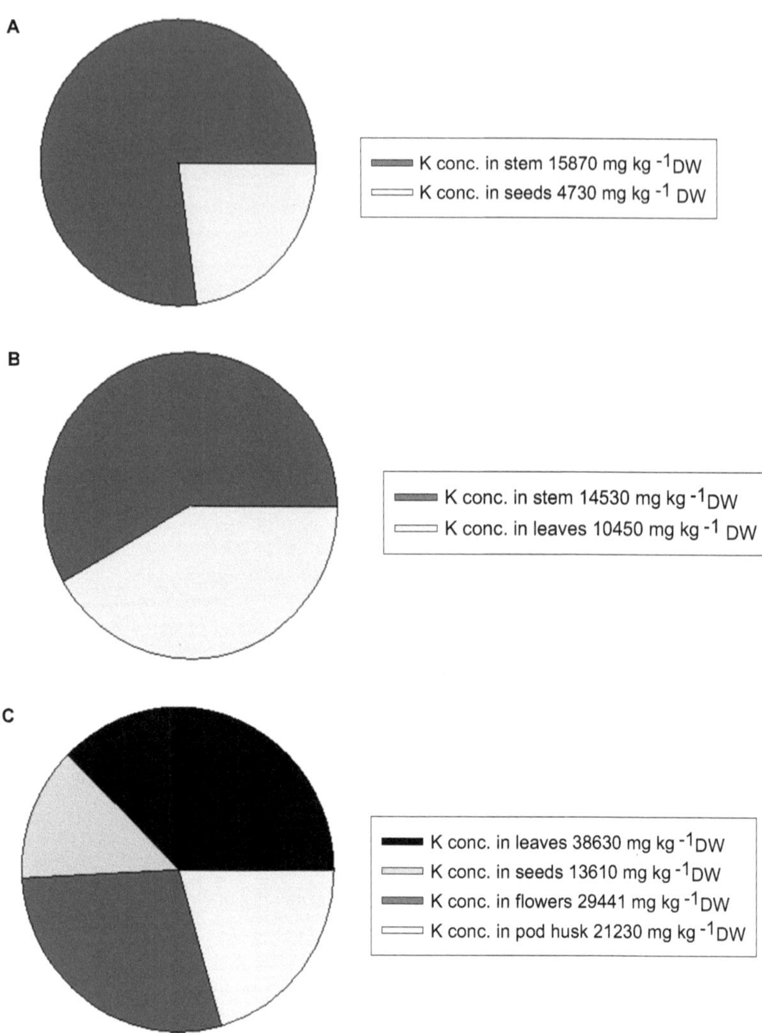

Fig. 12.2 K^+ concentration in various parts of (**a**) wheat grown in natural agricultural soil in Punjab; (**b**) maize grown in natural agricultural soil in Punjab, Pakistan; and (**c**) Indian mustard grown in shale rock-derived soil, Fort Collins, CO, USA

producing phytohormones which stimulate the development of root, modify root exudation and physiology, and help improve nutrient uptake and mobilization of minerals. The fixation and release of K mainly depend on the K concentration in the soil solution (Schneider et al. 2013). Co-inoculation of K-dissolving bacteria (KDB) and P-dissolving bacteria (PDB) in combination with P and K materials resulted an increase in P and K accumulation and plant growth (shoot and root growth) of corn grown in P- and K-limited soils (Abou-el-Seoud and Abdel-Mageed 2012). Potassium concentration in various plant parts is not similar with results of some field studies under various environmental conditions; we observed the K^+ concentration in various parts of wheat, corn, and Indian mustard plants with ICP-OES, through acid digestion method (Fig. 12.2). It was noticed that under aerobic condition, more K is released from feldspar, illite, and muscovite as compared to anaerobic condition (Sheng 2005; Badr 2006). In liquid medium *Bacillus edaphicus* showed better growth and greater K-releasing ability with illite as compared to feldspar (Sheng and He 2006).

12.5 K^+ Transporters

In most of the soils, the K^+ availability is generally low and fluctuating in various geographical areas. Under the K-limiting conditions, to preserve the vital functions and growth of plants, we have developed various strategies to take up more K^+. These strategies may include increased capacity for high-affinity K^+ uptake, ensuring cytosolic homeostasis, and modification of root system development (Cherel et al. 2014).

Epstein et al. (1963) reported that in plants, the transport of K^+ from soils into roots is through high- and low-affinity components, and K^+ uptake is mainly mediated by high-affinity K^+ uptake K m ~10–40 μM (Daniel et al. 1994). Plants may contain multiple K^+ transporters for high-affinity uptake, and a genome-wide survey revealed seven major families of cation transporters (total 75 genes) in *Arabidopsis* which mediate K^+ uptake in plants across the plasma membrane (Anschütz et al. 2014). These include putative K^+/H^+ anti-porters (6 genes), Shaker-type K^+ channels (9 genes), KUP/HAK/KT transporters (13 genes), two-pore K^+ channels (6 genes), cyclic nucleotide-gated channels (20 genes), glutamate receptors (20 genes), and HKT transporters (1 gene) (Shabala 2003; Véry and Sentenac 2002).

HKT1 protein (a membrane protein in roots) is important in plant nutrition which confers the ability to take up K^+ and could possibly contribute to environmental alkali metal toxicities (Daniel et al. 1994). Another one study by Kim et al. (1998) had identified a new family (AtKUP) of K transporters in *Arabidopsis thaliana* and proposed that the AtKUP family may provide important components of low- and high-affinity K^+ nutrition and uptake in different plant cell types. Li et al. (2014) demonstrated that in rice roots, the Os-AKT1 (the Shaker K^+ channel)-mediated K^+ uptake is modulated by the Os-CBL1-Os-CIPK23 complex (Os-CBL1 and Os-CIPK23 are upstream regulators of Os-AKT1 in rice). Moreover, K channels mediate facilitated diffusion of K^+ across the plasma membrane, and this opening and closing of K^+ channels is controlled by changes in plasma membrane voltages (Badr et al. 2006; Basak and Biswas 2009).

12.6 Effect of KSB on Growth and Yield of Crop Species

KSB inoculants having the ability to solubilize K from rocks and minerals can greatly influence the plants growth as well as they are beneficial for environment. Several studies have been conducted on K-solubilizing *Bacillus* spp. which mainly focused on potassium-solubilizing capability along with their impact on plant growth and K^+ uptake in the KSB-inoculated plants. Seedling treatment or seed inoculation with KSB usually resulted in increased germination, seedling strength, growth, and yield and improved uptake of K^+ by plant roots. Singh et al. (2010) studied the impact of *Azotobacter chroococcum*, *Bacillus mucilaginosus*, and *Rhizobium* spp. to solubilize K from waste mica by inoculating *Triticum aestivum* and *Zea mays* plants. Improved uptake of K^+ was observed in wheat and corn with increased biomass, protein, and chlorophyll content. *B. mucilaginosus* showed higher K solubilization ability than the other inoculants. Mikhailouskaya and Tcherhysh (2005) observed the effect of inoculation with KSB on eroded soils was comparable to the crop yield in moderately eroded soils without KSB inoculation, and results showed increase in yield of wheat up to 1.04 t/ha. Wheat inoculated with wild-type NBT strain of *B. edaphicus* resulted in enhanced growth of shoots and roots as well as increased NPK content in plants as compared to control plants (Sheng and He 2006). Similarly, the inoculation of soil with *B. circulans* for silicate and K solubilization from minerals resulted in increased organic matter as well as ~17 % increase in rice yield (Muralikannan and Anthomiraj 1998).

Bader et al. (2006) reported that the inoculation with KSMs in sorghum plants treated with K- and P-containing minerals resulted in enhanced uptake of K (79 %, 41 %, and 93 %) and P (116 %, 71 %, and 110 %) and increased dry matter content (58 %, 48 %, and 65 %) in calcareous, clay, and sandy soils, respectively. According to Sugumaran and Janarthanam (2007), increase in oil content (35 %) and dry matter (25 %) was observed in groundnut plants with increased availability of K and P in soil (86.57–99.60 and 6.24–9.38 mg/kg, respectively) when inoculated with *B. mucilaginosus*. Nayak (2001) reported that there is increase uptake of K and plant biomass of brinjalon potash mobilizer's treatment as compared to untreated control. According to Archana et al. (2008), KSB strain from rhizosphere of *Vigna radiata* and rock was found efficient in K solubilizers in acidic soils and further improved the growth of green gram. The effect of K-mobilizing *B. edaphicus* on cotton and chili was studied which resulted in increased available K and P content in plant biomass (Sheng et al. 2003). Sheng and Huang (2002) studied the effect of KSB and reported the increase in plant biomass and better uptake of K and P in sorghum plant as compared to control. The combination of K solubilizers with chemical fertilizer and bio-fertilizer resulted in increased K^+ uptake as well as increased yield in tapioca (*Manihot esculenta*) and yam (*Dioscorea* spp.) (Clarson 2004). These KSB like *B. edaphicus* increased the nutrient uptake and root and shoot growth in rapeseed and cotton in K-deficient soil, and K content was increased by ~26 % and 30 %, respectively (Sheng 2005). K-solubilizing *Bacillus* spp. improved the nutrition, growth, and yield of corn plant as compared to fertilizer control (Archana et al. 2008). Recently, Subhashini (2015) observed the impact of potassium-mobilizing bacterium *Frateuria aurantia* on the increased uptake of potassium in tobacco.

12.7 Concluding Remark

In conclusion, in soil K mobilization is affected by many biotic and abiotic environmental factors. *Bacillus* spp. can effectively improve K^+ uptake in different plant species by producing various organic acids in plant rhizosphere and stimulating plant root growth. Plant inoculation with KSB can greatly enhance product yields in terms of increased plant biomass and also improve efficacy of NPK fertilizers through enhanced solubilization and more uptake by plant roots.

References

Abdel-Salam M, Shams A (2012) Feldspar-K fertilization of potato (*Solanum tuberosum* L.) augmented by biofertilizer. J Agri Environ Sci 12:694–699

Abou-el-Seoud II, Abdel-Mageed A (2012) Impact of rock materials and biofertilizations on P and K availability for maize (*Zea maize*) under calcareous soil conditions. Saudi J Biol Sci 19:55–63

Aleksandrov V, Blagodyr R, Ilev I (1967) Liberation of phosphoric acid from apatite by silicate bacteria. Mikrobiol Z 29:111–114

Anschütz U, Becker D, Shabala S (2014) Going beyond nutrition: regulation of potassium homoeostasis as a common denominator of plant adaptive responses to environment. J Plant Physiol 171:670–687

Archana DS, Savalgi VP, Alagawadi AR (2008) Effect of potassium solubilizing bacteria on growth and yield of maize. Soil Biol Ecol 28:9–18

Archana DS, Nandish MS, Savalagi VP, Alagawadi AR (2012) Screening of potassium solubilizing bacteria (KSB) for plant growth promotional activity. Bioinfolet 9:627–630

Archana DS, Nandish MS, Savalagi V, Alagawadi A (2013) Characterization of potassium solubilizing bacteria (KSB) from rhizosphere soil. Bioinfolet 10:248–257

Badr MA (2006) Efficiency of K-feldspar combined with organic materials and silicate dissolving bacteria on tomato yield. J Appl Sci Res 2:1191–1198

Badr MA, Shafei AM, Sharaf El-Deen SH (2006) The dissolution of K and P-bearing minerals by silicate dissolving bacteria and their effect on sorghum growth. Res J Agri Biol Sci 2:5–11

Bagyalakshmi B, Ponmurugan P, Balamurugan A (2012) Impact of different temperature, carbon and nitrogen sources on solubilization efficiency of native potassium solubilizing bacteria from tea (*Camellia sinensis*). J Biol Res 3:36–42

Bajpai PD, Sundara R (1971) Phosphate solubilizing bacteria. Solubilization of phosphate in liquid culture by selected bacteria as affected by different pH values. J Soil Sci Plant Nutr 17:41–43

Basak BB, Biswas DR (2009) Influence of potassium solubilizing microorganism (*Bacillus mucilaginosus*) and waste mica on potassium uptake dynamics by sudan grass (*Sorghum vulgare* Pers.) grown under two Alfisols. Plant Soil 317:235–255

Biswas DR (2011) Nutrient recycling potential of rock phosphate and waste mica enriched compost on crop productivity and changes in soil fertility under potato-soybean cropping sequence in an Inceptisol of Indo-Gangetic plains of India. Nutr Cycl Agroecocyst 89:15–30

Cherel I, Lefoulon C, Boeglin M, Sentenac H (2014) Molecular mechanisms involved in plant adaptation to low $K^{(+)}$ availability. J Exp Bot 65:833–848

Clarson D (2004) Potash biofertilizer for ecofriendly agriculture. Agro-Clinic and Research Centre, Kottayam, pp 98–110

Daniel P, Schachtman DP, Schroeder JI (1994) Structure and transport mechanism of a high-affinity potassium uptake transporter from higher plants. Nature 370:655–658

Diep CN, Hieu TN (2013) Phosphate and potassium solubilizing bacteria from weathered materials of denatured rock mountain, Ha Tien, Kiên Giang province Vietnam. Am J Life Sci Res 1:88–92

Epstein E, Rains DW, Elzam OE (1963) Resolution of dual mechanisms of potassium absorption by barley roots. Proc Natl Acad Sci U S A 49:684–692

Gahoonia TS, Care D, Nielsen NE (1997) Root hairs and phosphorus acquisition of wheat and barley cultivars. Plant Soil 191:181–188

Goldstein AH (1994) Involvement of the quino-protein glucose dehydrogenase in the solubilization of exogenous mineral phosphates by gram-negative bacteria. In: Torriani-Gorini A, Yagil E, Silver S (eds) Phosphate in microorganisms: cellular and molecular biology. ASM Press, Washington, DC, pp 197–203

Gundala PB, Chinthala P, Sreenivasulu B (2013) A new facultative alkaliphilic, potassium solubilizing, *Bacillus* sp. SVUNM9 isolated from mica cores of Nellore District, Andhra Pradesh, India. Research and reviews. J Microbiol Biotechnol 2:1–7

Hu X, Chen J, Guo J (2006) Two phosphate-and potassium-solubilizing bacteria isolated from Tianmu mountain, Zhejiang, China. World J Microbiol Biotechnol 22:983–990

Keshavarz ZJ, Aliasgharzad N, Oustan S, Emadi M, Ahmadi A (2013) Isolation and characterization of potassium solubilizing bacteria in some Iranian soils. Arch Agron Soil Sci 59:1713–1723

Kim EJ, Kwak JM, Uozumi N, Schroeder JI (1998) AtKUP1: an Arabidopsis gene encoding high-affinity potassium transport activity. Plant Cell 10:5–62

Kumar P, Dubey R, Maheshwari D (2012) *Bacillus* strains isolated from rhizosphere showed plant growth promoting and antagonistic activity against phytopathogens. Microbiol Res 167:493–499

Kumar A, Bahadur I, Maurya BR, Raghuwanshi R, Meena VS, Singh DK, Dixit J (2015) Does a plant growth-promoting rhizobacteria enhance agricultural sustainability? J Pure Appl Microbiol 9(1):715–724

Li F, Li S, Yang Y, Cheng L (2006) Advances in the study of weathering products of primary silicate minerals, exemplified by mica and feldspar. Acta Petrol Mineral 25:440–448

Li J, Long Y, Qi GN, Li J, Xu ZJ, Wu WH, Wang Y (2014) The Os-AKT1 channel is critical for K^+ uptake in rice roots and is modulated by the rice CBL1-$CIPK_{23}$ complex. Plant Cell 26:3387–3402

Lian B, Fu P, Mo D, Liu C (2002) A comprehensive review of the mechanism of potassium releasing by silicate bacteria. Acta Mineral Sin 22:179–183

Liu D, Lian B, Dong H (2012) Isolation of *Paenibacillus* sp. and assessment of its potential for enhancing mineral weathering. Geomicrobiol J 29:413–421

Marschner H (1995) Mineral nutrition of higher plants. Academic, London

Maurya BR, Meena VS, Meena OP (2014) Influence of Inceptisol and Alfisol's potassium solubilizing bacteria (KSB) isolates on release of K from waste mica. Vegetos 27(1):181–187

Meena OP, Maurya BR, Meena VS (2013) Influence of K-solubilizing bacteria on release of potassium from waste mica. Agric Sustain Dev 1(1):53–56

Meena VS, Maurya BR, Bahadur I (2014a) Potassium solubilization by bacterial strain in waste mica. Bangladesh J Bot 43(2):235–237

Meena VS, Maurya BR, Verma JP (2014b) Does a rhizospheric microorganism enhance K+ availability in agricultural soils? Microbiol Res 169:337–347

Meena RK, Singh RK, Singh NP, Meena SK, Meena VS (2015a) Isolation of low temperature surviving plant growth-promoting rhizobacteria (PGPR) from pea (*Pisum sativum* L.) and documentation of their plant growth promoting traits. Biocatal Agric Biotechnol doi:10.1016/j.bcab.2015.08.006

Meena VS, Maurya BR, Verma JP, Aeron A, Kumar A, Kim K, Bajpai VK (2015b) Potassium solubilizing rhizobacteria (KSR): isolation, identification, and K-release dynamics from waste mica. Ecol Eng 81:340–347

Mikhailouskaya N, Tcherhysh A (2005) K-mobilizing bacteria and their effect on wheat yield. Latv J Agron 8:154–157

Muntz A (1890) Sur La decomposition des rockes et al. formation de la terraiable. Compt Rend Akad Sci Paris 110:1370–1372

Muralikannan N, Anthomiraj S (1998) Occurrence of silicate solubilizing bacteria in rice ecosystem. Madras Agric J 85:47–50

Nayak B (2001) Uptake of potash by different plants with the use of potash mobilizing bacteria *Frateuria aurantia*. M.Sc. (Agric) thesis, QUAT, Bhubaneswar

Parmar P, Sindhu S (2013) Potassium solubilization by rhizosphere bacteria: influence of nutritional and environmental conditions. J Microbiol Res 3:25–31

Prajapati K, Modi H (2012) Isolation and characterization of potassium solubilizing bacteria from ceramic industry soil. CIBTech J Microbiol 1:8–14

Prajapati K, Sharma MC, Modi HA (2013) Growth promoting effect of potassium solubilizing microorganisms on okra (*Abelmoschus esculentus*). Int J Agri Sci 23:181–188

Rajawat MVS, Singh S, Singh G, Saxena AK (2012) Isolation and characterization of K-solubilizing bacteria isolated from different rhizospheric soil. In: Proceeding of 53rd annual conference of association of microbiologists of India, pp 124

Rosa-Magri MM, Avansini SH, Lopes-Assad ML, Tauk-Tornisielo SM, Ceccato-Antonini SR (2012) Release of potassium from rock powder by the yeast *Torulaspora globosa*. Braz Arch Biol Technol 55:577–582

Sangeeth KP, Bhai RS, Srinivasan V (2012) *Paenibacillus glucanolyticus*, a promising potassium solubilizing bacterium isolated from black pepper (*Piper nigrum* L.) rhizosphere. J Spices Aromatic Crops 21:118–124

Schneider A, Tesileanu R, Charles R, Sinaj S (2013) Kinetics of soil potassium sorption-desorption and fixation. Commun Soil Sci Plant Anal 44:837–849

Shaaban E, El-Shamma M, El Shazly S, El-Gazzar A, Abdel-Hak RE (2012) Efficiency of rock-feldspar combined with silicate dissolving bacteria on yield and fruit quality of Valencia orange fruits in reclaimed soils. J Appl Sci Res 8:4504–4510

Shabala S (2003) Regulation of potassium transport in leaves: from molecular to tissue level. Ann Bot 92:627–634

Sheng XF (2005) Growth promotion and increased potassium uptake of cotton and rape by a potassium releasing strain of *Bacillus edaphicus*. Soil Biol Biochem 37:1918–1922

Sheng XF, He LY (2006) Solubilization of potassium-bearing minerals by a wild-type strain of *Bacillus edaphicus* and its mutants and increased potassium uptake by wheat. Can J Microbiol 52:66–72

Sheng XF, Huang WY (2002) Mechanism of potassium release from feldspar affected by the strain NBT of silicate bacterium. Acta Pedol Sin 39:863–871

Sheng XF, He LY, Huang W (2002) The conditions of releasing potassium by a silicate dissolving bacterial strain NBT. Agri Sci China 1:662–666

Sheng XF, Xia JJ, Chen J (2003) Mutagenesis of the *Bacillus edaphicus* strain NBT and its effect on growth of chili and cotton. Agri Sci China 2:400–412

Sheng XF, Zhao F, He LY, Qiu G, Chen L (2008) Isolation and characterization of silicate mineral-solubilizing *Bacillus globisporus* Q12 from the surfaces of weathered feldspar. Can J Microbiol 54:1064–1068

Singh G, Biswas DR, Marwaha TS (2010) Mobilization of potassium from waste mica by plant growth promoting rhizobacteria and its assimilation by maize (*Zea mays*) and wheat (*Triticumaestivum* L.): a hydroponics study under phytotron growth chamber. J Plant Nutr 33:1236–1251

Singh NP, Singh RK, Meena VS, Meena RK (2015) Can we use maize (*Zea mays*) rhizobacteria as plant growth promoter? Vegetos 28(1):86–99

Subhashini (2015) Growth promotion and increased potassium uptake of tobacco by potassium-mobilizing bacterium *Frateuria aurantia* grown at different potassium level in Vertisols. Commun Soil Sci Plant Anal 46:210–220

Sugumaran P, Janarthanam B (2007) Solubilization of potassium containing minerals by bacteria and their effect on plant growth. World J Agri Sci 3(3):350–355

Taha SM, Mahmoud SAZ, Halim El-Damaty A, Abd El-Hafez A (1969) Activity of phosphate-dissolving bacteria in Egyptian soils. Plant Soil 31:149–160

Ullman WJ, Kirchman DL, Welch SA, Vandevivere P (1996) Laboratory evidence for microbially mediated silicate mineral dissolution in nature. Chem Geol 132:11–17

Uroz S, Calvaruso C, Turpault MP, Pierrat JC, Mustin C, Frey-Klett P (2007) Effect of the mycorrhizosphere on the genotypic and metabolic diversity of the bacterial communities involved in mineral weathering in a forest soil. Appl Environ Microbiol 73:3019–3027

Uroz S, Calvaruso C, Turpault M-P, Frey-Klett P (2009) Mineral weathering by bacteria: ecology, actors and mechanisms. Trends Microbiol 17:378–387

Véry A-A, Sentenac H (2002) Cation channels in the *Arabidopsis* plasma membrane. Trends Plant Sci 7:168–175

Zakaria AAB (2009) Growth optimization of potassium solubilizing bacteria isolated from biofertilizer. B.S thesis, Faculty of Chemistry and Natural Resources, Engineering University, Malaysia, Pahang, pp 40

Zhang C, Kong F (2014) Isolation and identification of potassium-solubilizing bacteria from tobacco rhizospheric soil and their effect on tobacco plants. Appl Soil Ecol 82:18–25

Zhao F, Sheng XF, Huang Z, He L (2008) Isolation of mineral potassium-solubilizing bacterial strains from agricultural soils in Shandong Province. Biodivers Sci 16:593–600

Potassium-Solubilizing Microorganisms (KSMs) and Its Effect on Plant Growth Improvement

13

S.S. Sindhu, Priyanka Parmar, Manisha Phour, and Anju Sehrawat

Abstract

All the crops require nitrogen, phosphorus, and potassium macronutrients for proper plant growth, and addition of mineral fertilizers leads to impressive crop yield increases. Excessive application of chemical fertilizers increases the cost of crop production and causes environmental pollution. Many microorganisms inhabiting soil/rhizosphere play an important role in solubilization of bound form of soil minerals and enhance the availability of nutrients in the soil for plant growth and development. These plant growth-promoting rhizobacteria, including N_2-fixing, phosphate-/potassium-solubilizing bacteria, are being used as biofertilizers to minimize health hazards caused by the use of chemical fertilizers. About 90–98 % of total potassium in soil exists in unavailable mineral form. Silicate minerals such as K-feldspars and biotite are the most common minerals in the Earth's crust and are a source of inorganic nutrients in soils to provide optimal nutrition for crops. Recently, K-solubilizing bacteria/fungi have been isolated from rhizosphere soil of different crops, which cause potassium solubilization by production of organic/inorganic acids or by production of polysaccharides. Efficient K-solubilizing bacteria have been reported to enhance potassium uptake in plants leading to plant growth stimulation under pothouse and field conditions. These K-solubilizing bacteria (KSB) could be applied as potential biofertilizers along with application of rock K materials to provide a continuous supply of available potassium for increasing the crop yield.

Keywords

Potassium-solubilizing microorganisms (KSMs) • K-feldspars • Environmental pollution • Rhizosphere bacteria • Plant growth

S.S. Sindhu (✉) • P. Parmar • M. Phour • A. Sehrawat
Department of Microbiology, CCS Haryana Agricultural University, Hisar 125004, India
e-mail: sindhuss58@gmail.com

V.S. Meena et al. (eds.), *Potassium Solubilizing Microorganisms for Sustainable Agriculture*,
DOI 10.1007/978-81-322-2776-2_13

13.1 Introduction

Soil is the basis for crop production, and it provides mechanical support, water, and necessary nutritional elements for plant growth. To feed the ever-increasing population, agricultural productivity has to be increased significantly. However, the deficiencies of plant-available nitrogen (N), phosphorus (P), and potassium (K) are major limiting factor for food production in many agricultural soils (Xie 1998; Zord et al. 2014). With the progressive intensification of agriculture due to small landholdings and introduction of high-yielding crop varieties and hybrids during green revolution, the soils are getting depleted in these macronutrients including potassium at a faster rate. Moreover, available soil K levels have also dropped due to leaching, runoff, and erosion (Sheng and Huang 2002a). As a consequence, potassium deficiency is becoming one of the major constraints in crop production, and therefore, many crops do respond to K fertilization in soils.

Potassium is the most abundantly absorbed cation and major essential plant macronutrient in higher plants. Potassium has been reported to play important roles in activation of plant enzymes, maintains cell turgor, enhances photosynthesis, helps in transport of sugars and starches, helps in nitrogen uptake, and is also essential for protein synthesis. Potassium also helps in grain filling and kernel weight, strengthens straw, increases disease resistance against pest and diseases, and also helps the plant to withstand abiotic and biotic stresses. Potassium deficiency symptoms usually occur first on the lower leaves of the plant and progress toward the top as the severity of the deficiency increases. One of the most common signs of potassium deficiency is the yellow searching or firing (chlorosis) along the leaf margin. In severe cases of potassium deficiency, the fired margin of the leaf may fall out. Without adequate potassium, the plants will have poorly developed roots, grow slowly, produce small seeds, result in lodging of cereal crops, and have lower yields. Sometimes, severe losses in terms of yield and quality could be caused to the crop, long before the symptoms of K deficiency become visible (Khanwilkar and Ramteke 1993).

Due to intensive cultivation and imbalanced fertilizer application, potassium level has declined in different kinds of soils recently (Zord et al. 2014). India ranks fourth in consumption of potassium fertilizers in the world, after the USA, China, and Brazil (FAI 2007). K fertilizers are imported in India in the form of muriate of potash (KCl) and sulfate of potash (K_2SO_4), because, there is no reserve of K-bearing minerals in India for production of commercial K fertilizers. On an average, 1.7 million tonnes of K is being imported annually in India (Anonymous 2003). Soluble potassic fertilizers along with nitrogenous and phosphorus fertilizers are applied at high recommended doses to enhance crop yields, which has caused environmental pollution problems (Brady 1990; Akande et al. 2008). Therefore, direct application of rock phosphate and rock potassium materials in soils may be agronomically more useful and environmentally safer than application of soluble P and K applied as chemical fertilizers (Rajan et al. 1996). However, P and K nutrients from rock minerals are released slowly in the soil and therefore not readily available to the plant (Zapata and Roy 2004).

Eco-friendly agricultural system has emerged as an important thrust area globally for long-term soil environmental sustainability and to minimize the environmental pollution associated with indiscriminate use of chemical fertilizers. Therefore, plant growth-promoting rhizobacteria (PGPR) including nitrogen-fixing, phosphate-mobilizing, and potassium-solubilizing bacteria (KSB) are being used as biofertilizers to conserve our existing resources and to minimize environmental pollution hazards (Vessey 2003; Ekin 2010; Bahadur et al. 2014; Sindhu et al. 2014c; Meena et al. 2014a). The use of these PGPR including potassium-solubilizing bacteria as a biofertilizer could work as a sustainable solution to improve plant nutrient uptake and crop production (Meena et al. 2014b; Sindhu et al. 2014b). Furthermore, application of potassium containing rock materials may be agronomically more useful and environmentally more feasible than application of soluble K as fertilizer (Rajan et al. 1996). Slow release of nutrients from rock K materials and their adsorption by plant roots may cause significant yield increases of the various crops (Meena et al. 2014b, 2015).

13.2 Availability of Potassium in the Soil

Among the three major plant nutrients N, P, and K, potassium is the seventh most abundant element in the Earth's crust, and actual concentrations of this nutrient in the soil vary widely ranging from 0.04 % to 3.0 % (Sparks and Huang 1985). Three forms of potassium are found in the soil, viz., soil mineral, non-exchangeable, and available form. Mineral form makes up more than 90–98 % of soil potassium (Sparks and Huang 1985; Sparks 1987), which is tightly bound, and most of it is unavailable for plant uptake. In mineral soils, K occurs in the form of silicate minerals, viz., muscovite, orthoclase, biotite, feldspar, illite, mica, vermiculite, smectite, etc. Another non-exchangeable form of potassium makes up approximately 1–10 % of soil potassium and acts as a reserve to replenish potassium taken up by plants or lost from the soil solution (Memon et al. 1988; Sharpley 1989). When level of exchangeable and solution K is decreased by crop removal and/or by leaching, release of non-exchangeable K to the exchangeable form occurs by increase in microbial activity (Sparks 1987). The third form of available potassium in the soil constitutes only 1–2 %, which is found either in the solution or as part of the exchangeable cation held by negative charge of clay minerals and organic matter in soils. The rate and direction of reactions between solution and exchangeable forms of K determine whether applied K will be leached into lower horizons, taken up by plants, converted into unavailable forms, or released into available forms (Sparks 2000).

The state-wise analysis of available potassium status in India showed that 21 % districts are in low, 51 % are in medium, and 28 % soils are having high available potassium (Ghosh and Hasan 1980). The average total potassium content of these Indian soils was found to be 1.52 % (Mengel and Kirkby 1987). However, total K in the soil is poorly correlated with available K and is rarely used to describe K fertility status of a soil. Supply of K to the plants depends directly on the concentration of K in soil solution (Sparks and Huang 1985). With the uptake or removal of K, the equilibrium is disturbed and K will be drawn upon from the non-exchangeable and soil mineral fraction. Soil K reserve pool can be solubilized by microorganisms for uptake by plants and microbes through production of acids or exopolysaccharides (Ullman et al. 1996; Rogers et al. 1998; Parmar and Sindhu 2013).

13.3 Characterization of Potassium-Solubilizing Microorganisms (KSMs)

Potassium-solubilizing microorganisms play an important role in the natural K cycle (Parmar and Sindhu 2013; Meena et al. 2014a; Sindhu et al. 2014b). A wide range of bacteria including members of the genus *Pseudomonas*, *Burkholderia*, *Acidithiobacillus ferrooxidans*, *Bacillus mucilaginosus*, *Bacillus edaphicus*, and *B. circulans* have been reported to release potassium from potassium-bearing minerals (Sperberg 1958; Lian et al. 2002; Sheng 2005; Li et al. 2006; Zhang et al. 2013). Due to deficiency of potassium in significant areas of cultivated soils, application of K-solubilizing bacteria as biofertilizer has been a focus of recent research for improvement in agriculture productivity (Xie 1998; Sheng et al. 2003; Zarjani et al. 2013; Bahadur et al. 2014; Don and Diep 2014).

Several microorganisms like *Aspergillus niger*, *Bacillus extorquens*, and *Clostridium pasteurianum* were found to grow on muscovite, biotite, orthoclase, microclase, and micas under in vitro conditions (Muntz 1890; Reitmeir 1951).

Silicate-solubilizing bacteria *B. mucilaginosus* subsp. *siliceus* showed the release of potassium from feldspar and aluminosilicates (Norkina and Pumpyansakya 1956). Duff and Webley (1959) observed silicate-dissolving action of Gram-negative bacteria *Erwinia herbicola* and *Pseudomonas* strains. Similarly, *Pseudomonas* strain isolated from soil showed clearing zone in silicate medium (Webley et al. 1960). Aleksandrov et al. (1967) isolated different bacterial species which dissolved potassium, silica, and aluminum from insoluble minerals. Purushothaman et al. (1974) reported the occurrence of silicate-solubilizing bacteria (SSB) in marine environments and suggested that these bacteria play a significant role in cycling of silicon in seawater. Avakyan et al. (1981) characterized *B. mucilaginosus* strain that was found to solubilize insoluble silicates. Khudsen et al. (1982) isolated potassium-solubilizing bacteria from rock and mineral samples which showed higher activity in potassium release from acid leached soil.

K-solubilizing bacteria were isolated from soil, rock, and mineral samples (Li 1994), and based on morphological and physiological characters, bacterial isolate MCRCp1 was later identified as *B. mucilaginosus*. Similarly, silicate-solubilizing bacteria were isolated from rice rhizosphere and identified as *Bacillus* sp. (Muralikannan 1996). Silicate- and phosphate-solubilizing bacteria were enumerated from soil's tank sediments (Kannan and Raj 1998), and 3 out of 17 isolates were identified as *Bacillus* sp. based on biochemical characteristics. Liu (2001) reported isolation of silicate-solubilizing bacteria *B. mucilaginosus* strains CS1 and CS2 from soil, and these bacteria also showed growth inhibition of Gram-negative bacteria *E. coli*. Strain CS1 was identified as *B. mucilaginosus*. Lin et al. (2002) reported solubilization of silicates by slime-producing *B. mucilaginosus* strain, and slime production also contributed in the colonization of rhizosphere as well as nonrhizosphere soil. Raj (2004) isolated silicate-solubilizing bacteria from rice ecosystem in a medium containing 0.25 % insoluble magnesium trisilicate, and *Bacillus* spp. were found to solubilize silicate minerals more efficiently under in vitro conditions. Murali et al. (2005) isolated silicate solubilizers from soil samples collected from coconut palms using modified Bunt and Rovira medium, and majority of the silicate solubilizers were identified as *Bacillus* sp. and *Pseudomonas* sp.

Two phosphate- and potassium-solubilizing strains KNP413 and KNP414 were isolated from the soil of Tianmu Mountain, Zhejiang Province (China) (Hu et al. 2006). Strain KNP414 showed higher dissolution capacity even than *Bacillus mucilaginosus* AS1.153, the inoculant of potassium fertilizer widely used in China. These strains KNP413 and KNP414 were classified to the genus of *Paenibacillus*, i.e., *P. mucilaginosus*, based on G + C contents and 16S rRNA gene sequence similarity. Gram-positive, rod-shaped endospore former bacterium *Bacillus mucilaginosus* was characterized, which solubilized silicon from illite at 30 °C (Zhou et al. 2006). K-solubilizing bacteria were isolated from soil, rocks, and mineral samples, viz., microcline, orthoclase, and muscovite mica (Sugumaran and Janarthanam 2007), and among the isolates, *B. mucilaginosus* strain MCRCp1 was found to solubilize more potassium by producing slime in muscovite mica. The inoculation of this bacterium markedly improved the phosphorus and potassium nutritional status in the soil. Zhao et al. (2008) isolated bacterial strains with multiple beneficial activities such as potassium solubilization, indole acetic acid (IAA) production, and siderophore production. Seventy bacterial isolates were isolated from the rhizosphere of wheat (Parmar 2010) by using modified Aleksandrov medium plates (consisting of glucose, 5.0; $MgSO_4.7H_2O$, 0.5; $CaCO_3$, 0.1; $FeCl_3$, 0.006; Ca_3PO_4, 2.0; mica powder, 2.0; agar-agar, 20.0 g). Twenty rhizobacterial isolates were found to solubilize potassium from mica powder and formed potassium solubilization zone (Fig. 13.1).

Two fungi *Penicillium frequentans* and *Cladosporium cladosporioides* were found to cause weathering of unaltered sandstone, granite, and limestone (Argelis et al. 1993). Both the

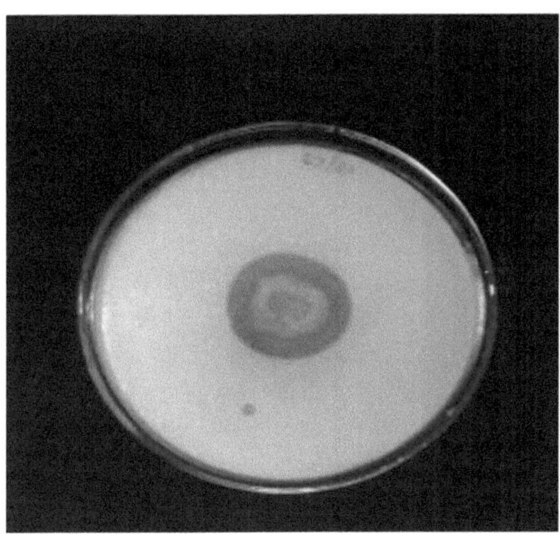

Fig. 13.1 Rhizobacterial isolate showing formation of K solubilization zone on Aleksandrov medium supplemented with mica powder

fungal species were found to produce large amounts of oxalic, citric, and gluconic acids in broth culture. The excretion of organic acids by the fungi caused extensive deterioration of clay silicates, of mica and feldspar from both sandstone and granite, and also of calcite and dolomite from limestone. Potassium-solubilizing fungi were isolated from ceramic industry soils, and a high ratio of clear zone on Aleksandrov agar supplemented with 0.5 % potassium aluminum silicate was observed with the four fungal isolates (Prajapati et al. 2012). In liquid medium containing potassium aluminum silicate, two fungal strains, i.e., KF1 and KF2, showed highest available potassium. These isolates were characterized as *Aspergillus niger* and *Aspergillus terreus*. More solubilization was reported by inoculation of *A. terreus* than *A. niger*, when grown in the presence of 1 % rock potassium (feldspar).

Mineral-solubilizing *Paenibacillus* strain KT was isolated from a soil in Henan Province, China (Liu et al. 2012). The release of water-soluble Al, Ca, and Fe from the potassium-bearing rock in active bacterial culture was higher after inoculation of this strain than from the control containing autoclaved inoculum. A potassium-solubilizing *B. circulans* strain Z_{1-3} was isolated from soil, which dissolved potassium from feldspar (Xiaoxi et al. 2012). Zarjani et al. (2013) obtained six isolates of potassium-solubilizing bacteria from some Iranian soils, which showed higher solubilization of K in biotite treatment. Five isolates belonged to *Bacillus megaterium* (JK3, JK4, JK5, JK6, JK7) and isolate JK2 belonged to *Arthrobacter* sp. *Bacillus megaterium* isolates were found as the most efficient potassium solubilizer under in vitro conditions.

Seven multifunctional strains were obtained from wheat field soil samples collected from Chang'an, Shanxi, China, and these strains caused the solubilization of both phosphate and potassium (Zhang et al. 2013). Most efficient CX-7 strain was identified as *Paenibacillus kribensis*. Solubilization study results showed that strain CX-7 released 71.60 mg/L and 5.18 mg/L water-soluble phosphorus and 3.44 mg/L potassium from phosphate powder, lecithin powder, and potassium feldspar powder, respectively. Antagonistic experiment showed that CX-7 strain had wide antagonism against pathogenic microorganisms. Don and Diep (2014) reported that 34 isolates out of 237 bacterial isolates (obtained from 118 sample soils/weathered granite rock of That So Mountain) were capable of dissolving both phosphate and potassium. Seven strains showed high phosphate and potassium dissolution capacity. Three strains including CA09 (*Agrobacterium tumefaciens*), CA29 (*Rhizobium tropici*), and K16B (*Azotobacter tropicalis*) were proposed as potential microbial inoculants for sustainable crop production in sandy acid soil in Vietnam. Recently, twenty seven strains were isolated from the tobacco rhizosphere, which solubilized K-feldspar powder in solid and liquid media (Zhang and Kong 2014). *Klebsiella variicola* was observed at the highest frequency with 18 potassium-solubilizing strains out of 27 strains isolated.

13.4 Potassium Solubilization: Mechanisms Involved

Naturally abundant soil microorganisms were found to cause solubilization of structural potassium compounds under in vitro conditions and pothouse and field conditions (Sheng and He 2006; Basak and Biswas 2010; Zarjani et al. 2013). The production of organic acids like citric, oxalic, and tartaric acids and the production of capsular polysaccharides by microorganisms were found to cause dissolution of minerals illite and feldspar to release potassium (Liu et al. 2006; Meena et al. 2014a; Sindhu et al. 2014b). Sheng and Huang (2002a) found that K release from the minerals was affected by pH, oxygen, and the bacterial strains used. Potassium solubilization was enhanced 84.8–127.9 % by inoculation of microbial strain as compared to uninoculated control. The extent of potassium solubilization by *Bacillus edaphicus* was reported higher in illite in broth culture than feldspar (Sheng and He 2006). Badr (2006) showed that silicate-solubilizing bacteria caused 4.96 mg/L potassium solubilization at pH 6.5–8.0. *B. mucilaginosus* was found to solubilize 4.29 mg/L of potassium in media supplemented with muscovite mica (Sugumaran and Janarthanam 2007).

Production of organic acids like citric, tartaric, and oxalic acids has been reported as predominant mechanism contributing to K solubilization (Table 13.1) (Hazen et al. 1991; Styriakova et al. 2003; Sheng and He 2006; Meena et al. 2014a). Potassium, silicon, and aluminum were obtained from dissolution of insoluble K-bearing minerals such as micas, illite, and orthoclases by excretion of organic acids by silicate bacteria, which either directly dissolved rock K or chelated silicon ions to bring K into the solution (Aleksandrov et al. 1967; Friedrich et al. 1991; Ullman et al. 1996; Bennett et al. 1998). Organic acids can either directly enhance dissolution by either a proton- or ligand-mediated mechanism or they can also indirectly enhance dissolution by the formation of complexes in solution with reaction products (Ullman and Welch 2002). The potassium is made available to plants when the minerals are slowly weathered or solubilized (Bertsch and Thomas 1985).

Solubilization of potassium from precipitated forms was demonstrated through production of inorganic and organic acids by *Thiobacillus*, *Clostridium*, and *Bacillus* (Berthelin 1983). Similarly, solubilization of feldspar by *B. mucilaginosus* and *B. edaphicus* was associated with the production of carboxylic acids like citric, tartaric, and oxalic acids (Malinovskaya et al. 1990; Sheng and Huang 2002b). The production of gluconate promoted dissolution of silicates like albite, quartz, and kaolinite by subsurface bacteria (Duff et al. 1963; Vandevivere et al. 1994). Potassium-solubilizing microorganisms were found to excrete different organic acids such as acetic, citric, lactic, propionic, glycolic, oxalic, malonic, succinic, fumaric, and tartaric acid (Wu et al. 2005). Production of tartaric acid was reported as the most frequent agent of mineral potassium solubilization (Prajapati

Table 13.1 Production of acids or polysaccharides (EPS/CPS) by potassium-solubilizing microorganisms

Organism	Acid/EPS/CPS production	References
B. mucilaginosus and *B. edaphicus*	Citric, tartaric, and oxalic acids	Malinovskaya et al. (1990) and Sheng and Huang (2002b)
B. edaphicus	Oxalic and tartaric acids and CPS	Sheng and He (2006)
B. mucilaginosus	Oxalate, citrate, and polysaccharides	Liu et al. (2006)
Penicillium frequentans and *Cladosporium cladosporioides*	Oxalic, citric, and gluconic acids	Argelis et al. (1993)
Paenibacillus mucilaginosus	Tartaric, citric, and oxalic acids	Liu et al. (2012)
Aspergillus niger and *Penicillium* spp.	Citric, glycolic, and succinic acids	Sperberg (1958)
Thiobacillus, *Clostridium*, and *Bacillus* spp.	Inorganic and organic acids	Berthelin (1983)
Pseudomonas aeruginosa	Acetic, citric, and oxalic acids	Badr et al. (2006)
Rhizobium tropici strain Q34	Tartaric, succinic, and citric acids and exopolysaccharides	Wang et al. (2015)

et al. 2012; Zarjani et al. 2013). The synthesis and release of organic acids by the microorganisms into the surrounding environment acidify the surrounding niche that ultimately leads to the release of K ions from the mineral K by protonation and acidification (Goldstein 1994; Meena et al. 2015).

In some cases, the ability to reduce the pH did not correlate with the ability to solubilize mineral potassium, indicating that acidification alone may not be the only mechanism of potassium solubilization (Zarjani et al. 2013). Bacteria were also found to enhance dissolution of silicates by the production of excess protons and organic ligands or by the production of hydroxyl anion, extracellular polysaccharides, and enzymes (Berthelin and Belgy 1979; Hieberk and Bennett 1992; Vandevivere et al. 1994; Barker et al. 1998).

Production of slime or acidic exopolysaccharides (EPS) also contributed to the release of potassium from silicates (Groudev 1987). The rate of plagioclase dissolution in solutions containing organic acids was found more in comparison to inorganic acids (Welch and Ullman 1993) and further showed that polysaccharides produced by the bacterium can combine with the minerals to form bacterial-mineral complexes, which lead to degradation of the minerals. Liu et al. (2006) demonstrated that polysaccharides of *B. mucilaginosus* strain strongly adsorbed the organic acids (oxalate and citrate) and the complex attached to the surface of the mineral, resulting in an area of high concentration of organic acids near the mineral. It was suggested that the extracellular polysaccharides adsorbed SiO_2, and this affected the equilibrium between the mineral and fluid phases and led to the reaction toward SiO_2 and K^+ solubilization. Several naturally occurring polymers were tested for their effect on mineral dissolution (Welch and Vandevivere 2009). Microbial EPS solutions extracted from the subsurface microbes increased the dissolution rate of feldspars, probably by forming complexes with framework ions in the solution. Solubilization and release of potassium from illite and feldspar by wild-type strain of *Bacillus edaphicus* and its four mutants were found due to the production of organic acids like oxalic acid and tartaric acids along with the production of capsular polysaccharides (CPS) (Sheng and He 2006). In liquid cultures, five bacterial strains showed better growth on Suzhou illite than on Nanjing feldspar. Oxalic acid was found more active for the solubilization of Nanjing feldspar, whereas oxalic and tartaric acids caused more solubilization of Suzhou illite.

Parmar (2010) showed that some rhizobacterial isolates caused solubilization of potassium in mica by acid production only, whereas other isolates caused K solubilization by production of CPS and EPS. Six rhizobacterial isolates caused solubilization by production of acid, CPS, and EPS. *Rhizobium tropici* strain Q34 showed significant increase in the production of tartaric acids and extracellular polysaccharides in the presence of feldspar, while better growth and more citric acid production were observed in the presence of biotite (Wang et al. 2015). Mineral dissolution experiments showed that organic acids and polysaccharides produced by strain Q34 promoted the release of Si and K from the minerals.

13.5 Influence of KSB on Growth and Yield of Different Crops

The use of beneficial soil microorganisms could reduce the amount of mineral fertilizer inputs by increasing the efficiency of nutrient availability and other plant growth-promoting activities. These plant growth-promoting bacteria associated with plant roots exert their beneficial effects on nutrition uptake and plant growth through a number of mechanisms such as N_2 fixation, production of phytohormones and siderophores, and transformation of nutrient elements such as phosphorus, potassium, and iron, when they are either applied to seeds or incorporated into the soil (Kloepper et al. 1989; Glick et al. 1999; Herridge et al. 2008; Sindhu et al. 2010). Some rhizosphere bacteria have also been found to suppress various plant diseases (Weller 2007; Haas and Defago 2005; Sindhu et al. 2011).

13.5.1 Inoculation Effect of Potassium-Solubilizing Bacteria on Plant Growth

Increase in the yield of maize and wheat was observed by application of organo-minerals inoculated with silicate bacteria (Aleksandrov 1958). Khudsen et al. (1982) showed that inoculation of potassium-solubilizing bacteria isolated from rock and mineral samples improved the growth of green gram's seedling. Similarly, an increased yield in rice crop was observed due to inoculation of silicate-solubilizing bacteria (Muralikannan 1996; Kalaiselvi 1999). Lin et al. (2002) observed 125 % increase in biomass, whereas K and P uptake were more than 150 % in tomato plant due to inoculation of silicate-dissolving bacteria *B. mucilaginosus* strain RCBC13 as compared to uninoculated plants. Inoculation of silicate-solubilizing bacteria *Bacillus edaphicus* on chilly and cotton increased the available P and K contents in plant biomass (Sheng et al. 2003). Beneficial effect of inoculation of potassium-solubilizing bacteria was reported in sorghum, which resulted in increased biomass and increased contents of P and K in plants than the control (Zhang et al. 2004). Increase in grain yield, silica content of rice, and available silica in soil was observed by inoculation of silicate-solubilizing *Bacillus* sp. (Raj 2004). Ramarethinam and Chandra (2005) recorded significantly increased brinjal yield, plant height, and K uptake in comparison to control due to inoculation of potash-solubilizing bacteria *Frateuria aurantia* in a field experiment.

The inoculation of *B. edaphicus* strain NBT on cotton and rape in K-deficient soil pot experiments resulted in increased root and shoot growth, and potassium content was increased by 30 % and 26 %, respectively (Sheng 2005). Christophe et al. (2006) reported that inoculation of *Burechulderia glathei* strains PMB (7) and PML1 (12) on pine roots significantly affected pine growth and root morphology, which was attributed to the release of K from the mineral. Sheng and He (2006) recorded an increased root and shoot growth of wheat plants and also showed significantly higher N, P, and K contents due to inoculation of *B. edaphicus* and its mutants in a yellow brown soil that had low available K. In the field experiment, increased yield in tomato crop was observed due to inoculation of silicate-dissolving bacteria *B. cereus* as a bioinoculant along with feldspar and rice straw (Badr 2006). Inoculation of bacteria combined with K- and P-bearing minerals caused increase in dry matter yield of sorghum plants along with P and K uptake and also improved fertility in three different soils, i.e., clay, sandy, and calcareous soils (Badr et al. 2006). In a field experiment, increased rice grain yield was observed due to inoculation of silicate-solubilizing bacteria that recorded 5218 kg/ha grain yield than the control yield of 4419 kg/ha (Balasubramaniam and Subramanian 2006).

Inoculation of groundnut with slime-forming *B. mucilaginosus* strain MCRCp1 showed increase in the plant dry matter by 125 % and in oil content by 35.4 % (Sugumaran and Janarthanam 2007). Available P and K increased from 6.24 to 9.28 mg/kg and 86.57 to 99.60 mg/kg, respectively, in soil due to inoculation of *B. mucilaginosus* MCRCp1 in comparison to uninoculated control plants. Basak and Biswas (2008) reported that inoculation of K-solubilizing *B. mucilaginosus* along with application of waste mica increased the biomass yield and uptake of K in Sudan grass (*Sorghum vulgare* var. *sudanensis*) grown under two Alfisols. Parmar (2010) showed that inoculation of K-solubilizing isolate HWP47 in wheat (*Triticum aestivum* L.) var. WH711 caused 51.46 % increase in root dry weight (RDW) in soil at 60 days after sowing in pots. Similarly, 44.28 % increase in shoot dry weight (SDW) was found in HWP47 inoculated plants. Addition of rock material along with inoculation of HWP47 isolate showed 22.35 % increase in RDW and 73.68 % increase in SDW. Isolates HWP15 and HWP47 also caused significant K uptake in the shoot tissues. Zhang and Kong (2014) showed that inoculation of tobacco seedlings with four KSB strains increased plant dry weight and uptake of both potassium and nitrogen. Higher increases were obtained on addition of K-feldspar powder along with inoculation of potassium-solubilizing bacteria.

13.5.2 Coinoculation of Potassium-Solubilizing Microbes with Other Beneficial Bacteria

Simultaneous inoculation of nitrogen-fixing bacteria (azotobacterin) and silicate bacteria showed 50–94 % increase in the yield of cotton (Ciobanu 1961). The increased K uptake coupled with increased yield in yam and tapioca was observed by treating the plants with potassium mobilizer in conjunction with biofertilizers and chemical fertilizers (Clarson 2004). Abou-el-Seoud and Abdel-Megeed (2012) reported that coinoculation of P- and K-dissolving bacteria *Bacillus megaterium* var. *phosphaticum* and potassium-dissolving bacteria (*Bacillus mucilaginous* and *B. subtilis*) along with direct application of rock P and K materials (R (P + K)) into the soil increased P and K availability and uptake and enhanced the shoot and root growth of maize plants grown on P- and K-limited calcareous soils. Similarly, Chandra et al. (2005) reported an increased yield by 15–20 % in yam and tapioca due to the application of potash solubilizer in combination with other biofertilizers like *Rhizobium*, *Azospirillum*, *Azotobacter*, *Acetobacter*, and PSM. Inoculation of K-solubilizing *B. mucilaginosus* strain along with P solubilizer *Bacillus megaterium* and N_2 fixer *Azotobacter chroococcum* showed significant increase in growth and nutrient uptake of maize crop (Wu et al. 2005). Soil properties such as organic matter content and total N in soil were also improved due to bacterial inoculation.

Han and Lee (2005) found that coinoculation of PSB (*B. megaterium*) and KSB (*B. mucilaginosus* strain KCTC3870) on eggplant in combination with direct application of rock P and K materials into the soil resulted in increased N, P, and K uptake, photosynthesis, and the yield of eggplant grown in nutrient-limited soil. The combined treatment resulted in increase of N, P, and K uptake in the shoot (14 %, 22 %, and 14 %, respectively) and in the root (11, 14, and 21 %). Combined treatment of both bacteria and mineral rocks further increased shoot dry weight by 27 % and root dry weight by 30 % over the control 30 days following planting. Similarly, coinoculation effect of both PSB (*B. megaterium* var. *phosphaticum*) and KSB (*B. mucilaginosus*) was evaluated on pepper and cucumber in nutrient-limited soil (Han et al. 2006). Coinoculation of both PSB and KSB, and fertilization with rock P and K, increased the N, P, and K uptake in shoot (21 %, 31 %, and 33 % for pepper and 29, 41, and 29 % for cucumber, respectively) and in root (16, 33, and 26 % for pepper; 29, 34, and 50 % for cucumber). The treatment including bacteria and mineral rocks further increased plant growth, i.e., 26 % in shoot and 29 % in root dry weight for pepper whereas 22 % in shoot and 27 % in root dry weight for cucumber plant in comparison to controls during 30 days after planting. The combined inoculation of *B. megaterium* var. *phosphaticum* and *B. mucilaginosus* on pepper and cucumber crops showed that coinoculation of PSB and KSB resulted in consistently higher P and K availability than the control (Vassilev et al. 2006).

Supanjani et al. (2006) reported that inoculation of phosphorus- and potassium-solubilizing bacteria on *Capsicum annuum* along with addition of P- and K-containing rock materials increased P availability from 12 % to 21 % and K availability from 13 % to 15 % in the soil as compared with control and subsequently improved N, P, and K uptake in the crop. The integration approach of rocks and bacteria also increased plant photosynthesis by 16 % and leaf area by 35 % in comparison to control. Coinoculation with potassium-solubilizing *B. mucilaginosus* and N_2-fixing *A. chroococcum* A-41 in Sudan grass along with waste mica resulted in highest biomass production and nutrient acquisition by Sudan grass (Basak and Biswas 2010). *Bacillus mucilaginosus* strain was found more effective K solubilizer than *Azotobacter chroococcum* strain A-41. Coinoculation of bacterial strains maintained consistently higher amounts of available K and N in soils even at 150 days of crop growth. Similarly, inoculation of maize and wheat plants with *Bacillus mucilaginosus*, *Azotobacter chroococcum*, and *Rhizobium* spp. significantly improved the biomass accumulation, potassium content, and uptake by plants (Singh et al. 2010). *B. mucilaginosus* resulted in significantly higher mobilization of potassium than inoculation of *A. chroococcum* and *Rhizobium*. Results revealed that inoculation with P- and K-solubilizing bacterial strains along with incorporation of P- and K-containing rock materials could be applied as a sustainable alternative to the use of chemical fertilizers. Thus, inoculation with PGPR including phosphate- and potassium-solubilizing bacteria (PSB and KSB) as biofertilizers could be a sustainable solution to improve plant nutrition and crop production (Vessey 2003).

13.6 Strategies to Improve Potassium Solubilization Efficiency for Its Use as Biofertilizer

Soil microorganisms play a pivotal role in various biogeochemical cycles and are responsible for the cycling of nutrients in the plant utilizable form (Wall and Virginia 1999). Microbes influence aboveground ecosystems by contributing to plant nutrition, plant health, soil structure, and soil fertility. Therefore, microorganisms offer an environment-friendly sustainable system and play a vital role in maintaining soil nutrient status. Many rhizosphere bacteria are well known for their capacity to confer plant growth promotion and also increase resistance toward various diseases as well as abiotic stresses (Dey et al. 2009; Sindhu et al. 2014c).

Potassium-solubilizing bacteria are an integral component of soil microbial community and play an important role in K cycle in soil by solubilizing the unavailable K to utilizable form for uptake by the plants (Marques et al. 2010; Sindhu et al. 2014b). These KSBs have enormous potential for making use of fixed K under soil systems with low K availability in tropical and subtropical developing countries. The mechanism of K solubilization by microorganisms has been studied in detail, but the K solubilization is a complex phenomenon affected by many factors such as microorganisms involved, nutritional status of soil, soil mineral type, amount of mineral, size of mineral, and environmental factors.

Moreover, the stability and survival of the KSBs after inoculation in soil is also important for K solubilization to benefit crop growth and development. These bacteria often fail to confer these beneficial effects when applied in the field, which is often due to insufficient rhizosphere colonization (Lugtenberg et al. 2001). However, Lin et al. (2002) showed that population of silicate-dissolving bacteria increased 70 % in the rhizosphere soil and 20 % in nonrhizosphere soil, respectively. Sugumaran and Janarthanam (2007) also reported that the number of K-solubilizing bacterium *B. mucilaginosus* strain MCRCp1 increased to about 10^6–10^7 cfu g^{-1} in soil after 90 days of inoculation, whereas the count of K-solubilizing bacteria was only 10^3 g^{-1} in the control soil. Novel bacterial traits conferring strain survival in the rhizosphere have been found and opened a way to better understand specific signaling and the regulatory processes governing the plant-beneficial bacterial association (Matilla et al. 2007). The use of molecular techniques in genetic modification of microbial and plant biological activities allows their better functioning in the rhizosphere (Ryan et al. 2009) leading to substantial improvement in the sustainability of agricultural systems.

Characterization of potassium-solubilizing bacterial isolates for other ancillary plant growth-promoting characters is also required to allow for selection of bacterial strains with many beneficial functions (Sindhu et al. 2002, 2014c) or to develop consortia of compatible strains with several benefits. Engineering the soil with inoculation of such beneficial bacterial strains along with enough organic matter supplementation is of paramount importance to maintain the plant life in soil. Thus, having a bacterial strain with multiple benefits would be a better choice as inoculants since they can perform other beneficial functions as well. The commercial formulations should also maintain high levels of quality control in terms of populations actually present in the formulation with high shelf life (Lian et al. 2007; Sindhu et al. 2014a).

Research work should also aim at selection of efficient environment-specific (high temperature, alkaline/acidic, or water-deficient soils) strains for effective K solubilization (Kirk et al. 2004; Naik et al. 2008). Bacterial strains capable of efficient solubilizing activity under high buffering conditions should be isolated and tested. Engineering the microbes for higher solubilizing activity is another area that needs attention while simultaneously raising the issue of the acceptability of such strains for field release. Mutation through chemical mutagenesis is still a strong tool to develop more efficient strains. The information on the genetics and regulation of the microbial strain phenotype needs to be generated. The knowledge generated on biodiversity and genetic manipulation of potassium-solubilizing bacteria will be useful to design strategies for the use of these strains as inoculants in organic agriculture. Strategies are required to clone genes from potassium-solubilizing bacteria and to transfer these genes into the bacterial strains having good colonization potential along with other beneficial characteristics such as nitrogen fixation, phosphate solubilization, and biocontrol activity (Zhang et al. 2013; Sindhu et al. 2014c). Metagenomic analysis aimed at the genes involved in synthesis and release of the acids is another area of research (Richardson et al. 2001; Ryan et al. 2001) to answer the analysis of the mineral solubilization phenomenon on a community basis.

Microbial interactions in the soil between KSB, PSB, other PGPR, the plant, and the environment are responsible for the variability observed in solubilization of bound form of nutrients and the uptake of solubilized nutrients leading to promotion of plant growth (Sindhu et al. 2014c). By developing better cultural practices and delivery systems, the efficacy of potassium-solubilizing bacteria can be improved further for their establishment in the rhizosphere (Sindhu and Dadarwal 2000). At present, the production of liquid biofertilizers is supposed to be the breakthrough in biofertilizer technology over conventional carrier-based technology. Therefore, special liquid formulations of mineral-solubilizing microbes will also find greater acceptance by farmers. Moreover, the applications of mixture of PGPRs with different beneficial activities including potassium and phosphate solubilization ability may be a more ecologically sound approach because it may result in better colonization and better adaptation to the environmental changes occurring throughout the growing season. Thus, the use of biotechnological approaches for manipulation of bacterial traits will improve the potassium solubilization efficiency, and their inoculation as potassic biofertilizer will enhance the plant growth and crop productivity for sustainable agriculture.

13.7 Conclusion

Application of locally available soil minerals such as mica, feldspar, and rock phosphate in combination with efficient potassium mobilizing bacterial strains as biofertilizers is urgently required to replace chemical fertilizers and for reducing the cost of crop cultivation. Although, many bacterial strains have been found to improve the growth of plants under pothouse conditions, the extent of growth stimulation by bacterial strains under field conditions usually remains unexplored. Therefore, effective potassium-solubilizing and plant growth-promoting bacterium-plant systems must be tested under field conditions with specific crop experimental designs, keeping in consideration of the soil type, plant types grown, and the environmental factors. In addition, plant variety has also been found to influence the root colonization ability of the inoculated strains. Thus, competitive and effective bacterial strains could be selected from the indigenous beneficial soil bacteria which could be adopted to the particular conditions of the inoculation site. The application of efficient strains of potassium-solubilizing bacteria may find their use in the amelioration of potassium-deficient soils, and further research could lead to an alternative mean of potassium nutrition for sustainable agriculture.

References

Abou-el-Seoud II, Abdel-Megeed A (2012) Impact of rock materials and biofertilizations on P and K availability for maize (*Zea maize*) under calcareous soil conditions. Saudi J Biol Sci 19:55–63

Akande MO, Adediran JA, Oluwatoyinbo FI, Makinde EA, Adetunji MT (2008) Suitability of poultry manure amended Sokoto rock phosphate on growth, nutrient uptake and yield of chilli pepper (*Capsicum fruitscens* L). Niger J Soil Sci 18:167–174

Aleksandrov VG (1958) Organo-mineral fertilizers and silicate bacteria. Dokl Akad Nauk 7:43–48

Aleksandrov VG, Blagodyr RN, Iiiev IP (1967) Liberation of phosphoric acid from apatite by silicate bacteria. Mikrobiyol Zh (Kiev) 29:111–114

Anonymous (2003) Agricultural statistics at a glance. Ministry of Agriculture Cooperation, New Delhi, pp 51–53

Argelis DT, Gonzala DA, Vizcaino C, Gartia MT (1993) Biochemical mechanism of stone alteration carried out by filamentous fungi living in monuments. Biogeochemistry 19:129–147

Avakyan ZA, Karavaiko GI, Mel'nikova EO, Krutsko VS, Ostroushhko YI (1981) Role of microscopic fungi in weathering of rocks from a pegmatite deposit. Mikrobiologiya 50:115–120

Badr MA (2006) Efficiency of K-feldspar combined with organic materials and silicate dissolving bacteria on tomato yield. J Appl Sci Res 2:1191–1198

Badr MA, Shafei AM, El-Deen Sharaf SH (2006) The dissolution of K and phosphorus bearing minerals by silicate dissolving bacteria and their effect on sorghum growth. Res J Agric Bio Sci 2:5–11

Bahadur I, Meena VS, Kumar S (2014) Importance and application of potassic biofertilizer in Indian agriculture. Int Res J Biol Sci 3:80–85

Balasubramaniam P, Subramanian S (2006) Assessment of soil test based potassium requirement for low land rice in udic haplustalf under the influence of silicon fertilization. Tamil Nadu Agricultural University, Tiruchirapalli, pp 621–712

Barker WW, Welch SA, Welch SC, Banfield F (1998) Experimental observations of the effects of bacteria on aluminosilicate weathering. Am Miner 83:1551–1563

Basak BB, Biswas DR (2008) Influence of potassium solubilizing microorganism (*Bacillus mucilaginous*) and waste mica on potassium uptake dynamics by sudan grass (*Sorghum vulgare* Pers) grown under two Alfisols. Plant Soil 317:235–255

Basak BB, Biswas DR (2010) Coinoculation of potassium solubilizing and nitrogen fixing bacteria on solubilization of waste mica and their effect on growth promotion and nutrient acquisition by a forage crop. Biol Fertil Soils 46:641–648

Bennett PC, Choi WJ, Rogera JR (1998) Microbial destruction of feldspars. Min Manag 8(62A):149–150

Berthelin J (1983) Microbial weathering processes. In: Krumbein WE (ed) Microbial geochemistry. Blackwell Scientific Publications, Oxford, pp 223–262

Berthelin J, Belgy G (1979) Microbial degradation of phyllosilicates during simulate podzolization. Geoderma 21(4):297–310

Bertsch PM, Thomas GW (1985) Potassium status of temperature region soils. In: Munson RD (ed) Potassium in agriculture. ASA, CSSA and SSSP, Madison, pp 131–162

Brady NC (1990) The nature and properties of soils. Macmillan, New York, pp 351–380

Chandra K, Greep S, Ravindranath P, Sivathsa RSH (2005) Liquid biofertilizers. Regional center for organic farming. Hebbal, Bangalore

Christophe C, Turpault MP, Freyklett P (2006) Root associated bacteria contribute to mineral weathering and to mineral nutrition in trees and budgeting analysis. Appl Environ Microbiol 72:258–1266

Ciobanu I (1961) Investigation of the efficiency of bacterial fertilizers applied to cotton. Cent Exp Ingras Bact Lucrari Stiint 3:203–214

Clarson D (2004) Potash biofertilizer for ecofriendly agriculture. Agro-clinic and Research Centre, Kottayam, pp 98–110

Dey C, Weinand T, Asch F (2009) Plant-rhizobacteria interactions alleviate abiotic stress conditions. Plant Cell Environ 32:1682–1694

Don NT, Diep CN (2014) Isolation, characterization and identification of phosphate- and potassium-solubilizing bacteria from weathered materials of granite rock mountain, that Son, an Giang province, Vietnam. Am J Life Sci 2:282–291

Duff RB, Webley DM (1959) 2-ketoglutaric acid and natural chelator produced by soil bacteria. Chemistry 44:1376–1377

Duff RB, Webley DM, Scott RO (1963) Solubilization of minerals and related materials by 2-ketogluconic acid producing bacteria. Soil Sci 5:105–114

Ekin Z (2010) Performance of phosphate solubilizing bacteria for improving growth and yield of sunflower (*Helianthus annuus* L.) in the presence of phosphorus fertilizer. Afr J Biotechnol 9:3794–3800

FAI (2007) Fertiliser statistics 2006–2007. The Fertilizer Association of India, New Delhi

Friedrich S, Platonova NP, Karavaiko GI, Stichel E, Glombitza F (1991) Chemical and microbiological solubilization of silicates. Acta Biotechnol 11:187–196

Ghosh AB, Hasan R (1980) Soil fertility map of India. Indian Agricultural Research Institute, New Delhi

Glick BR, Patten CL, Holguin G, Penrose DM (1999) Biochemical and genetic mechanisms used by plant growth-promoting bacteria. Imperial College Press, London

Goldstein AH (1994) Involvement of the quinoprotein glucose dehydrogenase in the solubilization of exogenous mineral phosphates by Gram negative bacteria. In: Torriani-Gorni A, Yagil E, Silver E (eds) Phosphate in microorganisms: cellular and molecular biology. ASM Press, Washington, DC, USA, pp 197–203

Groudev SN (1987) Use of heterotrophic microorganisms in mineral biotechnology. Acta Biotechnol 7:299–306

Haas D, Defago G (2005) Biological control of soil-borne pathogens by fluorescent pseudomonads. Nat Rev Microbiol 3:307–319

Han HS, Lee KD (2005) Phosphate and potassium solubilizing bacteria effect on mineral uptake, soil availability and growth of eggplant. Res J Agric Biol Sci 1:176–180

Han HS, Supanjani, Lee KD (2006) Effect of co-inoculation with phosphate and potassium solubilizing bacteria on mineral uptake and growth of pepper and cucumber. Plant Soil Environ 52:130–136

Hazen TC, Jimenez L, Victoria GL (1991) Comparison of bacteria from deep subsurface sediment and adjacent groundwater. Microbiol Ecol 22:293–304

Herridge DF, Peoples MB, Boddey RM (2008) Global inputs of biological nitrogen fixation in agricultural systems. Plant Soil 311:1–18

Hieberk FK, Bennett PC (1992) Microbial control of silicate weathering in organic-rich groundwater. Science 258:278–281

Hu XF, Chen J, Guo JF (2006) Two phosphate and potassium solubilizing bacteria isolated from Tianmu mountain, Zhejiang, China. World J Microbiol Biotechnol 22:983–990

Kalaiselvi P (1999) Influence of silicate solubilizing bacteria on dissolution of silicate and potassium in rice soil added with graded levels of potassium and rice residues. M. Sc. (Agric) thesis, Tamil Nadu Agricultural University, Coimbatore

Kannan NM, Raj SA (1998) Occurrence of silicate solubilizing bacteria in rice ecosystem. Madras Agric J 85:47–50

Khanwilkar SA, Ramteke JR (1993) Response of applied K in cereals in Maharashtra. Agriculture 4:84–96

Khudsen D, Peterson GA, Prov PF (1982) Lithium, sodium and potassium. In: Page AL (ed) Methods of soil analysis, part (2) agronomy monograph 9, 2nd edn. ASA and SSSA, Madison

Kirk JL, Beaudette LA, Hart M, Moutoglis P, Klironomus JN, Lee H, Trevors JT (2004) Methods of studying soil microbial diversity. J Microbiol Methods 58:169–188

Kloepper JW, Lifshitz R, Zablotowicz RM (1989) Free-living bacterial inocula for enhancing crop productivity. Trends Biotechnol 7:39–44

Li YF (1994) The characteristics and function of silicate dissolving bacteria fertilizer. Soil Fertil 2:48–49

Li FC, Li S, Yang YZ, Cheng LJ (2006) Advances in the study of weathering products of primary silicate minerals, exemplified by mica and feldspar. Acta Petrol Mineral 25:440–448

Lian B, Fu PQ, Mo DM, Liu CQ (2002) A comprehensive review of the mechanism of potassium release by silicate bacteria. Acta Mineral Sin 22:179–183

Lian B, Wang B, Pan M, Liu C, Tang HH (2007) Microbial release of potassium from K-bearing minerals by thermophilic fungus *Aspergillus fumigatus*. Geochem Cosmochim Acta 72:87–98

Lin QM, Rao ZH, Sun YX, Yao J, Xing LJ (2002) Identification and practical application of silicate-dissolving bacteria. Agric Sci China 1:81–85

Liu GY (2001) Screening of silicate bacteria with potassium releasing and antagonistic activity. China J Appl Environ Biol 7:66–68

Liu W, Xu X, Wu S, Yang Q, Luo Y, Christie P (2006) Decomposition of silicate minerals by *Bacillus mucilaginosus* in liquid culture. Environ Geochem Health 28:133–140

Liu D, Lian B, Dong H (2012) Isolation of *Paenibacillus* sp. and assessment of its potential for enhancing mineral weathering. Geomicrobiol J 29:413–421

Lugtenberg BJJ, Dekkers L, Bloemberg CV (2001) Molecular determinants of rhizosphere colonization by *Pseudomonas*. Annu Rev Phytopathol 39:461–490

Malinovskaya IM, Kosenko LV, Votselko SK, Podgorskii VS (1990) Role of *Bacillus mucilaginosus* polysaccharide in degradation of silicate minerals. Mikrobiologiya 59:49–55

Marques APGC, Pires C, Moreira H, Rangel AOSS, Castro ML (2010) Assessment of the plant growth promotion abilities of six bacterial isolates using *Zea mays* as indicator plant. Soil Biol Biochem 42:1229–1235

Matilla MA, Espinosa-Urgel M, Roderiguez-Herva JJ, Ramos JL, Ramos-Gonzalez MI (2007) Genomic analysis reveals the major driving forces of bacterial life in the rhizosphere. Genome Biol 8:R179

Meena VS, Maurya BR, Bahadur I (2014a) Potassium solubilization by bacterial strain in waste mica. Bangladesh J Bot 43(2):235–237

Meena VS, Maurya BR, Verma JP (2014b) Does a rhizospheric microorganism enhance K+ availability in agricultural soils? Microbiol Res 169:337–347

Meena VS, Maurya BR, Verma JP, Aeron A, Kumar A, Kim K, Vivek, Bajpai K (2015) Potassium solubilizing rhizobacteria (KSR): isolation, identification, and K-release dynamics from waste mica. Ecol Eng 81:340–347

Memon YM, Fergus IF, Hughes JD, Page DW (1988) Utilization of non-exchangeable soil potassium in relation to soil types, plant species and stage of growth. Aust J Soil Res 26:489–496

Mengel K, Kirkby EA (1987) Principles of plant nutrition. Internl Potash Inst Bern, Switzerland, pp 200–210

Muntz A (1890) Sur La decomposition des rockes et al formation de la terrarable. C R Akad Sci Paris 110:1370–1372

Murali G, Gupta A, Nair RV (2005) Variations in hosting beneficial plant associated microorganisms by root (wilt) diseased and field tolerant coconut palms of west coast tall variety. Curr Sci 89:1922–1927

Muralikannan M (1996) Biodissolution of silicate, phosphate and potassium by silicate solubilizing bacteria in rice ecosystem. M. Sc. (Agric) thesis, Tamil Nadu Agricultural University, Coimbatore

Naik PR, Raman G, Narayanan KB, Sakthivel N (2008) Assessment of genetic and functional diversity of phosphate solubilizing fluorescent pseudomonads isolated from rhizospheric soil. BMC Microbiol 8:230–243

Norkina SP, Pumpyansakya LV (1956) Certain properties of silicate bacteria. Crop Sci Soc Japan 28:35–40

Parmar P (2010) Isolation of potassium solubilizing bacteria and their inoculation effect on growth of wheat (*Triticum aestivum* L. em. Thell.). M. Sc. thesis submitted to CCS Haryana Agricultural university, Hisar

Parmar P, Sindhu SS (2013) Potassium solubilization by rhizosphere bacteria: influence of nutritional and environmental conditions. J Microbiol Res 3:25–31

Prajapati K, Sharma MC, Modi HA (2012) Isolation of two potassium solubilizing fungi from ceramic industry soils. Life Sci Leafl 5:71–75

Purushothaman A, Chandramohan D, Natarajan R (1974) Distribution of silicate dissolving bacteria in velar estuary. Curr Sci 43:282–283

Raj SA (2004) Solubilization on a silicate and concurrent release of phosphorus and potassium in rice ecosystem. In: Biofertilizer technology for rice based cropping system, India, pp 372–378

Rajan SSS, Watkinson JH, Sinclair AG (1996) Phosphate rock for direct application to soils. Adv Agron 57:77–159

Ramarethinam S, Chandra K (2005) Studies on the effect of potash solubilizing/mobilizing bacteria *Frateuria aurantia* on brinjal growth and yield. Pestology 11:35–39

Reitmeir RF (1951) Soil potassium. In: Norman AG (ed) Advances in agronomy, American Society of Agronomy, vol III. Academic Press, Int Publ, New York, pp 113–164

Richardson AE, Hadobas PA, Hayes JE (2001) Extracellular secretion of *Aspergillus* phytase from *Arabidopsis* roots enables plants to obtain phosphorus from phytate. Plant J 25:641–649

Rogers JR, Bennett PC, Choi WJ (1998) Feldspars as a source of nutrients for microorganisms. Amer Mineral 83:1532–1540

Ryan PR, Delhaise E, Jones DL (2001) Function and mechanism of organic anion exudation from plant roots. Annu Rev Plant Physiol Plant Mol Biol 52:527–560

Ryan PR, Dessaux Y, Thomashow LS, Weller DM (2009) Rhizosphere engineering and management for sustainable agriculture. Plant Soil 321:363–383

Sharpley AN (1989) Relationship between soil potassium forms and mineralogy. Soil Sci Soc Am J 52:1023–1028

Sheng XF (2005) Growth promotion and increased potassium uptake of cotton and rape by a potassium releasing strain of *Bacillus edaphicus*. Soil Biol Biochem 37:1918–1922

Sheng XF, He LY (2006) Solubilization of potassium bearing minerals by a wild type strain of *Bacillus edaphicus* and its mutants and increased potassium uptake by wheat. Can J Microbiol 52:66–72

Sheng XF, Huang WY (2002a) Study on the conditions of potassium release by strain NBT of silicate bacteria. Sci Agric Sin 35:673–677

Sheng XF, Huang WY (2002b) Mechanism of potassium release from feldspar affected by the strain NBT of silicate bacterium. Acta Pedol Sin 39:863–871

Sheng XF, Xia JJ, Chen J (2003) Mutagenesis of the *Bacillus edaphicus* strain NBT and its effect on growth of chilli and cotton. Agric Sci China 2:400–412

Sindhu SS, Dadarwal KR (2000) Competition for nodulation among rhizobia in *Rhizobium*-legume symbiosis. Indian J Microbiol 40:211–246

Sindhu SS, Suneja S, Goel AK, Parmar N, Dadarwal KR (2002) Plant growth promoting effects of *Pseudomonas* sp. on coinoculation with *Mesorhizobium* sp. *Cicer* strain under sterile and wilt sick soil conditions. Appl Soil Ecol 19:57–64

Sindhu SS, Dua S, Verma MK, Khandelwal A (2010) Growth promotion of legumes by inoculation of rhizosphere bacteria. In: Khan MS, Zaidi A, Musarrat J (eds) Microbes for legume improvement. Springer, Wien/New York/Berlin, USA, pp 195–235

Sindhu SS, Dua S, Sahu G (2011) Biological control of plant diseases. In: Rana MK (ed) Modern concepts of vegetable production. Biotech Books, Daryaganj/New Delhi, pp 470–517

Sindhu SS, Parmar P, Phour M (2014a) Nutrient cycling: potassium solubilization by microorganisms and improvement of crop growth. In: Parmar N, Singh A (eds) Geomicrobiology and biogeochemistry. Springer, Berlin/Heidelberg, pp 175–198

Sindhu SS, Parmar P, Phour M, Kumari K (2014b) Rhizosphere microorganisms for improvement in soil fertility and plant growth. In: Nagpal R, Kumar A, Singh R (eds) Microbes in the service of mankind: tiny bugs with huge impact. JBC Press, New Delhi, pp 32–94

Sindhu SS, Phour M, Choudhary SR, Chaudhary D (2014c) Phosphorus cycling: prospects of using rhizosphere microorganisms for improving phosphorus nutrition of plants. In: Parmar N, Singh A (eds) Geomicrobiology and biogeochemistry. Springer, Berlin/Heidelberg, pp 199–237

Singh G, Biswas DR, Marwah TS (2010) Mobilization of potassium from waste mica by plant growth promoting rhizobacteria and its assimilation by maize (*Zea mays*) and wheat (*Triticum aestivum* L.). J Plant Nutr 33: 1236–1251

Sparks DL (1987) Potassium dynamics in soils. Adv Soil Sci 6:1–63

Sparks DL (2000) Bioavailability of soil potassium. In: Summer ME (ed) Handbook of soil science. CRC Press, Boca Raton, pp D–38–D–52

Sparks DL, Huang PM (1985) Physical chemistry of soil potassium. In: Munson RD (ed) Potassium in agriculture. American Society of Agronomy Journal, Madison, pp 201–276

Sperberg JI (1958) The incidence of apatite solubilizing organisms in the rhizosphere and soil. Aust J Agric Res Econ 9:778–781

Styriakova I, Styriak I, Galko I, Hradil D, Bezdicka P (2003) The release of iron bearing minerals and dissolution of feldspar by heterotrophic bacteria of *Bacillus* species. Ceram Silicaty 47:20–26

Sugumaran P, Janarthanam B (2007) Solubilization of potassium containing minerals by bacteria and their effect on plant growth. World J Agric Sci 3 (3):350–355

Supanjani, Han HS, Jung SJ, Lee KD (2006) Rock phosphate potassium and rock solubilizing bacteria as alternative sustainable fertilizers. Agro Sustain Dev 26:233–240

Ullman WJ, Welch SA (2002) Organic ligands and feldspar dissolution. Geochem Soc 7:3–35

Ullman WJ, Kirchman DL, Welch SA (1996) Laboratory evidence by microbially mediated silicate mineral dissolution in nature. Chem Geol 132:11–17

Vandevivere P, Welch SA, Ullman WJ, Kirchman DJ (1994) Enhanced dissolution of silicate minerals by bacteria at near neutral pH. Microbiol Ecol 27:241–251

Vassilev N, Medina A, Azcon R, Vassilev M (2006) Microbial solubilization of rock phosphate as media containing agro industrial wastes and effect of the resulting products on plant growth and phosphorus uptake. Plant Soil 287:77–84

Vessey KJ (2003) Plant growth promoting rhizobacteria as biofertilizers. Plant Soil 25:557–586

Wall DH, Virginia RA (1999) Control of soil biodiversity- in sight from extreme environments. Appl Soil Ecol 13:137–150

Wang RR, Wang Q, He LY, Qui G, Sheng XF (2015) Isolation and the interaction between a mineral-weathering *Rhizobium tropici* Q34 and silicate minerals. World J Microbiol Biotechnol 31:747–753

Webley DM, Duff RB, Mitchell WA (1960) A plate method for studying the breakdown of synthetic and natural silicates by soil bacteria. Nature 188:766–767

Welch SA, Ullman WJ (1993) The effect of organic acids on plagioclase dissolution rates and stoichiometry. Geochim Cosmochim Acta 57:2725–2736

Welch SA, Vandevivere P (2009) Effect of microbial and other naturally occurring polymers on mineral dissolution. Geomicrobiol J 12:227–238

Weller DM (2007) *Pseudomonas* biocontrol agents of soilborne pathogens: looking back over 30 years. Phytopathology 97:250–256

Wu SC, Cao ZH, Li ZG, Cheung KC, Wong MH (2005) Effects of biofertilizer containing N-fixer, P and K solubilizers and AM-fungi on maize growth: a greenhouse trial. Geoderma 125:155–166

Xiaoxi Z, Liu X, Tang J, Hu S, Jiang P, Li W, Xu L (2012) Characterization and potassium solubilizing ability of *Bacillus circulans* Z_{1-3}. Adv Sci Lett 10:173–176

Xie JC (1998) Present situation and prospects for the world's fertilizer use. Plant Nutr Fertil Sci 4:321–330

Zapata F, Roy RN (2004) Use of phosphate rock for sustainable agriculture. FAO and IAEA, Rome

Zarjani JK, Aliasgharzad N, Oustan S, Emadi M, Ahmadi A (2013) Isolation and characterization of potassium-solubilizing bacteria in some Iranian soils. Arch Agron Soil Sci 59:1713–1723

Zhang C, Kong F (2014) Isolation and identification of potassium-solubilizing bacteria from tobacco rhizosphere soil and their effect on tobacco plants. Appl Soil Ecol 82:18–25

Zhang CJ, Tu GQ, Cheng CJ (2004) Study on potassium dissolving ability of silicate bacteria. Shaguan Coll J 26:1209–1216

Zhang AM, Zhao GY, Gao TG, Wang W, Li J, Zhang SF, Zhu BC (2013) Solubilization of insoluble potassium and phosphate by *Paenibacillus kribensis* CX-7: a soil microorganism with biological control potential. Afr J Microbiol Res 7:41–47

Zhao F, Sheng XF, Huang Z (2008) Isolation of mineral potassium solubilizing bacterial strains from agricultural soils in Shandong province. Biodivers Sci 16:593–600

Zhou H, Zeng XX, Liu FF, Qiu GZ, Hu YH (2006) Screening, identification and desilication of a silicate bacterium. J Cent South Univ Technol 13:337–341

Zord C, Senbayram M, Peiter E (2014) Potassium in agriculture – status and perspective. J Plant Physiol 171:656–659

Dynamics of Potassium and Their Bioavailability for Plant Nutrition

14

Brijesh Kumar Yadav and Ajaib Singh Sidhu

Abstract

Many physical, chemical, and mineralogical factors govern the release of K from micas by cation exchange reaction and dissolution processes. Mainly in Indian agricultural conditions, most of the farmers use only nitrogen and phosphorus and do not use the potassic fertilizer due to unawareness that the problem of K deficiency occurs in soils and it is reported that ~72 % of soils need immediate K fertilization for good crop production. The efficient soil microorganisms influence the availability of minerals in soil and play a major role in ion cycling and soil fertility. A number of microorganisms such as bacteria (*Bacillus mucilaginosus*, *B. edaphicus*, *B. circulans*, *B. cereus*, *B. subtilis*, *B. coagulans*, *B. amyloliquefaciens*, *B. megaterium*, *Enterobacter hormaechei*, *Flectobacillus* spp., *Acidithiobacillus ferrooxidans*, *Paenibacillus* spp.), fungal strains (*Aspergillus niger*, *A. fumigatus*, and *A. terreus*), and yeast (*Torulaspora globosa*) solubilize the insoluble K to soluble forms of K for plant uptake. The main mechanisms of KSMs are acidolysis, chelation, exchange reactions, complexolysis, and production of organic acids. Therefore, the efficient KSMs should be applied for solubilization of unavailable form of K to an available form of K that is taken up by the plant for their growth and development. Therefore, our aim is to elaborate on the studies of native K-solubilizing microbes to develop efficient microbial consortium and their uses as biofertilizers or a biocontrol agent, which helps to enhance the K availability in agricultural soils and increase crop production and environmental and agricultural sustainability.

Keywords

K release • K-solubilizing microorganisms (KSMs) • Bioavailability • K uptake

B.K. Yadav (✉) • A.S. Sidhu
Regional Research Station, Punjab Agricultural University, Bathinda 151001, Punjab, India
e-mail: bkyadav74@pau.edu; assidhu@pau.edu

V.S. Meena et al. (eds.), *Potassium Solubilizing Microorganisms for Sustainable Agriculture*,
DOI 10.1007/978-81-322-2776-2_14

14.1 Introduction

Potassium (K) was first isolated in 1807 by Sir Humphrey Davy. The word "potash" is often used to refer to potassium chloride (KCl), a common K fertilizer and derivative from "pot ashes" which refers to the earliest practice of applying the leachate of wood ashes as a source of K. Although K constitutes ~ 2.5 % of the lithosphere, in the soil the K concentration varies widely between 0.04 and 3.0 % (Sparks and Huang 1985). Igneous rocks such as granites and syenites (46–54 g K kg^{-1}), basalts (7 g K kg^{-1}), and peridotites (2 g K kg^{-1}) have higher K content than sedimentary rocks like clayey shales (30 g K kg^{-1}), whereas limestone has an average of only 6 g K kg^{-1} on the Earth's crust (Malavolta 1985). In general a mineral soil ranges between 0.04 and 3 % K. The upper and subsurface layers of the soil profile contain between 3 and 100 ton ha^{-1} of total K. Of this total K, about 98 % is bound in the mineral form, whereas only 2 % is in soil solution and exchangeable phases (Schroeder 1979; Bertsch and Thomas 1985). The amount of K in a soil depends upon nature of parent material, degree of weathering, addition of manures and fertilizers, and losses due to leaching, erosion, and crop removal. Soil test results for K fertility status among India's agricultural soils are categorized by Hasan (2002) as per nutrient index values suggested by Ramamurthy and Bajaj (1969) and indicate that 21 %, 51 %, and 28 % of districts are low, medium, and high K fertility status, respectively. Thus, ~72 % of soils need immediate K fertilization for good crop production as these soils are low and medium K status (Table 14.1). Available soil K was extracted with 1 N ammonium acetate (NH_4OAc, pH 7.0), and soils containing less than 130 kg K_2O ha^{-1} were categorized as low, between 130 and 335 kg K_2O ha^{-1} as medium, and above 335 kg K_2O ha^{-1} as high.

K is an essential plant nutrient for its growth and is extremely important for the productive farming in agriculture. It is required in high quantities and plays two major roles in the functioning of plant cells. First, it plays a unique part in the activation of enzymes which are fundamental to metabolic processes, mainly the production of proteins and sugars. Second, K is the "plant-preferred" ion for the turgor (rigidity) of each plant cell by maintaining the water content. More concentration of K in the cell sap creates conditions to move water into the cell (osmosis) through the cell wall. The turgid cells sustain the leaf's vigor which helps photosynthesis to proceed efficiently (Mengel and Kirkby 1987;

Table 14.1 Categorization of districts as low, medium, and high K fertility status

	Potassium status (kg K_2O ha^{-1})		
States	Less than 130	130–335	More than 335
Andhra Pradesh	2	14	3
Arunachal Pradesh	2	3	0
Assam	7	3	0
Bihar and Jharkhand	1	24	2
Chandigarh	0	1	0
Dadra and Nagar Haveli	0	1	0
Delhi	0	1	0
Goa	1	0	0
Gujarat	0	3	16
Haryana	0	2	9
Himachal Pradesh	6	4	3
Jammu and Kashmir	5	5	0
Karnataka	3	10	7
Kerala	4	6	0
Madhya Pradesh and Chhattisgarh	3	10	31
Maharashtra	0	12	13
Manipur	1	0	0
Meghalaya	1	0	0
Mizoram	1	0	0
Nagaland	5	0	0
Orissa	2	11	0
Punjab	0	9	3
Pondicherry	1	0	0
Rajasthan	0	23	0
Sikkim	0	4	0
Tamil Nadu	0	6	7
Tripura	3	0	0
Uttar Pradesh and Uttaranchal	26	23	7
West Bengal	2	13	1
Total districts, (%)	76 (21)	190 (51)	105 (28)

Marschner 1995; Valmorbida and Boaro 2007; Meena et al. 2013; Maurya et al. 2014).

The K controls the size of the stomata openings in response to ecological and internal plant conditions and influences the transpiration rates. Deficiencies of K in crops were felt generally in areas of high crop yields and more so in situations where continued nitrogen fertilization, with or without phosphorus, depletes the soils of their K. Hence, with the passage of time and intensification of agriculture, K availability in the soil tends to decline. At early stages of K deficiency, mostly the yield decreases. This early stage is called "hidden hunger" since plants show no specific symptoms at this stage. As the intensity of the deficiency increases, significant symptoms appear. The plant cannot distribute K to younger leaves, when its deficiency occurs in soil, as this is a mobile nutrient. So its deficiency symptoms first appear on older/lower leaves. Typical K deficiency symptoms in plants include brown scorching and curling of leaf tips as well as chlorosis (yellowing) between the leaf veins. Sometimes purple spots may also appear on the leaf undersides. Under insufficient supply of K, the plants have poorly developed root system, grow slowly, produce lesser seeds, and have lower yields (McAfee 2008; White and Karley 2010) which increased susceptibility to many diseases (Amtmann et al. 2008; Armengaud et al. 2010) and pest (Amtmann et al. 2006; Troufflard et al. 2010). In soil the K exists in different forms, viz., water-soluble, exchangeable, non-exchangeable, and lattice K. The water-soluble along with exchangeable K constitutes the plant-available K (Mishra et al. 1993). These different forms of soil K are in dynamic equilibrium, and any depletion in a given K form is possible to shift equilibrium in the direction to deplete it (Ramamoorthy and Paliwal 1976).

The K presents as abundant element in soil or is applied as natural or synthetic fertilizers, only 1–2 % is available to plants, and the rest is bound with other minerals and therefore unavailable to plants. The efficient soil microorganisms influence the availability of minerals in soil and play a major role in ion cycling and soil fertility (Bin et al. 2010). Alexander (1985) reported silicate bacteria to resolve K, silicon, and aluminum from insoluble minerals. Their uses as bio-inoculants or biocontrol agents will help to increase K availability in soil and increase crop production by fulfilling the K requirement of the plants (Meena et al. 2015b; Singh et al. 2015).

14.2 The Forms of K in Soil

Knowledge of various forms of K and their distribution in soils (Fig. 14.1) is a great relevance in assessing the long-term availability of K to crops and in framing a sound basis of fertilizer recommendations. About 98 % of the total K reserve in soils exists in inorganic combinations which can further be characterized as: water-soluble K, exchangeable K, non-exchangeable K, and lattice K. Depending on the type of soil and environmental conditions, K availability for plant may vary.

14.2.1 Readily Available K

Readily available K includes water-soluble K (dissolved in soil water) and exchangeable K (held on exchange sites of clay particles). Plants readily absorb the water-soluble K. As soon as the K concentration in soil water decreases, more K is released into this solution from the exchangeable K. The K attached to the exchange sites of the clay particles are more readily available for plant than the K trapped between the layers of the clay particles.

14.2.2 Slowly Available K

The K trapped between layers of clay minerals and frequently referred to as being fixed is known as slowly available K. The plants are unable to use the slowly available K during a single crop growing season and serve as a reservoir for readily available K. The quantity of K fixed varies with the type of clay minerals that dominates in the soil. Montmorillonite clays fix K when the

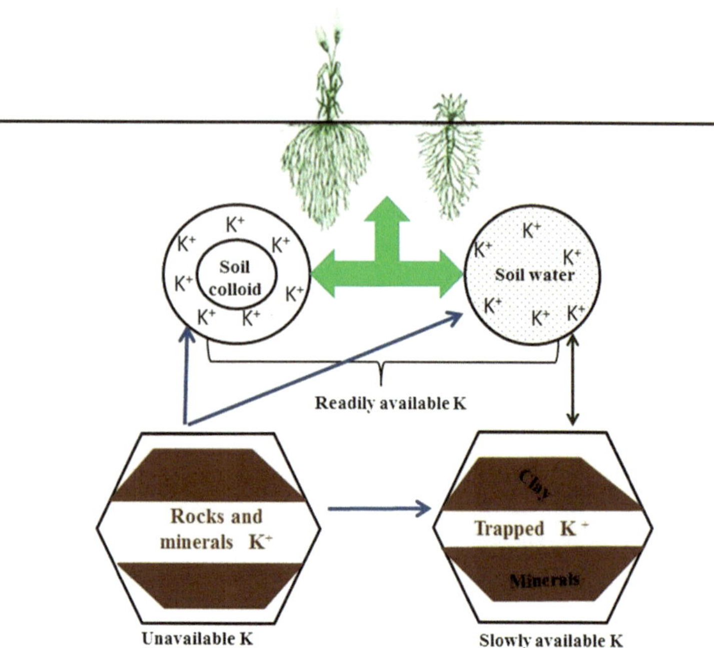

Fig. 14.1 Relationship among readily available, slowly available, and unavailable K in soil-plant system

soils become dry and released when the soil becomes wet because K is trapped between the layers in the clay mineral. Illite dominant clays also fix K between layers when they become dry, but do not release all of the fixed K when water is added in to the soil.

14.2.3 The Unavailable K

Plants are unable to use the K in these forms. Depending on soil type, ~90–98 % of total soil K is found in unavailable form. Crystalline structure of feldspars and micas which is part of the soils contains most of the unavailable K. These minerals take a long time to weather (break down) and small quantities of K released in to the soil solution, and the process is too slow to supply the full K needs of field crops.

14.3 The Behavior of K in Soil

The behavior of the K in soil is influenced by the type and amount of clay and soil organic matter. The type of clay depends on the parent rocks (igneous or sedimentary) and the degree to which the particles have undergone modification. The clay-sized particles are made up of many layers, each composed of interlocking silicon and oxygen atoms or aluminum and oxygen atoms. In the illite clay, there are spaces between the layers in which positive ions, like K, can migrate and be held thereby negative charges on the layers, whereas in kaolinitic clay, the layers are so tightly compressed that ions like K cannot enter between the layers. In both types of clay minerals, K can be held at the edges of the clay layers. Moreover, at the edges, the layers tend to open up, allowing ready entrance of K. The positive K ions held at the edges of the clay layers and toward the outer edge of any interlayer space can be replaced quite easily by other positively charged ions. When this process occurs, the K is released into the soil solution from where it is taken up by plant roots. The reverse process occurs when K concentration in soil solution increased by applying chemical fertilizers and manures (Maurya et al. 2014; Meena et al. 2014a, b, 2015b). Thus, reserves of K can be built up in those soils where there is separation between the layers that form the clay particle.

14.4 The K Cycle in Soil

The main pathway for K in the soil is presented in Fig. 14.2. In soil, K occurs as: water-soluble K, exchangeable K, non-exchangeable K, and lattice K. The water-soluble and exchangeable pools of K are inequilibrium with each other. The plants take up K completely as the K^+ ion, which is the only form that exists in the water-soluble form. Exchangeable K accounts only 1–2 % of soil K and refers to ions adsorbed to exchange sites on soil particles. When water-soluble K is removed by plant uptake, it is replenished by K released from the exchangeable pool.

Similarly, if the concentration of K exceeds in soil solution than in the exchangeable pool, K will adsorb to the exchange sites to ensure a stable equilibrium between pools of available K (water soluble and exchangeable). K fixation refers to entrapment of the K^+ ion in the structure of clay minerals. The fixed pool/lattice K is remarkable to release K at sufficient rates to meet the demands of growing crops. However, a portion of this pool is turning into available form as the K is depleted in available form. More than 90 % of the total K in soils exists apart of the parent minerals (mica and feldspar) and liberate of K at a rate far too slow to meet the needs of growing plants. There are mainly two ways such as leaching and erosion by which K can be lost from the system. K leaching can take place in coarse-textured soils that receive above-average precipitation, and erosion is a more important route of K losses from soil.

14.5 K Fixation and Release

The K availability in soils depends upon the phenomenon of K fixation. In the process of K fixation in soils, the added soluble K is converted to a form that cannot be extracted with a neutral salt solution, commonly employed to extract the plant-available form of soil K. As a result of K fixation, availability of added K decreases. The extent of K fixation depends on the type of clay mineral and its charge density, degree of interlayering, moisture content, concentration of K^+ ions in addition to the concentration of

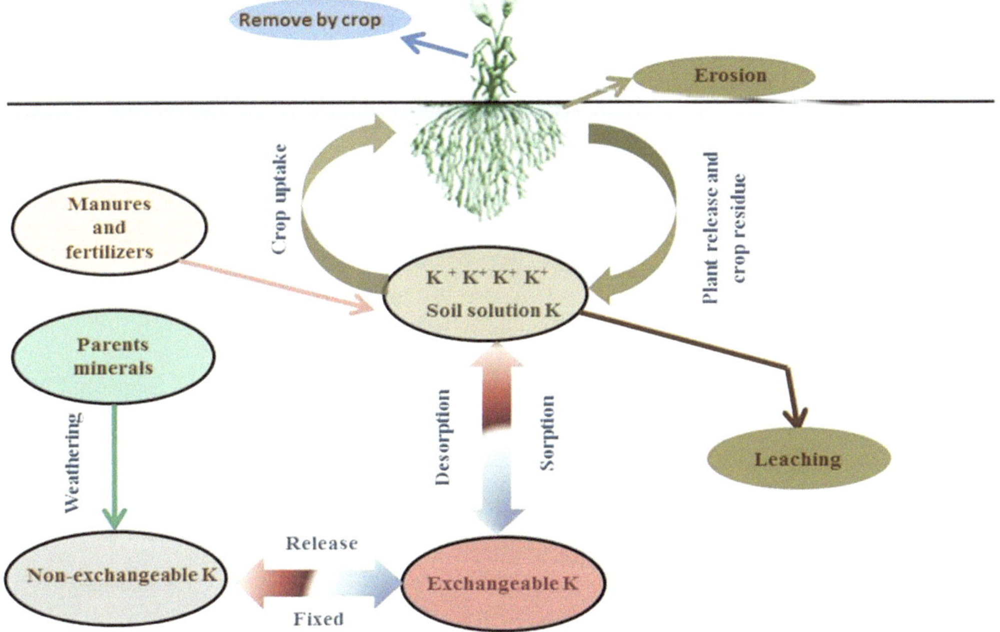

Fig. 14.2 Potassium dynamics in soil-plant system

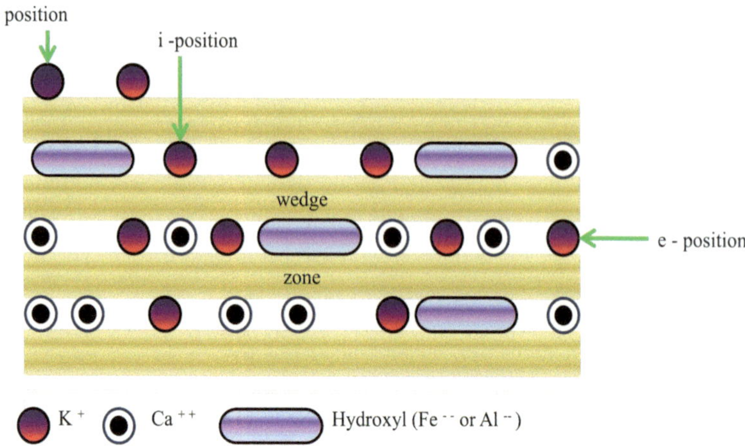

Fig. 14.3 Different adsorption sites of K in an expandable layer silicate with interlayers

opposing cations, and pH of the ambient solution bathing the clay or soil. The K fixation is almost deficient in soils dominated in kaolinite, chlorite, and unweathered micas, minor in montmorillonite, considerable in illite, and more in vermiculite-dominated soils. The degree of K fixation is strongly influenced with the charge density on the layer silicate and fixes more K with high charge density than those with low charge density. K^+ ions are adsorbed by clay minerals onto the binding sites which differ in their selectivity (Fig. 14.3). Three diverse adsorption sites can be distinguished in 2:1-type clay minerals (illite, vermiculites, and weathered mica) known as the planar surface (p-position), at the edges of the layers (e-position), and in interlayer space (i-position).

The all three binding sites are specific for K adsorption. The binding selectivity for K by organic matter and kaolinite are similar to the location sites. At this point, the K-bond is comparatively weak so that K adsorbed may easily be replaced by other cations like Ca^+ and Mg^+ ions. The interlayer space (i-position) has highest specificity for K^+ and mainly accounts for K^+ fixation in soils. The potassium added through manures and fertilizers initially increased the readily available K in soil. The saturated exchange complex with respect to K leads to the entry of K into wedge (partially opened) and interlayer spaces. This results in fixation of K in non-exchangeable form.

Mainly two processes are concerned in the release of K from micas: (1) the alteration of K-bearing micas to expansible 2:1 layer silicates by exchanging the K with hydrated cations and (2) the dissolution of the micas followed by the formation of weathering products. The phenomenon of fixation of exchangeable K as well as liberation of non-exchangeable K plays an important role in the dynamics of K in soil. The regular release of K from the mica lattice to form illite (hydrous mica) and ultimately vermiculite with simultaneous gain of water or H_3O^+ and swelling of K lattice is illustrated in Fig. 14.4. The low hydration energy of K^+ favors its entrapment. Due to crop removal and leaching, low concentration of K in soil solution favors the release of K. In the absence of external additions of K, plants are capable of taking up a very large amount of K without bringing about substantial decrease in exchangeable K. This means that K which is not initially in exchangeable form has changed into exchangeable form and becomes available to plants.

Many physical, chemical, and mineralogical factors such as tetrahedral rotation and cell dimensions, extent of tetrahedral tilting, hydroxyl orientation, chemical composition, particle size, structural imperfections, degree of K reduction, layer charge alterations and associated reactions, biological activity, inorganic cations, wetting, and drying govern the release of K from micas by cation exchange reaction and

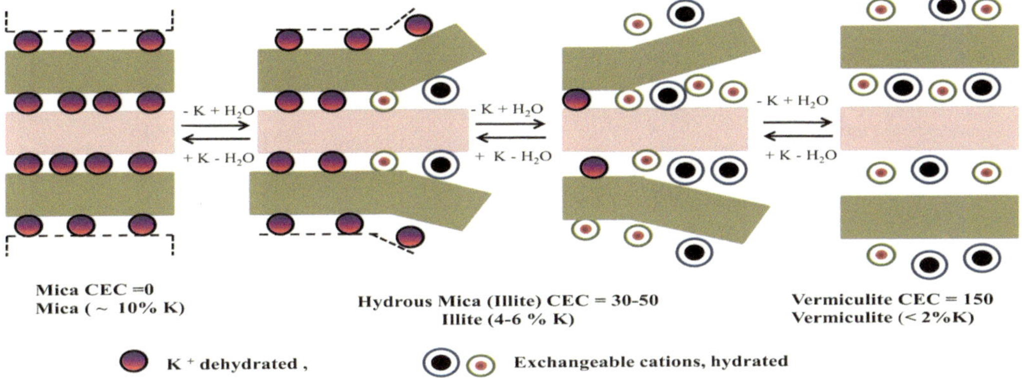

Fig. 14.4 Mechanism of K fixation and release

dissolution processes. The organisms reduce the K in soil solution, and their action may be similar to that of tetraphenylboron (TPB) in artificial weathering of micas. In addition, generally the action of organisms is more complex when organic acids are produced by them. A small but measurable amount of biochemical compounds such as organic acids are present in all soils. Moreover, the time required for soil formation can be extended over a period of centuries; the collective effect in a soil which has very little quantities of chelating agents will be considerable. Song and Huang (1988) studied the influence of oxalic and citric acids on the dynamics of K release from micas and feldspars and reported the order of K release from K-bearing minerals: biotite > microcline > orthoclase > muscovite.

14.6 The K-Solubilizing/K-Mobilizing Microorganisms

With intensive agriculture, availability of K in soils dropped due to crop removal, leaching, runoff, and erosion. Thus, it becomes urgent to explore the bio-activation of soil K reserves to improve the potash fertilizer shortage. Total soil K reserves are usually large in all soils, but the allocation of K forms differs from soil to soil depending on dominant soil minerals present (Mclean and Watson 1985). In all soils K is a plentiful element or is applied as natural or synthetic fertilizers, only 1–2 % is available to plants, and the rest being bound with other minerals and therefore unavailable to plants. Many soil microbes have been identified to play a key role in the K cycle, and therefore, KSMs present in the soil could provide a substitute technology to make K available for plant nutrition (Rogers et al. 1998). Various groups of soil microflora (Table 14.2) were reported to be involved in solubilization/mobilization of unavailable forms of K into available forms of K which is easily absorbed by plants' roots. These KSMs are economically viable and eco-friendly and are capable to dissolve K-bearing minerals and rocks, enhance plant growth, and improve the crop yield.

The first evidence of microbial contribution in solubilization of rock K was reported by Muentz (1890). Further, many researchers reported a wide range of the rhizospheric microorganisms as the K solubilizers include *B. mucilaginosus* (Raj 2004; Sugumaran and Janarthanam 2007; Zhao et al. 2008; Basak and Biswas 2009; Zarjani et al. 2013); *B. edaphicus* (Sheng 2005); *B. circulans* (Lian et al. 2002); *Burkholderia* and *A. ferrooxidans* (Sheng and Huang 2002; Sheng and He 2006); *Arthrobacter sp.* (Zarjani et al. 2013); *Enterobacter hormaechei* (Prajapati et al. 2013); *Paenibacillus mucilaginosus* (Hu et al. 2006; Liu et al. 2012); *P. frequentans* and *Cladosporium* (Argelis et al. 1993); *Aminobacter*, *Sphingomonas*, and *Burkholderia* (Uroz et al. 2007); and

Table 14.2 List of some potassium-solubilizing/potassium-mobilizing microorganisms

Microorganisms	References
Bacteria	
Bacillus amyloliquefaciens	Gundala et al. (2013)
Agrobacterium tumefaciens, *B. cereus*, *B. coagulans*. *B. megaterium*, *Flectobacillus* spp., *Microbacterium hominis*	Diep and Hieu (2013)
Bacillus mucilaginosus	Zhou et al. (2006)
Bacillus circulans	El-Hadad et al. (2011)
Bacillus subtilis	Abou and Abdal (2012)
Bacillus edaphicus	Sheng (2005)
Bacillus megaterium var. *phosphaticum*	Sugumaran and Janarthanam (2007)
Enterobacter hormaechei	Prajapati et al. (2012)
Burkholderia	Bagyalakshmi et al. (2012)
Paenibacillus spp.	Liu et al. (2012)
Paenibacillus glucanolyticus	Sangeeth et al. (2012)
Pseudomonas spp.	Bagyalakshmi et al. (2012) and Archana et al. (2013)
Fungal	
Aspergillus fumigatus	Bin et al. (2010)
Aspergillus niger, *A. terreus*	Prajapati et al. (2012)
Yeast	
Torulaspora globosa	Rosa-Magri et al. (2012)

Paenibacillus glucanolyticus (Sangeeth et al. 2012). Wu et al. (2005) recorded increase in K uptake by maize crop by applying two arbuscular mycorrhizal fungi (AMF) species, *Glomus mosseae* and *G. intraradices*, on soil weight basis. Lian et al. (2008) also reported some fungal strains are able to solubilize rock potassium and K-aluminum silicate.

Moreover, Prajapati et al. (2012) reported that K-solubilizing fungi such as *A. terreus* and *A. niger* solubilize insoluble K and showed the highest available K in liquid medium by using two various insoluble sources of K, i.e., feldspar and K-aluminum silicate. It is also reported that *A. terreus* shows the highest solubilization as well as acid production on both of the insoluble K sources.

Furthermore, symbiotic nitrogen-fixing rhizobia and *Pseudomonas*, along with fixing atmospheric nitrogen into ammonia and exporting to the host plants, have also shown K- and P-solubilizing activity. For instance, *Aspergillus* spp., *A. terreus* (Prajapati et al. 2013), *A. niger* (Lian et al. 2002; Prajapati et al. 2012), and *Penicillium* spp. (Sangeeth et al. 2012) enhanced K solubilization by mobilizing inorganic and organic K and liberate of K from rocks and minerals. Rosa-Magri et al. (2012) isolated the yeast *Torulaspora globosa*, from the sugarcane rhizosphere, and evaluated its capacity to solubilization of K from alkaline ultramafic rock powder and found 38 % more total K released in the medium with the yeast during a 15-day period of incubation.

14.7 The Mechanisms of K Solubilization/Mobilization

At present, insufficient information is available on K solubilization/mobilization by the rhizospheric microorganism as well as crop growth enhancement. The mechanism for KSMs to solubilize K involves lowering the pH of the medium and improving chelation of the cations bound to K and acidolysis of the adjoining area of microorganism. The release of organic acids and protons by K-solubilizing microorganisms lowers the pH of the medium (Uroz et al. 2009; Parmar and Sindhu 2013; Zarjani et al. 2013). Such acidolysis due to organic acids produced by the rhizospheric microorganisms directly breaks up the mineral K, slowly releases exchangeable K and readily available exchangeable K, and can chelate both Si and Al ions associated with K minerals (Romheld and Kirkby 2010). Thus, the synthesis and release of organic acids by microorganisms into adjoining surroundings acidify the microbe's cells and their immediate atmosphere that ultimately lead to release of K ions from the mineral K by protonation and acidification (Goldstein 1994). Sheng et al. (2008) detected some organic acids in microbial suspension. Further, organic acids formed by KSMs can be detected by high performance liquid chromatography (HPLC) and enzymatic methods (Archana et al. 2013; Zhang et al. 2013). The acidification is not only mechanism of K solubilization, as the ability to decrease the pH in some cases did not associate with the capacity to solubilize mineral K (Rosa-Magri et al. 2012; Zarjani et al. 2013). Furthermore, the chelating ability of organic acids is important, because it has been reported that the addition of 0.05 M EDTA in to the medium has similar solubilizing result as inoculation with *P. bilaii* (Kucey 1988; Liu et al. 2006; Sheng and He 2006). According to Sheng and Huang (2002), the K release from the K-bearing minerals was also affected by pH, oxygen, and the bacterial strains used.

The effectiveness of the K solubilization by different microorganisms depends on the nature of K-bearing minerals and aerobic conditions (Uroz et al. 2009). Sheng and He (2006) observed better growth on illite than feldspar due to greater solubilization of K by B. *edaphicus* in the liquid media. The K release affected pH aerobic conditions and soil mineral properties (Lian 1998; Chen et al. 2008; Lian et al. 2008; Bin et al. 2010). The organic and inorganic acids convert insoluble K (mica, muscovite, biotite, and feldspar) to the soluble K (soil solution form), resulting in the increase of the availability of K to the plants. Various types of organic acid produced by KSMs differed with different organisms. KSMs are capable to disintegrate phlogopite via aluminum chelation and acidic dissolution of the crystal network (Abou and Abdel 2012). The release of various types of organic acids was reflected by microorganisms to solubilize the insoluble K to an available form of K which is easily taken up by the plant.

14.8 The Effect of KSMs on Crop Growth and Yield

Many researchers showed significant improvement of germination percentage, seedling vigor, plant growth and yield, and K uptake by plants under glasshouse and field conditions due to seed inoculation and seedling treatments of plants with the KSMs (Awasthi et al. 2011; Zhang et al. 2013). Application of organo-minerals with combination of silicate bacteria for enhancing plant growth was first reported Aleksandrov (1985). Sheng (2005) examined plant growth-promoting effects and nutrient uptake on cotton and rape in K-deficient soil with *B. edaphicus* NBT and found increase in root and shoot growth of cotton and rape. Similarly, Han et al. (2006) conducted an experiment with PSB (*B. megaterium* var. *phosphaticum*) and KSB (*B. mucilaginosus*) along with pepper (*Capsicum annum* L.) and cucumber (*Cucumis sativus* L.) and reported that co-inoculation of PSB and KSB resulted in consistently higher P and K availability, Sheng and He (2006) conducted an experiment with low K status using wheat as a test crop and reported that after inoculation with bacterial strains, the root and shoot growth of wheat was significantly increased and they had higher NPK contents of plant components as compared to those un-inoculated. Similarly, Badar et al. (2006) reported that application of KSMs with K- and P-bearing minerals on sorghum enhanced dry matter yield by 48 %, 65 %, and 58 %; phosphorus uptake by 71 %, 110 %, and 116 %; and K uptake by 41 %, 93 %, and 79 % in clay, sandy, and calcareous soils, respectively. Supanjani et al. (2006) reported that combination of K rocks with inoculation of K-solubilizing bacteria increased K availability from 13 % to 15 %. Soil application of KSM plant has 16 % photosynthesis and 35 % higher leaf area to control. According to Sugumaran and Janarthanam

(2007), inoculation of *B. mucilaginosus* in groundnut crop increases the dry matter by 25 %, oil content by 35.4 %, and on an average 93.0 mg kg^{-1} K in soil as compared to control. Archana et al. (2008) isolated KSMs from rock and rhizosphere soils of green gram (*Vigna radiata* L.) and reported that these KSMs enhanced the solubilization of K in acid leached soil and increased seedling growth and yield of green gram. Basak and Biswas (2009) observed that biomass yield, uptake, and percent K improvement by Sudan grass were increased when mica was inoculated with K-solubilizing bacteria (*B. mucilaginosus*).

A hydroponics experiment was conducted by Singh et al. (2010) with *B. mucilaginosus*, *Azotobacter chroococcum*, and *Rhizobium* spp. using maize and wheat as the test crops under a phytotron growth chamber. They reported that among the *Rhizobacteria*, *B. mucilaginosus* resulted in significantly higher mobilization of K than *A. chroococcum* and *Rhizobium* inoculation. According to Archana et al. (2012), the efficient K-solubilizing bacteria *Bacillus* spp. significantly improved plant development, K uptake, and yield over absolute fertilizer control. Bagyalakshmi et al. (2012) used K-solubilizing strains as biofertilizers in tea plantation and reported that the influence of indigenous K-solubilizing bacteria upon K nutrient exhibited improvement in the productivity and nutrient uptake in plants and retained in soil including quality parameters in tea plantations. Abou and Abdel (2012) evaluated the synergistic effects of soil fertilization with rock K materials and co-inoculation K-dissolving bacteria (*B. mucilaginosus* and *B. subtilis*) using maize as a test crop under limited K soil conditions (calcareous soil). They observed that the co-inoculation of bacteria with K materials into the soil increased K availability, plant shoot and root growth, and K uptake by maize plant.

Prajapati et al. (2013) conducted an experiment with K-releasing bacterial *Enterobacter hormaechei* and fungal *A. terreus* strains to study their plant growth-promoting effects and nutrient uptake on Okra (*Abelmoschus esculantus*) in K-deficient soil. They observed that *E. hormaechei* enhance root and shoot growth and both microorganisms were able to mobilize K efficiently in plant. They also reported that in soils K content was increased with *E. hormaechei* and *A. terreus*.

14.9 Factors Affecting K Availability and Uptake

The availability and uptake of K by plants are frequently complicated by many interacting components. Plants vary in their capability to take up K depending on a number of factors. Factors that influence availability of K in soil and resulting plant uptake are generally classified as soil factors, plant factors, and fertilizer and management practices.

14.9.1 Soil Factors

The following factors are responsible for the K availability and uptake:

(a) Parent materials: It includes soil developing material, amount and type of clay minerals present, vegetation and climate under which soil developed, topography and drainage, and the time duration for soil forming. Soils which are inherently high in K minerals generally contain more K availability than those soils which are inherently little in K minerals.
(b) Cation exchange capacity (CEC): It reflects the soil's capacity to hold K as well as other cations and accumulate in soil for crop uptake. The available K present in soil as ionic (K^{+}) form and detained by negatively charged colloids (primarily clay minerals and organic matter) that build the cation exchange capacity (CEC) of the soil. Generally, soils with higher CEC have greater storage capacity and supplying power of K to the plants.
(c) Available K: The available K includes water-soluble K plus exchangeable K known as soil test K. High amount of soil

test K increases the available of K for plants, by increasing the quantity and balance of K comparative to other cations. When the intensity of soil test K decreases, the crop response to applied K fertilizers increases.

(d) K fixation capacity: Soils which have clay-type minerals can fix huge amounts of K from applied fertilizers or other sources and reduce the availability of K to the crops.
(e) Cation balance: A significant imbalance between soil available K and the other major cations like Ca^+, Mg^+, H^+, Al^+, and Na^+ present in soil affects the K availability to the crops.
(f) Moisture: Soil moisture involves a number of functions like mass-flow movement of K to the plant roots and diffusion of K to the roots that is absorbed by the roots. At sufficient moisture, diffusion occurs more quickly. On the other hand, drought stress reduces K availability in soil and uptake by the roots.
(g) Temperature: The unavailable K turn into available K more rapidly at higher temperature because warm temperature accelerates the release of K from K-bearing minerals. The root activity, plant functions, and physiological processes increase along with rise in soil temperature, and increase in physiological activity leads to increased K uptake. At low soil temperature, K availability and uptake by root are reduced.
(h) pH: The K availability is reduced as soil pH reduced. Soil acidity causes poor root growth due to Al^+ and Mn^+ toxicities, which hinders K uptake. When liming materials are added in acidic soils, exchangeable K increases due to increases in the cation exchange capacity. Again excessive amounts of Ca^+ and Mg^+ reduced the K saturation on cation exchange sites by increasing competition with cations like Ca^+ and Mg^+.
(i) Aeration: Oxygen is essential for root respiration and sufficient oxygen is required by plants for K uptake. As soil moisture increases, the root activity and subsequent K uptake decrease due to low oxygen levels in saturated soils.
(j) Tilth: The soil tilth is directly correlated to the soil friability and capacity to get air into the soil. The K availability is reduced in no-till or ridge-till planting systems due to restricted root growth along with a limited distribution of roots. Oxygen is required for root respiration and K uptake which improves as a result of tillage.

14.9.2 Plant Factors

The following are plant factors also responsible for the K availability and uptake:

(a) The crop: Each crop varies in their ability to take up K from soils, which is related with the type of root structure and surface area of the roots. For example, grasses have more ability to take up K from the soils due to their fibrous, branching roots.
(b) The crop variety: Crop genetics justify the differences among varieties or hybrids of a particular crop. They are generally related to the type of root system, root density, and metabolic activity that influences K uptake. The K uptake is mainly influenced by plant water relations and metabolic and physiological processes. K has very positive effect on root branching and density, and new varieties have higher yield potential which increases the demand to supply a sufficient amount of K in soil.
(c) Plant populations: The yields of crops are higher and demands of K increased as the plant population increased, although the crop yields will not increase with higher populations if K levels in soil are inadequate, from native and fertilizer sources.

(d) Crop yield: Total K uptake increases with increasing crop yields. The K uptake per unit of crop yield might be almost constant at optimum yield.

(f) Weed and insect control: Controlling of weeds and insects reduces competition for moisture and nutrients, so that the crop being produced has relatively more K available.

14.9.3 Fertilizer and Management Practices

The K availability and uptake influenced with kinds and methods of fertilizer application and management practices include:

(a) Increased use of nitrogen (N) and other limiting nutrients: Addition of nitrogen and/or phosphorus normally increases K uptake when sufficient K is available. In general, the uptake of K by crops is similar to N uptake and may be greater sometimes. Therefore, when limiting nutrients are added, the demands on soil K increased.

(b) Applications of manures and K fertilizers: The K availability in soils increased mainly by adding manures or crop residues and K fertilizers in adequate amounts. K is readily available from all these sources, when they are provided near the roots, where roots can absorb the K.

(c) Placement of K: Broadcast plow-down applications of K are more available than surface-applied disked-in K as K reached in root zone in prior method of application. Deep placement, drip irrigation, and gypsum application help down movement of K in very fine-textured soils.

(d) Conservation tillage: The soil K levels should be built to high before shifting to minimum or conservation tillage for improving K distribution within the plow layer. In many fine-textured soils, surface-applied K moves very little in the soil and has low availability, particularly under dry-land conditions.

(e) Drainage: It improves the aeration of the soil and improves the availability of soil K by draining excess moisture of soils and helps soil warm up earlier.

14.10 Concluding Remarks

K is the essential macronutrient for growth and development of the plant next to nitrogen and phosphorus. Plants acquire K from soil solution as K ions. The K availability in soil is generally very low (~1–2 %), and the major portion of K (~98 %) are unavailable to plants and establish as insoluble rocks, minerals, and other deposits. However, these unavailable forms constitute the major reservoirs of K in soil as, under suitable environment, they solubilized and turn into available for plant nutrition. The phenomenon of both the fixation of available K and release of unavailable K plays a vital role in K dynamics. KSMs have the ability to solubilize fixed forms of K in K-bearing minerals (micas, illite, and orthoclase) due to formation and secretion of organic acids such as citric, oxalic, and tartaric acids. The organic acids formed by KSMs make easy to weathering of K minerals by directly dissolving the K from K-bearing minerals or formation of metal-organic complexes by forming chelate together with silicon ions to bring the K in soil solution. Thus, KSMs may be used as cost-effective and environmental-friendly microbial consortia, which are essential for enhancing production of various crops along with improving the soil fertility for sustainable agriculture.

References

Abou-el-Seoud, Abdel-Megeed A (2012) Impact of rock materials and biofertilizations on P and K availability for maize (*Zea maize*) under calcareous soil conditions. Saudi J Biol Sci 19:55–63

Aleksandrov VG (1985) Organo-mineral fertilizers and silicate bacteria. Dokl Akad Nauk 7:43–48

Alexander M (1985) Introduction to soil microbiology. Wiley, New York, pp 382–385

Amtmann A, Hammond JP, Armengaud P, White PJ (2006) Nutrient sensing and signaling in plants: potassium and phosphorus. Adv Bot Res 43:209–257

Amtmann A, Troufflard S, Armengaud P (2008) The effect of potassium nutrition on pest and disease resistance in plants. Physiol Plant 133:682–691

Archana DS, Savalgi VP, Alagawadi AR (2008) Effect of potassium solubilizing bacteria on growth and yield of maize. Soil Biol Ecol 28:9–18

Archana DS, Nandish MS, Savalagi VP, Alagawadi AR (2012) Screening of potassium solubilizing bacteria (KSB) for plant growth promotional activity. Bioinfolet 9:627–630

Archana DS, Nandish MS, Savalagi VP, Alagawadi AR (2013) Characterization of potassium solubilizing bacteria (KSB) from rhizosphere soil. Bioinfolet 10:248–257

Argelis DT, Gonzala DA, Vizcaino C, Gartia MT (1993) Biochemical mechanism of stone alteration carried out by filamentous fungi living in monuments. Biogeochemistry 19:129–147

Armengaud P, Breitling R, Amtmann A (2010) Coronatine-intensive 1 (COII) mediates transcriptional responses of *Arabidopsis thaliana* to external potassium supply. Mol Plant 3:390–405

Awasthi R, Tewari R, Nayyar H (2011) Synergy between plants and P-solubilizing microbes in soils: effects on growth and physiology of crops. Int Res J Microbiol 2:484–503

Badar MA, Shafei AM, Sharaf El-Deen SH (2006) The dissolution of K and phosphorus bearing minerals by silicate dissolving bacteria and their effect on sorghum growth. Res J Agric Biol Sci 2:5–11

Bagyalakshmi B, Ponmurugan P, Balamurugan A (2012) Impact of different temperature, carbon and nitrogen sources on solubilization efficiency of native potassium solubilizing bacteria from tea (*Camellia sinensis*). J Biol Res 3:36–42

Basak BB, Biswas DR (2009) Influence of potassium solubilizing microorganism (*Bacillus mucilaginosus*) and waste mica on potassium uptake dynamics by sudan grass (*Sorghum vulgare* Pers.) grown under two Alfisols. Plant Soil 317:235–255

Bertsch PM, Thomas GW (1985) Potassium status of temperate region soils. In: Munson RD (ed) Potassium in agriculture. American Society of Agronomy Journal, Madison, pp 131–162

Bin L, Bin W, Mu P, Liu C, Teng HH (2010) Microbial release of potassium from K-bearing minerals by thermophilic fungus *Aspergillus fumigatus*. Geochim Cosmochim Acta 72:87–98

Chen S, Lian B, Liu CQ (2008) Bacillus *mucilaginosus* on weathering of phosphorite and primary analysis of bacterial proteins during weathering. Chin J Geochem 27:209–216

Diep CN, Hieu TN (2013) Phosphate and potassium solubilizing bacteria from weathered materials of denatured rock mountain, Ha Tien, Kiên Giang province Vietnam. Am J Life Sci 1:88–92

El-Hadad ME, Mustafa MI, Selim Sh M, El-Tayeb TS, Mahgoob AEA, Norhan H, Abdel A (2011) The nematicidal effect of some bacterial biofertilizers on meloidogyne incognita in sandy soil. Braz J Microbiol 42:105–113

Goldstein AH (1994) Solubilization of exogenous phosphates by gram negative bacteria. In: Silver S et al (eds) Cellular and molecular biology of phosphate and phosphorylated compounds in microorganisms. ASM, Washington, DC, pp 197–203

Gundala PB, Chinthala P, Sreenivasulu B (2013) A new facultative alkaliphilic, potassium solubilizing, *Bacillus* Sp. SVUNM 9 isolated from mica cores of Nellore District, Andhra Pradesh, India. Research and Reviews. J Microbiol Biotechnol 2:1–7

Han HS, Supanjani, Lee KD (2006) Effect of co-inoculation with phosphate and potassium solubilizing bacteria on mineral uptake and growth of pepper and cucumber. Plant Soil Environ 52:130–136

Hasan R (2002) Potassium status of soils in India. Better Crops Int 16:3–5

Hu X, Chen J, Guo J (2006) Two phosphate- and potassium-solubilizing bacteria isolated from Tianmu Mountain, Zhejiang, China. World J Microbiol Biotechnol 22:983–990

Kucey RMN (1988) Effect of *Penicillium biloji* on the solubility and uptake of P and micronutrients from soil by wheat. Can J Soil 68:261–267

Kumar A, Bahadur I, Maurya BR, Raghuwanshi R, Meena VS, Singh DK, Dixit J (2015) Does a plant growth- promoting rhizobacteria enhance agricultural sustainability? J Pure Appl Microbiol 9(1):715–724

Lian BA (1998) A study on how silicate bacteria GY92 dissolves potassium from illite. Acta Mineral Sin 18:234–238

Lian B, Fu PQ, Mo DM, Liu CQ (2002) A comprehensive review of the mechanism of potassium release by silicate bacteria. Acta Mineral Sin 22:179–183

Lian B, Wang B, Pan M, Liu C, Teng HH (2008) Microbial release of potassium from K-bearing minerals by thermophilic fungus *Aspergillus fumigatus*. Geochim Cosmochim Acta 72:87–98

Liu W, Xu X, Wu S, Yang Q, Luo Y, Christie P (2006) Decomposition of silicate minerals by *Bacillus mucilaginosus* in liquid culture. Environ Geochem Health 28:133–140

Liu D, Lian B, Dong H (2012) Isolation of *Paenibacillus sp.* and assessment of its potential for enhancing mineral weathering. Geomicrobiol J 29:413–421

Malavolta E (1985) Potassium status of tropical and subtropical region soils. In: Munson RD (ed) Potassium in agriculture. Am Soc Agron J, Madison, pp 163–200

Marschner H (1995) Mineral nutrition of higher plants, 2nd edn. Academic, London

Maurya BR, Meena VS, Meena OP (2014) Influence of Inceptisol and Alfisol's potassium solubilizing bacteria (KSB) isolates on release of K from waste mica. Vegetos 27(1):181–187

McAfee J (2008) Potassium, a key nutrient for plant growth. Department of Soil and Crop Sciences. http://jimmcafee.tamu.edu/files/potassium

Mclean EO, Watson ME (1985).Soil measurement of plant-available potassium. In: Munson RD (ed) Potassium in agriculture. Am Soc Agron J, Madison, pp 277–278

Meena OP, Maurya BR, Meena VS (2013) Influence of K-solubilizing bacteria on release of potassium from waste mica. Agric Sustain Dev 1(1):53–56

Meena VS, Maurya BR, Bahadur I (2014a) Potassium solubilization by bacterial strain in waste mica. Bangladesh J Bot 43(2):235–237

Meena VS, Maurya BR, Verma JP (2014b) Does a rhizospheric microorganism enhance K+ availability in agricultural soils? Microb Res 169:337–347

Meena RK, Singh RK, Singh NP, Meena SK, Meena VS (2015a) Isolation of low temperature surviving plant growth promoting rhizobacteria (PGPR) from pea (*Pisum sativum* L.) and documentation of their plant growth promoting traits. Biocatal Agric Biotechnol. doi:10.1016/j.bcab.2015.08.006

Meena VS, Maurya BR, Verma JP, Aeron A, Kumar A, Kim K, Bajpai VK (2015b) Potassium solubilizing rhizobacteria (KSR): isolation, identification, and K-release dynamics from waste mica. Ecol Eng 81:340–347

Mengel K, Kirkby EA (1987) Principles of plant nutrition, 4th edn. International Potash Institute, IPI, Bern, 685p

Mishra MK, Srivastava PC, Ghosh D (1993) Forms of potassium in relation to soil properties and clay mineralogy in some profiles of Chambal command area, Rajasthan. J Pot Res 9:187–194

Muentz A (1890) Surla decomposition desrochesetla formation de la terre arable. C R Acad Sci 110:1370–1372

Parmar P, Sindhu SS (2013) Potassium solubilization by rhizosphere bacteria: influence of nutritional and environmental conditions. J Microbiol Res 3:25–31

Prajapati K, Sharma MC, Modi HA (2012) Isolation of two potassium solubilizing fungi from ceramic industry soils. Life Sci Leafl 5:71–75

Prajapati K, Sharma MC, Modi HA (2013) Growth promoting effect of potassium solubilizing microorganisms on *Abelmoschus esculentus*. Int J Agric Sci 3:181–188

Raj SA (2004) Solubilization of silicate and concurrent release of phosphorus and potassium in rice ecosystem. In: Conference paper biofertilizers technology, Coimbatore, pp 372–378

Ramamoorthy B, Paliwal KV (1976) Potassium adsorption ratio for some paddy soils in relation to their potassium availability. Soil Sci 99:236–242

Ramamurthy B, Bajaj JC (1969) Soil fertility map of India. Indian Agril Res Inst, New Delhi

Rogers JR, Bennett PC, Choi WJ (1998) Feldspars as a source of nutrients for microorganisms. Am Mineral 83:1532–1540

Romheld V, Kirkby EA (2010) Research on potassium in agriculture: needs and prospects. Plant Soil 335:155–180

Rosa-Magri MM, Avansini SH, Lopes-Assad ML, Tauk-Tornisielo SM, Ceccato- Antonini SR (2012) Release of potassium from rock powder by the yeast *Torulaspora globosa*. Braz Arch Biol Technol 55:577–582

Sangeeth KP, Bhai RS, Srinivasan V (2012) *Paenibacillus glucanolyticus*, a promising potassium solubilizing bacterium isolated from black pepper (*Piper nigrum* L.) rhizosphere. J Spic Aromather Crops 21:118–124

Schroeder D (1979) Structure and weathering of potassium containing minerals. Proc Cong Int Pot Inst 11:43–63

Sheng XF (2005) Growth promotion and increased potassium uptake of cotton and rape by a potassium releasing strain of *Bacillus edaphicus*. Soil Biol Biochem 37:1918–1922

Sheng XF, He LY (2006) Solubilization of potassium-bearing minerals by a wild type strain of *Bacillus edaphicus* and its mutants and increased potassium uptake by wheat. Can J Microbiol 52:66–72

Sheng XF, Huang WY (2002) Mechanism of potassium release from feldspar affected by the strain NBT of silicate bacterium. Acta Pedol Sin 39:863–871

Sheng XF, Zhao F, He H, Qiu G, Chen L (2008) Isolation, characterization of silicate mineral solubilizing *Bacillus globisporus* Q12 from the surface of weathered feldspar. Can J Microbiol 54:1064–1068

Singh G, Biswas DR, Marwah TS (2010) Mobilization of potassium from waste mica by plant growth promoting rhizobacteria and its assimilation by maize (*Zea mays*) and wheat (*Triticum aestivum* L.). J Plant Nutr 33:1236–1251

Singh NP, Singh RK, Meena VS, Meena RK (2015) Can we use maize (*Zea mays*) rhizobacteria as plant growth promoter? Vegetos 28(1):86–99

Song SK, Huang PM (1988) Dynamics of potassium release from potassium-bearing minerals as influenced by oxalic and citric acids. Soil Sci Soc Am J 52:383–390

Sparks DL, Huang PM (1985) Physical chemistry of soil potassium. In: Munson RD (ed) Potassium in agriculture. Am Soc Agron J, Madison, pp 201–276

Sugumaran P, Janarthanam B (2007) Solubilization of potassium containing minerals by bacteria and their effect on plant growth. World J Agric Sci 3:350–355

Supanjani HHS, Jung SJ, Lee KD (2006) Rock phosphate potassium and rock solubilizing bacteria as alternative sustainable fertilizers. Agron Sustain Dev 26:233–240

Troufflard S, Mullen W, Larson TR, Graham IA, Crozier A, Amtmann A, Armengaud P (2010) Potassium deficiency induced the biosynthesis of oxylipins and glucosinolates in Arabidopsis thaliana. Plant Biol 10:172

Uroz S, Calvaruso C, Turpault MP, Pierrat JC, Mustin C, Frey-Klett P (2007) Effect of the mycorrhizosphere on the genotypic and metabolic diversity of the bacterial communities involved in mineral weathering in a forest soil. Appl Environ Microbiol 73:3019–3027

Uroz S, Calvaruso C, Turpault MP, Frey-Klett P (2009) Mineral weathering by bacteria: ecology, actors and mechanisms. Trends Microbiol 17:378–387

Valmorbida J, Boaro CSF (2007) Growth and development of *Mentha piperita* L. in nutrient solution as affected by rates of potassium. Braz Arch Biol Technol 50:379–384

White PJ, Karley AJ (2010) Potassium. In: Hell R, Mendel RR (eds) Cell biology of metals and nutrients, plant cell monographs, vol 17. Springer, Berlin, pp 199–224

Wu SC, Cao ZH, Li ZG, Cheung KC, Wong MH (2005) Effects of biofertilizer containing N fixer, P and K solubilizers and AM fungi on maize growth: a greenhouse trial. Geoderma 125:155–166

Zarjani JK, Aliasgharzad N, Oustan S, Emadi M, Ahmadi A (2013) Isolation and characterization of potassium solubilizing bacteria in some Iranian soils. Arch Agron Soil Sci 77:7569, http://dx.doi.org/10.1080/03650340.2012

Zhang A, Zhao G, Gao T, Wang W, Li J, Zhang S (2013) Solubilization of insoluble potassium and phosphate by *Paenibacillus kribensis* CX-7: a soil microorganism with biological control potential. Afr J Microbiol Res 7:41–47

Zhao F, Sheng X, Huang Z, He L (2008) Isolation of mineral potassium-solubilizing bacterial strains from agricultural soils in Shandong Province. Biodivers Sci 16:593–600

Zhou Hong BO, Zheng Xiao XI, Fei-Fei L, Guan-Zhou Q, HuYue H (2006) Screening, identification and desiccation of silicate bacterium. J Cent South Univ Techol 13:337–341

Potassium-Solubilizing Microorganisms: Mechanism and Their Role in Potassium Solubilization and Uptake 15

Ashish Sharma, Deepti Shankhdhar, and S.C. Shankhdhar

Abstract

K is the third most important plant macronutrient after nitrogen and phosphorus. It is absorbed from soil primarily in the form of K^+ and is required in the plants for early growth, production, and modification of proteins, maintenance of water use efficiency, stand persistence, longevity, etc. The total K content in soil exceeds 20,000 ppm which is primarily divided into unavailable, slowly available, and readily available fractions of which readily available fractions constitute only 1–2 % of the total K available in the soil. To mobilize and utilize these large reserves, a viable strategy is the identification and utilization of K-solubilizing microorganisms (KSMs). Soil microbes playing a key role in K cycling have been known since a long time now; therefore, identification of KSMs and their utilization are of prime importance to reduce the fertilizer usage and the effects caused by effective fertilizer usage. Although many bacteria like *Acidithiobacillus*, *Burkholderia*, and *Pseudomonas* have been identified as the potential K solubilizers, a clear cut mechanism has not been reported. However, most solubilization activities of bacteria and fungi have been attributed to the activities like acid hydrolysis of K from minerals present in the soil, chelation by production of some organic acids, etc. Therefore, the content presented in this manuscript explores the applicability of these KSMs, their mode of action, and their percent contribution in K solubilization and availability to the plants.

A. Sharma
Botany Department, DAV University, Sarmastpur, Jalandhar-Pathankot Highway, Jalandhar, Punjab 144011, India

D. Shankhdhar • S.C. Shankhdhar (✉)
Department of Plant Physiology, College of Basic Sciences and Humanities, G. B. Pant University of Agriculture and Technology, Pantnagar 263145, Uttarakhand, India
e-mail: shankhdhar.sc@rediffmail.com

V.S. Meena et al. (eds.), *Potassium Solubilizing Microorganisms for Sustainable Agriculture*,
DOI 10.1007/978-81-322-2776-2_15

Keywords
K solubilization • KSMs • K^+ transporters • Soil K status

15.1 Introduction

K is one of the most important essential nutrients as it is required by all plants, animals, and humans in large quantities for proper functioning of their cells. Plant-based foods are the major source of K for humans along with some animal-based foods. K is the seventh in most abundant element in the Earth's crust. Disintegration of rocks releases K along with some other nutrients, but this quantity is low and releases too slow to meet the requirements of crop plants. K fertilizers are generally used for the soil supplementation in order to achieve economically viable yields of crop plants and maintain soil fertility. The current demand for potash fertilizers in the world is ~30 MT which is expected to increase up to ~33 MT in 2016, of which ~72 % of the total fertilizer is expected to be used in Asia. Among the Asian countries, ~32 % of the growth in world demand for potash is expected in India, 25 % in China, 5 % in Indonesia, 3 % in Thailand, 2 % in Vietnam, 1 % in Malaysia, and the balance in the rest of Asia (FAO 2012).

In plants, animals, and humans, K always occurs as a positive ion (K^+) in every living cell. K is very important in the human body. Along with sodium, it regulates the water balance and the acid-base balance in the blood and tissues. K enters the cell more readily compared to sodium and instigates the brief Na-K exchange across the cell membranes. In the nerve cells, this Na-K flux generates the electrical potential that aids the conduction of nerve impulses. When K leaves the cell, it changes the membrane potential and allows the nerve impulse to progress. This electrical potential gradient, created by the "Na-K pump," helps generate muscle contractions and regulates the heartbeat. Another of the pump's most important functions is preventing the swelling of cells. If sodium is not pumped out, water accumulates within the cell causing it to swell and ultimately burst (Haas 2011). In plant nutrient classification also, it has been classified in the group IV (nutrients present in the plants in elemental form). It is also found as the K^+ ion in soils, many rocks, and seawater, but is generally present in the bound form which may be unavailable to the plants. K is extremely important for the plants, which is evident from the discussion that follows. K "activates" ~70 different enzymes involved in plant growth (Suelter 1970). The K changes the physical shape of the enzyme molecule, exposing the appropriate chemically active sites for reaction. K also neutralizes various organic anions and other compounds within the plant, helping to stabilize pH between 7 and 8, optimum for most enzyme reactions (Suelter 1970).

Such functions include the short-term maintenance of electrical potentials across membranes (Cheeseman and Hanson 1979) and turgor-related phenomena such as cell expansion (Dolan and Davies 2004), plant movements (Philippar et al. 1999; Moran 2007), pollen tube development (Mouline et al. 2002), and stomatal opening and closing (Humble and Hsiao 1969; Dietrich et al. 2001). Other well-characterized biological functions of K include the activation of numerous enzymes (Suelter 1970), the long-distance transport of nitrate (Ben-Zioni et al. 1970) and sucrose (Cakmak et al. 1994), and the charge stabilization of anions within the cell (Clarkson and Hanson 1980). When water supply is short, K is pumped out of the guard cells. The pores close tightly to prevent loss of water and minimize drought stress to the plant. If K supply is inadequate, the stomata become sluggish – slow to respond – and water vapor is lost. Closure may take hours rather than minutes and is incomplete. As a result, plants with an insufficient supply of K are much more susceptible to water stress. Accumulation of K in plant roots produces a gradient of osmotic pressure that draws water into the roots. Plants deficient in K are thus less able to absorb water and are more subject to stress when water is in short supply (Dietrich et al. 2001).

The role of K in photosynthesis is complex. The activation of enzymes by K and its involvement in adenosine triphosphate (ATP) production are probably more important in regulating the rate of photosynthesis than the role of K in stomatal activity. When the sun's energy is used to combine CO_2 and water to form sugars, the initial high-energy product is ATP. The ATP is then used as the energy source for many other chemical reactions. The electrical charge balance at the site of ATP production is maintained with K ions. When plants are K deficient, the rate of photosynthesis and the rate of ATP production are reduced, and all of the processes dependent on ATP are slowed down. Conversely, plant respiration increases which also contributes to slower growth and development. In some plants, leaf blades reorient toward light sources to increase light interception or away to avoid damage by excess light, in effect assisting to regulate the rate of photosynthesis. These movements of leaves are brought about by reversible changes in turgor pressure through movement of K into and out of specialized tissues similar to that described above for stomata (Moran 2007).

The plant's transport system uses energy in the form of ATP. If K is inadequate, less ATP is available, and the transport system breaks down. This causes photosynthesis to build up in the leaves, and the rate of photosynthesis is reduced. Normal development of energy storage organs, such as grain, is regarded as a result (Ben-Zioni et al. 1970). An adequate supply of K helps to keep all of these processes and transportation systems functioning normally. K also plays a major role in the transport of water and nutrients throughout the plant in the xylem. When K supply is reduced, translocation of nitrates, phosphates, calcium (Ca), magnesium (Mg), and amino acids is depressed. As with phloem transport systems, the role of K in xylem transport is often in conjunction with specific enzymes and plant growth hormones. An ample supply of K is essential to efficient operation of these systems (Ben-Zioni et al. 1970).

K is required for every major step of protein synthesis. The "reading" of the genetic code in plant cells to produce proteins and enzymes that regulate all growth processes would be impossible without adequate K. When plants are deficient in K, proteins are not synthesized despite an abundance of available nitrogen (N). Instead, protein "raw materials" (precursors) such as amino acids, amides, and nitrate accumulate. The enzyme nitrate reductase catalyzes the formation of proteins, and K is likely responsible for its activation and synthesis. The enzyme responsible for synthesis of starch (starch synthetase) is activated by K. Thus, with inadequate K, the level of starch declines, while soluble carbohydrates and N compounds accumulate. Photosynthetic activity also affects the rate of sugar formation for ultimate starch production. Under high K levels, starch is efficiently moved from sites of production to storage organs.

Increased evidence has shown that crop production is significantly restricted by biotic stresses. Oerke and Dehne (2004) estimated that weeds produce the highest potential loss (32 %), followed by animal pests (18 %), fungi and bacteria (15 %), and viruses (3 %). In many cases, K-deficient plants tend to be more susceptible to infection than those with an adequate supply of K. For example, the rate of rice borer infestation was greatest when there was no supply of K, but decreased rapidly as the K concentration increased (Sarwar 2012). Similar results were found with a *Discula destructiva* Redlin infection in *Cornus florida* L. (Holzmueller et al. 2007). Williams and Smith (2001) also reported that increased K fertilizer significantly reduces the disease incidence of stem rot and aggregate sheath spot (AgSS), and negative correlations were found between the percentage of K in leaf blades and disease severity. K fertilizer is widely reported to decrease insect infestation and disease incidence in many host plants. Perrenoud (1990) reviewed that the use of K significantly decreases the incidence of fungal diseases by 70 %, bacteria by 69 %, insects and mites by 63 %, viruses by 41 %, and nematodes by

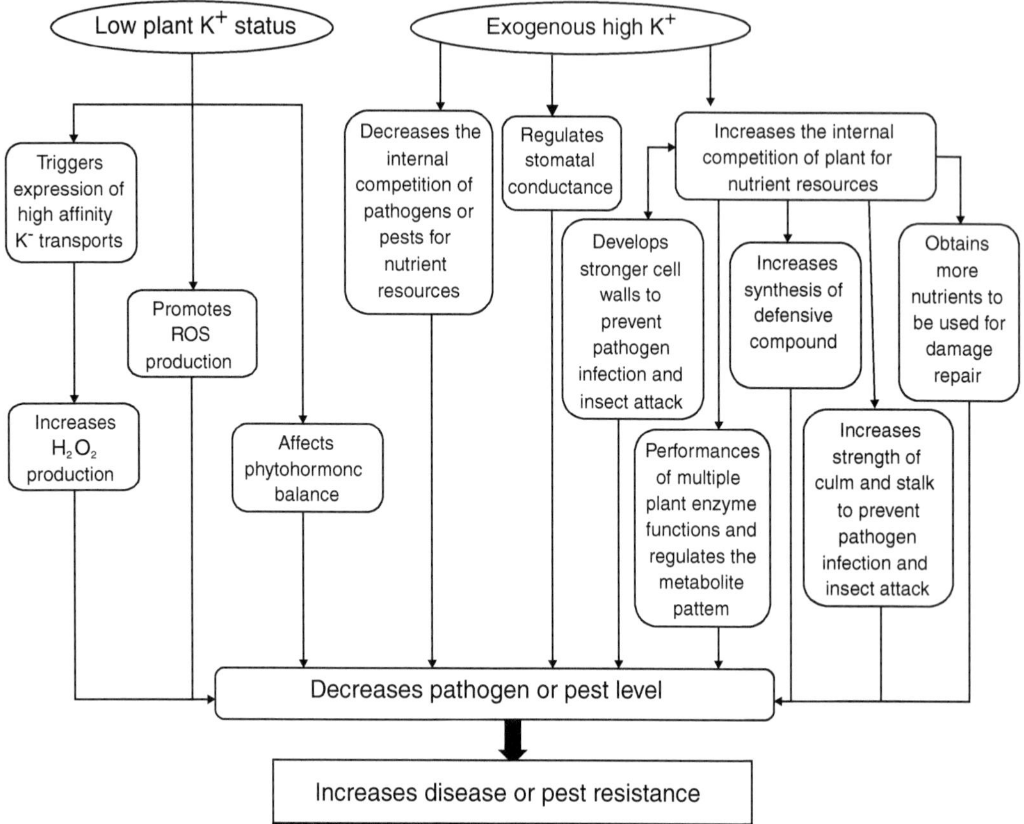

Fig. 15.1 Role of K under biotic stresses (Adapted from Wang et al. 2013)

33 %. Meanwhile, K increased the yield of plants infested with fungal diseases by 42 %, bacteria by 57 %, insects and mites by 36 %, viruses by 78 %, and nematodes by 19 % (Wang et al. 2013). The role of K in mitigating biotic stress can be demonstrated as below (Fig. 15.1). Besides biotic stress K can also regulate the plant's functioning during abiotic stresses like drought. The role that K plays in mitigating abiotic stress is shown as under (Fig. 15.2).

15.2 Potassium Status of Indian Soils

Different soil types that are found in India include alluvial, medium, and deep black soils, red and lateritic soils, etc. K status of these soils varies depending upon soil type, parent material, and texture and management practices. Black soils contain smectite as dominant clay mineral which results in the soil having high level of exchangeable K. On the other hand, alluvial soils with mica being the dominant mineral contain medium levels of exchangeable K and high levels of non-exchangeable K (Anonymous 2002). However, red and lateritic soils contain kaolinite as dominant clay mineral and are light in texture which results in low exchangeable and high non-exchangeable K content.

In rainfed agroecosystems, the soils are characterized by low to high in available K status, and its deficiency can be easily seen in coarse-textured alluvial soils, red and lateritic soils, shallow soils, and soils which supported continuous high yields without K addition. Vertisols show relatively high available K as compared to inceptisols and alfisols because of higher clay content. Mean K availability under different cropping systems varies (Table 15.1).

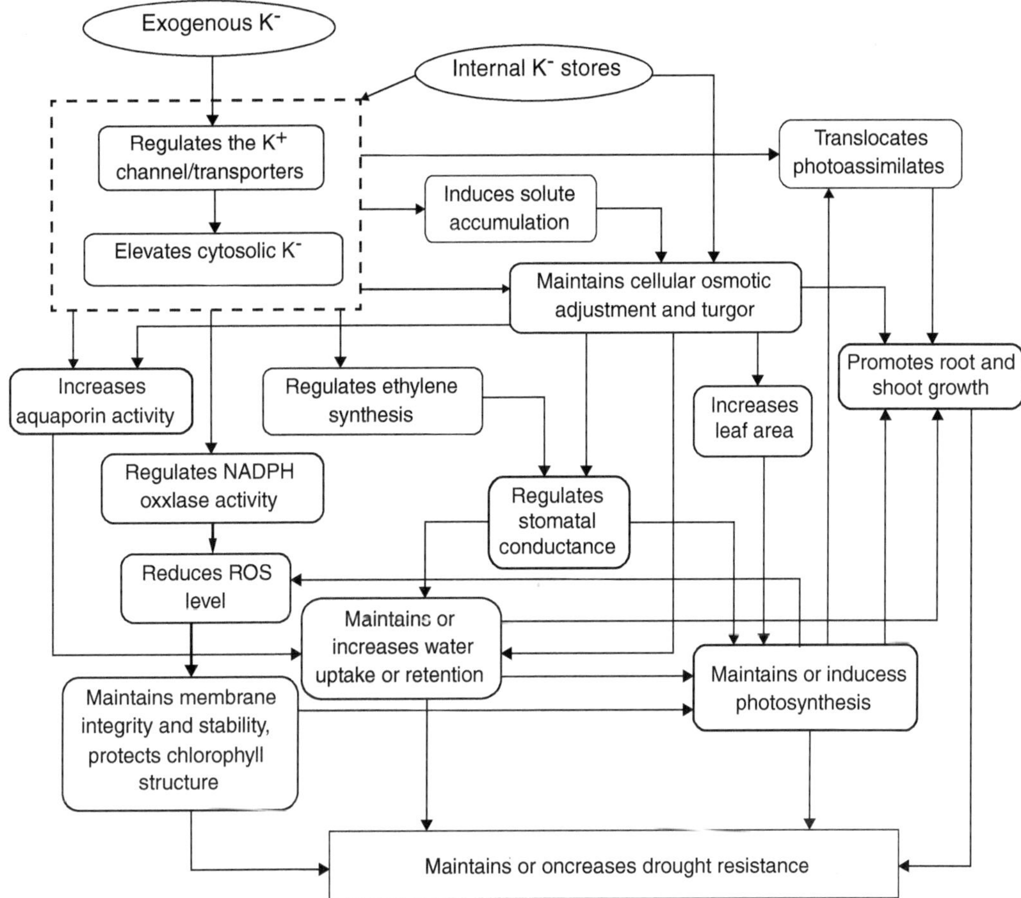

Fig. 15.2 Role of K in osmoregulation during drought stress (Adapted from Wang et al. 2013)

Table 15.1 Availability of potassium under different cropping systems

Cropping system	Potassium availability (kg ha^{-1})
Rice-based system	138.8–95.1
Groundnut-based system	129.2–188.8
Soybean-based system	322.2–407.5
Cotton-based system	76.7–272.2
Rabi sorghum system	365.4–500.4
Pearl millet-based system	85.1–163.1
Finger millet-based system	53
Maize-based system	55.6–109.4

15.3 Potassium Pools in Agricultural Soils

The total K content of soils frequently exceeds 20,000 ppm. Nearly all of this is in the structural component of soil minerals and is not available for plant growth. Three forms of K, unavailable, slowly available or fixed, and readily available or exchangeable (Fig. 15.3), exist in soils (Srinivasrao et al. 2011):

(a) *Unavailable potassium*: Depending on soil type, ~90–98 % of total soil K is found in this form. Feldspars and micas are minerals that contain most of the K. Plants cannot use

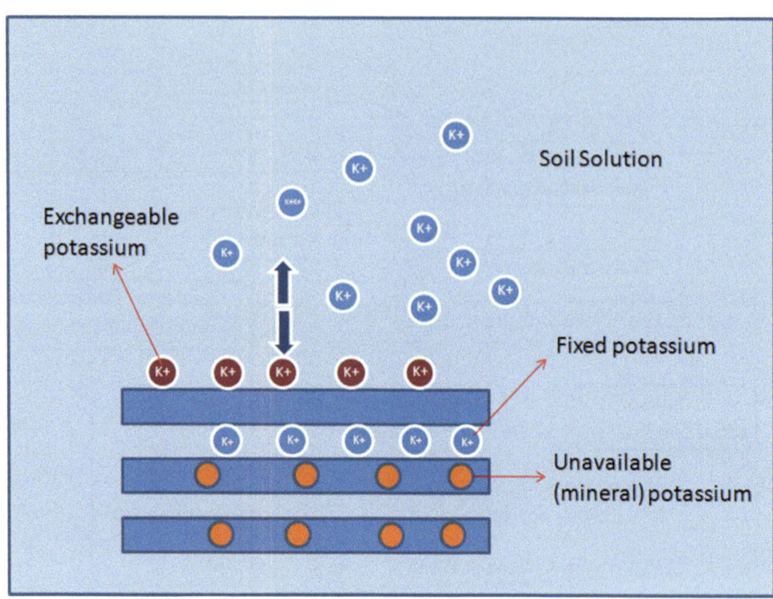

Fig. 15.3 Potassium fractions in soil (Adapted from www.smart-fertilizer.com/potassiuminsoil)

the K in this crystalline-insoluble form. Over long periods of time, these minerals weather (break down) and K is released. This process, however, is too slow to supply the full K needs of field crops. As these minerals weather, some K moves to the slowly available pool. Some also moves to the readily available pool (Srinivasrao et al. 2011).

(b) *Slowly available potassium*: This form of K is thought to be trapped between layers of clay minerals and is frequently referred to as being fixed. Growing plants cannot use much of the slowly available K during a single growing season. This slowly available K is not measured by the routine soil testing procedures. Slowly available K can also serve as a reservoir for readily available K. While some slowly available K can be released for plant use during a growing season, some of the readily available K can also be fixed between clay layers and thus converted into slowly available K. The amount of K fixed in the slowly available K varies with the type of clay that dominates in the soil. Montmorillonite clays are dominant in many of central and western Minnesota soils. These clays fix K when soils become dry because K is trapped between the layers in the clay mineral. This K, however, is released when the soil becomes wet. Illite clays are dominant in most of the soils in southeastern Minnesota. These clays also fix K between layers when they become dry, but do not release all of the fixed K when water is added. This fixation without release causes problems for management of potash fertilizers for crop production in the region (Srinivasrao et al. 2011).

(c) *Readily available potassium*: K that is dissolved in soil water (water soluble) plus that held on the exchange sites on clay particles (exchangeable K) is considered readily available for plant growth. The exchange sites are found on the surface of clay particles. This is the form of K measured by the routine soil testing procedure. Plants readily absorb the K dissolved in the soil water. As soon as the K concentration in soil water drops, more is released into this solution from the K attached to the clay minerals. The K attached to the exchange sites on the clay minerals is more readily available for plant growth than the K trapped between the layers of the clay minerals (Srinivasrao et al. 2011).

15.4 Factors Affecting K Uptake

K uptake by plants is affected by the following factors. Soil moisture: higher soil moisture ~75 % usually means greater availability of K. Increasing soil moisture increases movement of K to plant roots and enhances availability. (a) Soil aeration and oxygen level: air is necessary for root respiration and K uptake. Root activity and subsequent K uptake decrease as soil moisture content increases to saturation. Levels of oxygen are very low in saturated soils (15–25 %). (b) Soil temperature: root activity, plant functions, and physiological processes all increase as soil temperature increases. This increase in physiological activity leads to increased K uptake. Optimum soil temperature for uptake is 60–80 °F. K uptake is reduced at low (5 °C below normal surface temperature) soil temperatures. (c) Tillage system: availability of soil K is reduced in no-till and ridge-till planting systems. The exact cause of this reduction is not known. Results of research point to restrictions in root growth combined with a restricted distribution of roots in the soil (Srinivasrao et al. 2011).

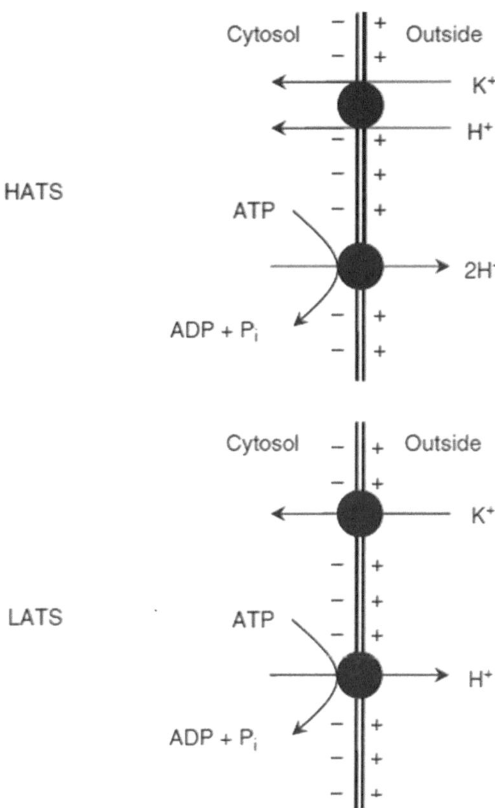

Fig. 15.4 General mechanisms proposed for K influx into plant cells (Adapted from Britto and Kronzuker 2008)

15.5 Mechanism of K Uptake in Plants

K is taken up by the plants either by high-affinity transport system (HATS) or by low-affinity transport system (LATS) (Fig. 15.4). In the HATS mechanism, the thermodynamically uphill flux of K^+ is driven by the downhill flux of H^+; charge balance is achieved by the outward pumping of two H^+ by the plasma membrane proton ATPase. In the LATS mechanism, by contrast, an electrogenic uniport of K^+ is electrically balanced by the ATP-driven efflux of one H^+ (Britto and Kronzucker 2008). Thermodynamically, channel activity relies on electrochemical potential gradients for K^+ to drive transport and is regulated by a wide range of agents: membrane voltage (Gassmann and Schroeder 1994), pH (Hoth and Hedrich 1999), cyclic nucleotides (Talke et al. 2003), CO_2-dependent light activation (Deeken et al. 2000), reactive oxygen species (Cakmak 2005), and K^+ itself (Johansson et al. 2006; Liu et al. 2006). The thermodynamic distinction between these two basic types of transporters forms the basis for their designation, in the recently devised transporter classification (TC) system, as class 1 (channel/pore type) for LATS and class 2 (electrochemical potential-driven transporter type) for HATS (Busch and Saier 2002).

K^+ acquisition from low external concentrations is usually considered to be an energy-demanding process, while that from high concentrations is energetically passive. On the surface, this view is supported by analyses of the electrochemical potential gradient for K^+ transport into the plant cell, which is defined primarily by the differences in K^+ concentration and electrical potential on either side of the plasma membrane (Cheeseman and Hanson

1980; Szczerba et al. 2006a). K^+ is usually the most abundant cation in the cytosol, with concentrations typically ranging from 40 to 200 mM (Leigh and Wyn Jones 1984; Walker et al. 1996; Kronzucker et al. 2003), a condition that could present an energetic obstacle to K^+ entry even under abundant K supply, were it not for the electrical charge separation across the membrane (inside negative), which greatly enhances the uptake of cations. However, in very dilute solutions, this electrical pull is insufficient to drive K^+ influx, and an active transport mechanism is thus postulated under such conditions (Cheeseman and Hanson 1980; Maathuis and Sanders 1994).

15.6 Potassium Channels and Transporters

K^+ transporters (Fig. 15.5 and Table 15.2) in the plants belong to either of the following classes.

15.6.1 Shaker Channels

Plant Shaker channels share similarities, both at the sequence and structure levels, with animal voltage-gated K^+ channels forming the so-called Shaker family (Jan and Jan 1997). Animal Shaker channels are made up of four subunits arranged around a central pore. The hydrophobic core of each subunit consists of six transmembrane segments (TMS), the fourth one with repetition of basic residues acting as a voltage sensor. A highly conserved membrane loop, located between the fifth and the sixth TMS and called the P (pore) domain, forms part of the selectivity filter of the ion-conducting pore. This loop contains a TxxTxGYGD motif, the hallmark of K^+-selective channels (Jan and Jan 1997). The plant Shaker cytosolic C-terminal region harbors regulatory domains comprising a putative cyclic nucleotide-binding site and, at the extreme C-terminus, the so-called KHA region (rich in hydrophobic and acidic residues), which

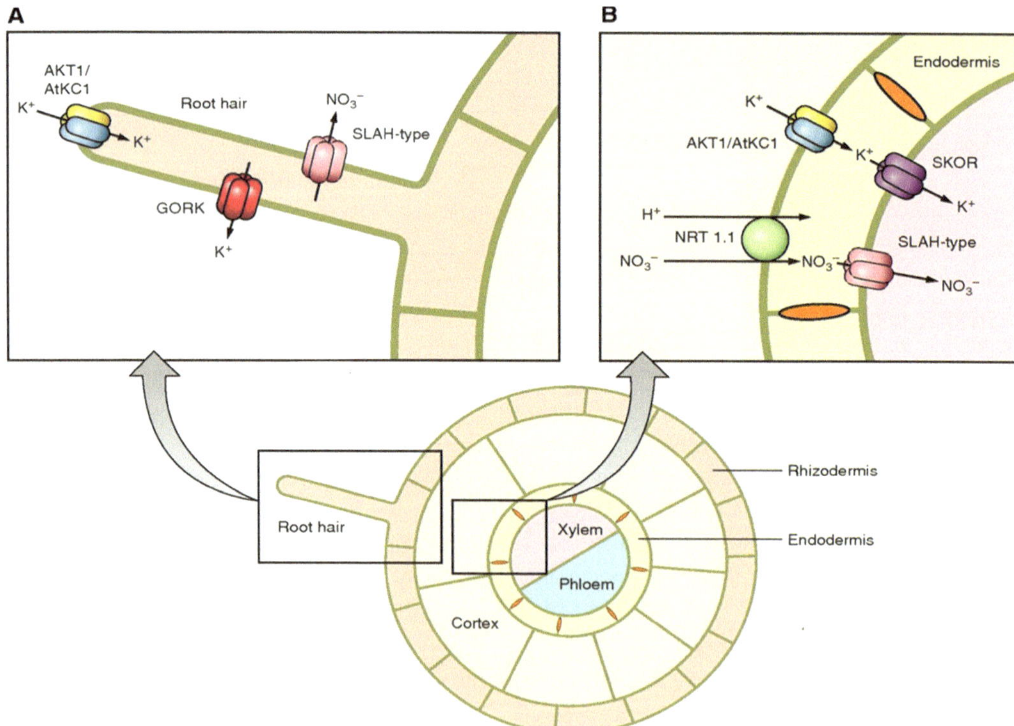

Fig. 15.5 (**a**) Working of root hair equipped with the inward-rectifying K^+-channel-unit AKT1/AtKC1, the outward K^+ rectifier GORK, and SLAH-type anion channels. (**b**) Polar transport across the endodermis cell layer (Adapted from Hedrich 2012)

Table 15.2 List of few K^+ transporters of plant with type and plant organ in which they are reported

Names	Plant discovered	Type	Organ	Reference
Shaker family				
KAT1	*Arabidopsis thaliana*	Inward rectifying	Guard cells	Anderson et al. 1992; Ichida et al. 1997
AKT1	*A. thaliana*	Inward rectifying	Root cortex, epidermis, and hairs	Basset et al. 1995; Dennison et al. 2001
SPIK	*A. thaliana*	Inward rectifying	Pollen	Mouline et al. 2002
AtKC1	*A. thaliana*	Silent	Root cortex, epidermis, and hairs	Pilot et al. 2003; Reintanz et al. 2002
SKOR	*A. thaliana*	Outward rectifying	Root pericycle, xylem parenchyma	Gaymard et al. 1998; Lacombe et al. 2001
GORK	*A. thaliana*	Outward rectifying	Guard cells, stem, root hairs	Ache et al. 2000; Ivashikina et al. 2001
LKT1	*Lycopersicon esculentum*	Inward rectifying	Root hairs	Hartje et al. 2000
MKT1	*Mesembryanthemum crystallinum*	Inward rectifying	Root cortex, epidermis, and stele	Su et al. 2001
SPICK2	*Samanea saman*	Inward rectifying	Pulvinus	Moshelion et al. 2002
KST1	*Solanum tuberosum*	Inward rectifying	Guard cells	Hoth et al. 2001
ZMK1	*Zea mays*	Inward rectifying	Coleoptiles	Philippar et al. 1999
KCO family				
KCO1	*A. thaliana*	Outward rectifying	Mesophyll, guard cells	Czempinski et al. 2002
SPOCK1	*Samanea saman*	Inward rectifying	Pulvinus	Moshelion et al. 2002
KUP/HAT/KT family				
AtKUP1	*A. thaliana*	Inward rectifying	Stem, root	Fu and Luan 1998
CnHAK1	*Cymodocea nodosa*	Inward rectifying	Root	Rubio et al. 2000
OsHAK1	*Oryza sativa*	Inward rectifying	Root	Banuelos et al. 2002
HKT family				
AtHKT1	*A. thaliana*	Outward rectifying	Root	Rus et al. 2001
OsHKT1	*Oryza sativa*	Inward rectifying	Root	Horie et al. 2001

might be involved in subunit tetramerization (Daram et al. 1997) and/or channel clustering in the membrane (Ehrhardt et al. 1997). An ankyrin domain, hypothesized to be a site of interaction with regulatory proteins (Sentenac et al. 1992), is present in most channels (e.g., in six of the nine *Arabidopsis* Shakers) between the putative cyclic nucleotide-binding site and the KHA region.

15.6.2 KCO Channels

KCO channels display a hydrophobic core composed of either four TMS and two P domains (KCO-2P family) or two TMS and one P domain (KCO-1P family). They do not possess any TMS that might be expected to behave as voltage sensors. Their pore domains bear a high K^+-permeability hallmark motif. Most of them possess putative Ca_2C-binding sites (one or two EF hands) in their cytosolic C-terminal region (Czempinski et al. 1999; Moshelion et al. 2002). K^+ channels with a hydrophobic core sharing structural homologies with KCO channels exist in the animal field. Functionally, 2TMS-1P animal K^+ channels are inward rectifiers (Doupnik et al. 1995; Jan and Jan 1997). Most 4TMS-2P animal channels have been described as leak-like channels (i.e., their open probability is weakly sensitive to voltage), with some of them gated by membrane stretch (Patel and Honore 2001).

15.6.3 KUP/HAK/KT Transporters

Plants possess a family of genes encoding polypeptides homologous to K^+ transporters that were first identified in *Escherichia coli* named KUP, for K^+ uptake (Schleyer and Bakker 1993), and the soil yeast *Schwanniomyces occidentalis* named HAK1, for high-affinity K^+ (Banuelos et al. 1995). The bacterial KUP system was reported to be a constitutive low-affinity K^+ uptake system with a preponderant role at low pH, likely to operate as an HC-KC symport (Trchounian and Kobayashi 1999; Zakharyan and Trchounian 2001). In *S. occidentalis* or *Neurospora crassa*, HAK1 mediates high-affinity K^+ uptake, probably acting as an H^+-K^+ symport, and is the major K^+ transport system at work in conditions of K^+ starvation (Navarro 2000). The plant homologues, called KUP, HAK, or KT (for KC transporter), form a large family, with 13 members in *Arabidopsis* (Maser et al. 2001) and at least 17 members in rice (Banuelos et al. 2002). Little is known about the structure of these transporters. Hydrophobicity profiles suggest that they might possess 12 TMS and a long cytosolic loop between the second and third TMS (Banuelos et al. 2002; Kim et al. 1998; Quintero and Blatt 1997; Rubio et al. 2000). No region involved in ion conduction has yet been identified.

15.6.4 HKT Transporters

Plant HKT transporters are related to the fungal Trk transporters and prokaryote KtrB and TrkH K^+ transporter subunits (Durell and Guy 1999; Navarro 2000). Both fungal and prokaryote K^+ transport systems of this family are believed to work as K^+ cotransporters, where the coupling ion is H^+ (Bihler et al. 1999) or Na^+ (Tholema et al. 1999), or as K^+: K^+ cotransport systems (Haro et al. 1999; Haro and Navarro 2002), depending on the transporter and possibly also on the ionic conditions (Navarro 2000). In fungi, Trk transporters are the major systems involved in K^+ uptake at micromolar to sub-millimolar K^+ concentrations [at least at neutral and basic pH (Navarro 2000)].

15.7 Potassium-Solubilizing Microorganisms (KSMs)

Introduction of high-yielding varieties of many crops and agricultural intensification has led to the depletion of K reserve of soils at a faster rate. Moreover, due to imbalanced fertilizer application, K deficiency is becoming one of the major constraints in crop production (Meena et al. 2013; Maurya et al. 2014). This emphasized the search to find an alternative indigenous source of K for plant uptake and to maintain K status in soils for sustaining crop production. Soil microbes have been reported to play a key role in the natural K cycle (Fig. 15.6), and therefore, indigenous KSMs present in the soil could provide an alternative technology to make K

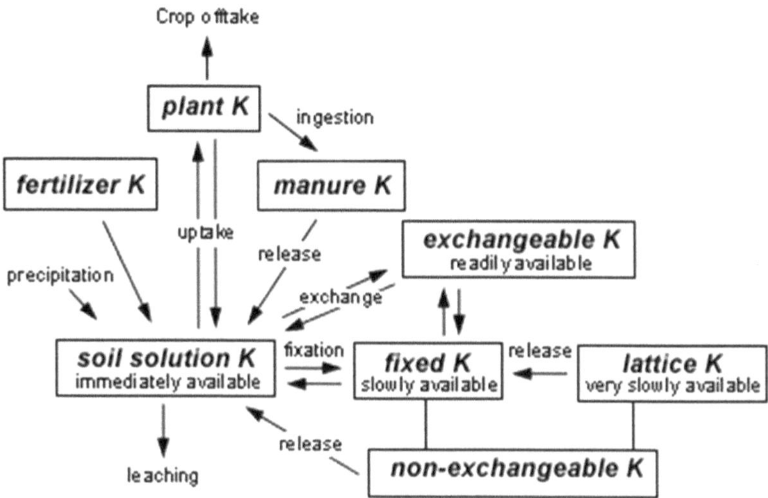

Fig. 15.6 K cycle operating in the nature (Adapted from Kraus and Jhonston 2002)

available for uptake by plants. Thus, identification of microbial strains capable of solubilizing K minerals quickly can conserve our existing resources and avoid environmental pollution hazards caused by heavy application of chemical fertilizers (Prajapati et al. 2012; Meena et al. 2015b; Singh et al. 2015).

A wide range of bacteria, namely, *Pseudomonas, Burkholderia, Acidithiobacillus ferrooxidans, Bacillus mucilaginosus, B. edaphicus, B. circulans*, and *Paenibacillus* spp. (Fig. 15.7), has been reported to release K in accessible form from K-bearing minerals in soils. These KSB were found to dissolve K, silicon, and aluminum from insoluble K-bearing minerals such as micas, illite, and orthoclases, by excreting organic acids which either directly dissolved rock K or chelated silicon ions to bring K into the solution (Meena et al. 2014a, 2015a). Reports are available about the beneficial effects of bacterial inoculation on K uptake and growth enhancement in cotton and rape, pepper and cucumber, sorghum, wheat, and Sudan grass (Parmar and Sindhu 2013).

KSMs that can dissolve K (Table 15.3) from rock minerals not only enhance plant growth but are also economically viable and eco-friendly. Since the first report of potassium solubilization by Muentz (1890), a large number of microbes mostly bacteria like *B. mucilaginosus, B. edaphicus, B. circulans, Paenibacillus* spp., *A. ferrooxidans, Pseudomonas* spp., and *Burkholderia* spp. (Lian et al. 2002; Sheng et al. 2008; Singh et al. 2010; Basak and Biswas 2012; Liu et al. 2012; Rajawat et al. 2012) have been reported to release K in accessible form from K-bearing minerals in soil (Meena et al. 2014b; Kumar et al. 2015).

15.8 Mechanism of K Solubilization

Currently, little information is available on the K solubilization by the rhizospheric microorganism; however, the most well-understood mechanisms of the K solubilization are by production of organic acid which can bind and extract K as well as other nutrients from non-soluble soil deposits for enhancement of crop growth. K release from different minerals is also a function of pH of the growing medium and available oxygen along with the bacterial strain used (Sheng and Huang 2002). The nature of K-bearing minerals also affects the efficiency of K solubilization by different KSMs (Uroz et al. 2009). In an experiment to compare the growth and K solubilization of *B. edaphicus*, it was observed that the bacteria was able to solubilize K better when illite was used as K mineral in the media in comparison to feldspar. Also, indigenous rhizospheric microorganisms have the potential to absorb and mobilize the fixed form of minerals from trace mineral sources.

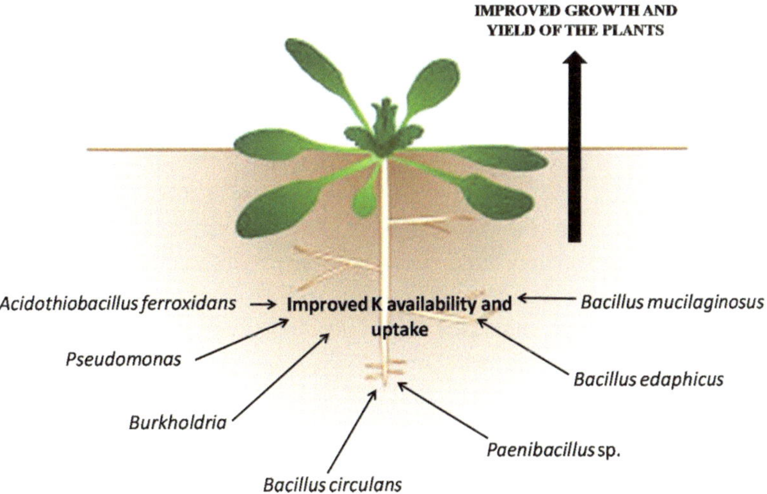

Fig. 15.7 Different bacterial species that help in K solubilization and uptake in the plants

Table 15.3 Microorganisms and minerals from which they can release potassium

Microorganisms	Source of K solubilization	Reference
A. niger	Rock powder	Assad et al. 2010
E. hormaechei	Feldspar	Prajapati et al. 2013
A. terreus	Feldspar	Prajapati et al. 2013
B. mucilaginosus	Waste mica	Basak and Biswas 2009
A. fumigatus	Rock powder	Lian et al. 2008
B. mucilaginosus	Microcline, muscovite, orthoclase	Sugumaran and Janartham 2007
B. edaphicus	Feldspar and illite	Sheng and He 2006
B. megaterium	Biotite and muscovite	Zarjani et al. 2013
Arthrobacter spp.	Biotite and muscovite	Zarjani et al. 2013
Pseudomonas spp.	Feldspar	Archana et al. 2013
Torulaspora globosa	Rock powder	Rosa-Magri et al. 2012
Burkholderia spp.	Feldspar	Sheng and Huang 2002
A. ferrooxidans	Feldspar	Sheng and He 2006
P. mucilaginosus	Illite	Liu et al. 2012
Cladosporium spp.	Mica	Argelis et al. 1993
Aminobacter spp.	Rock powder	Uroz et al. 2007
Sphingomonas spp.	Rock powder	Uroz et al. 2007
P. glucanolyticus	Mica	Sangeeth et al. 2012

For example, silicate bacteria can dissolve K, silica, and aluminum from insoluble minerals. Besides these factors, proton concentration of soil or soil solution is directly related to releases of K from minerals. In this respect, a study conducted by Badar (2006) showed that the extent of K solubilization by silicate-solubilizing bacteria was higher at a near neutral-basic pH (pH range 6.5–8.0) in comparison to the other pH tested. From the above discussion, it can be inferred that the K release from the insoluble minerals affects pH, aerobic conditions, and soil mineral properties (Lian 1998; Chen et al. 2008; Lian et al. 2008; Bin et al. 2010).

The most well-understood and widely studied mechanism of potassium solubilization is the production of various types of organic acids. These acids primarily bring about acidolysis, complexolysis, and exchange reactions when they interact with the K-containing minerals and for the key features attributed to the conversion of insoluble K into soluble form (Uroz et al. 2009). The organic acids convert insoluble K (mica, muscovite, biotite feldspar) to the soluble form of K (soil solution form) as a result availability of nutrients to the plants increases. The production and excretion of different organic acids vary with the organism. When these microorganisms are grown in a suspension culture, then organic acids can easily be detected with the help of chromatographic methods (Sheng et al. 2008). KSMs can also weather phlogopite via aluminum chelation and acidic dissolution of the crystal network (Leyval and Berthelin 1989; Abou-el-Seoud and Abdel-Megeed 2012).

The release of various types of organic acids by microorganisms is correlated to their ability to solubilize the insoluble K to an available form of K which can easily take up by the plants. Researchers have also suggested that organic acids produced by the KSMs, along with K solubilization, are also related to the direct plant growth promotion. Sheng and He (2006) reported that solubilization of illite and feldspar by KSMs is due to the production of organic acids like oxalic acid and tartaric acids, gluconic acid and 2-ketogluconic acid, oxalic acid, citric acid, malic acid, and succinic acid. Tartaric acid has also been found by other workers as one of the most frequent agents of mineral K solubilization (Prajapati et al. 2012; Prajapati and Modi 2012; Zarjani et al. 2013). Other organic acids, such as acetic, citric, lactic, propionic, glycolic, oxalic, malonic, succinic, fumaric, tartaric, etc., have also been identified among K solubilizers (Wu et al. 2005; Lian 2010).

The solubilization of structural K compounds by naturally abundant KSMs is common under in vitro conditions (Raj 2004; Sheng and He 2006; Sugumaran and Janarthanam 2007; Prajapati et al. 2012; Zarjani et al. 2013) and field and greenhouse condition (Basak and Biswas 2009; Singh et al. 2010; Prajapati et al. 2013; Parmar and Sindhu 2013). The indigenous rhizospheric microorganisms are effective in releasing K from structural K through solubilization and form exchangeable pools of total soil K by acidolysis, chelation, and solubilization by KSMs (Uroz et al. 2009). Biomasses of the rhizospheric microorganism in soil also contain a significant quantity of fixed K that is potentially available to plants (Jones et al. 2003; Girgis 2006).

Some other mechanisms for K solubilization by KSMs are by (i) lowering the pH, (ii) enhancing chelation of the cations bound to K, and (iii) acidolysis of the surrounding area of microorganism. The lowering in pH of the medium suggests the release of organic acids and protons by the K-solubilizing microorganisms (Uroz et al. 2009; Parmar and Sindhu 2013; Zarjani et al. 2013). Such acidolysis by organic acids produced by the rhizospheric microorganisms can either directly dissolve the mineral K as a result of slow releases of exchangeable K and readily available exchangeable K or can chelate both Si and Al ions associated with K minerals (Romheld and Kirkby 2010). Thus, the synthesis and discharge of organic acids by the microorganisms into the surrounding environment acidify the microbe's cells and their surrounding environment that ultimately lead to the release of K^+ from the mineral K by protonation and acidification (Goldstein 1994). Of the different organic acids involved in the solubilization of insoluble K, succinic, citric, gluconic, α-ketogluconic, and oxalic acids are the most prominent acids released by microbial strains.

Organic acids produced by the KSMs can be detected by high-performance liquid chromatography and enzymatic methods (Archana et al. 2012, 2013; Zhang et al. 2013). However, the acidification does not seem to be the only mechanism of solubilization, as the ability to reduce the pH in some cases did not correlate with the ability to solubilize mineral K (Rosa-Magri et al. 2012; Zarjani et al. 2013). Furthermore, the chelating ability of the organic acids is also important, as it has been shown that the addition of 0.05 M EDTA to the medium has the same solubilizing effect as inoculation with *Penicillium bilaii* (Kucey 1988; Sheng and He 2006; Liu et al. 2006).

15.9 Conclusions

K is one of the most widely available minerals in the Earth's crust and is of prime importance to the plants, but it is unavailable to the plants as it can form insoluble complexes with other minerals in the soil. Many different K channels are known and mechanism of K uptake in the plants has been elucidated in fair details. Using all this information now, we are in the position to identify new methods that will allow the plants to utilize phosphorus in a more efficient way. A promising technique is the use of K-solubilizing microorganisms that can solubilize the K reserves from soil and make it available to the plants. The use of KSMs has proved to be useful in the work done so far. Moreover, utilization of naturally occurring KSMs on agricultural and other soils can increase the ratio of exchangeable K in comparison to non-exchangeable K. Thus, application of KSMs will enhance the availability of K in soils which will be useful in the longer run and will form an important factor in the sustainable agriculture. Hence, it can be concluded that utilization of KSMs with judicious research will be useful and reduce the agricultural reliance on the fertilizers.

References

Abou-el-Seoud, Abdel-Megeed A (2012) Impact of rock materials and biofertilizations on P and K availability for maize (*Zea maize*) under calcareous soil conditions. Saudi J Biol Sci 19:55–63

Ache P, Becker D, Ivashikina N, Dietrich P, Roelfsema MR, Hedrich R (2000) GORK, a delayed outward rectifier expressed in guard cells of *Arabidopsis thaliana*, is a K^+-selective, K^+-sensing ion channel. FEBS Lett 486:93–98

Anderson JA, Huprikar SS, Kochian LV, Lucas WJ, Gaber RF (1992) Functional expression of a probable *Arabidopsis thaliana* potassium channel in *Saccharomyces cerevisiae*. Proc Natl Acad Sci U S A 89:3736–3740

Anonymous (2002) Potassium status of soils in India. Better Crop Int 16(2). 3–5, Nov 2002

Archana DS, Nandish MS, Savalagi VP, Alagawadi AR (2012) Screening of potassium solubilizing bacteria (KSB) for plant growth promotional activity. Bioinfolet 9:627–630

Archana DS, Nandish MS, Savalagi VP, Alagawadi AR (2013) Collection of potassium solubilizing bacteria (KSB) from rhizosphere soil. Bioinfolet 10:248–257

Argelis DT, Gonzala DA, Vizcaino C, Gartia MT (1993) Biochemical mechanism of stone alteration carried out by filamentous fungi living in monuments. Biogeo Chem 19:129–147

Assad ML, Avansini SH, Rosa MM, Carvalho JRP, Ceccato-Antonini SR (2010) The solubilization of potassium-bearing rock powder by *Aspergillus niger* in small-scale batch fermentations. Can J Microbiol 56:598–605

Badar MA (2006) Efficiency of K feldspar combined with organic material and silicate dissolving bacteria on tomato yield. J Applied Sci Res 2:1191–1198

Banuelos MA, Garciadeblas B, Cubero B, Navarro AR (2002) Inventory and functional characterization of the HAK potassium transporters of rice. Plant Physiol 130:784–794

Bin L, Bin W, Mu P, Liu C, Teng HH (2010) Microbial release of potassium from K-bearing minerals by thermophilic fungus Aspergillus fumigatus. Geochim Cosmochim Acta 72:87–98

Banuelos MA, Klein RD, Alexander-Bowman SJ, Navarro AR (1995) A potassium transporter of the yeast *Schwanniomyces occidentalis* homologous to the Kup system of *Escherichia coli* has a high concentrative capacity. EMBO J 14:3021–3027

Basak BB, Biswas DR (2012) Modification of waste mica for alternative source of potassium: evaluation of potassium release in soil from waste mica treated with potassium solubilizing bacteria (KSB). Lambert Academic Publishing, Saarbrücken. ISBN 978-3-659-29842-4

Basset M, Conejero G, Lepetit M, Foucroy P, Sentenac H (1995) Organization and expression of the gene coding for the potassium transport system AKT1 of *Arabidopsis thaliana*. Plant Mol Biol 29:947–958

Ben-Zioni A, Vaadia Y, Lips SH (1970) Correlations between nitrate reduction, protein synthesis and malate accumulation. Physiol Plant 23:1039–1047

Bihler H, Gaber RF, Slayman CL, Bertl A (1999) The presumed potassium carrier Trk2p in *Saccharomyces cerevisiae* determines an H^+-dependent, K^+-independent current. FEBS Lett 447:115–120

Biswas DR, Basak BB (2009) Influence of potassium solubilizing microorganism (*Bacillus mucilaginosus*) and waste mica on potassium uptake dynamics by sudan grass (*Sorghum vulgare*) grown under two Alfisols. Plant Soil Environ J 317:235–255

Britto DT, Kronzucker HJ (2008) Cellular mechanisms of potassium transport in plants. Physiol Plant 137:637–650

Busch W, Saier MH (2002) The transporter classification (TC) system. Crit Rev Biochem Mol Biol 37:287–337

Cakmak I (2005) The role of potassium in alleviating detrimental effects of abiotic stresses in plants. J Plant Nutr Soil Sci 168:521–530

Cakmak I, Hengeler C, Marschner H (1994) Changes in phloem export of sucrose in leaves in response to phosphorus, potassium and magnesium deficiency in bean plants. J Exp Bot 45:1251–1257

Cheeseman JM, Hanson JB (1980) Does active K1 influx to roots occur? Plant Sci Lett 18:81–84

Cheeseman JM, Hanson JB (1979) Energy-linked potassium influx as related to cell potential in corn roots. Plant Physiol 64:842–845

Chen S, Lian B, Liu CQ (2008) *Bacillus mucilaginosus* on weathering of phosphorite and primary analysis of bacterial proteins during weathering. Chin J Geochem 27:209–216

Clarkson DT, Hanson JB (1980) The mineral nutrition of higher plants. Annu Rev Plant Physiol 31:239–298

Czempinski K, Frachisse JM, Maurel C, Barbier-Brygoo-H, Mueller-Roeber B (2002) Vacuolar membrane localization of the *Arabidopsis* 'two-pore' K^+ channel KCO1. Plant J 29:809–820

Czempinski K, Gaedeke N, Zimmermann S, Mueller-Roeber B (1999) Molecular mechanisms and regulation of plant ion channels. J Exp Bot 50:955–966

Daram P, Urbach S, Gaymard F, Sentenac H, Cherel I (1997) Tetramerization of the AKT1 plant potassium channel involves its C-terminal cytoplasmic domain. EMBO J 16:3455–3463

Deeken R, Sanders C, Ache P, Hedrich R (2000) Developmental and light-dependent regulation of a phloem-localised K^+ channel of *Arabidopsis thaliana*. Plant J 23:285–290

Dennison KL, Robertson WR, Lewis BD, Hirsch RE, Sussman MR, Spalding EP (2001) Functions of AKT1 and AKT2 potassium channels determined by studies of single and double mutants of *Arabidopsis*. Plant Physiol 127:1012–1019

Dietrich P, Sanders D, Hedrich R (2001) The role of ion channels in light-dependent stomatal opening. J Exp Bot 52:1959–1967

Dolan L, Davies J (2004) Cell expansion in roots. Curr Opin Plant Biol 7:33–39

Doupnik CA, Davidson N, Lester HA (1995) The inward rectifier potassium channel family. Curr Opin Neurobiol 5:268–277

Durell SR, Guy HR (1999) Structural models of the KtrB, TrkH, and Trk1,2 symporters based on the structure of the KcsA K^+ channel. Biophys J 77:789–807

Ehrhardt T, Zimmermann S, Mueller-Reober B (1997) Association of plant K^+ in channels is mediated by conserved C termini and does not affect subunit assembly. FEBS Lett 409:166–170

FAO (2012) World fertilizer trends and outlook to 2018. Food and Agriculture Organization of the United Nations, Rome, pp 1–45

Fu HH, Luan S (1998) AtKuP1: a dual affinity K^+ transporter from *Arabidopsis*. Plant Cell 10:63–73

Gassmann W, Schroeder JI (1994) Inward-rectifying K^+ channels in root hairs of wheat – a mechanism for aluminum-sensitive low-affinity K^+ uptake and membrane potential control. Plant Physiol 105:1399–1408

Gaymard F, Pilot G, Lacombe B, Bouchez D, Bruneau D (1998) Identification and disruption of a plant shaker-like outward channel involved in K^+ release into the xylem sap. Cell 94:647–655

Girgis MGZ (2006) Response of wheat to inoculation with phosphate and potassium mobilizers and organic amendment. Ann Agric Sci Ain Shams Univ Cairo 51:85–100

Goldstein MAH (1994) Involvement of the quinoprotein glucose dehydrogenase in the solubilization of exogenous phosphates by gram-negative bacteria. In: Torriani-Gorini A, Yagil E, Silver S (eds) Phosphate in microorganisms: cellular and molecular biology. ASM Press, Washington, DC, pp 197–203

Haas EM (2011) Role of potassium in maintaining health. http://hkpp.org/patients/potassium-health

Haro R, Navarro AR (2002) Molecular analysis of the mechanism of potassium uptake through the TRK1 transporter of *Saccharomyces cerevisiae*. Biochim Biophys Acta 1564:114–122

Haro R, Sainz L, Rubio F, Navarro AR (1999) Cloning of two genes encoding potassium transporters in *Neurospora crassa* and expression of the corresponding cDNAs in *Saccharomyces cerevisiae*. Mol Microbiol 31:511–220

Hartje S, Zimmermann S, Klonus D, Mueller-Roeber B (2000) Functional characterization of LKT1, a K^+ uptake channel from tomato root hairs, and comparison with the closely related potato inwardly rectifying KC channel SKT1 after expression in *Xenopus* oocytes. Planta 210:723–731

Hedrich R (2012) Ion channels in plants. Physiol Rev 92:1777–1811

Holzmueller EJ, Jose S, Jenkins MA (2007) Influence of calcium, potassium, and magnesium on *Cornus florida* L. density and resistance to dogwood anthracnose. Plant Soil 290:189–199

Horie T, Yoshida K, Nakayama H, Yamada K, Oiki S, Shinmyo A (2001) Two types of HKT transporters with different properties of Na^+ and K^+ transport in *Oryza sativa*. Plant J 27:129–138

Hoth S, Geiger D, Becker D, Hedrich R (2001) The pore of plant K^+ channels is involved in voltage and pH sensing: domain-swapping between different K^+ channel alpha-subunits. Plant Cell 13:943–952

Hoth S, Hedrich R (1999) Distinct molecular bases for pH sensitivity of the guard cell K^+ channels KST1 and KAT1. J Biol Chem 274:11599–11603

Humble GD, Hsiao TC (1969) Specific requirement of potassium for light-activated opening of stomata in epidermal strips. Plant Physiol 44:230–234

Ichida AM, Pei ZM, Baizabal-Aguirre VM, Turner KJ, Schroeder JI (1997) Expression of a Cs^+-resistant guard cell K^+ channel confers Cs^+-resistant, light induced stomatal opening in transgenic *Arabidopsis*. Plant Cell 9:1843–1857

Ivashikina N, Becker D, Ache P, Meyerhoff O, Felle HH, Hedrich R (2001) K^+ channel profile and electrical properties of *Arabidopsis* root hairs. FEBS Lett 508:463–469

Jan LY, Jan YN (1997) Cloned potassium channels from eukaryotes and prokaryotes. Annu Rev Neurosci 20:91–123

Johansson I, Wulfetange K, Poree F, Michard E, Gajdanowicz P, Lacombe B, Sentenac H, Thibaud JB, Mueller-Roeber B, Blatt MR, Dreyer I (2006) External K^+ modulates the activity of the Arabidopsis potassium channel SKOR via an unusual mechanism. Plant J 46:269–281

Jones DL, Dennis PG, Owen AG, Van-Hees PAW (2003) Organic acid behavior in soils misconceptions and knowledge gaps. Plant Soil 248:31–41

Kim EJ, Kwak JM, Uozumi N, Schroeder JI (1998) AtKUP1: an *Arabidopsis* gene encoding high-affinity potassium transport activity. Plant Cell 10:51–62

Kraus A, Jhonston AE (2002) Assessing soil potassium, can we do better? 9th international congress of soil science, Faisalabad, Pakistan 18–20 March 2002

Kronzucker HJ, Szczerba MW, Britto DT (2003) Cytosolic potassium homeostasis revisited: 42K-tracer analysis in *Hordeum vulgare* L. reveals set-point variations in [K^+]. Planta 217:540–546

Kucey RMN (1988) Effect of *Penicillium biloji* on the solubility and uptake of P and micronutrients from soil by wheat. Can J Soil 68:261–267

Kumar A, Bahadur I, Maurya BR, Raghuwanshi R, Meena VS, Singh DK, Dixit J (2015) Does a plant growth-promoting rhizobacteria enhance agricultural sustainability? J Pure Appl Microbiol 9(1):715–724

Lacombe B, Becker D, Hedrich R, De-Salle R, Hollmann M (2001) The identity of plant glutamate receptors. Science 292:1486–1487

Leigh RA, Wyn Jones RG (1984) A hypothesis relating critical potassium concentrations for growth to the distribution and functions of this ion in the plant cell. New Phytol 97:1–13

Leyval C, Berthelin J (1989) Interaction between *Laccaria laccata*, *Agrobacterium radiobacter* and beech roots: influence on P, K, Mg and Fe mobilization from minerals and plant growth. Plant Soil 117:103–110

Lian B, Fu PQ, Mo DM, Liu CQ (2002) A comprehensive review of the mechanism of potassium release by silicate bacteria. Acta Mineral Sinica 22:179

Lian B, Wang B, Pan M, Liu C, Teng HH (2008) Microbial release of potassium from K-bearing minerals by thermophilic fungus *Aspergillus fumigatus*. Geochim Cosmochim Acta 70:87–98

Lian B, Wang B, Pan M, Liu C, Teng HH (2010) Microbial release of potassium from K-bearing minerals by thermophilic fungus *Aspergillus fumigatus*. Geochim Cosmochim Acta 72:87–98

Lian BA (1998) A study on how silicate bacteria GY92 dissolves potassium from illite. Acta Mineral Sin 18:234–238

Liu D, Lian B, Dong H (2012) Isolation of *Paenibacillus* sp. and assessment of its potential for enhancing mineral weathering. Geomicrobiol J 29:413–421

Liu W, Xu X, Wu S, Yang Q, Luo Y, Christie P (2006) Decomposition of silicate minerals by *Bacillus mucilaginosus* in liquid culture. Environ Geochem Health 28:133–140

Maathuis FJM, Sanders D (1994) Mechanism of high-affinity potassium uptake in roots of *Arabidopsis thaliana*. Proc Natl Acad Sci U S A 91:9272–9276

Maser P, Thomine S, Schroeder JI, Ward JM, Hirschi K (2001) Phylogenetic relationships within cation transporter families of *Arabidopsis*. Plant Physiol 126:1646–1667

Maurya BR, Meena VS, Meena OP (2014) Influence of Inceptisol and Alfisol's potassium solubilizing bacteria (KSB) isolates on release of K from waste mica. Vegetos 27(1):181–187

Meena OP, Maurya BR, Meena VS (2013) Influence of K-solubilizing bacteria on release of potassium from waste mica. Agric Sustain Dev 1(1):53–56

Meena RK, Singh RK, Singh NP, Meena SK, Meena VS (2015a) Isolation of low temperature surviving plant growth-promoting rhizobacteria (PGPR) from pea (*Pisum sativum* L.) and documentation of their plant growth promoting traits. Biocatal Agric Biotechnol. doi:10.1016/j.bcab.2015.08.006

Meena VS, Maurya BR, Bahadur I (2014a) Potassium solubilization by bacterial strain in waste mica. Bangladesh J Bot 43(2):235–237

Meena VS, Maurya BR, Verma JP (2014b) Does a rhizospheric microorganism enhance K+ availability in agricultural soils? Microbiol Res 169:337–347

Meena VS, Maurya BR, Verma JP, Aeron A, Kumar A, Kim K, Bajpai VK (2015b) Potassium solubilizing rhizobacteria (KSR): isolation, identification, and K-release dynamics from waste mica. Ecol Eng 81:340–347

Moran N (2007) Osmoregulation of leaf motor cells. FEBS Lett 581:2337–2347

Moshelion M, Becker D, Czempinski K, Mueller-Roeber-B, Attali B (2002) Diurnal and circadian regulation of putative potassium channels in a leaf moving organ. Plant Physiol 128:634–642

Mouline K, Very AA, Gaymard F, Boucherez J, Pilot G (2002) Pollen tube development and competitive ability are impaired by disruption of a shaker K^+ channel in *Arabidopsis*. Genes Dev 16:339–350

Muentz A (1890) Surla decomposition desroches etla formation de la terre arable. C R Acad Sci 110:1370–1372

Navarro AR (2000) Potassium transport in fungi and plants. Biochim Biophys Acta 1469:1–30

Oerke EC, Dehne HW (2004) Safeguarding production-losses in major crops and the role of crop protection. Crop Prot 23:275–285

Parmar P, Sindhu SS (2013) Potassium solubilization by rhizosphere bacteria: influence of nutritional and environmental conditions. J Microbiol Res 3:25–31

Patel AJ, Honore E (2001) Properties and modulation of mammalian 2P domain K^+ channels. Trends Neurosci 24:339–346

Perrenoud S (1990) Potassium and plant health, 2nd edn. International Potash Institute, Bern, pp 8–10

Philippar K, Fuchs I, Luthen H, Hoth S, Bauer CS (1999) Auxin-induced K^+ channel expression represents an essential step in coleoptile growth and gravitropism. Proc Natl Acad Sci U S A 96:12186–12191

Pilot G, Gaymard F, Mouline K, Cherel I, Sentenac H (2003) Regulated expression of *Arabidopsis* shaker K^+ channel genes involved in K^+ uptake and distribution in the plant. Plant Mol Biol 51:773–787

Prajapati K, Sharma MC, Modi HA (2012) Optimization of medium components for potassium solubilizing fungus *Aspergillus terreus* (KSF 1) by response surface methodology. Indian J Fundam Appl Life Sci 2:50–54

Prajapati K, Sharma MC, Modi HA (2013) Growth promoting effect of potassium solubilizing microorganisms on *Abelmoscus esculantus*. Int J Agric Sci 3:181–188

Prajapati KB, Modi HA (2012) Isolation and characterization of potassium solubilizing bacteria from ceramic industry soil. CIBTech J Microbiol 1:8–14

Quintero FJ, Blatt MR (1997) A new family of K^+ transporters from *Arabidopsis* that are conserved across phyla. FEBS Lett 415:206–211

Raj SA (2004) Solubilization of silicate and concurrent release of phosphorus and potassium in rice ecosystem. In: Book chapter – Conference Paper Biofertilizers Technology, Coimbatore, India, pp 372–378

Rajawat MVS, Singh S, Singh G, Saxena AK (2012) Isolation and characterization of K solubilizing bacteria isolated from different rhizospheric soil. In: Proceeding of 53rd annual conference of Association of Microbiologists of India 2012, p 124

Reintanz B, Szyroki A, Ivashikina N, Ache P, Godde M (2002) AtKC1, a silent *Arabidopsis* potassium channel alpha-subunit modulates root hair K^+ influx. Proc Natl Acad Sci U S A 99:4079–4084

Romheld V, Kirkby EA (2010) Research on potassium in agriculture: needs and prospects. Plant Soil 335:155–180

Rosa-Magri MM, Avansani SH, Lopes-Assad MS, Tornisielo SM, Antonini SR (2012) Release of potassium from rock powder by the yeast *Torulaspora globosa*. Braz Arch Biol Technol 55:577–582

Rubio F, Santa-Maria GE, Navarro AR (2000) Cloning of *Arabidopsis* and barley cDNAs encoding HAK potassium transporters in root and shoot cells. Physiol Plant 109:34–43

Rus A, Yokoi S, Sharkhuu A, Reddy M, Lee BH (2001) AtHKT1 is a salt tolerance determinant that controls Na^+ entry into plant roots. Proc Natl Acad Sci U S A 98:14150–14155

Sangeeth KP, Bhai RS, Srinivasan V (2012) *Paenibacillus glucanolyticus*, a promising potassium solubilizing bacterium isolated from black pepper (*Piper nigrum* L.) rhizosphere. J Spic Aromat Crops 21:118–124

Sarwar M (2012) Effects of potassium fertilization on population build up of rice stem borers (lepidopteron pests) and rice (*Oryza sativa* L.) yield. J Cereals Oilseeds 3:6–9

Schleyer M, Bakker EP (1993) Nucleotide sequence and 3′-end deletion studies indicate that the K^+-uptake protein Kup from *Escherichia coli* is composed of a hydrophobic core linked to a large and partially essential hydrophilic C terminus. J Bacteriol 175:6925–6931

Sentenac H, Bonneaud N, Minet M, Lacroute F, Salmon JM (1992) Cloning and expression in yeast of a plant potassium ion transport system. Science 256:663–665

Sheng XF, He LY (2006) Solubilization of potassium-bearing minerals by a wild-type strain of *Bacillus edaphicus* and its mutants and increased potassium uptake by wheat. Can J Microbiol 52:66–72

Sheng XF, Huang WY (2002) Mechanism of potassium release from feldspar affected by the strain NBT of silicate bacterium. Acta Pedol Sin 39:863–871

Sheng XF, Zhao F, He LY, Qiu G, Chen L (2008) Isolation and characterization of silicate mineral solubilizing *Bacillus globisporus* Q12 from the surfaces of weathered feldspar. Can J Microbiol 54:1064–1068

Singh NP, Singh RK, Meena VS, Meena RK (2015) Can we use maize (*Zea mays*) rhizobacteria as plant growth promoter? Vegetos 28(1):86–99

Singh S, Pandey A, Kumar B, Palni LMS (2010) Enhancement in growth and quality parameters of tea [*Camellia sinensis* (L.) O. Kuntze] through inoculation with arbuscular mycorrhizal fungi in an acid soil. Biol Fertil Soils 46:427–433

Srinivasrao CH, Satyanarayana T, Venkateswarulu B (2011) Potassium mining in Indian agriculture: input and output balance. Karnataka J Agric Sci 24:20–28

Su H, Golldack D, Katsuhara M, Zhao C, Bohnert HJ (2001) Expression and stress dependent induction of potassium channel transcripts in the common ice plant. Plant Physiol 125:604–614

Suelter CH (1970) Enzymes activated by monovalent cations. Science 168:789–795

Sugumaran P, Janartham B (2007) Solubilization of potassium minerals by bacteria and their effect on plant growth. World J Agric Sci 3:350–355

Szczerba MW, Britto DT, Kronzucker HJ (2006) Rapid, futile K^+ cycling and pool-size dynamics define low-affinity potassium transport in barley. Plant Physiol 141:1494–1507

Talke IN, Blaudez D, Maathuis FJM, Sanders D (2003) CNGCs: prime targets of plant cyclic nucleotide signalling? Trends Plant Sci 8:286–293

Tholema N, Bakker EP, Suzuki A, Nakamura T (1999) Change to alanine of one out of four selectivity filter glycines in KtrB causes a two orders of magnitude decrease in the affinities for both K^+ and Na^+ of the Na^+ dependent K^+ uptake system KtrAB from *Vibrio alginolyticus*. FEBS Lett 450:217–220

Trchounian A, Kobayashi H (1999) Kup is the major K^+ uptake system in *Escherichia coli* upon hyper-osmotic stress at a low pH. FEBS Lett 447:144–148

Uroz S, Calvaruso C, Turpault M-P, Frey-Klett P (2009) Mineral weathering by bacteria: ecology, actors and mechanisms. Trends Microbiol 17:378–387

Uroz S, Calvaruso C, Turpault MP, Pierrat JC, Mustin C, Frey-Klett P (2007) Effect of the mycorrhizosphere on the genotypic and metabolic diversity of the bacterial communities involved in mineral weathering in a forest soil. Appl Environ Microbiol 73:3019–3027

Walker NA, Sanders D, Maathuis FJ (1996) High-affinity potassium uptake in plants. Science 273:977–979

Wang M, Zheng Q, Shen Q, Guo S (2013) The critical role of potassium in plant stress response. Int J Mol Sci 14:7370–7390

Williams J, Smith SG (2001) Correcting potassium deficiency can reduce rice stem diseases. Better Crops 85:7–9

Wu SC, Cao ZH, Li ZG, Cheung KC, Wong MH (2005) Effects of biofertilizer containing N fixer, P and K solubilizers and AM fungi on maize growth: a greenhouse trial. Geoderma 125:155–166

Zakharyan E, Trchounian A (2001) K^+ influx by Kup in *Escherichia coli* is accompanied by a decrease in H^+ efflux. FEMS Microbiol Lett 204:61–64

Zarjani JK, Aliasgarzad N, Oustan S, Emadi M, Ahmadi A (2013) Isolation and characterization of some potassium solubilizing bacteria in some Iranian soils. Arch Agron Soil Sci 59:1713–1723

Zhang A, Zhao G, Gao T, Wang W, Li J, Zhang S (2013) Solubilization of insoluble potassium and phosphate by *Paenibacillus kribensis* CX-7: a soil microorganism with biological control potential. Afr J Microbiol Res 7:41–47

KSM Soil Diversity and Mineral Solubilization, in Relation to Crop Production and Molecular Mechanism

16

Manoj Shrivastava, P.C. Srivastava, and S.F. D'Souza

Abstract

Bio-weathering and biomineralization are geochemical processes of the decomposition of rocks and minerals which are mediated by the living organisms. These processes play a fundamental role in the release of nutrients from rocks and minerals. Microorganisms, higher plants, and animals can weather rock aggregates through biomechanical and biochemical attack. Among such microbes, the information especially on phosphate solubilizers has been well documented; however, there are very few studies on the use of potassium-solubilizing microbes (KSMs) for the release of the native soil K. The strains of *Pseudomonas*, *Bacillus*, *Aspergillus*, *Penicillium*, etc. are some known K solubilizers. The principal mechanism for the microbe-mediated release of potassium from minerals is through the production of organic acids. A short overview of the KSMs and their effect of K uptake and crop growth are presented herein. K dynamics in soils and its availability to plants, metabolic pathways effecting the release of organic acids by KSMs, are covered. The aspects of immobilization of KSMs for the ease of application and the role of VAM in K mobilization have also been explored.

Keywords

Bio-weathering • Biomineralization • Microorganisms • K solubilization • K dynamics

M. Shrivastava
Centre for Environment Science and Climate Resilient Agriculture, Indian Agricultural Research Institute, New Delhi 110 012, India

P.C. Srivastava
Department of Soil Science, College of Agriculture, G.B. Pant University of Agriculture and Technology, Pantnagar 263 145, India

S.F. D'Souza (✉)
Bioscience Group, Bhabha Atomic Research Centre, Trombay, Mumbai 400 085, India
e-mail: stanfdsouza@gmail.com

V.S. Meena et al. (eds.), *Potassium Solubilizing Microorganisms for Sustainable Agriculture*,
DOI 10.1007/978-81-322-2776-2_16

16.1 Introduction

Potassium (K) is an essential primary nutrient which is required by all living forms including crop plants. In crop plants, it plays a significant role in the activation of several enzymes and controls important anabolic processes like photosynthesis and protein synthesis (Mengel and Kirkby 1987; Marschner 1995). It has also a role in imparting resistance against both abiotic and biotic stresses like drought incidence, disease, and insect attacks. Under intensive agriculture, the native bioavailable soil K pool which represents only a small fraction of the total K reserve in the soil is quickly depleted owing to crop removal, leaching, runoff, and erosion losses. In many developing countries, the lack of awareness about the balanced fertilization among farmers coupled with higher prices for imported potassic fertilizers and negative K balance has been noted for several crop rotations (Singh et al. 2004). In view of the present scenario, the research efforts to improve the efficiency of K fertilizers and bio-activation of soil K reserves should go hand in hand for sustainable food production and also for reducing the cost of cultivation in developing countries. The latter strategy sounds promising as it involves only little cost and avoids the heavy use of chemical fertilizers in agriculture. The present chapter examines the dynamics of K in soils and its availability to plants, K-solubilizing microbes, the mechanism of K solubilization, and the related aspects.

16.2 Potassium Dynamics in Soil and Availability to Plants

The total K content of soils is usually quite large as it is a constituent of many primary and secondary soil minerals. Total K content in surface (0–15 cm) soil layer may range between 0.13 % and 4.46 %. The K content of soil depends on both the nature of the parent material of a given soil and the courses of pedogenesis occurred. During the course of chemical weathering, K is released by water and weak acids formed in the nature, and the content of K in the parent material is reduced. Therefore, less weathered young soils from volcanic material are likely to be rich in K as compared to the highly weathered sandy soils. During the course of weathering, K^+ held as the interlayer cation in a dehydrated state within 2:1 clay minerals are replaced by other hydrated cations such as Ca^{2+}, Na^+, Mg^{+2}, etc., and this causes an expansion of the interlayer space. This release of interlayer K is diffusion controlled and favored by low pH and low K^+ concentration in the soil solution. The nature of soil minerals also influences the release of K due to their varying susceptibility to weathering. In soil, K is present in different fractions as released upon weathering from soil minerals enter in soil solution:

(i) K^+ in soil solution
(ii) K^+ as exchangeable cation on the exchange complex of soil colloids
(iii) Non-exchangeable K in the interlayer space of 2:1 clay minerals
(iv) K^+ as structural cation in the lattice of soil minerals

All these fractions of soil K are in a state of dynamic equilibrium (Fig. 16.1). K^+

These K^+ ions are absorbed by soil colloids (aluminosilicate clays, sesquioxides, and organic matter) and/or absorbed by plant or lost through leaching. Exchange sites on soil organic matter and 1:1 clay minerals retain hydrated K^+ ions in the exchangeable form, while on clay minerals (2:1 type), they are retained by the exchange sites of different selectivities. (a) Exchange sites of planar surfaces (p-site) hold hydrated K^+ easily releases K (organic matter and 1:1 clay minerals). (b) Edge surface site (e-site) of high selectivity holds K^+ with higher tenacity and releases K^+ less easily than p-sites, and (c) interlayer sites (i-site) of very high selectivity hold dehydrated K^+ very tightly and release K^+ least easily.

Both soil solutions K^+ and exchangeable K^+ are in equilibrium. Any depletion of K^+ from the soil solution as a result of plant uptake or leaching allows a rapid and easy replenishment

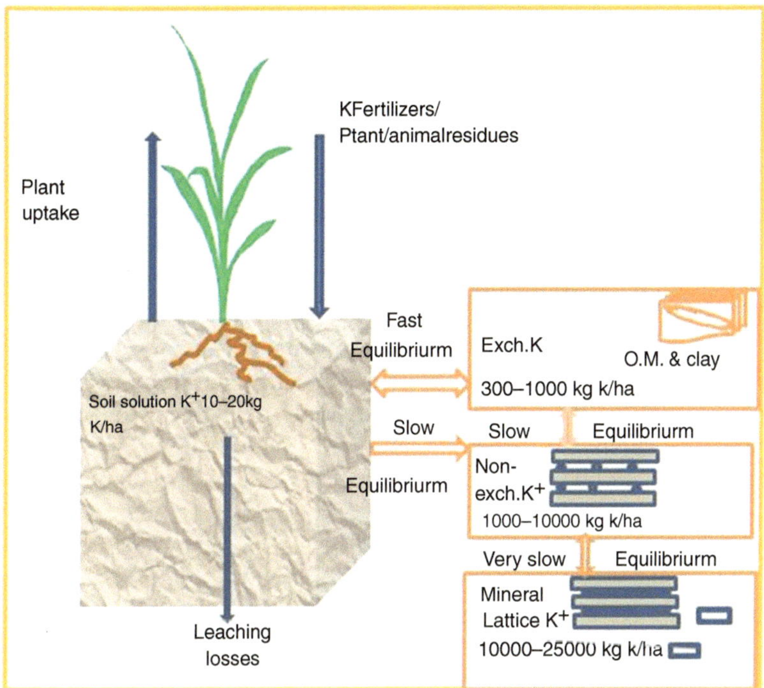

Fig. 16.1 Dynamics of K in soil–plant–microbes environmental system

from the K^+ present on the exchange sites of low selectivity (p-sites). Soils having a higher buffering capacity replenish K in the soil solution for longer periods as compared to soils of poor buffering capacity. Exchangeable and non-exchangeable fractions of K are also in equilibrium; however, the establishment of equilibrium is relatively slower. Upon exhaustion of the exchangeable K present on exchange sites, non-exchangeable K may release K for plant uptake. In this process, the interlayer K^+ ions are replaced by other cations of larger size such as Ca, Na, and Mg, causing an expansion of interlayer space for 14 Å. This creates a wedge zone from where replaced K^+ diffuses out of mineral (Jackson and During 1979). The minimum threshold concentration at which the release of fixed K starts varies with the type of mineral. For example, it is 1 mmol L^{-1} for biotite to 0.1 mmol K^+ L^{-1} for muscovite. Some 2:1-type clay minerals like mica, illite, and vermiculite have high selectivity for K^+ as compared to the replacing divalent cations. Any addition of K^+ to the solution, through application of soluble K fertilizers, results in a strong irreversible retention of K in their interlayer space. This process of conversion of added K to relatively unavailable form is also known as K fixation. The clay minerals of 2:1 type which maintain a high charge density in the silicon tetrahedral layer exhibit this characteristic. The hexagonal O network in the tetrahedral layer provides a cavity to allow just fitting of two K^+ ions. High charge density in the tetrahedral layer of two unit cells allows a strong mutual electrostatic attraction of layers resulting collapse of the interlayer space to lock K^+ ions there.

The phenomenon of K^+ fixation in soils is influenced by different soil and management factors. It is usually increased with the increase in soil clay content, and, therefore, K fixation is much higher in the subsurface layer as compared to surface soil. Among 2:1 clay minerals, the order of K fixation is: vermiculite > illite > montmorillonite. Dioctahedral micas retain interlayer K more strongly than trioctahedral micas. The reduction of ferric iron in the octahedral layer of clay crystal also enhances K^+ fixation (Chen et al. 1987). The presence of wedge zone in the 2:1 clay minerals increases K fixation.

Weathered micas, vermiculite, and illite fix K under both dry and wet, while smectite fixes K^+ only under dry state. In acidic soils (pH <4.5), K fixation is small in magnitude as compared to a limed or neutral soil because H_3O^+ competes for interlayer exchange position. Under acidic soil conditions, hydrated Al and Fe and their polymers occupy the K-selective binding sites. The presence of hydroxyls Al–Fe groups in the interlayer also prevents the collapse of interlayer space or physical entrapment of K^+ ions. Air-drying increases K fixation in soils having vermiculite or other expandable micas such as beidellite or illite with high exchangeable K^+ level, while in soils of low exchangeable K content, air-drying increases exchangeable potassium content due to cracking of clay edges exposing interlayer K^+ for replacement. Ammonium ions (NH_4^+) being similar in change and ionic size to K^+ effectively compete for fixation in expandable 2:1 clay minerals. The presence of any of the two ions blocks the release of the other ion.

16.2.1 Soil Factors Affecting K Availability to Plants

The availability of K in soils is governed by different soil properties. In general, fine-textured soils of high CEC have a high level of exchangeable K as compared to sandy soils. The mineralogy of soil clay fraction also influences K availability in the soils. Soils having vermiculite or montmorillonite clays are generally high in K supply as compared to highly weathered soils (Ultisols) which have higher content of kaolinite, a 1:1 clay mineral. Potassium availability is poor in acidic soils due to K leaching, and the inhibitory effect of Al^{+3} on root growth and liming practice increases the retention and availability of K in acidic soils. In calcareous soils, high contents of exchangeable Ca^{2+} and Mg^{2+} reduce the availability of K to crops due to the antagonistic effect on K uptake by plants. Among soil physical properties, soil compaction and reduced porosity impair K uptake by crop plants due to restricted root growth and poor O_2 supply. Low-temperature and low soil moisture conditions also slow down the root growth, diffusion of K^+ in soil, and the rate of K uptake by plants. Thus, K deficiency is expected in acidic sandy soils, leached sandy and lateritic soils, organic and eroded soils, and also soils rich in K-fixing minerals.

Among the processes of K mobilization from soil to the growing plant roots, diffusion plays a major role in supplying K^+, while mass flow assumes importance only when soils have high concentrations of K^+ in soil solution either naturally or due to liberal K application. Among crops, cereals having a fibrous root system with intense lateral branching absorb K^+ at a much faster rate as compared to dicots having only tap root system. Kraffezyk et al. (1984) observed that low K supply increased root exudation by maize plants with a higher proportion of citric and oxaloacetic acids. Cao et al. (1991) studied the distribution of exchangeable and slowly available (fixed) K in the rhizosphere of wheat and noted that exchangeable K was depleted in the vicinity of roots. A slight depletion of slowly available K^+ (fixed K) was also recorded in the close proximity of the root. The release of protons from plant roots could affect the release of non-exchangeable K and increase the diffusion coefficient of K in the rhizosphere (Kuchenbuch 1987). The release of non-exchangeable K from feldspar has also been attributed to exudation of acids, especially citric and oxalic acids, from roots (Song and Huang 1988; Wang et al. 2000; Moritsuka et al. 2004). Hinsinger et al. (1992, 1993) reported that phlogopite, a trioctahedral mica, was disintegrated in the rhizosphere of ryegrass and rape due to proton release from plant roots and vermiculitized after plant root induced the release of interlayer. K. Hinsinger and Jaillard (1993) also observed that ryegrass (*Lolium multiflorum*) could cause vermiculitization (release of non-exchangeable K) of phlogopite, a trioctahedral mica, by decreasing K concentration in the rhizosphere below a threshold of ~ 80 mol K dm^{-3}. Rhizosphere microflora including endomycorrhiza and nonsymbiotic microflora also promote the mobilization and

Table 16.1 Diversity of potassium-solubilizing microbes

KSMs	K source used	Reference
B. mucilaginosus	Mica and soil	Basak and Biswas 2010
B. edaphicus	K-deficient soil	Sheng 2005
B. mucilaginosus	Soil	Han and Lee 2005
B. megaterium	Muscovite and biotite	Zarjani et al. 2013
Arthrobacter spp.	Muscovite or biotite micas	Zarjani et al. 2013
B. circulanscan	Silicate mineral	Lian et al. 2002
A. terreus and A. niger	Feldspar	Prajapati et al. 2012
Penicillium spp.	Muscovite	Crawford et al. 2000
E. hormaechei (KSB-8)	Feldspar	Prajapati et al. 2013
Paenibacillus mucilaginosus	F-feldspar, kaolinite, montmorillonite	Liu et al. 2012; Hu et al. 2006
P. frequentans, Cladosporium	K-bearing minerals	Argelis et al. 1993
Aminobacter, Sphingomonas, Burkholderia	K-bearing minerals	Uroz et al. 2007
Paenibacillus glucanolyticus	K-bearing minerals	Sangeeth et al. 2012
A. fumigates	K-bearing minerals	Lian et al. 2008
M. hominis, Flectobacillus spp.	Kaolinite	Diep and Hieu 2013
B. globisporus	Feldspar, muscovite, and biotite	Sheng et al. 2008

absorption of insoluble K from biotite by maize plants (Berthelin and Leyral 1982). These reports opened up a possibility of using these minerals as a supplementary source of K for crop plants (Coroneos et al. 1996; Hinsinger et al. 1996). Indeed, several studies demonstrated that such clay minerals, even in subsidiary quantities, increase effective soil K availability (Barré et al. 2008).

16.3 Potassium-Solubilizing Microbes (KSMs)

The KSMs can solubilize the insoluble K to soluble form for the ready utilization by growing plants. The KSMs are ubiquitous whose numbers are likely to vary from soil to soil.

16.3.1 Diversity of KSMs in Agricultural Soil

The diverse groups of microorganisms, isolated from soils, minerals, rocks, and rhizosphere of various plants such as *Pseudomonas*, *Burkholderia*, *Bacillus*, *Aspergillus*, and *Penicillium* spp., are known to release K from K-bearing minerals and unavailable pools of soil potassium (Table 16.1). These KSMs dissolve silicon and aluminum along with K from insoluble K-bearing minerals such as micas, illite, feldspar, and orthoclases by various mechanisms (Meena et al. 2013, 2014a, Shanware et al. 2014). Application of these KSMs in conjunction with insoluble K sources like mica to soil has recently gained a great deal of attention. It is still unclear what drives microbial selectivity on mineral surfaces (Meena et al. 2013, 2014b; Maurya et al. 2014). However, the chemical composition of the mineral is likely to be very important, which may facilitate the nutritional requirements of microbes to govern microbial community composition (Hutchens 2009).

16.3.2 Role of Mycorrhiza in Potassium Uptake by Plants

Mycorrhiza is a mutually beneficial biological association between species of the fungal phylum *Glomeromycota* and roots of higher plants. The two common types of fungi involved in such association are arbuscular mycorrhizae (AM) and ectomycorrhizae (ECM). The AM are the most abundant fungi present in agricultural soils. They form a symbiotic association with both the terrestrial and aquatic plants (Willis et al. 2013). ~80 % of terrestrial plants, including most of the agricultural, horticultural, and hardwood crop species, are able to establish such mutualistic association (Giovannetti et al. 2006). These fungi penetrate into the root cortical cells and form haustoria-like structure called arbuscule that serves as a mediator for the exchange of metabolites between fungus and host cytoplasm (Oueslati 2003). The AM fungal hyphae also proliferate into the soil matrix (Bethlenfalvay and Linderman 1992) and help plants to acquire mineral nutrients and water from the soil and also contribute to improving soil structure (Javaid 2009). The AM fungi also help nutrient cycling in the ecosystems (Yaseen et al. 2012). The growth and productivity of several field crops involve root colonization by mycorrhizal fungi (Nunes et al. 2010). Guo et al. (2010) reported that mycorrhizal roots could explore larger soil volume by extending their extra metrical hyphae to facilitate the absorption and translocation of the nutrients as compared to non-mycorrhizal plants. Using a multitracer technique, Suzuki et al. (2001) recorded the uptake and transport of sodium (Na), zinc (Zn), selenium (Se), rubidium (Rb), and strontium (Sr) by AM fungal hyphae to the host plants. Under control conditions, Marschner and Dell (2006) noted that the AM symbiosis could account for up to 60 % of copper (Cu) and 10 % of K requirement of plants.

The primary function of mycorrhizal association especially in the nutrient-depleted soils is nutrient uptake of plant uptake, particularly of phosphorus (Bolan 1991) and other nutrients like K, Ca, Mg, Zn, Cu, Fe, etc. (Clark and Zeto 2000). Ectomycorrhizal-forming fungi (EMF) also play an important role in mineral weathering and the partitioning of the products of weathering in soils under nutrient-deficient conditions. The arbuscular mycorrhizal fungi enhance the solubility of mineral form of K by releasing protons (H^+) or CO_2 and organic acid anions such as citrate, malate, and oxalate (Balogh-Brunstad et al. 2008b). This also leads to the increase in the contents of nitrogen, potassium, calcium, and iron in the plant leaves and fruits (Veresoglou et al. 2011). The inoculation of the two arbuscular mycorrhizal fungi (AMF) species, *G. mosseae* and *G. intraradices*, has been demonstrated to increase the uptake of K by maize crop (Wu et al. 2005). Alves et al. (2010) reported that the mycorrhizal colonization of *Eucalyptus* plants improves the plant height, shoot dry weight, root length, and phosphorus and potassium contents as compared to the control. In AM switch grass grown in acid soil, the acquisition of K was especially enhanced (Clark et al. 1999). In *Eucalyptus dunnii* seedlings, the ectomycorrhizal fungi increase the efficiency of alkaline breccia as a source of P and K to the seedlings (Alves et al. 2010). Ectomycorrhizal fungi actively forage for minerals and act as biosensors that discriminate between different grain sizes and minerals (Leake et al. 2008). Further, ectomycorrhizal fungus, such as *Scleroderma citrinum*, has been reported to provide a niche for the mineral weathering by rhizosphere bacteria (*Burkholderia*) (Koele et al. 2009). In a greenhouse experiment, tomato (*Lycopersicon esculentum*), co-inoculation with *Glomus* spp., *P. putida*, *A. chroococcum*, and *A. lipoferum* enhanced the lycopene content, antioxidant activity, and shoot and fruit K content (Ordookhani et al. 2010).

16.3.3 Plant Growth Promotion by KSMs

The beneficial effects of KSMs applied to the soil to enhance the K uptake of plants have been reported in several crops including cotton, oilseed rape, pepper, cucumber, and Sudan grass (Sheng et al. 2003; Han and Lee 2005; Han et al. 2006). The KSMs have a plant growth-promoting effect (Table 16.2). Plant growth-promoting effect of KSMs implies that the exudates of these microorganisms can effectively enhance the release of K from clay minerals, which in turn increases the K uptake and plant growth. In a number of incubation trials, the application of inoculated feldspars in soil enhanced about 40–60 % of K solubility and K uptake by plants (Basak and Biswas 2009; Abou-el-Seoud and Abdel-Megeed 2012; Meena et al. 2015b; Singh

Table 16.2 Potassium solubilizers having plant growth-promoting effect on agricultural crops

KSMs	Crop species	References
B. edaphicus	Cotton and rape	Sheng 2005
B, mucilaginosus	Eggplant	Han and Lee 2005
B. megaterium, B. mucilaginosus	Pepper and cucumber	Han et al. 2006
B. edaphicus	Wheat	Sheng and He 2006
B. mucilaginosus	Tobacco	Li et al. 2007
P. syringae, E. aerogenes	Maize	Nadeem et al. 2007
B. mucilaginosus, A. chroococcum A-41	Sudan grass	Basak and Biswas 2010
P. putida, A. chroococcum	Tomato	Ordookhani et al. 2010
B. mucilaginosus	Maize and wheat	Singh et al. 2010
B. mucilaginosus	Potato–soybean	Biswas 2011
P. glucanolyticus	Black pepper	Sangeeth et al. 2012
P. putida	Tea	Bagyalakshmi et al. 2012
E. hormaechei, A. terreus	Okra	Prajapati et al. 2013

et al. 2015). *B. mucilaginosus* has been reported to promote the growth of tobacco by dissolving potassium from feldspar (Li et al. 2007). Inoculation with KSMs alone or co-inoculation with other plant growth-promoting microorganisms (PGPMs) like N fixers, P solubilizers, and mycorrhiza in conjunction with or without amendment of the respective K materials increased the availability of K in soil, enhanced K uptake, and promoted growth of crops (Table 16.2). The inoculation with PGPMs also regulates the plant nutrition by enhancing K^+ uptake over Na^+ in plants, even under salt stress conditions (Nadeem et al. 2007; Kumar et al. 2015; Meena et al. 2015a). Therefore, the application of K-solubilizing microbes as biofertilizer can reduce the use of agrochemicals for crop production and support an eco-friendly sustainable agriculture.

The use of KSMs in soils having a suboptimal supply of K is likely to release unavailable K for the utilization by the plants and could stimulate anabolic processes in plants to bring a significant improvement in plant growth. Besides this, the enhancement in plant growth may partly be attributed to the ability of some KSMs to produce growth hormones, especially indole-3-acetic acid (IAA). To be a successful plant growth promoter inoculants, the bacteria must colonize the root system rapidly during the growing season of the plant (Defreitas and Germida 1992). There are several greenhouse evidences to indicate significant improvements in plant growth due to the inoculation of KSMs. Sheng (2005) noted a significant increase in shoot and root dry yield and uptake of K by cotton and rape due to application of illite mineral inoculated with a potassium releasing strain, *B. edaphicus* NBT. The inoculated bacterium established larger population on the root and the rhizosphere of cotton and rape up to 5 weeks after sowing. Lin et al. (2002) and Egamberdiyeva and Hoflich (2003) also reported that the bacterial inoculation resulted in the growth promotion and higher K contents in the plant parts. In a pot culture experiment with groundnut, Sugumaran and Janathanam (2007) reported that the inoculation of *B. mucilaginosus* (MCRCp1), a slime-forming bacteria, increases the available P from 6.24 to 9.28 mg kg^{-1} soil and available K from 88.6 to 99.6 mg kg^{-1} soil and improves root length from 16.3 to 35.0 cm, plant dry matter from 0.49 to 1.10 g pot^{-1}, number of pods from 7.1 to 13.4, and oil content from 26.0 % to 35.4 %. Basak and Biswas (2009) also showed that the inoculation of *B. mucilaginosus* increases the biomass yields, K uptake of Sudan grass (*Sorghum vulgare* Pers. var. sudanensis), and percent K recoveries from mica in two Alfisols. Prajapati et al. (2013) reported that the inoculation of KSB-8 and application of rock K brought a significant increase in shoot and root length and dry weights of okra under greenhouse conditions.

A synergistic effect of the conjoint inoculation with both KSMs and PSMs along with soil application of their insoluble sources on plant growth has been reported by some workers. In a greenhouse trial with eggplant, Han and Lee (2005) noted that the conjunctive inoculation of *B. megaterium* (P solubilizers) and *B. mucilaginosus* (K solubilizers) with the application of their respective rock P or K materials increased the availability of P by ~25 and that of K by 15 % and also enhanced the photosynthetic rate; the uptake of N, P, and K; and the growth of eggplants. A similar effect had also been reported with pepper and cucumber (Han et al. 2006). Zhang and Kong (2014) reported that among different KSMs isolates, XF11 shows the highest beneficial effect on the growth and nutrient uptake of tobacco seedlings. These works indicated that a combination of inoculation of KSB and addition of K-feldspar powder could become a promising practice to substitute the use of costly commercial K fertilizer and may help to achieve economy in the expenditure on crop production.

16.3.4 Immobilization of KSMs for Soil Application

One of the major limitations associated with the applications of microbial consortium into soil is ensuring their survival under both biotic and abiotic stresses. Soil moisture content, salinity, heavy metal toxicity, temperature, pH, texture, oxygen availability, rate of oxygen diffusion, and nutrient availability have been suggested as some important abiotic factors controlling survival of introducing bacteria in soils. Biological factors include predation by protozoans and phages, a lower level of starvation resistance of the introduced bacteria, and lack of suitable soil which limit the extended cell survival. In this context, the immobilized cells are being investigated as an alternative to ensure the survival of desired microbial communities for a variety of environmental applications in agriculture, biocontrol, pesticide application, and pollutant (pesticide/xenobiotics) degradation in contaminated soils (D'Souza 2002).

The immobilization of microbes through encapsulation or entrapment in certain well-defined polymers has been shown to afford protection to cells under adverse conditions. The encapsulation provides not only protection, but a more stable microenvironment for the entrapped microbial cells in the adverse environment (Cassidy et al. 1996; Joshi and D'Souza 1999). Immobilization techniques may also help in devising new future strategies for the introduction of microbes into the soils. Based on this approach, it is now feasible to provide a more congenial microenvironment to the microbial consortium. The immobilized cells can also act as synthetic inoculation carriers for the slow release of plant growth-related organisms into soils. In such case, microenvironment in the immobilized matrix initially protects the cells from the adverse soil microenvironment. The microorganisms are released after necessary adaptation to the prevailing environmental conditions. This may also enable cells to overcome the changing conditions in soil and increase the survival of microbial cells, and their metabolic activities can be maintained for extended periods. The immobilization can provide additional benefits, especially for commercial purposes, in terms of ease of storage and transportation and also in terms of biosafety features that limit contamination and bioaerosol formation. Some immobilization studies have already been conducted with the plant growth-promoting microbes (Ahuja and D'Souza 2009; Shrivastava et al. 2008: Shrivastava 2008); however, to the best of our knowledge, no published report is available on the immobilization of KSMs. Sincere efforts need to be attempted on the immobilization of KSMs cells as an alternate technology, especially for the problem soils.

16.4 Mechanism of Potassium Solubilization

KSMs are capable of producing some low molecular weight organic acids such as citric, oxalic, and tartaric acids, which release protons for the displacement of insoluble K from some insoluble

K minerals like micas, illite, feldspar, and orthoclase. These organic acids can also accelerate the weathering of K minerals through the formation of metal–organic complexes (specially with Fe^{2+}, Al^{3+}, and Ca^{2+}) which may destabilize the lattice structures and release K into soil solution (Song and Huang 1988; Friedrich et al. 1991; Ullaman et al. 1996; Bennett et al. 1998). The carboxylic acids and capsular polysaccharide produced by *B. mucilaginosus* and *B. edaphicus* (Lin et al. 2002; Sheng and Huang 2002) are responsible for the solubilization of feldspar. Malinovskaya et al. (1990) opined that a mixture of polymers and low molecular weight ligands had a synergistic effect on mineral weathering by *B. mucilaginosus*. Bennett et al. (2001) reported that the metabolic by-products, extracellular enzymes, chelates, and both simple and complex organic acids produced by microorganisms helped in the dissolution of feldspar and increased the availability of K to plants. Basak and Biswas (2009) noted that the *B. mucilaginosus* strain effects the dissolution of mica at the edges to release K and the dissolution of mica was demonstrated by a considerable decrease in the integral width (full length at half maximum) of 10 Å peaks of treated mica.

Besides bacteria, a mold *A. niger* which is known to solubilize RP due to the production of organic acid, especially citric acid and gluconic acid (Nahas et al. 1990; Vassileva et al. 1998; Shrivastava and D'Souza 2014) had been verified to solubilize K as well from two rock K sources (alkaline ultramafic and phlogopite) after 21 DAI (Lopes-Assad et al. 2006). Rosa-Magri et al. (2012) reported that a yeast species, *Torulaspora globosa*, released about 38 % of the total potassium from an alkaline ultramafic rock during a 15 DAI. These workers also demonstrated that the agitation during incubation (which increased oxygen availability) resulted in ~20 % more bio-solubilization of the alkaline ultramafic rock as compared to the static culture. Their data indicated that this yeast has good potential for bio-solubilization of K and K biofertilizer production.

Some mycorrhizal fungi also help in mobilizing unavailable K to host plants. Leyval and Berthelin (1989) noted that the ectomycorrhizal fungi in beech solubilized K and other minerals from phlogopite. The process is likely to operate through the removal of nutrients and the addition of acids by mycorrhiza. Some ectomycorrhizal fungi release low molecular weight organic anions from the hyphal tips and form microscopic tunnels within feldspar and hornblende mineral grains and accelerate weathering rates in soil minerals (van Scholl et al. 2008). The other plausible mechanism for the release of unavailable K from soil K minerals is through the formation of biofilm by some bacteria in rhizosphere soil (Balogh-Brunstad et al. 2008a). These biofilms besides offering protections for microbial community against abiotic and biotic stress also facilitate the activity of KSB in the rhizosphere and accelerate the weathering of biotite and anorthite.

Some microbes bring changes in the root architecture and the chemical exudation behavior of the plant roots to affect K solubilization. Calvaruso et al. (2006) showed that the weathering of biotite was doubled by *Burkholderia glathei* PML1(12) with a promotion of plant growth. These workers also noted a significant increase in the number of lateral roots and root hairs and promoted growth of pine. Bakker et al. (2004) opined that the following changes could promote the effect of K solubilization from minerals: (i) the fragmentation of the minerals due to root activity with a consequent increase in their reactive surfaces, (ii) the root exudates provide the substrates for the bacteria to produce metabolites, (iii) and the production of growth hormones by the bacteria and organic acids to stimulate root development and more root exudation, to effect weathering of K mineral and removal of nutrients by plant uptake (Gahoonia et al. 1997).

16.5 Molecular Mechanism of K Solubilization

Microorganisms produce various extracellular polymers (primarily proteins and polysaccharides) that serve as attachment structures to mineral or rock surface. These microbes also synthesize exopolymers to form thick gel layers or biofilms, which create controllable microenvironments around microbial cells for weathering (Christensen and Characklis 1990; Meena et al. 2014b). Biofilm formation on aluminosilicate increases the residence time of water as compared to the residence time at the bare rock or mineral surface and enhances the mineral weathering. Additionally, polysaccharide coatings on mineral grains also help to maintain diffusion pathways as water potential decreases in soils (Chenu and Roberson 1996). KSMs have the capability of degrading potassium-containing silicate minerals, but the molecular mechanisms of microbial mineral weathering are not yet clear. KSB secreted certain proteins, related to the weathering of potassium minerals, which are induced under bacterial nutritional deficiency, particularly, by K and also the presence of minerals and of K-bearing rocks. This suggests a direct evidence of the metabolic changes of extracellular enzymes in bacteria during the process of weathering of potassium minerals. It was speculated that these secreted proteins, together with extracellular polymers like polysaccharides, may accelerate the weathering of potassium minerals, resulting in the release of K^+ needed for the bacterial growth and plant nutrition as well (Xiao et al. 2012). The production of organic acids is another mechanism for solubilization of mineral phosphate and potassium by microorganism. Gluconic acid is the principal organic acid produced by most of the potassium and phosphate solubilizers like *Pseudomonas* spp., *Bacillus* spp., and *Arthrobacter* spp.; mainly, two genes are involved in gluconic acid production, viz., pqq and gab Y (Shrivastava et al. 2010; Meena et al. 2014a).

16.6 Concluding Remarks and Future Prospects

The efficacy of most of the KSMs has been tested either under laboratory or greenhouse conditions. Currently, very meager information is available on the field application of KSMs. In order to evaluate the potential of the KSMs for agricultural production systems, there is a need to conduct well-planned field studies to evaluate their effect on both soil properties and crop growth under different agroclimatic conditions. Further, efforts to immobilize these microbes on suitable carrier(s) could enhance their utility under various abiotic and biotic stress conditions. The molecular mechanism of KSMs has not been very well understood and a question still remains, why microbes solubilize the insoluble K sources? Studies on the extracellular polymer structure released by KSMs and its distribution are also very important and need to be explored. Microenvironments adjacent to microbial colonies need to be examined to evaluate the microbial impact on mineral dissolution rates to elucidate the mechanism of rock and mineral weathering by KSM and the consequent release of potassium to the soil solution.

References

Abou-el-Seoud, Abdel-Megeed A (2012) Impact of rock materials and biofertilizations on P and K availability for maize (*Zea mays*) under calcareous soil conditions. Saudi J Biol Sci 19:55–63

Ahuja A, D'Souza SF (2009) Bioprocess for solubilization of rock phosphate on starch based medium by *Paecilomyces marquandii* immobilized on polyurethane foam. Appl Biochem Biotechnol 152:1–5

Alves L, Oliveira VL, Filho GNS (2010) Utilization of rocks and ectomycorrhizal fungi to promote growth of eucalypt. Braz J Microbiol 41:676–684

Argelis DT, Gonzala DA, Vizcaino C, Gartia MT (1993) Biochemical mechanism of stone alteration carried out by filamentous fungi living in monuments. Biogeochemistry 19:129–147

Bagyalakshmi B, Ponmurugan P, Marimuthu S (2012) Influence of potassium solubilizing bacteria on crop productivity and quality of tea (*Camellia sinensis*). Afr J Agric Res 7:4250–4259

Bakker MR, George E, Turpault MP, Zhang J, Zeller B (2004) Impact of Douglas-fir and Scots pine seedlings on plagioclase weathering under acidic conditions. Plant Soil 266:247–259

Balogh-Brunstad Z, Keller CK, Dickinson JT, Stevens F, Li CY, Bormann BT (2008a) Biotite weathering and nutrient uptake by ectomycorrhizal fungus *Suillus tomentosus*, in liquid culture experiments. Geochim Cosmochim Acta 72:2601–2618

Balogh-Brunstad Z, Keller CK, Gill RA, Bormann BT, Li CY (2008b) The effect of bacteria and fungi on chemical weathering and chemical denudation fluxes in pine growth experiments. Biogeochemistry 88: 153–167

Barré P, Montagnier C, Chenu C, Abbadie L, Velde B (2008) Clay minerals as a soil potassium reservoir: observation and quantification through X-ray diffraction. Plant Soil 302:213–220

Basak BB, Biswas DR (2009) Influence of potassium solubilizing microorganism (*Bacillus mucilaginous*) and waste mica on potassium uptake dynamics by sudan grass (*Sorghum vulgare* Pers) grown under two Alfisols. Plant Soil 317:235–255

Basak BB, Biswas DR (2010) Co-inoculation of potassium solubilizing and nitrogen fixing bacteria on solubilization of waste mica and their effect on growth promotion and nutrient acquisition by a forage crop. Biol Fertil Soils 46:641–648

Bennett PC, Choi WJ, Rogera JR (1998) Microbial destruction of feldspars. Mineral Manag 8:149–150

Bennett PC, Rogers JR, Choi WJ (2001) Silicates, silicate weathering, and microbial ecology. Geomicrobiol J 18:3–19

Berthelin J, Leyval C (1982) Ability of symbiotic and non-symbiotic rhizospheric microflora of maize (*Zea mays*) to weather micas and to promote plant growth and plant nutrition. Plant Soil 68:369–377

Bethlenfalvay GJ, Linderman RG (1992) Mycorrhizae in sustainable agriculture. ASA special publication no 54. Madison 8–13

Biswas DR (2011) Nutrient recycling potential of rock phosphate and waste mica enriched compost on crop productivity and changes in soil fertility under potato–soybean cropping sequence in an inceptisol of Indo-gangetic plains of India. Nutr Cycl Agroecosyst 89: 15–30

Bolan NS (1991) A critical review on the role of mycorrhizal fungi in the uptake of phosphorus by plants. Plant Soil 134:189–207

Calvaruso C, Turpault MP, Frey-Klett P (2006) Root-associated bacteria contribute to mineral weathering and to mineral nutrition in trees: a budgeting analysis. Appl Environ Microbiol 72:1258–1266

Cao YP, Xu YT, Li XL (1991) The distribution of potassium is rhizosphere of wheat. Acta Agric Univ Pekin 17:69–74

Cassidy MB, Lee H, Trevors JT (1996) Environmental applications of immobilized microbial cells: a review. J Ind Microbiol 16:79–101

Chen SZ, Low PF, Roth CB (1987) Relation between potassium fixation and the oxidization state of octahedral iron. Soil Sci Soc Am J 51:82–86

Chenu C, Roberson EB (1996) Diffusion of glucose in microbial extracellular polysaccharide as affected by water potential. Soil Biol Biochem 28:877–884

Christensen BE, Characklis WG (1990) Physical and chemical properties in biofilms. In: Characklis WG, Marshall KC (eds) Biofilms. Wiley, New York, pp 93–130

Clark RB, Zeto SK (2000) Mineral acquisition by arbuscular mycorrhizal plants. J Plant Nutr 23: 867–902

Clark RB, Zobel RW, Zeto SK (1999) Effects of mycorrhizal fungus isolate on mineral acquisition by *Panicum virgatum* in acidic soil. Mycorrhiza 9:167–176

Coroneos C, Hinsinger P, Gilkes RJ (1996) Granite powder as a source of potassium for plants: a glasshouse bioassay comparing two pasture species. Fertil Res 45:143–152

Crawford RH, Floyd M, Li CY (2000) Degradation of serpentine and muscovite rock minerals and immobilization of cations by soil Penicillium spp. Phyton (Horn, Austria) 40:315–322

D'Souza SF (2002) Trends in immobilized enzyme and cell technology. Indian J Biotechnol 1:321–338

Defreitas JR, Germida JJ (1992) Growth promotion of winter wheat by fluorescent Pseudomonas under field conditions. Soil Biol Biochem 24:1137–1146

Diep CN, Hieu TN (2013) Phosphate and potassium solubilizing bacteria from weathered materials of denatured rock mountain, Ha Tien, Kiên Giang province, Vietnam. Am J Life Sci 3:88–92

Egamberdiyeva D, Hoflich C (2003) Influence of growth promoting bacteria on the growth of wheat in different soils and temperatures. Soil Biol Biochem 35:973–978

Friedrich S, Platonova NP, Karavaiko GI, Stichel E, Glombitza F (1991) Chemical and microbiological solubilization of silicates. Acta Biotechnol 11: 187–196

Gahoonia TS, Care D, Nielsen NE (1997) Root hairs and phosphorus acquisition of wheat and barley cultivars. Plant Soil 191:181–188

Giovannetti M, Avio L, Fortuna P, Pellegrino E, Sbrana C, Strani P (2006) At the root of the wood wide web. Plant Signal Behav 1:1–5

Guo Y, Ni Y, Huang J (2010) Effects of rhizobium, arbuscular mycorrhiza and lime on nodulation, growth and nutrient uptake of lucerne in acid purplish soil in China. Trop Grasslands 44:109–114

Han HS, Lee KD (2005) Phosphate and potassium solubilizing bacteria effect on mineral uptake, soil availability and growth of eggplant. Res J Agric Biol Sci 1:176–180

Han HS, Supanjani, Lee KD (2006) Effect of co-inoculation with phosphate and potassium solubilizing bacteria on mineral uptake and growth of pepper and cucumber. Plant Soil Environ 52:130–136

Hinsinger P, Jaillard B (1993) Root-induced release of interlayer potassium and vermiculation of phlogopite as related to potassium depletion in the rhizosphere of ryegrass. J Soil Sci 44:525–534

Hinsinger P, Jaillard B, Dufey JE (1992) Rapid weathering of a trioctahedral mica by the roots of ryegrass. Soil Sci Soc Am J 65:977–982

Hinsinger P, Elsass F, Jaillard B, Robert M (1993) Root induced irreversible transformation of a trioctahedral mica in the rhizosphere of rape. J Soil Sci 44:535–545

Hinsinger P, Bolland MDA, Gilkes RJ (1996) Silicate rock powder: effect on selected chemical properties of a range of soils from Western Australia and on plant growth as assessed in a glasshouse experiment. Fertil Res 45:69–79

Hu X, Chen J, Guo J (2006) Two phosphate- and potassium-solubilizing bacteria isolated from Tianmu Mountain, Zhejiang, China. World J Microbiol Biotechnol 22:983–990

Hutchens E (2009) Microbial selectivity on mineral surfaces: possible implications for weathering processes. Fungal Biol Rev 23:115–121

Jackson BLJ, During C (1979) Studies of slowly available potassium in soils of New Zealand. I. Effect of leaching, temperature and potassium depletion of the equilibrium concentration of potassium in solution. Plant Soil 51:197–204

Javaid A (2009) Arbuscular mycorrhizal mediated nutrition in plants. J Plant Nutr 32:1595–1618

Joshi NT, D'Souza SF (1999) Immobilization of activated sludge for the degradation of phenol. J Environ Sci Health-A 34:1689–1700

Koele N, Turpault MP, Hildebrand EE, Uroz S, Frey-Klett P (2009) Interactions between mycorrhizal fungi and mycorrhizosphere bacteria during mineral weathering: budget analysis and bacterial quantification. Soil Biol Biochem 41:1935–1942

Kraffezyk I, Trolldenies G, Beringer H (1984) Soluble root exudates of maize influence of potassium supply and rhizosphere microorganisms. Soil Biol Biochem 16:315–322

Kuchenbuch RO (1987) Potassium dynamics in the rhizosphere and potassium availability. In: Methodology in soil-K research. Proc Coll Int Potash Inst, International Potash Institute, Horgen, Switzerland, p 215–234

Kumar A, Bahadur I, Maurya BR, Raghuwanshi R, Meena VS, Singh DK, Dixit J (2015) Does a plant growth-promoting rhizobacteria enhance agricultural sustainability? J Pure Appl Microbiol 9(1):715–724

Leake JR, Duran AL, Hardy KE, Johnson I, Beerling DJ, Banwart SA, Smits MM (2008) Biological weathering in soil: the role of symbiotic root-associated fungi biosensing minerals and directing photosynthate-energy into grain-scale mineral weathering. Mineral Mag 72:85–89

Leyval C, Berthelin J (1989) Interactions between *Laccaria laccata*, *Agrobacterium radiobacter* and beech roots: influence on P, K, Mg and Fe mobilization from minerals and plant growth. Plant Soil 117: 103–110

Li X, Wu Z, Li W, Yan R, Li L, Li J (2007) Growth promoting effect of a transgenic *Bacillus mucilaginosus* on tobacco planting. Appl Microbiol Biotechnol 74:1120–1125

Lian B, Fu PQ, Mo DM, Liu CQ (2002) A comprehensive review of the mechanism of potassium release by silicate bacteria. Acta Mineral Sin 22:179–183

Lian B, Wang B, Pan M, Liu C, Teng HH (2008) Microbial release of potassium from K-bearing minerals by thermophilic fungus *Aspergillus fumigatus*. Geochim Cosmochim Acta 72:87–98

Lin QM, Rao ZH, Sun YX, Yao J, Xing LJ (2002) Identification and practical application of silicate-dissolving bacteria. Agric Sci China 1:81–85

Liu D, Lian B, Dong H (2012) Isolation of *Paenibacillus* sp. and assessment of its potential for enhancing mineral weathering. Geomicrobiol J 29:413–421

Lopes-Assad ML, Rosa MM, Erler G, Ceccato-Antonini SR (2006) Solubilização de pó-de-rocha por *Aspergillus niger*. Espaço e Geogr 9:1–16

Malinovskaya IM, Kosenko LV, Votselko SK, Podgorskii VS (1990) Role of *Bacillus mucilaginosus* polysaccharide in degradation of silicate minerals. Mikrobiologiya 59:49–55

Marschner H (1995) Mineral nutrition of higher plants, 2nd edn. Academic, London

Marschner H, Dell B (2006) Nutrient uptake in mycorrhizal symbiosis. Plant Soil 159:89–102

Maurya BR, Meena VS, Meena OP (2014) Influence of Inceptisol and Alfisol's potassium solubilizing bacteria (KSB) isolates on release of K from waste mica. Vegetos 27(1):181–187

Meena OP, Maurya BR, Meena VS (2013) Influence of K-solubilizing bacteria on release of potassium from waste mica. Agric Sustain Dev 1(1):53–56

Meena VS, Maurya BR, Bahadur I (2014a) Potassium solubilization by bacterial strain in waste mica. Bangladesh J Bot 43(2):235–237

Meena VS, Maurya BR, Verma JP (2014b) Does a rhizospheric microorganism enhance K+ availability in agricultural soils? Microbiol Res 169:337–347

Meena RK, Singh RK, Singh NP, Meena SK, Meena VS (2015a) Isolation of low temperature surviving plant growth-promoting rhizobacteria (PGPR) from pea (*Pisum sativum* L.) and documentation of their plant growth promoting traits. Biocatal Agric Biotechnol. doi:10.1016/j.bcab.2015.08.006

Meena VS, Maurya BR, Verma JP, Aeron A, Kumar A, Kim K, Bajpai VK (2015b) Potassium solubilizing rhizobacteria (KSR): isolation, identification, and K-release dynamics from waste mica. Ecol Eng 81: 340–347

Mengel K, Kirkby EA (1987) Principles of plant nutrition, 4th edn. International Potash Institute, Bern

Moritsuka N, Yanai J, Kosaki T (2004) Possible processes releasing nonexchangeable potassium from the rhizosphere of maize. Plant Soil 258:261–268

Nadeem SM, Zahir ZA, Naveed M, Arshad M (2007) Preliminary investigations on inducing salt tolerance in maize through inoculation with rhizobacteria containing ACC deaminase activity. Can J Microbiol 53:1141–1149

Nahas E, Banzatto DA, Assis LC (1990) Fluorapatite solubilization by *Aspergillus niger* in vinasse medium. Soil Biol Biochem 22:1097–1101

Nunes JLD, DeSouza PVD, Marodin GAB, Fachinello JC (2010) Effect of arbuscular mycorrhizal fungi and indole butyric acid interaction on vegetative growth of 'Aldrighi' peach rootstock seedlings. Cienc Agrotecnol 34:80–86

Ordookhani K, Khavazi K, Moezzi A, Rejali F (2010) Influence of PGPR and AMF on antioxidant activity, lycopene and potassium contents in tomato. Afr J Agric Res 5:1108–1116

Oueslati O (2003) Allelopathy in two durum wheat (*Triticum durum* L.) varieties. Agric Ecosyst Environ 96:161–163

Prajapati K, Sharma MC, Modi HA (2012) Isolation of two potassium solubilizing fungi from ceramic industry soils. Life Sci Leafl 5:71–75

Prajapati K, Sharma MC, Modi HA (2013) Growth promoting effect of potassium solubilizing microorganisms on okra (*Abelmoschus esculentus*). Int J Agric Sci Res 3:181–188

Rosa-Magri MM, Avansini SH, Lopes-Assad ML, Tauk-Tornisielo SM, Ceccato-Antonini SR (2012) Release of potassium from rock powder by the yeast *Torulaspora globosa*. Braz Arch Biol Technol 55: (http://dx.doi.org/10.1590/S1516-89132012000400013)

Sangeeth KP, Bhai RS, Srinivasan V (2012) *Paenibacillus glucanolyticus*, a promising potassium solubilizing bacterium isolated from black pepper (*Piper nigrum L.*) rhizosphere. J Spic Aromat Crops 21:118–124

Shanware AS, Kalkar SA, Trivedi MM (2014) Potassium solublisers: occurrence, mechanism and their role as competent biofertilizers. Int J Curr Microbiol App Sci 3:622–629

Sheng XF (2005) Growth promotion and increased potassium uptake of cotton and rape by a potassium releasing strain of *Bacillus edaphicus*. Soil Biol Biochem 37:1918–1922

Sheng XF, He LY (2006) Solubilization of potassium bearing minerals by a wild type strain of *Bacillus edaphicus* and its mutants and increased potassium uptake by wheat. Can J Microbiol 52:66–72

Sheng XF, Huang WY (2002) Study on the conditions of potassium release by strain NBT of silicate bacteria. Sci Agric Sin 35:673–677

Sheng XF, Xia JJ, Chen J (2003) Mutagenesis of the *Bacillus edaphicus* strain NBT and its effect on growth of chilli and cotton. Agric Sci China 2: 400–412

Sheng XF, Zhao F, He LY, Qiu G, Chen L (2008) Isolation and characterization of silicate mineral-solubilizing *Bacillus globisporus* Q12 from the surfaces of weathered feldspar. Can J Microbiol 54: 1064–1080

Shrivastava M (2008) Biochemical and agronomic evaluation of the ability of phosphate solubilizing microorganisms for enhancement of bioavailability of phosphorus and plant growth. Ph.D. thesis, University of Mumbai

Shrivastava M, D'Souza SF (2014) Bio-solubilization of rock phosphate and plant growth promotion by *Aspergillus nige*r TMPS1 in ultisol and vertisol. In: LK Heng, K Sakadevan, G Dercon, ML Nguyen (eds) Proceeding-international symposium on managing soils for food security and climate change adaptation and mitigation. Food and Agriculture Organization of the United Nations, Rome 73–77

Shrivastava S, D'Souza SF, Desai PD (2008) Production of indole-3-acetic acid by immobilized actinomycete (*Kitasatospora* spp.) for soil applications. Curr Sci 94: 1595–1604

Shrivastava M, Rajpurohit YS, Misra HS, D'Souza SF (2010) Survival of phosphate-solubilizing bacteria against DNA damaging agents. Can J Microbiol 56: 822–830

Singh B, Singh Y, Imas P, Xie JC (2004) Potassium nutrition of the rice-wheat cropping system. Adv Agron 81:203–259

Singh G, Biswas DR, Marwah TS (2010) Mobilization of potassium from waste mica by plant growth promoting rhizobacteria and its assimilation by maize (*Zea mays*) and wheat (*Triticum aestivum* L.). J Plant Nutr 33: 1236–1251

Singh NP, Singh RK, Meena VS, Meena RK (2015) Can we use maize (*Zea mays*) rhizobacteria as plant growth promoter? Vegetos 28(1):86–99

Song SK, Huang PM (1988) Dynamics of potassium release from potassium-bearing minerals as influenced by oxalic and citric acids. Soil Sci Soc Am J 52: 383–390

Sugumaran P, Janathanam B (2007) Solubilization of potassium containing minerals by bacteria and their effect on plant growth. World J Agric Sci 3:350–355

Suzuki H, Kumagai H, Oohashi K, Sakamoto K, Inubushi K, Enomoyo S (2001) Transport of trace elements through the hyphae of an arbuscular mycorrhizal fungus into marigold determined by the multitracer technique. Soil Sci Plant Nutr 47:131–137

Ullman WJ, Kirchman DL, Welch SA (1996) Laboratory evidence by microbially mediated silicate mineral dissolution in nature. Chem Geol 132:11–17

Uroz S, Calvaruso C, Turpault MP, Pierrat JC, Mustin C, Frey Klett P (2007) Effect of the mycorrhizosphere on the genotypic and metabolic diversity of the bacterial communities involved in mineral weathering in a forest soil. Appl Environ Microbiol 73:3019–3027

van Schöll L, Kuyper TW, Smits MM, Landeweert R, Hoffland E, van Breemen N (2008) Rock-eating mycorrhizas: their role in plant nutrition and biogeochemical cycles. Plant Soil 303:35–47

Vassileva M, Vassilev N, Azcon R (1998) Rock phosphate on olive cake-based medium and its further application in a soil-plant system. World J Microb Biotechnol 14:281–284

Veresoglou SD, Mamolos AP, Thornton B, Voulgari OK, Sen R, Veresoglou S (2011) Medium-term fertilization of grassland plant communities masks plant species-linked effects on soil microbial community structure. Plant Soil 344:187–196

Wang JG, Zhang FS, Cao YP, Zhang XL (2000) Effect of plant types on release of mineral potassium from gneiss. Nutr Cycl Agroecosyst 56:37–44

Willis A, Rodriguesb BF, Harrisa PJC (2013) The ecology of arbuscular mycorrhizal fungi. Crit Rev Plant Sci 32: 1–20

Wu SC, Cao ZH, Li ZG, Cheung KC, Wong MH (2005) Effects of biofertilizer containing N-fixer, P and K solubilizers and AM fungi on maize growth: a greenhouse trial. Geoderma 125:155–166

Xiao B, Lian B, Shao W (2012) Do Bacterial secreted proteins play a role in the weathering of potassium-bearing rock powder? Geomicrobiol J 29:497–505

Yaseen T, Burni T, Hussain F (2012) Effect of arbuscular mycorrhizal inoculation on nutrient uptake, growth and productivity of chickpea (*Cicer arietinum*) varieties. Int J Agron Plant Prod 3:334–345

Zarjani JK, Aliasgharzad N, Oustan S, Emadi M, Ahmadi A (2013) Isolation and characterization of potassium solubilizing bacteria in some Iranian soils. Arch Agro Soil Sci 59:1713–1723

Zhang C, Kong F (2014) Isolation and identification of potassium-solubilizing bacteria from tobacco rhizospheric soil and their effect on tobacco plants. Appl Soil Ecol 82:18–25

Potassium and Its Role in Sustainable Agriculture

17

Jyoti Rawat, Pankaj Sanwal, and Jyoti Saxena

Abstract

Apart from two major components nitrogen and phosphorous, potassium is the third essential macronutrient required for the growth and metabolism of plant, and its deficiency in plants causes poorly developed roots, slow growth, low resistance to disease, delayed maturity, small seed production and lower yields. The concentration of soluble K in soil is very small as maximum part of K exists in insoluble form. Silt, clay and sand are important components of soil in earth and biggest reservoir of potassium. Most common deposits of potassium are feldspar and mica. The available K level in soil dropped in the last decade due to rapid development of agriculture and application of imbalanced fertilizers. Potassium is released when these minerals are slowly weathered, or, alternatively, it can be solubilized by some beneficial microorganisms and made available for plants. Several bacterial and fungal strains have been identified for their ability of high potassium solubilization. Various species of *Pisolithus*, *Cenococcum*, *Piloderma*, *Bacillus*, *Paenibacillus*, *Acidithiobacillus*, *Pseudomonas*, *Burkholderia*, *Aspergillus* and *Clostridium* have been reported to release large amount of potassium from different minerals and enhance the productivity of many crops. Co-inoculation of PSMs and KSMs in conjunction with direct application of rock P and K minerals into the soil has been reported to increase N, P and K uptake, photosynthesis and the yield of plants grown in P- and K-limited soils. Thus, identification of microbial strains capable of solubilizing potassium minerals can rapidly conserve our existing resources and escape environmental pollution hazards caused by heavy application of chemical fertilizers.

J. Rawat (✉) • P. Sanwal • J. Saxena
Biochemical Engineering Department, B. T. Kumaon Institute of Technology, Dwarahat 263653, Uttarakhand, India
e-mail: jyotisaxena2000@yahoo.co.in

V.S. Meena et al. (eds.), *Potassium Solubilizing Microorganisms for Sustainable Agriculture*,
DOI 10.1007/978-81-322-2776-2_17

Keywords
Potassium • Sustainable agriculture • Biotic and abiotic stress • KSMs

17.1 Introduction

Potassium (K) is a soft, silver-white metal, light in its pure form that reacts very violently with water. It is commonly called potash (K_2O), a term that has been derived from an early production technique where K was leached from wood ashes and concentrated by evaporating the leachate in large iron pots (Mikkelsen and Bruulsema 2005). Soils contain varying quantities of K-bearing minerals that constitute a major K reserve. The K reservoir in the earth crust is associated with primary alumina silicates that are the most abundant K-bearing minerals such as K feldspar, mica, biotite, muscovite and nepheline. The secondary alumina silicates, however, comprise hydrous mica (illite) as well as a continuum of micaceous weathered or inherited products, viz. mixed-layer phyllosilicates (Bertsch and Thomas 1985). McAfee (2008) described feldspar and mica ranging from 90 % to 98 % as the most common soil components of potassium. The potassium content of Indian soils varies from less than 0.5 % to 3.00 % (Mengel and Kirkby 1987). Some of the mineral sources available in India containing 3–14 % K_2O are glauconitic sand, feldspar, muscovite and nepheline syenite (Indian Minerals Yearbook 2011). Out of various feldspars, potassium feldspar is the most common one and contains up to 13 % K_2O (Rao et al. 1998). Micas are important for plant nutrition because they represent a major source of K, whereas the K in biotite acts as a good fertilizer for plants (Arnold 1963).

17.1.1 Forms of Potassium

Potassium is available in four forms in the soil which are K^+ ions in the soil solution, an exchangeable cation tightly held on the surfaces of clay minerals and organic matter, potassium fixed by weathered micaceous minerals and potassium present in the lattice of certain K-containing primary minerals. All forms of potassium are linked with each other. The amounts of exchangeable and readily available potassium do not provide the rate at which the non-exchangeable or fixed potassium can move into the exchangeable form. These are the reasons why some soils may have a relatively low level when tested and yet supply enough potassium for relatively high crop yields. It has been hypothesized that the lack of crop response on these soils may be attributed to K release from non-exchangeable soil K, particularly from K feldspars (Rehm and Sorensen 1985).

17.1.1.1 Soluble Potassium

Soluble potassium is the most available form but its contribution to total K is very small. According to a report of During (1984), it ranged from 3 to 30 μg/ml in most soils of New Zealand. This type of potassium does not form any chelates, complexes or ion pairs in the soil. Plants take up most of their potassium directly in this form and so deplete it very rapidly in soil.

17.1.1.2 Exchangeable Potassium

Exchangeable K is defined as the fraction that occupies sites in the soil colloidal complex (Malavolta 1985). It is a major bioavailable form of K in the soil, usually 0.1–2 % of total potassium, i.e. between 10 and 400 ppm (Schroeder 1974). The amount of K^+ held by clay minerals at exchange sites depended on kinetic as well as thermodynamic factors (Parfitt 1992).

Release of non-exchangeable K to the exchangeable form occurred when levels of exchangeable and soluble K were decreased due to crop uptake or leaching and perhaps by increase in microbial activity (Sparks et al. 1980; Sparks 2000). Hence, the amount of the exchangeable K concentration determines the effectiveness of resupply. The exchangeable K is more related to the type of clay and its net negative charge. It was found that the exchangeable K levels of allophanic soils were relatively small, whereas soils with large amounts of vermiculite or mica had greater amounts of exchangeable K (Parfitt 1992). Potassium in this form is mostly contained by minerals as feldspar and mica. For optimal nutrition of a crop, the replenishment of a K-depleted soil solution was affected predominantly by the release of exchangeable K from clay minerals (Sheng and Huang 2002).

17.1.1.3 Fixed Potassium

Fixed K is held between the layers of micaceous clay minerals and not readily accessible for exchange with other solution cations. The K^+ present in the wedge, edge, step and crack positions can be referred to as fixed K. Potassium in this form is temporarily trapped between the expanding layers of some clay minerals and most likely between structural layers in the soil of micas, intergraded hydrous micas (e.g. illites) and vermiculites or in the wedge zones during edge weathering of the micas (Kirkman et al. 1994). A fixed form of potassium in soils which is high in feldspars and volcanic glass where the K is structurally bonded is slowly available to plants over time and cannot be replaced by ordinary cation exchange process. This would contribute little to plant growth, because of the low levels or absence of micaceous 2:1 clay minerals (Fieldes and Swindale 1954).

17.1.1.4 Structural Potassium

Potassium within a soil also exists as structural potassium and is variously known as mineral K, unweathered K, native K, matrix K or inert K. It constitutes the largest amount of the total K in most soils (Metson 1980). It is mostly bound covalently within the crystal structure of the K-bearing minerals such as micas (biotite and muscovite), feldspars (orthoclase and microcline) and volcanic glasses (acidic and basic) (Metson 1968), and it only becomes available upon long-term weathering. Biotite and basic volcanic glasses were found to weather easily, whereas feldspars and acidic volcanic glass weathered slowly (Fieldes and Swindale 1954). Generally, highly weathered soils from humid and tropical areas have much less structural potassium remaining than newer soils or soils from cold arid climates.

17.2 Potassium as Fertilizer

Glauconite sand contains around 4–8 % K_2O and has been used as a source of potash fertilizer worldwide (Majumder et al. 1995; Yadav et al. 2000). The need for potash in fertilizer can be determined by plant analysis and soil testing.

17.2.1 Potassium Chloride

It is a common source of fertilizer, and for the most part, it is mined as sylvite ore (KCl, the potassium analogue of halite, or rock salt, NaCl), mixed with NaCl, beneficiated to remove some of the contaminants and sold. It is highly soluble; hence, excessive rate can cause salt damage to plants. Further, KCl may also be the result of crystallization from brine, either from solution mining of KCl ore or precipitation from hypersaline waters, e.g. Searles Lake and Dead Sea, often with carnallite ($KClMgCl_2$) as an intermediate precipitate. The final form of KCl from crystallization processes is white crystals. Typical fertilizer analysis of KCl consists of 60–63 % K_2O and 46 % Cl.

17.2.2 Potassium Sulfate

Potassium sulfate has ~5 % market share and basically constitutes 48–53 % K_2O and 17–18 % S. It is found rarely in pure form. This fertilizer is used particularly for horticultural crops in which chloride uptake is a problem, as in tobacco plant.

17.2.3 Potassium Magnesium Sulfate

This is the primary K fertilizer produced by German and French mines from langbeinite ore and historically was called 'double manure salts'. The analysis comprises 20–22 % K_2O with secondary nutrients like S (21–22 %) and Mg (10–11 %). Potassium magnesium sulfate is a good source of K when there is also a need for magnesium and sulfate in plant nutrition, though excess K is expected to induce magnesium deficiency.

17.2.4 Potassium Nitrate

In this fertilizer the formulated forms of potassium are KCl and nitric acid with 44 % K_2O and 13 % N. It is expensive for agronomic use, mostly applied for horticulture and greenhouses. It was also known as saltpetre to the ancients and recognized by Glauber in the seventeenth century as the 'principle of vegetation'. Besides above-cited fertilizers, potassium phosphate (KH_2PO_4) and potassium thiosulfate ($K_2S_2O_3$) are also available. Kelp meal, plant residues and wood ash, which contain K mainly as a carbonate (K_2CO_3), hydroxide (KOH) and rock powder, e.g. granite and minerals such as alunite, orthoclase, microcline, etc., are some other sources of potassium sold as K fertilizer.

17.3 Functions of Potassium in Plants

Potassium is a major nutrient element found in the soil which is required by plants in greater amounts. Being essential or vital nutrient for plant growth, potassium (K^+) plays an important role in plant regulatory development including osmoregulation, plant-water relation and internal cation/anion balance. It also has substantial effect on enzyme activation involved in the formation of organic substances, protein and starch synthesis, respiratory and photosynthetic metabolism (Lauchli and Pfluger 1979; Wyn Jones et al. 1979; Marschner 2010), stomatal movement and water relations (turgor regulation and osmotic adjustment) by increasing protein production in plants (Marschner 1995). Enzyme activation is also needed to metabolize carbohydrates for the manufacture of amino acids and proteins and tolerance of external stress such as frost, drought, heat and light. Potassium also functions in triggering the growth of young tissues and cell enlargement and, hence, improves early growth. Besides, potassium is important during plant ontogeny and in improving plant quality and oil content in plants. Hence, large amount of potassium is required to maintain plant health but it often receives less attention than N and P in many crop production systems. The crop yield can be positively and negatively influenced by favourable and unfavourable environmental conditions. Unfavourable environmental conditions which would create potentially damaging physiological changes within plants are known as stresses (Shao et al. 2008). Potassium has also been involved in various physiological functions related to plant health and resistance to biotic and abiotic stresses such as diseases, pests, drought, salinity, cold, frost and waterlogging (Wang et al. 2013) However, suitable application of potassium can improve insect and disease resistance in plants. The protective role of K^+ in plants suffering from drought stress has been well documented (Pier and Berkowitz 1987; Sen Gupta et al. 1989). There are major challenges for agriculture to enhance crop yields and to stabilize plant development and yield formation under biotic and abiotic stress conditions (Reynolds et al. 2011).

17.3.1 Biotic Stress Resistance

Biotic components such as weeds caused the highest potential loss (~32 %), followed by animal pests (18 %), fungi, bacteria (15 %) and viruses (3 %) (Oerke and Dehne 2004). Recently, it was seen in K-deficient soil that plants were more susceptible to infection than those plants having adequate supply of K

(Wang et al. 2013) and also the rate of rice borer infestation was greatest when there was no supply of K, but decreased rapidly as the K concentration increased (Sarwar 2012). It was found that the use of K in fields extensively decreased the occurrence of fungal diseases by 70 %, bacterial by 69 %, viral by 41 %, insect and mite infestation by 63 % and nematodes by 33 % (Perrenoud 1990). The synthesis of high-molecular-weight compounds (such as proteins, starch and cellulose) was markedly increased in K-sufficient plants, thus lowering the concentrations of low-molecular-weight compounds, such as soluble sugars, organic acids, amino acids and amides in the plant tissues which were generally responsible for the development of infections and insect infestations. Thereby, plants were protected from diseases and pest attacks in K-sufficient plants (Marschner 2012). Adequate K also increased phenol concentrations, which could also play a significant role in plant resistance (Prasad et al. 2010).

17.3.2 Abiotic Stress Resistance

Abiotic stresses can cause major damage to the crops as compared to yield losses from biotic stress (Oerke 2006). Major abiotic factors are drought stress (low moisture, aquaporins and water uptake, osmotic and stomatal regulation, detoxification of reactive oxygen species, etc.), salt stress, cold stress and waterlogging stress. The use of potassium triggers many plant activities, whereas depletion of potassium uptake can cause problem for plant growth. Potassium to a great extent contributes to the survival of the plants exposed to various abiotic stresses. It was documented by Wang et al. (2013) that abiotic stress factors such as heat, cold, drought and salinity had a huge impact on world agriculture, and they might reduce average yields by ~50 % for most major crop plants. Increased application of K^+ has been shown to enhance photosynthetic rate, plant growth and yield and drought resistance in different crops under water stress conditions (Sharma et al. 1996; Tiwari et al. 1998; Yadav et al. 1999; Egilla et al. 2001).

17.3.2.1 Drought and Low Moisture Stress

Association between K stress and plant drought resistance has been demonstrated by many workers. Drought can be defeated by plants by inducing deeper rooting, larger absorption surfaces and greater water retention in plant tissues, which can also be overcome by application of K fertilizers with other nutrients like phosphorus and nitrogen (Kirkby et al. 2009). Sufficient amounts of K enhanced the total dry mass accumulation of crop plants under drought stress/low-moisture conditions in comparison to lower K concentrations (Egilla et al. 2001). Similar findings were documented by Lindhauer (1985) that not only plant dry mass was increased but also leaf area and water retention in plant tissues under drought-stressed conditions improved. It was found that plants that were continuously exposed to drought stress could form reactive oxygen species, which caused leaf damage (Cakmak et al. 2005; Foyer et al. 2002; Oerke and Dehne 2004). Wang et al. (2013) found a close relation of K in physiological and molecular mechanisms of plant drought resistance. For the period of drought stress, root growth and the rate of K^+ diffusion in the soil towards the root got restricted and depressed the plant resistance as well as K absorption.

Aquaporins and Water Uptake

Potassium is also involved in plant-water relations by regulating the osmotic potential and hydraulic conductivity of membranes and altering water permeability (Heinen et al. 2009; Maurel and Chrispeels 2001). Aquaporin is water along with K^+ ions which moves through specific channel protein present in the plasma and intracellular membranes of cells (Maathuis et al. 1997; Steudle 2000). K^+ is found in the plant cell in two distinct compartments, the cytosol and the vacuole (Leigh 2001), hence transported through plant cell membranes with the help of specific protein channels (Maathuis et al. 1997). Under drought stress conditions, aquaporin gene expression could be regulated (Tyerman et al. 2002; Lian et al. 2004) to help the plants to maintain their water balance (Kaldenhoff et al. 2008). During water stress, roots regulated their water and ion uptake capacities by modifying PIPs (plasma membrane intrinsic proteins) and K^+ channel at the transcription level to cope with the water deficiency (Smart et al. 2001; Galmes et al. 2007; Cuellar et al. 2010). Kanai et al. (2011) also observed close coupling between aquaporin activities and K channel transporters. They found that aquaporin activities might have been suppressed by K deficiency and resulted in a reduction of root hydraulic conductance and water supply to the growing stem for diameter expansion and the leaf for transpiration.

Osmotic and Stomatal Regulation

Under drought stress condition, the maintenance of enough water levels is important for plant survival; hence, osmotic regulation is the most important trait involved in maintaining high cellular turgor potential and water retention in response to drought stress. An adequate amount of K may help osmotic adjustment, which maintains higher turgor pressure, relative water content and lower osmotic potential, thus improving the ability of plants to tolerate drought stress (Kant and Kafkafi 2002; Egilla et al. 2005). Stomata are essential to control plant water loss via transpiration and quick stomatal closures are needed to survive during stressed conditions. Stomatal closure is preceded by rapid release of K^+ from the guard cells into the leaf, thus improving the ability of plants to tolerate drought stress (Kant and Kafkafi 2002). When K^+ is deficient, the stomata cannot function properly and water losses from plant may reach damaging levels (Gething 1990).

Detoxification of Reactive Oxygen Species (ROS)

Generally, drought stress or K-deficient plants induce ROS, e.g. superoxide radical ($O^{\bullet}_2$), hydrogen peroxide (H_2O_2) and hydroxyl radical ($^{\bullet}HO$), production (Cakmak 2005). Production of ROS is mainly responsible for impairment of cellular functions and growth depression in stress conditions. K requirement for drought-stressed plants could be related to the role of K in enhancing photosynthetic CO_2 fixation and transporting photosynthates into sink organs and inhibiting the transfer of photosynthetic electrons to O_2, thus reducing ROS production (Cakmak 2005). Besides the photosynthetic electron transports, nicotinamide adenine dinucleotide phosphate (NADPH)-dependent oxidase activation represents another major source for production of ROS in plant cells by a number of biotic and abiotic stress factors. Furthermore, abscisic acid has also been shown to be effective in increasing H_2O_2 and O_2^- accumulations in roots or leaves (Lin and Kao 2001; Jiang and Zhang 2001). Hence, maintaining an adequate K nutritional status critical for plant osmotic adjustment and for mitigating ROS damage as induced by drought stress.

17.3.2.2 Salinity Stress

Salinity is another major abiotic stress that affects major part of the total land area on earth. Saline soils generally had higher concentrations of Na^+ than K^+ and Ca^{2+} which resulted in passive accumulation of Na^+ in root and shoot (Bohra and Doerffling 1993). Due to the accumulation of salt, water uptake by plant roots in soil became difficult thus disturbing water balance, while high concentrations of salts in plant tissue were found to be toxic (Wang et al. 2013). Salinity inhibited seed germination and plant growth, affected the leaf anatomy and physiology of plants and, thereby, influenced plant-water relations, photosynthesis, protein synthesis, energy production and lipid metabolism (Parida and Das 2005).

17.3.2.3 Cold Stress

Cold stress inhibited plant growth and development and resulted in reduced crop productivity (Wang et al. 2013). Devi et al. (2012) noticed that in *Panax ginseng*, a high K^+ concentration activated the plant's antioxidant system and increased the levels of ginsenoside-related secondary metabolite transcripts, which are associated with cold tolerance. Thus, cold stress might have destroyed photosynthetic processes and reduced the efficacy of antioxidant enzymes, resulting in ROS accumulation (Mittler 2002; Xiong et al. 2002; Suzuki and Mittler 2006). Potassium improved plant survival under cold stress by increasing antioxidant levels and reducing ROS production (Cakmak 2005; Devi et al. 2012).

High concentrations of K^+ protect plant cell against freezing by lowering the freezing point of the cell solution. The plasma membrane is the prime site which is mainly affected when changes in temperature occur, and membrane fluidity is reduced by cold stress resulting in changes of fatty acid unsaturation and the lipid-protein composition of the cell membrane (Wang et al. 2006). Maximum growth response and chilling resistance in tomato, eggplant and pepper plants with the addition of K^+ were associated with increase in phospholipids, membrane permeability and improvement in biophysical and biochemical properties of cell (Hakerlerker et al. 1997). Hence, higher K tissue concentrations reduced chilling damage and increased cold resistance, thereby increasing yield production (Mengel 2001; Kant and Kafkafi 2002).

17.3.2.4 Waterlogging Stress

Waterlogging, a serious obstacle for sustainable agriculture development, affected approximately 10 % of the global land area (Setter and Waters 2003). The main biological consequence that plant encounters during waterlogging stress is the respiration of roots, and microorganisms further deplete the residual oxygen, hence making the environment hypoxic (i.e. oxygen levels limit mitochondrial respiration) and later anoxic (i.e. respiration is completely inhibited). It resulted in a severe decline in energy status of root cells which affected important metabolic processes of plants (Bailey-Serres and Voesenek 2008; Wegner 2010). To overcome this difficulty, application of K could efficiently offset the adverse effects of waterlogging on plants. It was reported that K supplement under waterlogging not only increased plant growth, photosynthetic pigments and photosynthetic capacity but also improved plant nutrient uptake as a result of higher K^+, Ca^{2+}, N, Mn^{2+} and Fe^{2+} accumulation (Ashraf et al. 2011). Exogenous application of K in soil and as foliar spray alleviated the adverse effects of waterlogging on cotton and many other plants such as corn (Welch and Flannery 1985; Csatho 1991), rice (Datta and Mikkelsen 1985), wheat (Beaton and Sekhon 1985; Khurana and Bhaya 1990) and oilseed rape (Sharma and Kolte 1994). There are major challenges for agriculture to enhance crop yields and to stabilize plant development and yield formation under biotic and abiotic stress conditions (Reynolds et al. 2011).

17.4 Deficiency of Potassium

There are several factors that lead to the insufficient supply of nutrients in soil and as a result plant has to face deficiency. In addition, the presence of extreme amounts of reduced substances in poorly drained soils is also responsible for retarded root growth and reduced K uptake (Fairhurst et al. 2007). Fundamentally, K^+ is water soluble and highly mobile and transported to the xylem in plants (Lack and Evans 2005). It acts as regulator in plants since it is constituent of 60 different enzyme systems of drought tolerance and water-use efficiency. K deficiency, also known as potash deficiency, occurs widely in plants and has a strong impact on plant metabolism. A plenty of deficiency symptoms have been reported which include chlorosis, poor growth, reduced yield and poor fibre quality with the increased susceptibility to diseases (Amtmann et al. 2008) and pests (Amtmann et al. 2006; Troufflard et al. 2010). It also affects photosynthesis process and plant growth; hence, purple spots may also appear on the leaf undersides. Besides, older leaves change from yellow to brown, leaf tips and margins dry up, root oxidation power declines, younger leaves decolourize, and root length and density reduce causing reduction of nutrient uptake and cytokine production in roots. Moreover, K-deficient plants are highly light sensitive and very rapidly become chlorotic and necrotic when exposed to high light intensity (Cakmak 2005). Potassium deficiency also inhibits evaporation which causes the temperature in the leaves to rise and results in burning of cells, which mainly occurs on the edges of the leaves. Stressed plants may be more sensitive to the cold injury. Sometimes rust-brown spots may appear in the leaf if K deficiency is severe, and leaves often turn or curl radially on the top and entire leaves become necrotic and eventually fall off. Flowering will be severely inhibited which ultimately affects crop production.

The reasons for potassium deficiency could be insufficient fertilization, excessive 'table salt' (sodium) in the root environment, unfavourable soil structure (e.g. sandy soils) and formation of depletion zones around roots (Kayser and Isselstein 2005; Moody and Bell 2006; Andrist-Rangel et al. 2007). Effects of K deficiency on plants depended on specific crop type such as corn becoming small in size and showing low yield and tomatoes exhibiting uneven fruit ripening, whereas leaves of cotton plants turned reddish brown and then black, appeared scorched and eventually fell, and yield and quality of forage crops were adversely affected (www.ncagr.gov). To overcome the deficiency, the prevention and cure can be achieved at great extent by adding potassium-specific fertilizer, e.g. potassium carbonate, rock potash, potassium nitrate, potassium sulfate, etc. Thus, study of potassium uptake and strategies to enhance uptake in plants is necessary.

17.5 Pathways and Mechanisms for Potassium Uptake

Plant cannot acquire nutrient directly unless it is present in available or dissolved form or released by weathering. There are several processes that contribute to the availability of potassium in the soil. Besides, the release of K from the rocks and minerals requires weathering over long periods of time, although calcinations of rocks can break the K out of the structure but makes the K more expensive than that from the evaporite ores. In addition to plant uptake of K from the soil solution, some of the exchangeable potassium on the soil colloids is also absorbed directly by roots. A plant root possesses a negative charge and attracts the positively charged potassium (K^+) which is held on the clay mineral surfaces and edges. However, potassium uptake of plants can be increased by using potassium solubilizers as bioinoculants which further increased the crop productivity (Shanware et al. 2014). Lian et al. (2008) reported that there were three major reaction pathways utilized by *Aspergillus fumigatus* to release potassium from potassium minerals which were acid hydrolysis, secretion of insoluble macromolecules and polymers bound in the cell membrane and direct biophysical forces which could split mineral grains.

It was demonstrated that potassium-solubilizing microorganisms have the capacity to dissolve K from insoluble minerals (Alexander 1985). Potassium-solubilizing bacteria are able to solubilize rock K mineral powder, such as micas, illite and orthoclases through various processes such as acidolysis, enzymolysis, capsule absorption and complexation by extracellular polysaccharides (Avakyan 1984; Rozanova 1986; Malinovskaya 1988; Malinovskaya et al. 1990; Friedrich et al. 1991; Ullman et al. 1996; Welch et al. 1999). Mechanisms involved in degradation of potassium-bearing minerals can be divided into direct (bacterial cell wall) and indirect (bioleaching, mineral weathering, microbial weathering and mechanical fragmentation) mechanisms.

17.5.1 Bacterial Cell Wall

Prokaryotes have various cell wall types which secrete one or more metabolic products that react with ions or compounds in the environment resulting in the deposition of mineral particles. Bacterial surfaces such as cell walls or polymeric materials (exopolymers) exuded by bacteria, including slimes, sheaths or biofilms, and even dormant spores, acted as important sites for the adsorption of ions and mineral nucleation and growth (Konhauser 1998; Beveridge 1989; Banfield and Zhang 2001; Bäuerlein 2003). In gram-negative bacteria, the layers external to the bacterial cell wall that may be involved in mineral nucleation include S-layers, capsules, slimes and sheaths. S-layers are acidic and possess a net negative charge, thereby having an affinity for metal cations (Southam 2000). In some cases, capsules are also known to form in response to the presence of metal ions (Appanna and Preston 1987).

17.5.2 Bioleaching

The process of removal of soluble material from the rocks in solution by percolating water is termed as leaching. Halite (NaCl) and sylvite (KCl) are highly water soluble, while carbonates and sulfates are sparingly soluble. Bioleaching microbes are mainly *Thiobacillus ferrooxidans*, *T. thiooxidans* and *Leptospirillum ferrooxidans*.

17.5.3 Mineral Weathering

The weathering reactions contribute to nutrient cycling (Huntington et al. 2000). Chemical weathering of bedrock releases inorganic nutrients such as Ca, Mg, K, Fe and P, which are then cycled through the saprolite, soil and vegetation. Inorganic nutrients were cycled by microorganisms via uptake, release, biomineralization, oxidation and reduction (Berner and Berner 1996). Microorganisms also contributed to mineral weathering and soil formation by secreting organic acids and other ligands such as siderophores as indicated by many researchers (Richter and Markewitz 1995; Kalinowski et al. 2000; Liermann et al. 2000, 2005; Richter and Oh 2002). Within the rooting zone, plants take up mineral nutrients, which are recycled back into the soil when plants get decomposed.

During weathering, physical, chemical and biological forces act on the parent materials and break them down into finer fractions, largely sand-, silt- and clay-size particles. This breakdown results in the release of several chemical elements, including potassium, and the formation of different clay minerals. Most of the total potassium inherited from the parent material during the soil-forming processes will be in the non-exchangeable and exchangeable forms. The relative amounts of sand, silt and clay fractions found in a soil depend on the kind of parent material (sandstone, limestone, shale or mica) from which the soil was derived. Potassium fixation and release is greatly influenced by the relative amounts of these fractions and the kinds of clay minerals present in the soil. Mineral dissolution studies with cultures of bacteria and fungi showed a dramatic increase in the dissolution rates of feldspar, biotite, quartz, apatite and other minerals (Berthelin and Belgy 1979; Callot et al. 1987; Thorseth et al. 1995; Ullman et al. 1996; Barker et al. 1997; Paris et al. 1996).

17.5.4 Microbial Weathering

The known and potential mechanisms of microbial weathering included redox reactions through the production of organic acids which led to weakening of chemical bonds in minerals for promoting mineral dissolution (Banfield et al. 1999; Harley and Gilkes 2000) and chelating molecules for mineral degradation (Uroz et al. 2007, 2009; Lian et al. 2008). Bacteria produced a wide range of low-molecular-weight organic acids such as citric, malic, oxalic, succinic and tartaric acid (Jones 1998; Neaman et al. 2005). Han and Lee (2005) concluded in a study that KSB solubilized potassium rock through production and secretion of organic acids. Similar were the observations of Prajapati and Modi (2012), which attributed the solubilization to reduction in pH due to organic acids. Few other reports are also available for feldspar solubilization by *Bacillus mucilaginosus* and *Bacillus edaphicus* due to acid production (Malinovskaya et al. 1990; Sheng and Huang 2002). In addition to organic acids, microorganisms (such as bacteria, algae, fungi and protozoa) used carbonic acid formed from carbon dioxide to attack the mineral surface, promoting the chemical weathering of rocks and minerals (Gadd 2007; Park et al. 2009).

Microbes played a key role in the weathering of major type of rocks, releasing various elements they needed as nutrients (Calvaruso et al. 2006). Many rock-inhabiting fungi were melanized, and melanin pigmentation conferred extra-mechanical strength to the hyphae to penetrate the rock surface and crevices (Dornieden et al. 1997; Sterflinger and Krumbein 1997) and also offered protection from metal toxicity (Gadd 1993). Fungi have been reported from a wide range of rock types, including rocks from extreme environments (Staley et al. 1983; Nienow and Fridman 1993; Sterflinger 2000; Etienne and Dupont 2002; Gorbushina 2007). In an experimental study conducted by Puente et al. (2004), fluorescent pseudomonads and bacilli were found to weather igneous rock, limestone and marble.

17.5.5 Biofilm

Biofilm helps to accelerate weathering of minerals like biotite and anorthite. Biotite weathering occurs in two stages: by oxidation within the rindlet zone and by alteration to kaolinite within the saprolite. Biofilms and biocrusts were normally supposed to cause higher weathering rates due to biodegradation (Warscheid and Braams 2000). It was accepted that the microbial biofilms not only accelerated the weathering process but also regulated denudation losses by acting as a protective layer covering the mineral-water-hyphal/root hair interface in the mycorrhizosphere and rhizosphere of vascular plants. Besides, biofilm formation on mineral surface promoted the corrosion of potassium-rich shale and the release of K, Si and Al in the bacteria-mineral contact model (Li-yang et al. 2014).

17.5.6 Mechanical Fragmentation

Fragmentation of the mineral caused by root activity increases the reactive surfaces, so having direct positive effect of the bacteria on mineral weathering. Mechanical fragmentation of rock particles also occurred when there was extension of hyphae (unique for fungal organisms) into the interior of minerals to acquire nutrition (Jongmans et al. 1997).

17.6 Potassium-Solubilizing Microorganisms (KSMs)

Potassium is the third major essential macronutrient for plant growth and development. It constitutes ~2.5 % of the lithosphere but actual soil concentrations of this nutrient vary widely ranging from 0.04 % to 3.0 % (Sparks and Huang 1985). Plants absorb K only in soluble form from soil, and its availability to crop plants is generally as low as 90–98 % of total K in soil in the unavailable mineral forms (Sparks 1987) such as feldspar and mica (McAfee 2008). The addition of chemical fertilizers causes environmental pollution and has many deteriorating impacts such as global warming, alteration of soil microbial diversity, etc. Moreover, they also influence soil-plant dynamics with its microbial distribution (Meena et al. 2013; Maurya et al. 2014).

Microorganisms on solid and liquid medium were tested, and it was observed that bacterium *Enterobacter hormaechei* (KSB-8) was more viable in liquid broth as compared to solid medium containing carrier lignite, while fungus *Aspergillus terreus* (KSF-1) was good at both liquid and solid medium (Kalawati and Modi 2014). Rhizosphere microorganisms play an important role in solubilization of bound form of soil minerals and enhancing the availability of plant nutrients in the soil. Increasing the bioavailability of phosphorus (P) and potassium in soils with inoculation of plant growth-promoting rhizobacteria (PGPR) singly or in consortium with or without rock materials has been reported by many researchers (Lin et al. 2002; Sahin et al. 2004; Girgis 2006; Eweda et al. 2007; Jha et al. 2012; Meena et al. 2015b; Singh et al. 2015), which may lead to increasing P uptake and plant growth. The mould *A. niger* has been well documented for its ability to solubilize P in rocks due to organic acid production, especially citric acid (Nahas et al. 1990; Vassileva et al. 1998; Jain et al. 2014).

Different microorganisms are used to supply different kind of nutrients in the soil such as symbiotic and non-symbiotic nitrogen-fixing bacteria which can supply nitrogen to plants by fixing the atmospheric nitrogen and converting the nitrogen into ammonium ion. *Bacillus megaterium* and *Pseudomonas* sp. are the most common phosphate-solubilizing bacteria that are used as a biofertilizer to solubilize phosphorus in soil. Solubilization of potassium from aluminosilicate minerals has also been observed by some fungi (Wallander and Tonie 1999; Glowa et al. 2003). It was proposed by Yuan et al. (2000) that ectomycorrhizae could mobilize potassium from clay minerals and thus enhanced its uptake by plants. Further, Yuan et al. (2004) studied the effect of four fungal strains, *Pisolithus* XC1, *Pisolithus* sp., *P. microcarpus* and *Cenococcum geophilum* SIV, collected from the roots of eucalyptus on the degradation of phlogopite and vermiculite; the results revealed that all four strains were able to weather the mineral phases and release elemental K. Moreover, Glowa et al. (2003) evaluated the ability of fungus *Piloderma* in extracting potassium from biotite, microcline and chlorite and found that the fungus species was able to acquire potassium from all three minerals, out of which biotite was more biodegradable (Meena et al. 2014a, 2015a).

Muentz (1890) showed the first evidence of microbial involvement in solubilization of rock potassium. Since then a diverse group of soil microflora has been reported to be involved in the solubilization of insoluble and fixed forms of K into available forms, which can easily be absorbed by plants (Li et al. 2006; Gundala et al. 2013; Zarjani et al. 2013). A wide range of potassium-solubilizing bacteria (KSB), namely, *Bacillus edaphicus*, *B. circulans*, *Paenibacillus* spp., *Acidithiobacillus ferrooxidans*, *Pseudomonas* spp., *Burkholderia* spp., etc., have been reported to release potassium from K-bearing minerals in soil (Sheng et al. 2008; Lian et al. 2002; Rajawat et al. 2012; Liu et al. 2012; Basak and Biswas 2012; Singh et al. 2010).

KSB such as *Bacillus mucilaginosus* solubilized potassium rock and stimulated plant growth through synthesis of growth-promoting substances via their biological activities. Similarly, silicate-solubilizing bacteria were found to dissolve potassium, silicon and aluminium from insoluble minerals (Aleksandrov et al. 1967). KSB have capacity to dissolve K from insoluble minerals (Alexander 1985). Many microorganisms in the soil are able to solubilize 'unavailable' forms of K-bearing minerals, such as micas, illite and orthoclases, by excreting organic acids which either directly dissolved rock K or chelated silicon ions to bring the K into solution (Groudev 1987; Friedrich et al. 1991; Ullman et al. 1996; Bennett et al. 1998). Therefore, the application of K-solubilizing microorganisms (KSMs) is a promising approach for increasing K availability in soils (Zahra et al. 1984; Vandevivere et al. 1994; Barker et al. 1998; Meena et al. 2014b; Kumar et al. 2015).

KSMs can be isolated from many sources in in vitro conditions using different media. Prajapati and Modi (2012) isolated 14 bacterial strains from samples collected from ceramic industry using feldspar on a solid media, out of which five strains showed higher potassium solubilization. Recently, Parmar and Sindhu (2013) used Aleksandrov medium supplemented with mica to isolate 137 K-solubilizing bacteria from soil samples collected from wheat rhizosphere. Among isolated strains, 20 strains were found to solubilize potassium from mica. The amount of K released by the strains ranged from 15 to 48 mg L^{-1}. Further examination for optimization of conditions for K release revealed that maximum solubilization occurred with glucose as carbon source at 25 °C temperature and 7.0 pH. Potassium solubilization was maximum when KCl was used as potassium source, followed by K_2SO_4, and least solubilization was found in mica powder. It was suggested that efficient potassium-solubilizing bacterial strains could be further exploited for plant growth improvement under field conditions. Sheng (2005) used sucrose minimal salt medium with illite as K source for the isolation of *B. edaphicus* strain NBT from rhizosphere soil of cotton. Wu et al. (2005) used the same medium with glass powder for the growth of *B. mucilaginosus*, a vigorous K solubilizer. Further, it was found to have great K-releasing capability and could promote the release of potassium through weathering of silicate minerals (Hu et al. 2006; Lian 1998; Zhao et al. 2006, 2008). Sangeeth et al. (2012) identified potassium-solubilizing bacterium, *Paenibacillus glucanolyticus* IISRK2, isolated from rhizosphere of black pepper plant. It was further evaluated for plant growth, and after the studies, it was documented that the strain efficiently promoted the shoot and dry matter of wheat plants promoting the overall plant growth. Commercially available microbial inoculants that are able to dissolve K from minerals and rocks not only enhance plant growth and yield but are also eco-friendly; microbes *Aspergillus niger*, *Bacillus extorquens* and *Clostridium pasteurianum* were found to grow on muscovite, biotite, orthoclase microclase and mica in vitro (Archana et al. 2013).

17.7 KSM Role in Sustainable Agriculture

Potassium is an essential and major nutrient for crop production (Alfaro et al. 2003; Zhang et al. 2011). The role of potassium is well known for improving shelf life of crops and disease resistance (Khawilkar and Ramteke 1993). Potassium is useful in agriculture land to increase crop yield as proper amount of potassium in soil can enhance root growth, improve drought resistance, activate many enzyme systems, maintain turgor pressure, reduce water loss and wilting, aid in photosynthesis and food formation, reduce respiration, prevent energy losses, enhance translocation of sugars and starch, produce grain rich in starch, increase protein content of plants, reduce waterlogging and retard crop diseases. Therefore, it is essential for the growth and metabolism of plants; the deficiency of potassium in plants causes poorly developed roots, slow growth and low resistance to disease, delayed maturity, small seeds and lower yields. With rapid development of agriculture or due to application of imbalanced fertilizers, available K level in soils has dropped. The concentration of soluble K is very small and the maximum part of K exists in insoluble form. Silt, clay and sand are biggest reservoirs of potassium and important component of soil. The most common component of potassium is feldspar and mica, and fortunately, India has the largest deposits of mica mines distributed in some districts of Bihar and Jharkhand. In such conditions the application of KSMs can be an alternative approach for increasing K availability (Krishnamurti and Huang 1988; Prajapati and Modi 2014; Zhang and Kong 2014).

It has been studied that by introducing potassium-solubilizing bacteria *B. mucilaginosus* and phosphate-solubilizing bacteria *Bacillus megaterium* var. *phosphaticum*, simultaneously, macronutrient (nitrogen, phosphate and potassium) uptake was increased in eggplant, pepper and cucumber leading to higher yields (Han et al. 2006). *B. edaphicus* strain was also examined for the growth promotion and increased potassium uptake on cotton and rape plants, and increased plant growth was observed in the soil treated with insoluble potassium with strain NBT. The shoot and root dry weight increased from 25 % to 33 % (cotton) and from 24 % to 27 % (rape), whereas K content was increased from 31 % to 34 % ($P < 0.05$) (cotton) and 28–31 % (rape) when the soil with insoluble K source and bioinoculant was compared to the uninoculated soil (Sheng 2005). Experiments conducted with tobacco seedlings, inoculated with *Klebsiella variicola* strains JM3, XF4 and XF11, showed greater height and dry weight than uninoculated seedlings. The GL7 significantly increased plant dry weight. Inoculation with the strains JM3, GL7, XF4 and XF11 significantly increased seedling absorption of N, P and K. Seedlings exposed to K feldspar absorbed significantly more N and K than those not exposed to added K feldspar (Jhang and Kong 2014). Krishnamurthy (1990) reported that the potassium content in tobacco leaf was strongly and positively correlated with available K status in sandy and sandy loam soils.

Nowadays, biofertilizer is a substitute to chemical fertilizer to increase soil fertility and crop production in sustainable farming. The use of plant growth-promoting microorganisms, including phosphate-solubilizing and potassium-mobilizing bacteria as biofertilizers, was suggested as a possible solution to improve plant nutrient and production (Vessey 2003). Beneficial microorganisms in the form of biofertilizers are applied on seeds/roots or in soil which mobilizes the availability of nutrients, especially N-P-K by their biological activity, thereby helping in the build-up of positive microflora and enhancing the soil health. The use of bioinoculants can also improve the physical properties and enhance water-holding capacity of soil. Moreover, microorganisms that are applied as biofertilizer can prevent nutrient leaching and lead to soil enrichment with nutrients. They are low in cost, compatible with long-term sustainability, and eco-friendly as compared to chemical fertilizers. Besides, the nutrient supply is constant and sustainable through these microorganisms' activities. Though the use of biofertilizer has gained momentum in recent years since chemical fertilizers are high in cost and can cause hazardous effect (Aseri et al. 2008), unfortunately not much attention has been given to the manufacturing of K biofertilizers.

17.8 Concluding Remarks

Potassium being the third major nutrient for plants is vital for plant growth. However, majority of potassium in soils is available in insoluble forms. Therefore, potassium-solubilizing microorganisms, the component of soil microbial community, play an important role in K solubilization to provide available form to plants. K solubilization benefits crop growth and improves soil fertility in an eco-friendly manner. Potassium-solubilizing strains are able to colonize the rhizosphere, promote crop yield and enhance plant stress response during stress conditions and K uptake. Unfortunately, very little attention has been paid to K-solubilizing microorganisms and K biofertilizers as most of the researches are focused on nitrogen and phosphorus biofertilizers. Hence, it is requisite to study the efficient K-solubilizing microorganisms to improve sustainable agriculture and to keep the soil productive.

References

Aleksandrov VG, Blagodyr RN, Iiiev IP (1967) Liberation of phosphoric acid from apatite by silicate bacteria. Mikrobiol Zh (Kiev) 29:111–114

Alexander M (1985) Introduction to soil microbiology. Wiley, New York, pp 382–385

Alfaro MA, Jarvis SC, Gregory PJ (2003) The effect of grassland soil managements on soil potassium availability. J Soil Sci Plant Nutr 3(2):31–41

Amtmann A, Hammond JP, Armengaud P, White PJ (2006) Nutrient sensing and signalling in plants: potassium and phosphorus. Adv Bot Res 43:209–257

Amtmann A, Troufflard S, Armengaud P (2008) The effect of potassium nutrition on pest and disease resistance in plants. Physiol Plant 133:682–691

Andrist-Rangel Y, Edwards AC, Hillier S, Oborn I (2007) Long-term K dynamics in organic and conventional mixed cropping systems as related to management and soil properties. Agric Ecosyst Environ 122:413–426

Appanna VD, Preston CM (1987) Manganese elicits the synthesis of a novel exopolysaccharide in an arctic *Rhizobium*. FEBS Lett 215:79–82

Archana DS, Nandish MS, Savalagi VP, Alagawadi AR (2013) Characterization of potassium solubilizing bacteria (KSB) from rhizosphere soil. Bioinfolet 10:248–257

Arnold PW (1963) Potassium in partially weathered soils. In: Proceedings of the first regional conference of the International Potash Institute, Berne, pp 11–17

Aseri GK, Jain N, Panwar J, Rao AV, Meghwalc PR (2008) Biofertilizers improve plant growth, fruit yield, nutrition, metabolism and rhizosphere enzyme activities of pomegranate (*Punica granatum* L.) in Indian Thar Desert. Sci Hortic 117(2):130–135

Ashraf MA, Ahmad MSA, Ashraf M, Al-Qurainy F, Ashraf MY (2011) Alleviation of waterlogging stress in upland cotton (*Gossypium hirsutum* L.) by exogenous application of potassium in soil and as a foliar spray. Crop Pasture Sci 62:25–38

Avakyan ZA (1984) Silicon compounds in solution bacteria quartz degradation. Mikrobiology 54:301–307

Bailey-Serres J, Voesenek LACJ (2008) Flooding stress: acclimations and genetic diversity. Annu Rev Plant Biol 59:313–339

Banfield JF, Zhang H (2001) Nanoparticles in the environment. Rev Mineral Geochem 44:1–58

Banfield J, Barker W, Welch S, Taunton A (1999) Biological impact on mineral dissolution: application of the lichen model to understanding mineral weathering in the rhizosphere. Proc Natl Acad Sci 96:3404–3411

Barker WW, Welch SA, Banfield JF (1997) Geomicrobiology of silicate mineral weathering. Min Soc Am Rev Mineral 35:391–428

Barker WW, Welch SA, Chu S, Banfield JF (1998) Experimental observations of the effects of bacteria on aluminosilicate weathering. Am Min 83:1551–1563

Basak B, Biswas D (2012) Modification of waste mica for alternative source of potassium: evaluation of potassium release in soil from waste mica treated with potassium solubilizing bacteria (KSB). LAMBERT Academic Publishing, Germany, ISBN-13:978-3659298424

Bauerlein E (2003) Biomineralization of unicellular organisms: an unusual membrane biochemistry for the production of inorganic nano- and microstructures. Angew Chem Int 42:614–641

Beaton JD, Sekhon GS (1985) Potassium nutrition of wheat and other small grains. In: Munson RD (ed) Potassium in agriculture. ASA, Madison

Bennett PC, Choi WJ, Rogera JR (1998) Microbial destruction of feldspars. Mineral Manag 8 (62A):149–150

Berner EK, Berner RA (1996) Global environment: water, air, and geochemical cycles. Prentice Hall, Upper Saddle River, N.J: Prentice Hall

Berthelin J, Belgy G (1979) Microbial degradation of phyllosilicates during simulated podzolization. Geoderma 21:297–310

Bertsch PM, Thomas GW (1985) Potassium status of temperate region soils. In: Munson RD (ed) Potassium in agriculture. American Society of Agronomy, Crop Science Society of America, and Soil Science Society of America, Madison, pp 131–162

Beveridge TJ (1989) Role of cellular design in bacterial metal accumulation and mineralization. Annu Rev Microbiol 43:147–171

Bohra JS, Doerffling K (1993) Potassium nutrition of rice (*Oryza sativa* L.) varieties under NaCl salinity. Plant Soil 152:299–303

Cakmak I (2005) The role of potassium in alleviating detrimental effects of abiotic stresses in plants. J Plant Nutr Soil Sci 168:521–530

Callot G, Maurette M, Pottier L, Dubois A (1987) Biogenic etching of microfractures in amorphous and crystalline silicates. Nat (Lond) 328:147–149

Calvaruso C, Turpault MP, Frey-Klett P (2006) Root-associated bacteria contribute to mineral-weathering and to mineral nutrition in trees, a budgeting analysis. Appl Environ Microbiol 72:1258–1266

Csatho P (1991) Effect of NPK-fertilization and split application of nitrogen on lodging due to windstorm and harvestable grain yield of maize. Acta Agron Hung 40(3–4):281–294

Cuellar T, Pascaud F, Verdeil JL, Torregrosa L, Adam-Blondon AF, Thibaud JB, Sentenac H, Gaillard I (2010) A grapevine Shaker inward K^+ channel activated by the calcineurin B-like calcium sensor 1-protein kinase CIPK23 network is expressed in grape berries under drought stress conditions. Plant J 61:58–69

Datta SK, Mikkelsen DS (1985) Potassium nutrition of rice. In: Munson RD (ed) Potassium in agriculture. ASA, Madison

Devi BSR, Kim YJ, Selvi SK, Gayathri S, Altanzul K, Parvin S, Yang DU, Lee OR, Lee S, Yang DC (2012) Influence of potassium nitrate on antioxidant level and secondary metabolite genes under cold stress in *Panax ginseng*. Russ J Plant Physiol 59:318–325

Dornieden T, Gorbushina AA, Krumbein WE (1997) Changes in physical properties of marble by fungal growth. Int J Restor Build Mon 3:441–456

During C (1984) Fertilizers and soils in New Zealand, 3rd revised edn. Government Printer, Wellington, pp 355

Egilla JN, Davies FT, Drew MC (2001) Effect of potassium on drought resistance of *Hibiscus rosa-sinensis* cv. Leprechaun: plant growth, leaf macro- and micronutrient content and root longevity. Plant Soil 229:213–224

Egilla JN, Davies FT, Boutton TW (2005) Drought stress influences leaf water content, photosynthesis, and water-use efficiency of *Hibiscus rosa-sinensis* at three potassium concentrations. Photosynthetica 43:135–140

Etienne S, Dupont J (2002) Fungal weathering of basaltic rocks in a cold oceanic environment (Iceland): comparison between experimental and field observations. Earth Surf Proc Land 27:737–748

Eweda WE, Selim SM, Mostafa MI, Dalia A, El-Fattah A (2007) Use of *Bacillus circulans* as bio-accelerator enriching composted agricultural wastes É- identification and utilization of the Aust. J Basic Appl Sci 2 (1):68–81

Fairhurst TH, Witt C, Buresh RJ, Dobermann A (2007) Rice: a practical guide to nutrient management. International Rice Research Institute, International Plant Nutrition Institute, and International Potash Institute, pp 89

Fieldes M, Swindale LD (1954) Chemical weathering of silicates in soil formation. N Z J Sci Tech B36:140–154

Foyer CH, Vanacker H, Gomez LD, Harbinson J (2002) Regulation of photosynthesis and antioxidant metabolism in maize leaves at optimal and chilling temperatures: review. Plant Physiol Biochem 40:659–668

Friedrich S, Platonova NP, Karavaiko GI, Stichel E, Glombitza F (1991) Chemical and microbiological solubilization of silicates. Acta Biotechnol 11:187–196

Gadd GM (1993) Interactions of fungi with toxic metals. New Phytol 124:25–60

Gadd GM (2007) Geomycology: biogeochemical transformations of rocks, minerals, metals and radionuclides by fungi, bioweathering and bioremediation. Mycol Res 111:3–49

Galmes J, Pou A, Alsina MM, Tomas M, Medrano H, Flexas J (2007) Aquaporin expression in response to different water stress intensities and recovery in Richter-110 (*Vitis* sp.): relationship with ecophysiological status. Planta 226:671–681

Gething PA (1990) Potassium and water relationships. In: Potash facts. IPI, Bern Girgis, MGZ

Girgis MGZ (2006) Response of wheat to inoculation with phosphate and potassium mobilizers and organic amendment. Ann Agric Sci Ain Shams Univ Cairo 51 (1):85–100

Glowa KR, Arocena JM, Massicotte HB (2003) Extraction of potassium and/or magnesium from selected soil minerals by *Piloderma*. Geomicrobiol J 20:99–112

Gorbushina AA (2007) Life on rocks. Environ Microbiol 9(7):1613–1631

Groudev SN (1987) Use of heterotrophic microorganisms in mineral biotechnology. Acta Biotechnol 7:299–306

Gundala PB, Chinthala P, Sreenivasulu B (2013) A new facultative alkaliphilic, potassium solubilizing, *Bacillus sp*. SVUNM9 isolated from mica cores of Nellore District, Andhra Pradesh. India Res Rev J Microbiol Biotechnol 2(1):1–7

Hakerlerker H, Oktay M, Eryuce N, Yagmur B (1997) Effect of potassium sources on the chilling tolerance of some vegetable seedlings grown in hotbeds. In: Proceedings of Regional Workshop of IPI, held at Bornova, Izmir, Turkey. IPI, Basel, pp 353–359

Han HS, Lee KD (2005) Phosphate and potassium solubilizing bacteria effect on mineral uptake, soil availability and growth of eggplant. Res J Agric Biol Sci 1(2):176–180

Han HS, Supanjani E, Lee KD (2006) Effect of co-inoculation with phosphate and potassium solubilizing bacteria on mineral uptake and growth of pepper and cucumber. Plant Soil Environ 52:130–136

Harley AD, Gilkes RJ (2000) Factors influencing the release of plant nutrient elements from silicates rock powder: a geochemical overview. Nutr Cycl Agroecosyst 56:11–36

Heinen RB, Ye Q, Chaumont F (2009) Role of aquaporins in leaf physiology. J Exp Bot 60:2971–2985

Hu X, Chen J, Guo J (2006) Two phosphate and potassium-solubilizing bacteria isolated from Tianmu Mountain, Zhejiang, China. World J Microbiol Biotechnol 22:983–990

Huntington TG, Hooper RP, Johnson CE, Aulenbach BT, Cappellato R, Blum AE (2000) Calcium depletion in a south eastern United States forest ecosystem. Soil Sci Soc Am J 64:1845–1858

Indian Minerals Yearbook (2011) (Part-II) 50th Edition, Potash, Indian Bureau of Mines

Jain R, Saxena J, Sharma V (2014) Differential effects of immobilized and free forms of phosphate solubilizing fungal strains on the growth and phosphorus uptake of mung bean plants. Ann Microbiol 64(4):1523–1534

Jha A, Sharma D, Saxena J (2012) Effect of single and dual phosphate solubilizing bacterial strain inoculations on overall growth of mung bean plants. Arch Agron Soil Sci 58:967–981

Jiang M, Zhang J (2001) Effect of abscisic acid on active oxygen species, antioxidative defence system and oxidative damage. Plant Cell Physiol 42:1265–1273

Jones DL (1998) Organic acids in the rhizosphere – a critical review. Plant Soil 205:25–44

Jongmans AG, Breemen N, Lundstrom US, van Hees PAW, Finlay RD, Srinivasan M, Unestam T, Giesler R, Melkerud PA, Olsson M (1997) Rock eating fungi. Nature 389:682–683

Kaldenhoff R, Ribas-Carbo M, Flexas J, Lovisolo C, Heckwolf M, Uehlein N (2008) Aquaporins and plant water balance. Plant Cell Environ 31:658–666

Kalinowski BE, Liermann LJ, Givens S, Brantley SL (2000) Rates of bacteria-promoted solubilization of Fe from minerals: a review of problems and approaches. Chem Geol 169:357–370

Kanai S, Moghaieb RE, El-Shemy HA, Panigrahi R, Mohapatra PK, Ito J, Nguyen NT, Saneoka H, Fujita K (2011) Potassium deficiency affects water status and photosynthetic rate of the vegetative sink in green house tomato prior to its effects on source activity. Plant Sci 180:368–374

Kant S, Kafkafi U (2002) Potassium and abiotic stresses in plants. In: Pasricha NS, Bansal SK (eds) Potassium for sustainable crop production. Potash Institute of India, Gurgaon, pp 233–251

Kayser M, Isselstein J (2005) Potassium cycling and losses in grassland systems: a review. Grass Forage Sci 60:213–224

Khawilkar SA, Ramteke JR (1993) Response of applied K in cereals in Maharashtra. Agric: 5;84–96

Khurana GP, Bhaya GP (1990) Effect of potash on wheat irrigated with nitrate waters. Indian J Agron 35:429–431

Kirkby EA, LeBot J, Adamowicz S, Romheld V (2009) Nitrogen in physiology – an agronomic perspective and implications for the use of different nitrogen forms. International Fertiliser Society, Cambridge

Kirkman JH, Basker A, Surapaneni A, MacGregor AN (1994) Potassium in the soils of New Zealand – a review. N Z J Agric Res 37:207–227

Konhauser KO (1998) Diversity of bacterial iron mineralization. Earth Sci Rev 43:91–121

Krishnamurti GSR, Huang PM (1988) Kinetics of manganese released from selected manganese oxide minerals as influenced by potassium chloride. Soil Sci 146:326–334

Kumar A, Bahadur I, Maurya BR, Raghuwanshi R, Meena VS, Singh DK, Dixit J (2015) Does a plant growth-promoting rhizobacteria enhance agricultural sustainability? J Pure Appl Microbiol 9(1):715–724

Lack AJ, Evans DE (2005) Instant notes in plant biology, 2nd edn. Taylor and Francis, Oxford

Lauchli A, Pflueger R (1979) Potassium transport through plant cell membranes and metabolic role of potassium in plants. In: Proceedings of the 11th congress on potassium research – review and trends, pp 111–163

Leigh RA (2001) Potassium homeostasis and membrane transport. J Plant Nutr Soil Sci 164:193–198

Li FC, Li S, Yang YZ, Cheng LJ (2006) Advances in the study of weathering products of primary silicate minerals, exemplified by mica and feldspar. Acta Petrol Min 25:440–448

Lian BA (1998) A study on how silicate bacteria GY92 dissolves potassium from illite. Acta Min Sin 18 (2):234–238

Lian B, Fu PQ, Mo DM, Liu CQ (2002) A comprehensive review of the mechanism of potassium release by silicate bacteria. Acta Min Sin 22:179–183

Lian HL, Yu X, Ye Q, Ding XS, Kitagawa Y, Kwak SS, Su WA, Tang ZC (2004) The role of aquaporin RWC3 in drought avoidance in rice. Plant Cell Physiol 45:481–489

Lian B, Wang B, Pan M, Liu C, Teng HH (2008) Microbial release of potassium from K-bearing minerals by thermophilic fungus *Aspergillus fumigatus*. Geochim Cosmochim Acta 72(1):87–98

Liermann LJ, Kalinowski BE, Brantley SL, Ferry JG (2000) Role of bacterial siderophores in dissolution of hornblende. Geochim Cosmochim Acta 64:587–602

Liermann LJ, Guynn RL, Anbar A, Brantley SL (2005) Production of a molybdophore during metal-targeted dissolution of silicates by soil bacteria. Chem Geol 220(3–4):285–302

Lin CC, Kao CH (2001) Abscisic acid induced changes in cell wall peroxidase activity and hydrogen peroxide level in roots of rice seedlings. Plant Sci 160:323–329

Lin QM, Rao ZH, Sun YX, Yao J, Xing LJ (2002) Identification and practical application of silicate dissolving bacteria. Agric Sci China 1:81–85

Lindhauer MG (1985) Influence of K nutrition and drought on water relations and growth of sunflower (*Helianthus-annuus* L.). J Plant Nutr Soil Sci 148:654–669

Liu D, Lian B, Dong H (2012) Isolation of *Paenibacillus* sp. and assessment of its potential for enhancing mineral weathering. Geomicrobiol J 29:413–421

Li-yang M, Xiao-yan CAO, De-si S (2014) Effect of potassium solubilizing bacteria-mineral contact mode on decomposition behavior of potassium-rich shale. C J Non Ferr Metal 24:1099–1109

Maathuis JM, Ichida AM, Sanders D, Schroeder JI (1997) Role of higher plant K+ channels. Plant Physiol 114:1141–1149

Majumder AK, Govindarajan B, Sharma T, Ray HS, Rao TC (1995) An empirical model for chloridising-roasting of potassium in glauconitic sandstone. Int J Miner Process 43:81–89

Malavolta E (1985) Potassium status of tropical and subtropical region soils. In: Munson RD (ed) Potassium in agriculture. American Society of Agronomy, Crop Science Society of America, and Soil Science Society of America, Madison, pp 163–200

Malinovskaya IM (1988) A method for the determination of the ability of bacterial polysaccharides to sorb silicic acid ions. Mikrobiol Zhurn 57:84–86

Malinovskaya IM, Kosenko LV, Votselko SK, Podgorskii VS (1990) Role of *Bacillus Mucilaginosus* polysaccharide in degradation of silicate minerals. Mikrobiology 59:49–55

Marschner H (1995) Mineral nutrition of higher plants, 2nd edn. Academic, San Diego

Marschner H (2010) Mineral nutrition of higher plants. Academic, London

Marschner P (2012) Marschner's mineral nutrition of higher plants, 3rd edn. Academic, London, pp 178–189

Maurel C, Chrispeels MJ (2001) Aquaporins: a molecular entry into plant water relations. Plant Physiol 125:135–138

Maurya BR, Meena VS, Meena OP (2014) Influence of Inceptisol and Alfisol's potassium solubilizing bacteria (KSB) isolates on release of K from waste mica. Vegetos 27(1):181–187

McAfee J (2008) Potassium, a key nutrient for plant growth. Department of Soil and Crop Sciences: http://jimmcafee.tamu.edu/files/potassium

Meena OP, Maurya BR, Meena VS (2013) Influence of K-solubilizing bacteria on release of potassium from waste mica. Agric Sustain Dev 1(1):53–56

Meena VS, Maurya BR, Bahadur I (2014a) Potassium solubilization by bacterial strain in waste mica. Bangladesh J Bot 43(2):235–237

Meena VS, Maurya BR, Verma JP (2014b) Does a rhizospheric microorganism enhance K+ availability in agricultural soils? Microbiol Res 169:337–347

Meena RK, Singh RK, Singh NP, Meena SK, Meena VS (2015a) Isolation of low temperature surviving plant growth-promoting rhizobacteria (PGPR) from pea (*Pisum sativum* L.) and documentation of their plant growth promoting traits. Biocatal Agric Biotechnol doi:10.1016/j.bcab.2015.08.006

Meena VS, Maurya BR, Verma JP, Aeron A, Kumar A, Kim K, Bajpai VK (2015a) Potassium solubilizing rhizobacteria (KSR): isolation, identification, and K-release dynamics from waste mica. Ecol Eng 81:340–347

Mengel K (2001) Principles of plant nutrition, 5th edn. Kluwer Academic Publishers, Dordrecht, pp 481–509

Mengel K, Kirkby EA (1987) Principles of plant nutrition. International Potash Inst, Bern, pp 200–210

Metson AJ (1968) Potassium in: soils of New Zealand. Part 2, New Zealand soil bureau bulletin 26. Government Printer, Wellington, pp 82–95

Metson AJ (1980) Potassium in New Zealand soils. Department of Scientific and Industrial Research. New Zealand Soil Bureau report no. 38, pp 61. New Zeland

Mikkelsen RL, Bruulsema TW (2005) Fertilizer use for horticultural crops in the US during the 20th century. Hort Technol 15:24–30

Mittler R (2002) Oxidative stress, antioxidants and stress tolerance. Trends Plant Sci 7:405–410

Moody PW, Bell MJ (2006) Availability of soil potassium and diagnostic soil tests. Aust J Soil Res 44:265–275

Muentz A (1890) Surla decomposition desroches etla formation de la terre arable. C R Acad Sci 110:1370–1372

Nahas E, Banzatto DA, Assis LC (1990) Fluorapatite solubilization by *Aspergillus niger* in vinasse medium. Soil Biol Biochem 22:1097–1101

Neaman A, Chorover J, Brantley SL (2005) Implication of the evolution of organic acid moieties for basalt weathering over ecological time. Am J Sci 305:147–185

Nienow JA, Friedman EI (1993) Terrestrial lithophytic (rock) communities. In: Friedman EI (ed) Antarctic microbiology, Wiley series in ecological and applied microbiology (Wiley series in ecological and applied microbiology). Wiley-Liss, pp 342–412, 644 pp

Oerke EC (2006) Crop losses to pests. J Agric Sci 144:31–43

Oerke EC, Dehne HW (2004) Safeguarding production-losses in major crops and the role of crop protection. Crop Prot 23:275–285

Parfitt RL (1992) Potassium-calcium exchange in some New Zealand soils. Aust J Soil Res 30:145–158

Parida AK, Das AB (2005) Salt tolerance and salinity effects on plants: a review. Ecotoxicol Environ Safe 60:324–349

Paris F, Botton B, Lapeyrie F (1996) In vitro weathering of phlogopite by ectomycorrhizal fungi. II. The effect of K^+ and Mg^{2+} deficiency and N sources on accumulation of oxalate and H^+. Plant Soil 179:141–150

Park J, Sanford RA, Bethke CM (2009) Microbial activity and chemical weathering in the Middendorf Aquifer, South Carolina. Chem Geol 258:232–241

Parmar P, Sindhu SS (2013) Potassium solubilization by rhizosphere bacteria: influence of nutritional and environmental conditions. J Microbiol Res 3(1):25–31

Perrenoud S (1990) Potassium and plant health, 2nd edn. International Potash Institute, Bern, pp 8–10

Pier PA, Berkowitz GA (1987) Modulation of water stress effects on photosynthesis by altered leaf K+. Plant Physiol 85:655–661

Prajapati KB, Modi HA (2012) Isolation and characterization of potassium solubilizing bacteria from ceramic industry soil. CIB Tech J Microbiol 1:8–14

Prajapati K, Modi HA (2014) The study of shelf life of potassium solubilizing microorganisms for liquid biofertilizer. Paripex Ind J Res 3:13–14

Prasad D, Singh R, Singh A (2010) Management of sheath blight of rice with integrated nutrients. Indian Phytopathol 63:11–15

Puente ME, Bashan Y, Li CY, Lebsky VK (2004) Microbial populations and activities in the rhizoplane of rock-weathering desert plants. I. Root colonization and weathering of igneous rocks. Plant Biol 6:629–642

Rajawat MVS, Singh S, Singh G, Saxena AK (2012) Isolation and characterization of K-solubilizing bacteria isolated from different rhizospheric soil. In: Proceeding of 53rd annual conference of association of microbiologists of India, pp 124

Rao SC, Bansal SK, Subba Rao A, Takkar PN (1998) Potassium desorption kinetics of major benchmark soil series of India. J Ind Soc Soil Sci 46(3):357–362

Rehm GW, Sorensen RC (1985) Effects of potassium and magnesium applied for corn grown on an irrigated sandy soil. Soil Sci Soc Am J 49:1446–1450

Reynolds M, Bonnett D, Chapman SC, Furbank RT, Manes Y, Mather DE (2011) Raising yield potential of wheat. I. Overview of a consortium approach and breeding strategies. J Exp Bot 62:439–452

Richter DD, Markewitz D (1995) How deep is soil? Bioscience 45:600–609

Richter DD, Oh NH (2002) Biological factors affecting saprolite formation. Abstracts with Programs – Geological Society of America, Paper No. 48-4

Rozanova EP (1986) Leaching of glass and during microbiological oxidation of oil. Microbiology 55:787–791

Sahin F, Cakmakci R, Kantar F (2004) Sugar beet and barley yields in relation to inoculation with N -fixing and phosphate solubilizing bacteria. Plant Soil 2 (265):123–129

Sangeeth KP, Bhai RS, Srinivasan V (2012) *Paenibacillus glucanolyticus*, a promising potassium solubilizing bacterium isolated from black pepper (*Piper nigrum* L.) rhizosphere. J Spic Aromat Crops 21(2):118–124

Sarwar M (2012) Effects of potassium fertilization on population build up of rice stem borers (lepidopteron pests) and rice (*Oryza sativa* L.) yield. J Cereal Oilseeds 3:6–9

Schroeder D (1974) Relationships between soil potassium and the potassium nutrition of the plant. In: Potassium research and agricultural production. Proceedings of the 10th congress of the international Potash Institue, pp 53–63

Sen Gupta A, Berkowitz GA, Pier PA (1989) Maintenance of photosynthesis at low leaf water potential in wheat. Plant Physiol 89:1358–1365

Setter TL, Waters I (2003) Review of prospects for germplasm improvement for waterlogging tolerance in wheat, barley and oats. Plant Soil 253:1–34

Shanware AS, Kalkar SA, Trivedi MM (2014) Potassium solublisers: occurrence, mechanism and their role as competent biofertilizers. Int J Curr Microbiol Appl Sci 3(9):622–629

Shao HB, Chu LY, Jaleel CA, Zhao CX (2008) Water-deficit stress-induced anatomical changes in higher plants. C R Biol 331:215–225

Sharma SR, Kolte SJ (1994) Effect of soil-applied NPK fertilizers on severity of black spot disease (*Alternaria brassicae*) and yield of oilseed rape. Plant Soil 167:313–320

Sharma KD, Nandwal AS, Kuhad MS (1996) Potassium effects on CO_2 exchange, ARA and yield of cluster bean cultivars under water stress. J Pot Res 12:412–423

Sheng XF (2005) Growth promotion and increased potassium uptake of cotton and rape by a potassium releasing strain of *Bacillus edaphicus*. Soil Biol Biochem 37:1918–1922

Sheng XF, Huang WY (2002) Mechanism of potassium release from feldspar affected by the strain NBT of silicate bacterium. Acta Pedol Sin 39(6):863–871

Sheng XF, Zhao F, He H, Qiu G, Chen L (2008) Isolation, characterization of silicate mineral solubilizing *Bacillus globisporus* Q12 from the surface of weathered feldspar. Can J Microbiol 54:1064–1068

Singh G, Biswas DR, Marwah TS (2010) Mobilization of potassium from waste mica by plant growth promoting rhizobacteria and its assimilation by maize (*Zea mays*) and wheat (*Triticum aestivum* L.). J Plant Nutr 33:1236–1251

Singh NP, Singh RK, Meena VS, Meena RK (2015) Can we use maize (*Zea mays*) rhizobacteria as plant growth promoter? Vegetos 28(1):86–99

Smart LB, Moskal WA, Cameron KD, Bennett AB (2001) MIP genes are down-regulated under drought stress in *Nicotiana glauca*. Plant Cell Physiol 42:686–693

Southam G (2000) Bacterial surface-mediated mineral formation. In: Lovley DR (ed) Environmental microbe-mineral interactions. ASM Press, Washington, DC, pp 257–276

Sparks DL (1987) Potassium dynamics in soils. Adv Soil Sci 6:1–63

Sparks DL (2000) Bioavailability of soil potassium. In: Sumner ME (ed) Handbook of soil science. CRC Press, Boca Raton, pp D38–D52

Sparks DL, Huang PM (1985) Physical chemistry of soil potassium. In: Munson RD et al (eds) Potassium in agriculture. ASA, Madison, pp 201–276

Sparks DL, Martens DC, Zelazny LW (1980) Plant uptake and leaching of applied and indigenous potassium in Dothan soils. Agron J 72:551–555

Staley JT, Jackson MJ, Palmer FE, Adams JB, Borns DJ, Curtiss B, Taylor-George S (1983) Desert varnish coatings and microcolonial fungi on rocks of the Gibson and Great Victorian deserts, Australia. BMR J Aust Geol Geophys 8:83–87

Sterflinger K (2000) Fungi as geologic agents. Geomicrobiol J 17:97–124

Sterflinger K, Krumbein WE (1997) Dematiaceous fungi as a major agent for biopitting Mediterranean marbles and limestones. Geomicrobiol J 14(3):219–230

Steudle E (2000) Water uptake by plant roots: an integration of views. Plant Soil 226:45–56

Suzuki N, Mittler R (2006) Reactive oxygen species and temperature stresses: a delicate balance between signaling and destruction. Physiol Plantarum 126:45–51

Thorseth IH, Furnes H, Tumyr O (1995) Textual and chemical effects of bacterial activity on basaltic glass: an experimental approach. Chem Geol 119:139–160

Tiwari HS, Agarwal RM, Bhatt RK (1998) Photosynthesis, stomatal resistance and related characters as influenced by potassium under normal water supply and water stress conditions in rice (*Oryza sativa* L.). Indian J Plant Physi 3(4):314–316

Troufflard S, Mullen W, Larson TR, Graham IA, Crozier A, Amtmann A, Armengaud P (2010) Potassium deficiency induces the biosynthesis of oxylipins and glucosinolates in *Arabidopsis thaliana*. BMC Plant Biol 10(1):172

Tyerman SD, Niemietz CM, Bramley H (2002) Plant aquaporins: multifunctional water and solute channels with expanding roles. Plant Cell Environ 25:173–194

Ullman WJ, Kirchman DL, Welch SA (1996) Laboratory evidence by microbially mediated silicate mineral dissolution in nature. Chem Geol 132:11–17

Uroz S, Calvaruso C, Turpault MP, Pierrat JC, Mustin C, Frey-Klett P (2007) Effect of the mycorrhizosphere on the genotypic and metabolic diversity of the bacterial communities involved in mineral weathering in a forest soil. Appl Environ Microbiol 73:3019–3027

Uroz S, Calvaruso C, Turpault MP, Frey-Klett P (2009) Mineral weathering by bacteria: ecology, actors and mechanisms. Trends Microbiol 17:378–387

Vandevivere P, Welch SA, Ullman WJ, Kirchman DL (1994) Enhanced dissolution of silicate minerals by bacteria at near-neutral pH. Microb Ecol 27:241–251

Vassileva M, Vassilev N, Azcon R (1998) Rock phosphate solubilization by *Aspergillus niger* on olive cake-based medium and its further application in a soil-plant system. World J Microb Biot 14:281–284

Vessey JK (2003) Plant growth-promoting rhizobacteria as biofertilizers. Plant Soil 255:571–586

Wallander H, Tonie W (1999) Biotite and microcline as potassium sources in ectomycorrhizal and non-mycorrhizal *Pinus sylvestris* seedlings. Mycorrhiza 9:25–32

Wang XM, Li WQ, Li MY, Welti R (2006) Profiling lipid changes in plant response to low temperatures. Physiol Plantarum 126:90–96

Wang M, Zheng Q, Shen Q, Guo S (2013) The critical role of potassium in plant stress response. Int J Mol Sci 14:7370–7390

Warscheid T, Braams J (2000) Biodeterioration of stone: a review. Int Biodeterior Biodegrad 46:343–368

Wegner L (2010) Oxygen transport in waterlogged plants. In: Mancuso S, Shabala S (eds) Waterlogging signaling and tolerance in plants. Springer, Berlin/Heidelberg, pp 3–22

Welch LF, Flannery RL (1985) Potassium nutrition of corn. In: Munson RD (ed) Potassium in agriculture. ASA, Madison

Welch SA, Barker WW, Barfield JF (1999) Microbial extracellular polysaccharides and plagioclase dissolution. Geochim Cosmochim Acta 63:1405–1419

Wu SC, Cao ZH, Li ZG, Chung KC, Wong MH (2005) Effects of biofertilizer containing N-fixer, P and K solubilizers and AM fungi on maize growth: a greenhouse trial. Geoderma 125:155–166

Wyn Jones RG, Brady CJ, Speirs J (1979) Ionic and osmotic relations in plant cells. In: Laidman DC, Wyn Jones RG (eds) Recent advances in the biochemistry of cereals. Academic, New York, pp 63–103

Xiong LM, Schumaker KS, Zhu JK (2002) Cell signaling during cold, drought, and salt stress. Plant Cell 14: S165–S183

Yadav DS, Goyal AK, Vats BK (1999) Effect of potassium in *Eleusine coracana* (L.) Gaertn under moisture stress conditions. J Potassium Res 15(1–4):131–134

Yadav VP, Sharma T, Saxena VK (2000) Dissolution kinetics of potassium from glauconitic sandstone in acid lixiviant. Int J Miner Process 60:15–36

Yuan L, Fang DH, Wang ZH, Shun H, Huang JG (2000) Bio-mobilization of potassium from clay minerals: I. By ectomycorrhizas. Pedosphere 10:339–346

Yuan L, Huang JG, Li XL, Christie P (2004) Biological mobilization of potassium from clay minerals by ectomycorrhizal fungi and eucalypt seedling roots. Plant Soil 262:351–361

Zahra MK, Monib MS, Abdel-Al I, Heggo A (1984) Significance of soil inoculation with silicate bacteria. Zentralbl Mikrobiol 139(5):349–357

Zarjani JK, Aliasgharzad N, Oustan S, Emadi M, Ahmadi A (2013) Isolation and characterization of potassium solubilizing bacteria in some Iranian soils. Arch Agro Soil Sci 77:7569

Zhang C, Kong F (2014) Isolation and identification of potassium-solubilizing bacteria from tobacco rhizospheric soil and their effect on tobacco plants. Appl Soil Ecol 82:18–25

Zhang QC, Wang GH, Feng YK, Qian P, Schoe-nau JJ (2011) Effect of potassium fertilization on soil potassium pools and rice response in an intensive cropping system in China. J Plant Nutr Soil Sci 174:73–80

Zhao TJ, Sun S, Liu Y, Liu JM, Liu Q, Yan YB, Zhou HM (2006) Regulating the drought responsive element (DRE)-mediated signaling pathway by synergic functions of trans-active and transinactive DRE binding factors in *Brassica napus*. J Biol Chem 281:10752–10759

Zhao F, Sheng X, Huang Z, He L (2008) Isolation of mineral potassium-solubilizing bacterial strains from agricultural soils in Shandong Province. Biodivers Sci 16:593–600

www.ncagr.gov

Towards the Soil Sustainability and Potassium-Solubilizing Microorganisms

18

Indra Bahadur, Bihari Ram Maurya, Ashok Kumar, Vijay Singh Meena, and Richa Raghuwanshi

Abstract

Potassium (K) is needed in adequate quantities for a crop to achieve its maximum yield. The function of potassium in plant growth has not been clearly defined but it is associated with movement of water, nutrients and carbohydrates in plant tissue. When potassium is not adequate, the plants will have poor root development and will grow slowly, and their seeds will become small and have lower yields. About 5 million tonnes of potassic fertilizer requirement would be fulfilled through imports because India does not have commercial-grade sources of potash reserve. India is totally dependent on the import of potassic fertilizers. On the other hand, India has the largest reserve for low potassium-containing minerals. The depletion of potassium in soil has been started, and in future this will aggravate. In most of the soils, about 90–98 % of total K exists in relatively unavailable minerals such as feldspar, orthoclase and the micas (muscovite, biotite, phlogopite, etc.). These minerals are very resistant to decomposition and probably supply relatively smaller quantity of potassium to growing crops. Potassium in soil is present in four forms: water-soluble (solution K), exchangeable, non-exchangeable and structural or mineral forms. The fixed form of K minerals is solubilized by K solubilizers, and then acquisition or accumulation of potassium by crop plants certainly will be enhanced. For evergreen agriculture, production can only be fulfilled when the environment, its caretakers and surrounding communities are healthy, for this application of KSMs holds a key

I. Bahadur (✉) • B.R. Maurya
Department of Soil Science and Agricultural Chemistry, Institute of Agricultural Sciences, Banaras Hindu University, Varanasi 221005, India
e-mail: ibm07025@gmail.com

A. Kumar • R. Raghuwanshi
Department of Botany, MMV, Banaras Hindu University, Varanasi 221005, Uttar Pradesh, India

V.S. Meena
Department of Soil Science and Agricultural Chemistry, Institute of Agricultural Sciences, Banaras Hindu University, Varanasi 221005, India

Indian Council of Agricultural Research – Vivekananda Institute of Hill Agriculture, Almora 263601, Uttarakhand, India

V.S. Meena et al. (eds.), *Potassium Solubilizing Microorganisms for Sustainable Agriculture*, DOI 10.1007/978-81-322-2776-2_18

approach for K availability in soils. KSB increased K availability in soils and increased mineral nutrient specially K uptake by plant.

Keywords
KSMs • Biodiversity • Genetic modification • Agriculture • Soil sustainability

18.1 Introduction

Ensuring food security for burgeoning population requires additional food grain production from the same land and this is obviously fulfilled by the green revolution. Food grain production potential of the soil increases. Which is very impressive but with insufficient concern for environment and soil sustainability. It is a great challenge in date to search for sustainable strategies to alleviate detrimental effects of intensive farming practices based on chemical input. Use of soil microorganisms for sustainable agriculture has increased in various parts of the world because of the negative environmental impact of chemical fertilizers and their increasing costs during the last couple of decades. Now the government of India has been trying to promote an improved practice involving the use of biofertilizers/bio-organics together with chemical fertilizers. Nowadays, in a developing country like India, where land-person ratio is rapidly narrowing, the only means of meeting the needs of agricultural produce is through increasing productivity without harming the environment and agricultural sustainability. Fertilizers, therefore, assume a great significance and constitute one of the key inputs for achieving high productivity (Meena et al. 2013; Singh et al. 2015). Nowadays, chemical fertilizer has started to show side effects on human, soil, as well as environment due to their imbalanced application in crops. Among the chemical fertilizers, potassium is an important primary element which is utilized for many cereals, vegetables and fruits and being added to the soil as potassium sulphate or potassium chloride (Grayston et al. 1996; Khan et al. 2009; Li et al. 2003; Lopes-Assad et al. 2010).

Soils are store house of microorganisms of multiple natures and have been benefiting the vegetation without any care by human being. Many of them such as N_2 fixers, P solubilizers and organic matter decomposers have been explored for beneficial use in agriculture for plant nutrient acquisition. There are so many potassium-solubilizing bacteria that are in soil which vary in their nature and activities (Gromov 1957; Norkina and Pumpyansakya 1956). Indian farmer's economic condition is not so good. Sustainable agriculture is the process in which economically viable, environment-friendly and socially acceptable technology is used for enhancing agricultural productivity and soil fertility. For sustainability, agriculture can do in long run when the farmers and surrounding communities are healthy without degrading environment.

A good soil means it feels soft and crumby, it has high nutrient-supplying capacity, it is rich in nutrient, it has good water storage and drainage capacity, it has resistance to erosion and nutrient loss, it has good amount of soil microorganisms, its productivity is good, and it produces healthy and quality crops (Doran 1994).

Extensive use of chemical fertilizers caused ecological imbalance, environmental pollution and hazards to soil health. Microbial inoculants are now becoming more popular in India and abroad as they are inexpensive and simple to use and have no side effects. Therefore, the use of efficient strains of biofertilizer microorganisms plays a vital role in agriculture and ecosystems. Enhancement in productivity of cereals, pulses and vegetables has been reported by Maurya et al. (2014), Meena et al. (2014a) and Verma et al. (2013).

In India, the total potassium fertilizer requirement is nearly about 5 million tonnes and totally made through imports, because India does not have commercial-grade rock for K fertilizer making (Goteti et al. 2013). Nowadays, the use of efficient rhizospheric microorganisms may offer plant growth promotion and agronomic, pathogenic and environmental benefits for intensive agricultural systems. The plant growth-promoting microorganisms (PGPMs) exhibit a gradual increase in demand in the world market. One possible mechanism for the effectiveness of biofertilizers, such as mobilization of sparingly available potassium (Meena et al. 2013; 2015a; Maurya et al. 2014), is its capacity to produce plant growth-promoting substances, which enhanced/induced resistance to environmental stress. In some cases, it reduces the direct or indirect infection of plant pathogenic microorganisms. Ecofriendly agricultural system has emerged as an important priority area globally in view of the growing demand for safe and healthy food and long-term soil as well as environmental sustainability and concerns on environmental pollution associated with the indiscriminate use of agrochemicals (Milic et al. 2004). These efficient microorganisms play a very important role as the component of the biological soil phase and also as indicator of soil fertility and soil degradation (Kim et al. 1998).

India is promising to have the world's enormous accumulation of mica distributed in Bihar, Jharkhand, Rajasthan and Andhra Pradesh. During the dressing of raw mica mined from mica mines located in these districts, a huge amount of waste mica are produced (about 75 % of total mined mica), and it contains 8–12 % K_2O that may be used in agriculture as a source of potassium which increases the crop production significantly (Nishanth and Biswas 2008; Meena et al. 2015b).

18.2 Need of Potassium-Solubilizing Microorganisms (KSMs) in K Nutrition

Plants can take up potassium only from the soil solution. Its availability is totally dependent upon the K dynamics as well as on total K content. Most of the Indian soils are rich in potassium. However, as a result of increase in crop yield due to rapid development and the use of modern intensive agriculture in the world and India, in particular, soil nutrient levels have dropped due to mining through crop removal without replenishing soil through fertilizer. Soil potassium and fertilizer deficiency is the major problem in the future for the development of Indian agriculture (Sheng and Huang 2002). Overuses of chemical fertilizers in crop production are costly as well as different unpropitious consequences like soil degradation and inconsistency in plant nutrition. Now is the time to develop reasonable, effective and ecofriendly nutrient sources which work without disturbing the environment (Aleksandrov 1958; Barre et al. 2008).

Bio-intervention of waste mica with potassium-solubilizing microorganism could be the alternative and viable technology to solubilize insoluble K in mica into plant-available pool and used efficiently as a source of K fertilizer for sustaining crop production and maintaining soil K (Basak and Biswas 2009). Research carried out earlier reported that some microorganisms in the soil are able to solubilize unavailable forms of K-bearing minerals such as micas, illite and orthoclases by excreting organic acids (Friedrich et al. 1991; Ullaman et al. 1996; Bennett et al. 1998).

Right now satisfied species of microorganism are extensively used which have unique properties to provide natural products and can be used as chemical fertilizer substitutes. Injudicious use of chemical fertilizer increases the cost and decreases the efficiency of K fertilizer, ruining the environment (Zhang et al. 2013). The alternative of the chemical/mineral potassium fertilizer is necessary for the evergreen agriculture/sustainable agriculture. In India, it is estimated that by 2020, for fulfilling the targeted production of about 325 million tonnes of food grain, the requirement of nutrient will be about 29 million tonnes, while their availability will be only about 21.6 million tonnes, having a deficit of about 7.4 million tonnes (Uroz et al. 2009). Therefore, the application of potassium-solubilizing microorganisms may be a promising approach for increasing K availability in soils.

Their use in agriculture can reduce the use of chemical fertilizer and support ecofriendly crop production (Meena et al. 2014a; Berthelin and Leyval 1982; Bennet et al. 2001; Deshwal and Kumar 2013).

Besides the above facts, the long-term use of biofertilizers is cheap, ecofriendly, effective, productive and approachable to farmers over chemical fertilizers (Subba 2001). Potassium constitutes about 2.5 % of the lithosphere, and concentration in soil of these nutrients varies widely ranging from 0.04 % to 3.0 %: only 1–2 % of this is available to plants and the rest were bound with minerals present in soil and that is why they are unavailable to plants. Potassium availability depends on the K dynamics and its total K content. Plant can uptake only from soil solution (Girgis 2006; Goteti et al. 2013; Kawalekar 2013).

The role of efficient rhizospheric microorganisms in silicate mineral mineralization was known in the nineteenth century. After that considerable studies were done on mineral potassium mineralization by naturally abundant rhizospheric microbes. K solubilization is carried out by a large number of bacteria like *B. mucilaginosus*, *B. edaphicus*, *B. circulans*, *Acidithiobacillus ferrooxidans*, *Paenibacillus* spp. and *Aspergillus* spp. Major amounts of K-containing minerals (muscovite, orthoclase, biotite, feldspar, illite, mica) are present in the soil as a fixed form which is not directly utilize by the crop plant (Meena et al. 2014b; Kumar et al. 2015).

Nowadays, most of the farmers use injudicious application of chemical fertilizers for achieving maximum productivity (Sparks and Huang 1987). The KSMs are globally distributed but numbers vary from one soil to another. KSMs are present in rhizospheric soils in large number as compared with nonrhizospheric soils, and they are also metabolically active than non-rhizosphere-isolated microbes (Maurya et al. 2014). KSMs obtained from rhizosphere are more tolerant to salt, pH and temperature. When KSMs are inoculated in a solid medium containing insoluble/mineral K, they are detected by the formation of clear halo zone around the colony, and methods for the isolation of KSMs were developed by many researchers (Maurya et al. 2014; Bahadur et al. 2015; Meena et al. 2015a).

Microbes can enhance mineral dissolution rate by producing and excreting metabolic by-products that interact with the mineral surface. Complete microbial respiration and degradation of particulate and dissolved organic carbon can elevate carbonic acid concentration at mineral surfaces in soils and in ground water (Barker et al. 1998) which can lead to an increase in the rates of mineral weathering by a proton-promoted dissolution mechanism. Therefore, dissolution of soil K minerals by silicate-dissolving bacteria was purified by repeated inoculation and maintained for further genetical and morphological study. And efficient KSMs were selected on the basis of ability of K release and plant growth promotion activities. Finally, efficient isolates were selected and are used for making the inoculants and tested in field and pot experiments in various crop, and the strains were established as potassic biofertilizers (Purushothaman and Natarajan 1974; Rajawat et al. 2012; Prajapati et al. 2013).

18.3 Biodiversity of KSMs

Research issue of the use of microorganisms for plant growth and control of plant pests is very quickly expanding. Various amalgamations of amino acids (lysine, arginine, leucine, isoleucine, valine, glycine, proline, cysteine, etc.), organic acids (citric acid, oxalic acid, malic acid, acetic acid, pyruvic acid, formic acid, butyric acid, glycolic acid, succinic acid, etc.), sugars (glucose, fructose, galactose, ribose, xylose, niacin, raffinose, oligosaccharides, etc.), vitamins (biotin, thiamine, pantothenate, niacin, riboflavin, etc.), purines, adenine, guanine, cytidine, uridine, enzymes (phosphatase, invertase, amylase, protease, urease, etc.) and some gaseous molecules (OH^-, HCO_3^-, $CO_2^-H_2$, etc.) in root exudates of different plant species (Dakora and Phillips 2002) help microbes survive in various condition. Microbes produce many root exudates which are very helpful in plant nutrition and growth.

These amalgamations were decisive for the applicability of microbes as bio-inoculants/biofertilizers for acquisition of nutrient for plant growth. A wide range of bacteria is able to solubilize potassium from mineral K-containing soil that enhanced the availability of potassium in soil solution which is easily utilized by crop plant, resulting in healthy growth of plant and improved yield. Plant replenishes the automated platform, facilitating uptake of nutrient and water, and plant roots also secrete a wide range of compounds (Walker et al. 2003). These exudates released by roots of plant attract the wide range of microbes which is heterogeneous, diverse and metabolically active soil microbial communities. These root exudates act as binding material/cementing agent of soil and, thus, improve soil structure and regulate and maintain the microbial population near the root surface. That is why rhizosphere reaches in microbial communities and its population.

Microbial activity near the root surface plays an important role in the development and rooting pattern of the plant. A fraction of these plant-derived small organic molecules are further metabolized by microorganisms in the vicinity as carbon and nitrogen sources, and some microbe-oriented molecules are subsequently retaken up by plants for growth and development (He and Sheng 2006). Several bacterial species known as KSB assist plant growth by mobilizing insoluble K. Population of KSB vary from soil to soil. A variety of bacteria discovered by many scientists as K solubilizers includes *Bacillus mucilaginosus*, *B. edaphicus*, *B. circulans*, *Arthrobacter* spp., etc.

18.4 The Nature of K Biofertilizers

Rhizospheric topsoils are rich in microbes and contain about 400 kg per acre of earthworms, 1088.622 kg per acre of fungi, 680 kg per acre of bacteria, 60 kg per acre of protozoa and 403 kg per acre of arthropods and algae, and even in small mammals in some cases (Verma et al. 2009). That is why soil can be viewed as a living community rather than an inert body. A decomposed part of organic matter called as humus contains dead organisms and plant organic materials in different decomposition stages. Humus and organic matters are reservoir of plant nutrient element; they also help in soil structure formation and provide some other benefits (Welch and Ullman 1993).

In fact, topsoil is the most biologically diverse part of the earth. Soil-dwelling organisms release bound-up minerals, converting them into plant-available forms that are then taken up by the plants growing on the site (Deshwal and Kumar 2013; Kawalekar 2013).

K-solubilizing microorganisms are morphologically diverse in characteristics such as colony form, margin, elevation, colour, slime production and gram's reaction. In many experiments, KSMs have been found in circular form, entire margin and cream-coloured colony. They are both in gram-negative and gram-positive rods that varied in length from short to long rods, cocci, etc. KSMs were observed for production of slime in different amount, i.e. high, medium and low. Extracellular production of slime is the main feature of potassium solubilizers. They survive in various climatic conditions as well as different types of soil that have been tested in different temperature as well as pH ranges (US Department of Agriculture 1998; Vandevivere et al. 1994; Xie 1998; Sheng 2002; Zhao et al. 2008).

18.4.1 Potassium-Solubilizing Microorganisms (KSMs)

The potassium-solubilizing microorganisms (KSMs) are rhizospheric microorganisms which solubilize the insoluble potassium (K) to soluble forms of K for plant growth and yield. K solubilization is carried out by a wide range of microorganisms (*B. mucilaginosus*, *B. edaphicus*, *B. circulans*, *B. subtilis*, *B. pumilus*, *Agrobacterium tumefaciens*, *Flavobacterium* spp., *Rhizobium* spp., etc.) and fungal strains (*Aspergillus* spp.) (Li et al. 2003; Meena et al. 2014a, b; Maurya et al. 2015; Zarjani et al. 2013; Gundala et al. 2013). Major amounts of potassium-containing minerals (muscovite, orthoclase, biotite, feldspar, illite, mica) are present in the soil as a fixed form which is not directly taken up by the plant. These insoluble sources are solubilized by many specific types of microorganisms called KSMs (Meena et al. 2014a, b; 2015b). This available K can be easily taken up by the plant for growth and development. These areas of the research are less focused or unidentified, but nowadays the growth of the research in these areas is enhanced. The KSB/KSR is isolated from different crop rhizospheres and from minerals, and these microbes are able to dissolve potassium from mineral soils that enhanced the crop growth, yield and soil sustainability (Maurya et al. 2014; Meena et al. 2015b). The evidence of solubilization of mineral potassium is studied by many scientists starting from the nineteenth century; Berthelin and Leyval (1982) suggested the regulation of silicon and its cycling in sea water, and they also reported the production of different organic and inorganic acids by these organisms. Groudev (1987) stated that silicate mineral dissolution was enhanced by exopolysaccharides, extrapolysaccharides and mucilaginous compound.

18.4.2 Search for Potassium-Solubilizing Microorganisms (KSMs)

The KSMs will be isolated on Aleksandrov medium using serial dilution followed by pour plate or streaking (Aleksandrov et al. 1967) from rhizospheric soil, nonrhizospheric soil and overburden samples near the mica deposit area (Basak and Biswas 2010; Maurya et al. 2014) (Fig. 18.1).

Once the efficient KSMs are screened, then it is tested for solubilization of potassium in Aleksandrov broth medium containing insoluble source of potassium mineral. For screening the KSMs, considering the high K solubilization capacity, efficient potassium-solubilizing isolates were selected for in vivo study of K solubilization capacity in soil mixed with K mineral at different time intervals. Ultimately efficient potassium solubilizers are used for pot culture and field authentication using various crops (Archana et al. 2013). The KSMs are isolated/obtained from different rhizospheric soils of various plants such as wheat (Parmar and Sindhu 2013; Zhang et al. 2013), feldspar (Sheng et al. 2002), potato-soybean cropping sequence (Biswas 2011), Iranian soils (Zarjani et al. 2013), ceramic industry soil (Prajapati and Modi 2012), mica core of Andhra Pradesh (Gundala et al. 2013), common bean (Kumar et al. 2012), biofertilizers (Zakaria 2009), sorghum and maize (Archana et al. 2013).

18.5 KSM: A Promising Approach in Sustainable Agriculture

The currently increasing human population, industrialization and urbanization cause the shrinkage of agricultural land and food crisis. It is a misconception that Indian soils are rich in potassium. In the future, deficiency of potassium in soil certainly would be a serious problem. These K solubilizers play a significant role in the solubilization of the part of mineral K of soil (90 %) to partially cater the plants' need of K (Fig. 18.2).

It is estimated that ~50–60 % of potash chemical fertilizers usage can be reduced by using *Frateuria aurantia*, a new bacterial species (species conformation by IMTECH Chandigarh) as a bio-inoculants. These new bacteria belonging to the family *Pseudomonadaceae* have the extra ability to mobilize K in almost all types of soils especially low K content soils and soils of pH 5–11, and they survive in a temperature of up to 42 °C. This potash-mobilizing biofertilizers can be applied in combination with *Rhizobium*, *Azospirillum*, *Azotobacter*, *Acetobacter*, PSM, etc. Potash-mobilizing bacterial-based product containing *Frateuria aurantia* produces plant growth-promoting substances which offer plant a multifaceted benefit in terms of growth, by mobilizing potash and making it available to crops. It also enhances the efficiency of chemical fertilizer (Patel 2011). Soil microorganisms enhance the potassium availability and production of PGPS (plant growth-promoting substances) (Barre et al. 2008).

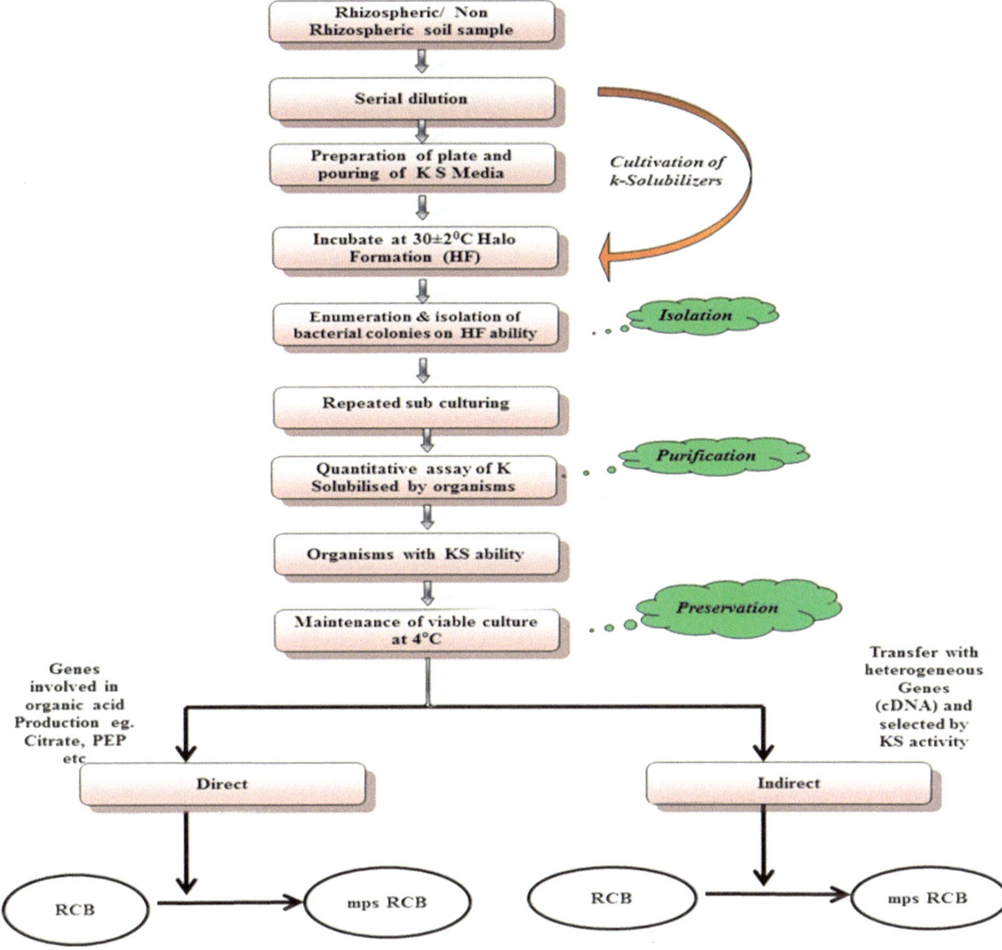

Fig. 18.1 Figure showing potassium-solubilizing microorganism (KSM) isolation, characterization and genetic modification

18.6 Methods of KSM Applications

KSB inoculants used as a seed treatment are cheap and common and the easiest means of inoculation. When properly applied, this method ensures that each seed receives the introduced KSB micophos. It is also used as a seedling treatment. One packet of inoculants (200 g) is mixed with 500 ml of water to make slurry (Bahadur et al. 2014). The seeds required are mixed in the slurry to have a uniform coating of the inoculants over the seeds and then shade dried for 30 min. The shade-dried seeds should be sown within 24 h. One packet of inoculants (200 g) is sufficient to treat 10 kg of seeds (Subba 2001). However, the use of a sticker solution of *gum acacia* improves the adherence of the inoculant KSMs on the seed. Thus, in accordance with these considerations, two approaches can be applied for KSM inoculation: firstly, the single-culture approach (SCA), where K solubilizers can be used alone, and, secondly, the multiple or mixed culture approach (MCA), often called co-inoculation, where KSMs are used along with other beneficial rhizosphere microorganisms.

The application of potassic biofertilizer is also used as a seedling root dip; this method is used for transplanted crops, and it is also used with

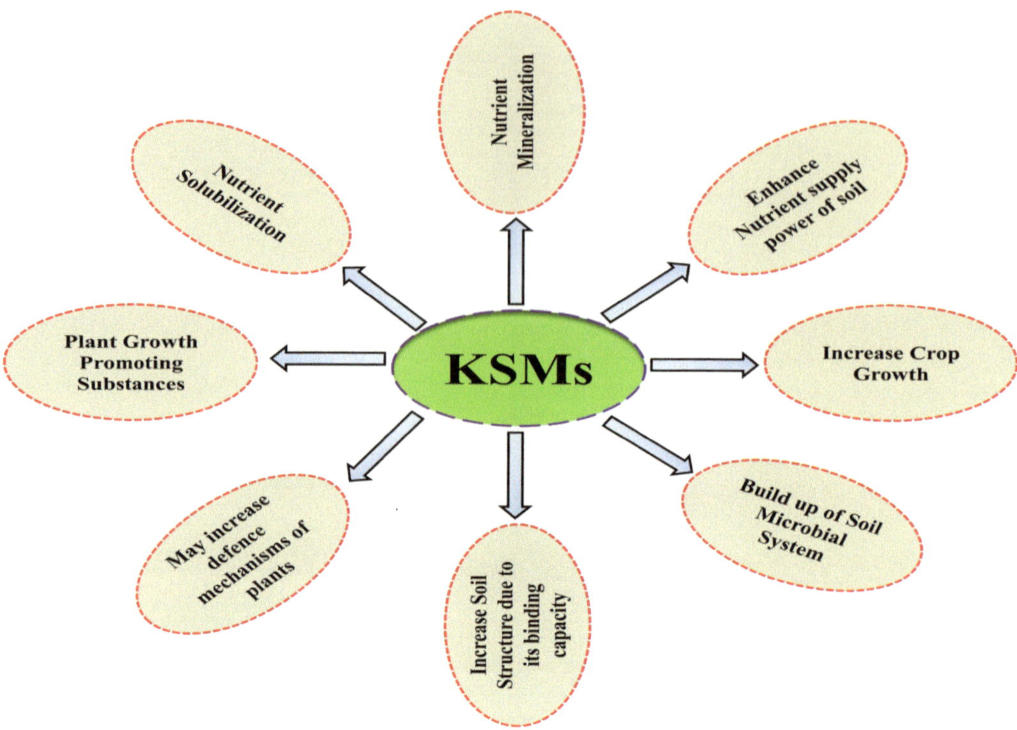

Fig. 18.2 How potassium-solubilizing microorganisms (KSMs) help in soil sustainability for a system development

farmyard manure by broadcasting in the main field just before transplanting/sowing of seeds. There are, however, certain situations where seed applications may be an ineffective means of application, e.g. with seeds dressed with pesticides incompatible with PSMs. Under such circumstances, soil application may be followed. Inoculants applied to the soil have the following advantages: greater population of KSM per unit area, minimized direct contact with chemically treated seeds, elimination of seed mixing and apparent ability to withstand low moisture conditions better than the powder form.

18.7 Factors Affecting KSM Inoculants

Indigenous microbes residing in soil contest for nutrition and moisture with applied microbial inoculants and frequently do not confess their productive establishment in the soil with the inoculated population. Sometimes poor or inefficient organic matter and moisture status in the inoculated field might limit the growth and multiplication of KSMs. The population declines during the off season but may increase after planting of crop. Using proper inoculation technique, survival of potassium-solubilizing microorganisms is enhanced. KSMs live in a microbial culture so they require careful handling, storage and transportation facility.

18.8 Effects of KSMs on Crop Response

Bio-tampering of K mineral (waste mica) with KSMs perhaps substitutes a feasible technology to solubilize mineral potassium present in mica into plant useable nutrient efficiently used as a source of K nutrition for sustaining crop production and maintaining soil potassium (Basak and Biswas 2009; Meena et al. 2015a, b; Maurya et al. 2015).

Inoculation with KSMs have been described strive valuable resultance on growth of many crops were established by many worker cotton and mustard (Sheng 2005) pepper and cucumber (Han et al. 2006), banana (Hassan et al. 2010), sorghum (Badr 2006), wheat (Sheng and He 2006), tomato (Lian et al. 2008), chilli (Ramarethinam and Chandra 2005) and sudan grass (Basak and Biswas 2010). Similarly, Zahra et al. (1984) reported that soil inoculated with silicate-dissolving bacteria *B. circulans* for solubilization of potassium and silicate from various silicate minerals showed significant increase of organic matter and ~17 % yield of rice. Augmentation of wheat yield up to 1.04 t per hectare was reported. According to Badar (2006), the co-inoculation of KSMs in both phosphorus- and potassium-bearing minerals on sorghum was recorded to enhance dry matter yield and nutrient uptake (~48 %, 65 % and 58 %), P (~71 %, 110 % and 116 %) and K (~41 %, 93 %, and 79 %) from three uptakes in three distinct soils, respectively. Archana et al. (2008) reported that the KSMs were isolated from rock and rhizosphere soils of *Vigna radiata* and reported that these KSMs enhance the solubilization of K in acid-leached soil as well as increase seedling growth and yield.

Many investigators have recorded the existence of soil microbes capable of solubilizing insoluble mineral resources to plant-available forms by excreting several metabolic by-products that interact with mineral surfaces and release nutrients. Silicophilic and aluminophilic bacteria are potential agents to release K from potassic minerals, and consequently they bring K solution for plant use. In vitro condition It was observed that K-solubilizing bacterial colonies characteristically produce slime intensively that provides a tool for the isolation of K-solubilizing bacteria. Microbial inoculants exhibit their response by improvement in soil fertility and/or increase in grain yield.

18.9 Reasons Why KSM Inoculants Do Not Respond to the Crop Species

In India, biofertilizer industry increased strangely in the last two and a half decades, but they are still in shortage of biofertilizer and are far away from their potential. And a very few private industries are engaged with K biofertilizers. Limited nutrient mobilization potential compared to their chemical counterparts and slow impact on crop growth are the major constraints. Inconsistent responses in the field under varied agroecological niches and cropping systems have also contributed to their low acceptance by farmers. Besides these, there are some technological constraints, which restrict the fast growth of biofertilizer industry. Some of the major constraints and limitations of the industry are as follows:

(a) Susceptibility of strains to high chemical fertilizer use.
(b) Less interest in scientific community on the development of K biofertilizer technologies.
(c) Culture collection banks not yet developed for KSMs due to this loss of efficient strains developed by scientists.
(d) Deficiency in technology in respect to carrier suitability and product formulations.
(e) Lack of automation in product handling.
(f) Liquid inoculants are coming up as solution, but the technology is still immature and not available in public domain and more or less it is costly.
(g) Distribution channels through government agencies are not effective, which leads to cut throat competition among bidders, resulting in a low-cost and poor-quality inoculant production.

18.10 Conclusions

The development of sustainable agricultural system requires a new technique to use less amount of chemical fertilizer while maintaining proper crop yields. The application of biological resources to exploit nutrient present in soil may hold promises for the future. Now facts are established that microbes are useful in increasing plant growth in many ways, like nutrient acquisition, solubilization, mobilization and secretion of root exudates which help in plant growth promotion, disease prevention and suppuration and stress control. Chemical fertilizer gradually commenced their side effect on human being and environment; however, the use of KSMs as a biofertilizer can improve available plant nutrient and production of crop in a sustainable way. It is very important to make a successful research work done for the recognition of an elite microbial strain capable of solubilizing potassium minerals quickly in large quantity which can conserve our existing resources and avoid environmental pollution hazards caused by excessive/injudicious use of chemical fertilizers. This communication highlighted the contributions of rhizospheric microorganisms especially potassium-solubilizing bacteria which can enhance the productivity of agricultural crops without disturbing the environment. This type of microbial consortium is cost-effective and ecofriendly for enhancing the sustainable agriculture. Application of waste mica could be a substitute for chemical fertilizer, and it is an important technology for the solubilization of potassium from insoluble sources in an ecofriendly way for sustainable crop production and supply of K. In addition, to safeguard the quantity as well as quality of food in developing countries like India, China, Canada, etc. in the long run, there is an indispensable urgency for the sustainable intensification of agricultural production system for productivity and revenue origination. In this situation, a viable, innovative, unique, genetically modified, soil and location-specific KSM biotechnology is the ultimate tool for use in the farmers' field in a short time to mitigate the potassium loss.

Acknowledgements Authors are thankful to the head of the Department of Soil Science and Agricultural Chemistry, Institute of Agricultural Sciences, Banaras Hindu University, Uttar Pradesh-05, India, for providing the critical suggestion, encouragement and support.

References

Aleksandrov VG (1958) Organo-mineral fertilizers and silicate bacteria. Dokl Akad SKh Nauk 7:43–48

Aleksandrov VG, Blagodyr RN, Iiiev IP (1967) Liberation of phosphoric acid from apatite by silicate bacteria. Mikrobiyol Zh (Kiev) 29:111–114

Archana DS, Savalgi VP, Alagawadi AR (2008) Effect of potassium solubilizing bacteria on growth and yield of maize. Soil Biol Ecol 28(1–2):9–18

Archana DS, Nandish MS, Savalagi VP, Alagawadi AR (2013) Characterization of potassium solubilizing bacteria (KSB) from rhizosphere soil. Bioinfolet 10: 248–257

Badar MA (2006) Efficiency of K-feldspar combined with organic material and silicate dissolving bacteria on tomato yield. J App Sci Res 2(12):1191–1198

Badr MA (2006) Efficiency of K-feldspar combined with organic materials and silicate dissolving bacteria on tomato yield. J Appl Sci Res 2:1191–1198

Bahadur I, Meena VS, Kumar S (2014) Importance and application of potassic biofertilizer in Indian agriculture. Int Res J Biol Sci 12:80–85

Bahadur I, Maurya BR, Kumar S, Dixit J, Chauhan AS, Manjhi BK, Meena VS, Narayan SRP (2015) The novel potassic bio-fertilizers: a promising approach for evergreen agriculture. Int J Microbiol Res 7(5): 692–697

Barker WW, Welch SA, Chu S, Banfield F (1998) Experimental observations of the effects of bacteria on aluminosilicate weathering. Am Miner 83:1551–1563

Barre P, Montagnier C, Chenu C, Abbadie L, Velde B (2008) Clay minerals as a soil potassium reservoir: observation and quantification through X-ray diffraction. Plant Soil 302:213–220

Basak BB, Biswas DR (2009) Influence of potassium solubilizing microorganism (Bacillus mucilaginosus) and waste mica on potassium uptake dynamics by sudangrass (*Sorghum vulgare Pers.*) grown under two Alfisols. Plant Soil 317:235–255

Basak BB, Biswas DR (2010) Co-inoculation of potassium solubilizing and nitrogen fixing bacteria on solubilization of waste mica and their effect on growth promotion and nutrient acquisition by a forage crop. Biol Fertil Soils Can J Microbiol 52:66–72

Bennett PC, Choi WJ, Rogers JR (1998) Microbial destruction of feldspars. Miner Mag 8(62A):149–150

Bennet PC, Rogers JR, Choi WJ, Hiebert FK (2001) Silicates, silicate weathering, and microbial ecology. Geomicrobiol J 18:3

Berthelin J, Leyval C (1982) Ability of symbiotic and non-symbiotic rhizospheric microflora of maize (*Zea mays*) to weather micas and to promote plant growth and plant nutrition. Plant Soil 68:369–377

Biswas DR (2011) Nutrient recycling potential of rock phosphate and waste mica enriched compost on crop productivity and changes in soil fertility under potato–soybean cropping sequence in an Inceptisol of Indo-Gangetic Plains of India. Nutr Cycl Agroecosyst 89:15–30

Dakora F, Phillips DA (2002) Root exudates as mediators of mineral acquisition in low-nutrient environments. Plant Soil 245:35–47

Deshwal VK, Kumar P (2013) Production of Plant growth promoting substance by Pseudomonas. J Acad Indus Res: pp 221–225

Doran JW (1994) Defining soil quality for a sustainable environment. Soil Science Society of America, Madison

Friedrich S, Plantonova NP, Karavaiko GI, Stichel E, Glombitza F (1991) Chemical and microbial solubilization of silicates. Acta Biotechnol 11:187–196

Girgis MGZ (2006) Response of wheat to inoculation with phosphate and potassium mobilizers and organic amendment. Ann Agric Sci Ain Shams Univ Cairo 51(1):85–100

Goteti PK, Leo DAE, Desai S, Mir Hassan Ahmed S (2013) Prospective zinc solubilising bacteria for enhanced nutrient uptake and growth promotion in Maize (Zea mays L.). Int J Microbiol doi:10.1155/2013/869697

Grayston SJ, Vaugha D, Jones D (1996) Rhizosphere carbon flow in trees in comparison with annual plants: the importance of root exudation and its impact on microbial activity and nutrient availability. Appl Soil Ecol 5:29

Gromov BV (1957) The microflora of rocks and primitive soil in some northern regions of the USSR. Mikrobiologiya 26:52–54

Groudev SN (1987) Use of heterotrophic microorganisms in mineral biotechnology. Acta Biotechnol 7:299–306

Gundala PB, Chinthala P, Sreenivasulu B (2013) A new facultative alkaliphilic, potassium solubilizing, Bacillus Sp. SVUNM9 isolated from mica cores of Nellore District, Andhra Pradesh, India. Research and reviews. J Microbiol Biotechnol 2(1):1–7

Han HS, Supanjani, Lee KD (2006) Effect of co-inoculation with phosphate and potassium solubilizing bacteria on mineral uptake and growth of pepper and cucumber. Plant Soil Environ 52:130 136

Hassan EA, Hassan EA, Hamad EH (2010) Microbial solubilization of phosphate–potassium rocks and their effect on khella (*Ammi visnaga*) growth. Ann Agric Sci 55(1):37 53

He LY, Sheng XF (2006) Solubilization of potassium-bearing minerals by a wild-type strain of Bacillus edaphicus and its mutants and increased potassium uptake by wheat. Can J Microbiol 52(1):66–72

Kawalekar JS (2013) Role of biofertilizers and biopesticides for sustainable agriculture. J Bio Innov 2(3):73–78

Khan AA, Jilani G, Akhtar MS, Naqvi SMS, Rasheed M (2009) Phosphorus solubilizing bacteria: occurrence, mechanisms and their role in crop production. J Agric Biol Sci 1(1):48–58

Kim KY, Jordan D, McDonald GA (1998) *Enterobacter agglomerans*, phosphate solubilizing bacteria, and microbial activity in soil: effect of carbon sources. Soil Biol Biochem 30:995–1003

Kumar P, Dubey RC, Maheshwari DK (2012) Bacillus strains isolated from rhizosphere showed plant growth promoting and antagonistic activity against phytopathogens. Microbiol Res 67:493–499

Kumar A, Bahadur I, Maurya BR, Raghuwanshi R, Meena VS, Singh DK, Dixit J (2015) Does a plant growth-promoting rhizobacteria enhance agricultural sustainability? J Pure Appl Microbiol 9(1):715–724

Li DX (2003) Study on the effects of silicate bacteria on the growth and fruit quality of apples. J Fruit Sci 20:64–66

Li J, Ovakim DH, Charles TC, Glick BR (2003) An ACC deaminase minus mutant of Enterobacter cloacae UW4 no longer promotes root elongation. Curr Microbiol 41:101–105

Lian B, Wang B, Pan M, Liu C, Teng HH (2008) Microbial release of potassium from K-bearing minerals by thermophilic fungus *Aspergillus fumigatus*. Geochim Cosmochim Acta 72(1):87–98

Lopes-Assad ML, Avansini SH, Erler G, Rosa MM, Porto de Carvalho JR, Ceccatop- Antonini SR (2010) Rock powder solubilization by *Aspergillus niger* as a source of potassium for agroecological systems. In: 19th World Congress of Soil Science, soil solutions for a changing world 1–6 August 2010. Brisbane, pp 219

Maurya BR, Meena VS, Meena OP (2014) Influence of inceptisol and alfisol's potassium solubilizing bacteria (KSB) isolates on release of K from waste mica. Vegetos 27(1):181–187

Maurya BR, Kumar A, Raghuwanshi R, Bahadur I, Meena VS (2015) Effect of phosphate solubilizing isolates on growth, yield and phosphate acquisition by rice and wheatcrops. Afr J Microbiol Res 9(12): 1367–1375

Meena OP, Maurya BR, Meena VS (2013) Influence of K-solubilizing bacteria on release of potassium from waste mica. Agric Sustain Dev 1(1):53–56

Meena VS, Maurya BR, Bahadur I (2014a) Potassium solubilization by bacterial strain in waste mica. Bangladesh J Bot 43(2):235–237

Meena VS, Maurya BR, Verma JP (2014b) Does a rhizospheric microorganism enhance K+ availability in agricultural soils? Microbiol Res 169:337–347

Meena RK, Singh RK, Singh NP, Meena SK, Meena VS (2015a) Isolation of low temperature surviving plant growth-promoting rhizobacteria (PGPR) from pea (*Pisum sativum* L.) and documentation of their plant growth promoting traits. Biocatal Agric Biotechnol. doi:10.1016/i.bcab.2015.08.006

Meena VS, Maurya BR, Verma JP, Aeron A, Kumar A, Kim K, Bajpai VK (2015b) Potassium solubilizing rhizobacteria (KSR): isolation, identification, and K-release dynamics from waste mica. Ecol Eng 81: 340–347

Milic VM, Jarak N, Mrkovacki N, Milosevic M, Govedarica S, Đuric, Marinkovic J (2004) Microbiological fertilizer use and study of biological activity for soil protection purposes. Field Veg Crop Res 40: 153–169

Nishanth D, Biswas DR (2008) Kinetics of phosphorus and potassium release from rock phosphate and waste mica enriched compost and their effect on yield and nutrient uptake by wheat (Triticumaestivum). Biores Technol 99:3342–3353

Norkina SP, Pumpyansakya LV (1956) Certain properties of silicate bacteria dokl. Crop Sci Soc Japan 28: 35–40

Parmar P, Sindhu SS (2013) Potassium solubilization by rhizosphere bacteria: influence of nutritional and environmental conditions. J Microbiol Res 3(1): 25–31

Patel BC (2011) Advance method of preparation of bacterial formulation using potash mobilizing bacteria that mobilize potash and make it available to crop plant. WIPO Patent Application WO/2011/154961

Prajapati K, Modi HA (2012) Isolation of two potassium solubilizing fungi from ceramic industry soils. Life Sci Leaflets 5:71–75

Prajapati K, Sharma MC, Modi HA (2013) Growth promoting effect of potassium solubilising microorganisms on Abelmoschus esculentus. Int J Agric Sci 3(1):181–188

Purushothaman DA, Natarajan RC (1974) Distribution of silicate dissolving bacteria in velar astury. Curr Sci 43(9):282–283

Rajawat MVS, Singh S, Singh G, Saxena AK (2012) Isolation and characterization of K-solubilizing bacteria isolated from different rhizospheric soil. In: Proceeding of: annual conference of microbiologists of India, At Punjab University, Punjab, India

Ramarethinam S, Chandra K (2005) Studies on the effect of potash solubilizing/mobilizing bacteria Fratewicia Aurania on brinjal growth and yield. Pestol 11:35–39

Sheng XF (2002) Study on the conditions of potassium release by strain NBT of silicate bacteria scientia. Agric Sin 35(6):673–677

Sheng XF (2005) Growth promotion and increased potassium uptake of cotton and rape by a potassium releasing strain of *Bacillus edaphicus*. Soil Biol Biochem 37:1918–1922

Sheng XF, He LY (2006) Solubilization of potassium bearing minerals by a wild type strain of *Bacillus edaphicus* and its mutants and increased potassium uptake by wheat. Can J Microbial 52(1):66–72

Sheng XF, He LY, Huang WY (2002) The conditions of releasing potassium by a silicate-dissolving bacterial strain NBT. Agric Sci China 1:662–665

Sheng XF, Huang WY (2002) Mechanism of potassium release from feldspar affected by the strain NBT of silicate bacterium. Acta Pedol Sin 39:863–871

Singh NP, Singh RK, Meena VS, Meena RK (2015) Can we use maize (*Zea mays*) rhizobacteria as plant growth promoter? Vegetos 28(1):86–99

Sparks DL, Huang PM (1987) Physical chemistry of soil potassium. In: Munson RD (ed) Potassium in agriculture. American Soc Agron J, Madison, WI.:201–276

Subba RNS (2001) An appraisal of biofertilizers in India. In: Kannaiyan S (ed) The biotechnology of biofertilizers. Narosa Pub House, New Delhi

Ullaman WJ, Kirchman DL, Welch WA (1996) Laboratory evidence by microbially mediated silicate mineral dissolution in nature. Chem Geol 132:11–17

Uroz S, Calvaruso C, Turpault MP, Freyklett P (2009) Mineral weathering by bacteria: ecology, actors and mechanisms. Trends Microbiol 17:378–387

US Department of Agriculture (1998) Soil biodiversity; soil quality information sheet, soil quality resource concerns 2. National Soil Survey Center, NRCS, USDA in cooperation with the Soil Quality Institute, and the National Soil Tilth Laboratory, Agricultural Research Service, USDA

Vandevivere P, Welch SA, Ullman WJ, Kirchman DJ (1994) Enhanced dissolution of silicate minerals by bacteria at near neutral pH. Microb Ecol 27:241–251

Verma JP, Yadav J, Tiwari KN (2009) Effect of *Mesorhizobium* and plant growth promoting rhizobacteria on nodulation and yields of chickpea. Biol Forum An Int J 1(2):11–14

Verma JP, Yadav J, Tiwari KN, Kumar A (2013) Effect of indigenous Mesorhizobium spp. and plant growth promoting rhizobacteria on yields and nutrients uptake of chickpea (Cicerarietinum L.) under sustainable agriculture. Ecol Eng 51:282–286

Walker TS, Bais HP, Grotewold E, Vivanco JM (2003) Root exudation and rhizosphere. Biol Plant Physiol 132:44–51

Welch SA, Ullman WJ (1993) The effect of organic acids on plagioclase dissolution rates and stoichiometry. Geochim Cosmochim Acta 57:2725–2736

Xie JC (1998) Present situation and prospects for the world's fertilizer use. Plant Nutr Fertil Sci 4:321–330

Zahra MK, Monib MS, Abdel-Al I, Heggo A (1984) Significance of soil inoculation with silicate bacteria. ZentralblMikrobiology 139(5):349–357

Zakaria AAB (2009) Growth optimization of potassium solubilizing bacteria isolated from biofertilizer. Bachelor of Chem. Eng. (Biotech.), Fac. of Chem., Natural Resources Eng. Univ., Malaysia Pahang, p. 40.

Zarjani JK, Aliasgharzad N, Oustan S, Emadi M, Ahmadi A (2013) Isolation and characterization of potassium solubilizing bacteria in some Iranian soils. Arch Agron Soil Sci 59(12):1713–1723. doi:10.1080/03650340.2012.756977

Zhang A, Zhao G, Gao T, Wang W, Li J, Zhang S (2013) Solubilization of insoluble potassium and phosphate by *Paenibacillus kribensis* a soil microorganism with biological control potential. Afr J Microbiol Res 7(1): 41–47

Zhao F, Sheng X, Huang Z, He L (2008) Isolation of mineral potassium-solubilizing bacterial strains from agricultural soils in Shandong Province. Biodivers Sci 16:593–600

Potassium Uptake by Crops as Well as Microorganisms

19

M.L. Dotaniya, V.D. Meena, B.B. Basak, and Ram Swaroop Meena

Abstract

Potassium (K) is one of the essential major plant nutrients. Its importance in agriculture has increased with intensive agriculture as well as deficiency occurring in crop plants. Global population is increasing at a quantum rate, which pushes the targeted yield to higher levels for mitigating the food demand of hungry mouths. Production of more food material from limited land is a challenge for the researcher and it aggravates nutrient deficiency, due to more uptake of plant nutrients, especially K, by high-yielding crop varieties. Among plant nutrients, deficiency of K limits the crop growth and reduces the crop yield. The source of K is native or via various replenishment paths, i.e. crop residue, microbial biomass and a range of waste materials. Researchers mentioning the wide gap between the addition of K and removal during crop production are highlighting the challenge to maintain a sustainable crop yield. The nutrient balance in the soil system is also affected by the quantity of nutrient that is taken up, raising nutrient storage in the soil–plant–microbe system, and how much is recycled by crop residues. The use of potassium-solubilizing microorganisms (KSMs) can increase the K level in the soil solution and ultimately increase plant growth and development. The main mechanisms of KSMs are acidolysis, chelation, exchange reactions, complexolysis and organic acid production in the soil. K fertilizers cost much more than other fertilizers, so use of KSMs in agricultural crop production can be a sustainable option for enhancing in situ K availability from the fixed sink of agricultural soils.

M.L. Dotaniya (✉) • V.D. Meena
ICAR-Indian Institute of Soil Science, Berasia Road, Nabibagh, Bhopal 462 038, India
e-mail: mohan30682@gmail.com

B.B. Basak
ICAR-Directorate of Medicinal and Aromatic Plant Research, Boriavi, Anand 387 310, India

R.S. Meena
Department of Agronomy, Institute of Agricultural Sciences, Banaras Hindu University, Varanasi 221005, Uttar Pradesh, India

V.S. Meena et al. (eds.), *Potassium Solubilizing Microorganisms for Sustainable Agriculture*, DOI 10.1007/978-81-322-2776-2_19

Keywords
Crop uptake • Plant nutrition • Potassium • Potassium solubilizing microorganisms

19.1 Introduction

Potassium is an element, represented by K. It was first isolated from potash, from which its name is derived. K is absorbed by plants in larger amounts than any other mineral element except nitrogen (N) and, in some cases, calcium (Ca). The K uptake is often equal to or more than that of nitrogen. Mostly, K values in fertile soil with K in soil solution vary between 100 and 500 μM (Claassen 1990) and, generally, in sandy soils, may reach values even above 1000 μM. It is a univalent cation found in the largest concentration (100–200 m*M* K) in the plant cell sap and so it is called a 'master cation'. K is ionic (K^+), free and mobile in plants. Besides its water balance role in plants (Rao and Brar 2002), it is important for plant metabolism and growth activities because of the great number of enzymes it activates (Rosolem 1982). It helps in the building of protein, photosynthesis, fruit quality, reduction of diseases and insect pest attack, and tolerance to cold and frost. Soil contains a huge amount of K-bearing minerals, thus fulfilling the crop K requirement during the crop growth periods. The mineralization of K from fixed sites into an exchangeable and available form often satisfies the crop demand, but in a few cases, there is low K content in minerals, showing poor availability in soil solution and leading to K deficiency in crops. Greater amounts of K-bearing rock are available in the Indo-Gangatic Plain (IGP) of India. The farmers of the IGP seldom apply K fertilizers for crop production. In recent years, K deficiency has been observed across the world in many agricultural crops and ultimately there has been a reduction in crop yield. The soil solution K is taken up by the crop as per its requirement, or sometimes more than its requirement is taken, and this is called luxury consumption. The excess K that is absorbed does not increase the crop yield to any appreciable extent. Such types of wasteful luxury consumption occur especially in forage crops. The decreased concentration of K in soil solution is in equilibrium with the exchangeable K in the soil and dissolution mechanisms are involved. In the modern scientific research era, many soil microorganisms have been identified, which actively participate in K dissolution mechanisms and enhance the soil solution K (Diep and Hieu 2013; Meena et al. 2013; Maurya et al. 2014).

The K-solubilizing bacteria are capable of freeing K from fixed sites of clay minerals (Parmar and Sindhu 2013). In addition to K solubilization, K-solubilizing bacteria (KSB) are also helpful for plant growth by improving plant nutrients and soil structure, and suppressing pathogens. For example, few bacteria can weather silicate clay minerals and release various elements, i.e. K, Si and Al, and in addition various types of organic acids and plant growth promoters are helpful to enhance plant growth. These bacteria are extensively utilized in biological K fertilizer and associated natural K leaching (Lian et al. 2002; Meena et al. 2015b; Singh et al. 2015).

19.2 Deficiency Token in Plants

K deficiency cannot be immediately identified from visual symptoms (Fig. 19.1). At first, the plant growth rate decreases, and in later stages, deficiency symptoms occur. Its deficiency might cause abnormalities in plants; usually they are growth related.

The common symptoms are chlorosis with scorching of plant leaves and yellowing of the margins of the leaf. This is one of the first symptoms of K deficiency in the middle and lower levels, with slow or stunted growth—as

Fig. 19.1 Potassium deficiency symptoms of different crop species

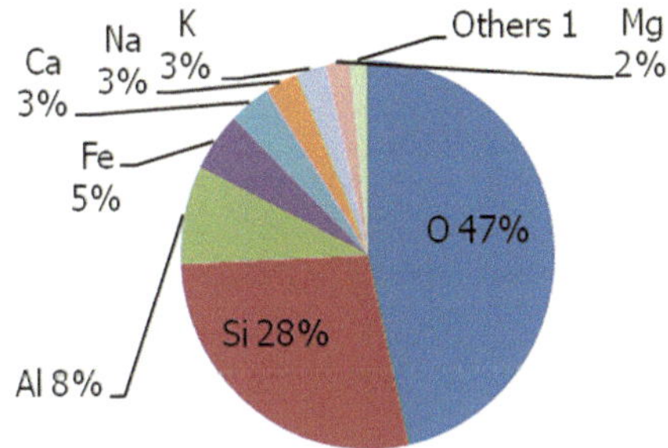

Fig. 19.2 Composition of the earth's crust

K is a vital growth catalyst in plants—and K-deficient plants will be unhealthy-looking with strongly reduced flowering and defoliation, as leaves turn yellow and then brown, and eventually fall off one by one. Left unattended, K deficiency in plants results in plants dying sooner than they should. This process might become even faster if plants are exposed to drought or high temperatures, as poor K uptake reduces plants' resistance to temperature changes and to drought, with the consequence of less water flow in the plants, and other effects such as uneven ripening of fruits, poor resistance to pests, and weak and unhealthy roots.

19.3 Potassium Chemistry in Soil

In the earth's crust, K accounts for ~2.6 % of the total mass after oxygen (46.6 %), silicon (27.7 %), aluminium (8.1 %), iron (5 %), calcium (3.6 %) and sodium (2.8 %) (Fig. 19.2). It is an important constituent of alkali K feldspars. In soil, key K-bearing silicate minerals are micas (~5 % of the crust) or silicate minerals. Biotite and muscovite are the most important micas and they both contain K. The most important K-bearing sedimentary mineral is sylvite (KCl) in nature.

Soil K is divided into solution, exchangeable, fixed, and structural K (Sparks and Huang 1985). K is supplied to plants by soil minerals, organic materials and fertilizer. Fixed K is non-exchangeable, non-extracted neutral molar ammonium acetate (NH_4OAc), which is commonly used to measure exchangeable K prior to application of K fertilizers, to determine the K fixation capacity of the soil (Denoroy et al. 2004), so that the applied fertilizer can help achieve better crop growth and yield. The availability of K soil solution to crop plants is governed by the four major pools in the soil: exchangeable, soluble, structural and fixed (non-exchangeable) (Romheld and Kirkby 2010). Among the K pools, soluble and exchangeable phases exist in major soils, but the structural and fixed pools provide negative charge sites on silicate clay mineral surfaces and organic matter. The fixed or non-exchangeable phase exists only in micaceous-type clays (2:1 layers) i.e. illite, vermiculite and other silicate clays from this group (Zhang et al. 2010). The main forms in which K is held in soils and the changes it undergoes as it is cycled through the soil–plant–animal system; equilibrium exists between these diverse pools and the correlation between them (Fig. 19.3). Most of the K in soil (90–98 %) is held in the primary minerals like micas and feldspars, but the release of this form to the soil solution during the process of weathering is very slow. As primary minerals

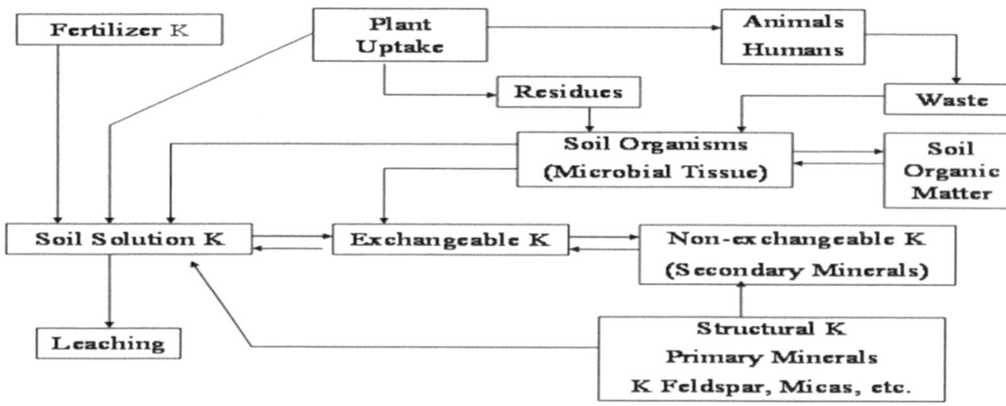

Fig. 19.3 The K cycle in agricultural soils (Adapted from Brady 1990)

weather, their rigid lattice structure becomes increasingly unstable. Over a period, K held in interlayer spaces in 2:1-type minerals (mica) is made more mobile, firstly through non-exchangeable sites, but slowly available form and lastly through readily exchangeable comes in soil solution. The release of elements from the two micas phlogopite and biotite is slow, but the release of K can be accelerated through biologically induced activities (Berthelin 1983). Some of the important sources of K are as follows: (a) mica is one of the important K-bearing minerals, mainly in the form of muscovite and biotite; these are not used in agriculture, though they contain significant amounts of K (8–12 % K_2O), and are dumped near mica mines; (b) feldspars of soil serve as an important source of available K and include orthoclase, microcline and sanidine; (c) zeolite is a larger group of hydrated alumino-silicates from which K cannot be removed without the destruction of the structure; another important mineral is glauconite; (d) -potassium-taranakite is considered to be a slow-releasing K source; and (e) vermiculite and illite are secondary clay minerals, contain significant amounts of K and are relatively easily available.

Intensive agriculture practice systems have reduced the opportunities for major K fixation, due to continuous requirements of crops, and are influenced by the soil parent materials, degree of weathering and nutrient balance in soil systems (Bertsch and Thomas 1985; Simonsson et al. 2007).

19.4 Potassium Released from Fixed Sites

The phenomena of both fixation-exchangeable K and the release of non-exchangeable K play a vital role in the dynamics of soil K. In the soil, more than 90–98 % of total soil K is found in an unavailable form, which is not easily taken up by crop plants. In the process of weathering, K is released into the soil solution from fixed sites. The available K is converted to a form that cannot be extracted with a neutral salt solution commonly used to extract the plant-available form of soil K. The fixation is nearly absent in kaolinite, chlorite and unweathered micas; slight in montmorillonite-type clay minerals; and considerable in illite- and vermiculite-dominant soils. The release of K from mica is mediated by soil water. The gradual release of K from trapped positions of the mica lattice to form illite (hydrous mica) happens due to the addition of water molecules and swelling of the K lattice (Fig. 19.4). The release of K from an unavailable form is also affected by the hydration energy of the K ion; lower hydration energy promotes K entrapment and vice versa. The concentration of soil solution K is decreased due to leaching and crop uptake; the substantial decrease in available K is replenished by exchangeable K. The releases of K from fixed sites to soil solution are at planar surfaces (p-position), at the edges of the layers (e-position), and in the interlayer space

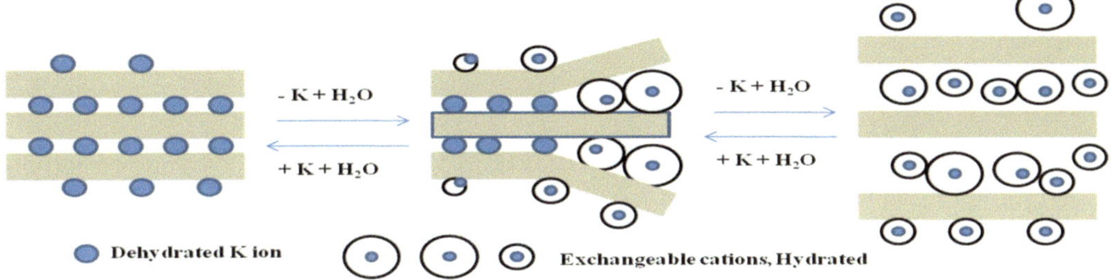

Fig. 19.4 Mechanism of K release and fixation (Modified from Rao and Brar 2002)

(i-position). The quantity and intensity relationship of K also determines both K availability and fixation phenomena in soil. The intensity factor is a measure of K in soil solution, which is immediately available for absorption by plant roots, and the quantity factor means the capacity of the soil to maintain the level of K in soil solution over a long period. The presence of other cations (Ca^{2+} and Mg^{2+} ions) in soil solution affects the intensity factor and also K availability (Beckett 1964). The capacity of various cations in saturated resins to replace the non-exchangeable K from soils has been found to be as follows: H^+ resin > Ca^{2+} resin > Na^+ resin > NH_4^+ resin.

19.5 Potassium Uptake by Crop Species

Figure 19.3 shows the soil–plant–microbe system for K uptake from soil in an easy way. The plant roots take up K from the soil solution only, i.e. there is no direct transfer of solid bound K (exchangeable and interlayer K) to the root surface. The crop plants absorb the K from the soil solution in the ionic form via roots through the plasmalemma or the cell membrane. However, the plant roots absorb the soil solution K, reducing the K concentration at the root surface, and consequently there is a concentration gradient build-up and K transport to the root by diffusion. Apart from diffusion K, also transport to roots by mass flow is usually low because of the relatively low concentration in soil solution. The part of the soil solution reduced due to crop plant uptake promotes the release of K from the solid phase, primarily by decrease from the exchangeable fraction and afterward from the interlude of non-exchangeable K.

The K requirement of various crops is expressed by total K uptake (Table 19.1). The sum of K in the soil solution depends on the type of soil, target yield, retention or removal of crop residues and the applied K fertilizer or K contained in irrigation water (Dotaniya et al. 2013; Foloni et al. 2008). Imbalance in fertilizer application, especially N fertilizers with no K, leads to K deficiency in crops after a certain period of continuous cropping, which is very common. There are some basic differences between N and K fertilizers (Table 19.2). During the estimation of K uptake, root K uptake is often neglected as the root K contributes only slightly (~10 %) to the total K uptake. The K uptake is often equal to or more than that of nitrogen. Tuber crops like potato, leafy vegetables like cauliflower and cabbage, forage crops like alfalfa, and fruits like banana and pineapple are heavy feeders on K. Soil fertility status also affects K uptake by crops; the level of nitrogen in soil enhances the K uptake in crops.

K availability in soil can be enhanced by the addition of crop rotations with vigorous, deep-root system crops, which is important in mounting a net of biopores in the soil profile for root exudates and growth (Williams and Weil 2004). Of the grasses with profuse root growth, which make the lower zone K in the surface zone, ruzigrass is very effective in taking up K from

Table 19.1 Potassium uptake by agricultural crops under field conditions (Tandon 1991)

Crop name	Yield (t ha^{-1})	Total uptake (kg K ha^{-1})
Wheat (*Triticum aestivum* L.)	3.90	137
Rice (*Oryza sativa* L.)	5.14	180
Chickpea (*Cicer arietinum* L.)	1.50	49
Pigeonpea (*Cajanus cajan*)	1.20	16
Groundnut (*Arachis hypogaea*)	2.54	95
Mustard (*Brassica juncea*)	2.60	133
Soybean (*Glycine max*)	2.50	101
Sunflower (*Helianthus annuus* L.)	2.38	141
Sugarcane (*Saccharum officinarum* L.)	87.60	270
Tea (*Camellia* spp.)	1.00	37
Tobacco (*Nicotiana tabacum* L.)	2.85	65
Alfalfa (*Medicago sativa* L.)	91.90	669
Potato (*Solanum tuberosum* L.)	29.50	119
Banana (*Musa paradisiaca* L.)	38.00	1053
Pineapple (*Ananas sativa*)	84.00	440
Papaya (*Carica papaya* L.)	150.00	415

Table 19.2 Comparison of different properties of nitrogen and potassium in soil and plants

Property	Nitrogen	Potassium
Soil composition	A component of organic matter, anion and cation forms, solid, dissolved and gas forms	Inorganic cation or component of soil minerals
Soil reactions	Complex chemical reactions in the soil, involving microorganisms	Simple chemical reactions in soils
Plant-uptake mechanism	Taken up in two main ionic forms: NH_4^+ and NO_3^-	Taken up only as K^+
Plant physiological role	A major component of proteins, nucleic acid and numerous other organic compounds	An osmotic regulator

the soil and making it available for corn (Garcia et al. 2008). The K balance of an agricultural system is calculated as per the formula below:

$$\begin{aligned}&\textbf{K Balance}\\&= (\text{Fertilizer K} + \text{Manure K} + \text{Rain K}\\&+ \text{Irrigation water K} +\\&+ \text{K In seedling and seeds})\\&-(\text{K uptake} + \text{Losses of K})\end{aligned}$$

In south Asia, the rice–wheat cropping system is one of the dominant cropping systems. It stretches over India, Pakistan, Nepal and Bangladesh; more than 10 Mha in India is occupied by this cropping system (Singh and Paroda 1994; Yadav et al. 1998). Earlier this area was known for sufficient K in soil for crop production, but in the last few decades, it has shown a K deficiency in agricultural crops. During 1960–1990, genetic improvements leading to the development of high-yielding rice and wheat varieties with a good fertilizer response resulted in a dramatic rise in productivity and crop yield. Both crops are heavy feeders, and the double cropping system speedily depleted the soil fertility level, especially K. A diagnostic survey showed that IGP farmers seldom apply K fertilizers during crop production (Yadav et al. 2000).

The general recommendations for K application are 25 kg K ha^{-1} in Punjab and up to 50 kg K ha^{-1} in the Uttar Pradesh and West Bengal parts of India for rice. For wheat in the IGP belt the recommendation is 21–58 kg K ha^{-1} (Tiwari 2000). Higher application of nitrogenous

Table 19.3 Commercial potassic fertilizers and K_2O percentages

Fertilizer name	Main compound(s)	K_2O (%)
Potassium chloride (40 %)	KCl	38–48
Potassium chloride (50 %)	KCl	48–52
Potassium chloride (60 %)	KCl	60
Potassium sulphate	K_2SO_4	50
Potassium magnesium sulphate	K_2SO_4, $MgSO_4$	48–52
Potassium nitrate	KNO_3	38–42
Potassium phosphates	KH_2PO_4, K_2HPO_4	45

fertilizers in IGP has increased the K deficiency in soils. The heavy subsidy on N fertilizers is also one of the major factors for high application of N fertilizers, but reductions in subsidies on phosphate and K fertilizers in India has adversely affected their consumption. This has led to a continued imbalance in fertilizer use in agricultural crop production. Various types of K fertilizers are available in the market and the K_2O percentage varies from 38 % to 60 % (Table 19.3). The chloride concentration in K fertilizers has affected the crop growth of tobacco, grapes, fruit trees, cotton, sugarcane, potatoes, tomatoes, onions and cucumber, whereas oil palm and coconuts conversely appear to be chloride-loving crops. The K fertilizers are generally broadcast or spread on the surface and mixed with the surface soil. In low–available K or high–K fixation capacity soils, band placement is recommended. Other than this, a few crops have performed better with a split application during the growth period, i.e. sugarcane, banana, papaya tea and pineapple.

19.6 Potassium Uptake and Solubilization by Microbes

Some microorganisms in the soil are able to solubilize 'unavailable' forms of K-bearing minerals, such as micas, illite and orthoclase, by excreting organic acids, which either directly dissolve rock K or chelate silicon ions to bring the K into solution (Bennett et al. 1998; Barker et al. 1998). These microorganisms, commonly known as KSMs, includes mainly bacteria and some fungi, but bacteria are mostly considered as K solubilizers and are widely known as K-solubilizing bacteria (KSB), K-dissolving bacteria or silicate-dissolving bacteria (SDB). Different kinds of possible mechanisms exist for mobilization of soil K minerals by KSB (Fig. 19.5). It has been demonstrated through laboratory and field experiments that microbes can accelerate weathering reactions of K minerals, especially when in direct contact with mineral surfaces by either (i) production of organic acids; (ii) production of metal-complexing ligands; (iii) formation of biofilms; or (iv) synergistic effects of microbes and root activities. These mechanisms are very complex, and none of these mechanisms occur in isolation.

The principal mechanism of K mineral solubilization involves low molecular acid production (Table 19.4) by soil microorganisms (Huang and Keller 1972; Leyval and Berthelin 1989). The protons associated with organic acid molecules decrease the pH of the solution and, therefore, induce the releasing capacity of cations such as iron, K and magnesium (Friedrich et al. 1991; Bennett et al. 1998; Sheng et al. 2003). Many workers have opined that production of carboxylic acids and capsular polysaccharide or extracellular polysaccharides (EPS) and enzymes are thought to accelerate the dissolution of a variety of silicates by application of KSMs, viz. *Bacillus mucilaginosus* and *B. edaphicus* (Lin et al. 2002). Production of metal-complexing ligands by K-mobilizing by microbes is another important mechanism of K mobilization. KSMs can increase the mineral dissolution rate by producing and excreting various organic ligands during their metabolism that interact with the mineral surface (Bennett et al. 2001). Microbial-produced organic ligands include metabolic by-products, extracellular enzymes,

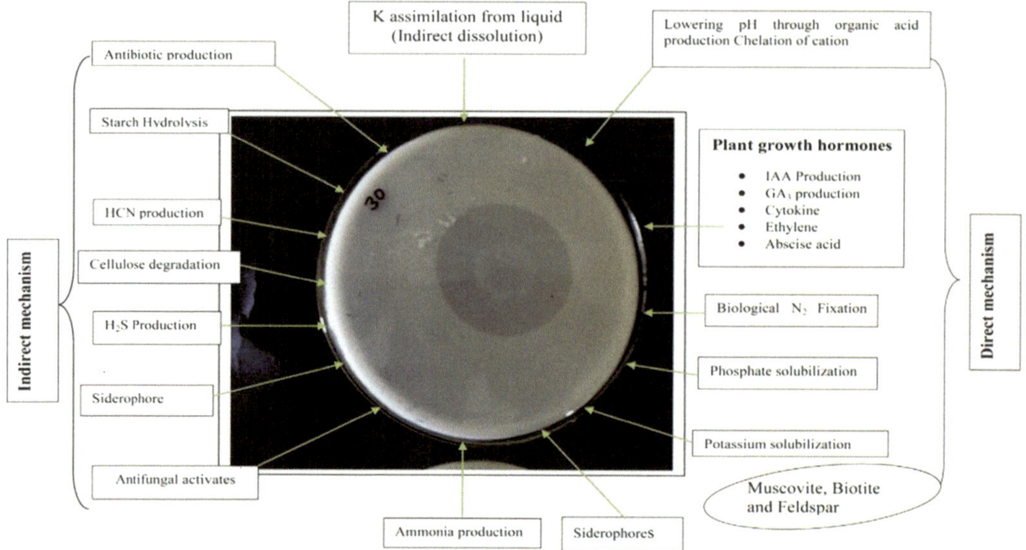

Fig. 19.5 Direct and indirect mechanisms of plant growth–promoting properties of KSMs and the K-solubilizing ability of mica (K-bearing mineral) on Aleksandrov medium (Meena et al. 2014b)

Table 19.4 KSMs produce various organic acids in different strains, which help in solubilization of insoluble K to soluble K

Organism	Predominant acid produced	Reference
Penicillium frequentans, *Clasdosporium* spp.	Oxalic, citric, gluconic	Argelis et al. (1993)
Paenibacillus mucilaginosus	Tartaric, citric, oxalic	Liu et al. (2012)
Aspergillus niger, *Penicillium* spp.	Citric, glycolic, sucinnic	Sperberg (1958)
B. megaterium, *Pseudomonas* spp.	Lactic, malic, oxalic, lactic	Taha et al. (1969)
B. megaterium, *E. freundii*	Citric, gluconic	Taha et al. (1969)
Arthrobacter, *Bacillus* spp.	Lactic, citric	Bajpai and Sundara (1971)
A. fumigatus, *A. candidus*	Oxalic, tartaric, citric, oxalic	Banik and Dey (1982)
P. aeruginosa	Acetate, citrate, oxalate	Badar et al. (2006)
B. mucilaginosus	Oxalate, citrate	Sheng and He (2006)
Pseudomonas spp.	Tartaric, citric	Krishnamurthy (1989)

chelates and both simple and complex organic acids, which help in dissolution of K minerals by decreasing the pH of the environment. Chelating molecules might increase the dissolution rates of cations by forming strong bonds with them or with mineral surfaces. The heterogeneous matrices of microorganisms held together and tightly bound to underlying surfaces by EPS are considered a biofilm. In this microenvironment (biofilm), bacteria extract inorganic nutrients and energy directly from the mineral matrix, thereby helping in mineral weathering. Thus, biofilms are known to help in the mobilization of K from K minerals (Meena et al. 2014a, 2015a).

19.7 Application of KSM for Plant Nutrition

Some effort has already been made to improve K nutrition of plants through application of K-bearing minerals in combination with potassium-solubilizing microorganisms (KSMs). K solubilizers enhance mineral uptake by plants

Table 19.5 The effect of KMB on wheat yield (t ha^{-1}) depends on the erosion degree of Luvisol sandy loam soil (2003–2004)

	NPK+FYM		
Soil	Control	Strain K-81	Response to bacteria
Not eroded	5.22	6.26	1.04
Moderately eroded	5.15	6.03	0.88
Severally eroded	4.67	5.95	1.28
Dealluvial	4.77	5.67	0.90

LSD (5 %) for soil: F_A–0.53
LSD (5 %) for bacterization: F_B–0.38

through solubilizing insoluble K from silicate minerals in soil. Bacteria inoculation can improve K availability in soils by producing organic acids and other chemicals, stimulating growth and mineral uptake by plants (Alexander 1977). Kloepper et al. (1991) reported that wheat yield increased by up to 30 % with *Azotobacter* inoculation and by up to 43 % with *Bacillus* inoculation. Xue-Zhiyong (1996) reported an increase in the yield of sweet potato by the effect of silicate bacterium on release of the K. Mikhailouskaya and Tchernysh (2005) demonstrated the effect of K-mobilizing bacteria (KMB) on releasing K from micas (biotite and muscovite) and hydromicas (glauconite and hydromuscovite) and their effect on wheat yield in a field experiment on Luvisol sandy loam soil characterized by different degrees of water erosion. Different mobility forms of K in muscovite were evaluated as a K source for bacteria (Hu et al. 2006; Kumar et al. 2015).

Application of native KMB on severely eroded soil resulted in higher wheat yields, which were comparable to those on moderately eroded soil without bacteria inoculation. The data revealed that indigenous KMB favours wheat nutrition and results in the formation of high grain yields (Table 19.5) on eroded soils. Moreover, application of active natural strains of KMB is environmentally sound and appropriate. The data thus suggest that use of biological fertilizers with KMB may be a prospective additional technique for conservation of fertility status and improvement of productivity in eroded soils. Sheng and He (2006) studied the effect of the wild type of bacterial strain *B. edaphicus* NBT and its four mutants on growth and uptake of K by wheat grown in pots using a K-deficient yellow-brown soil in Nanjing, China. The results showed that the root and shoot growth of wheat was significantly increased by *B. edaphicus* NBT and their mutants MPs (+) and MPs (+) 1. Bacterial inoculation also resulted in significantly higher uptake of NPK by plant components. Wu et al. (2005) studied the effects of co-inoculation of biofertilizer containing N fixer (*A. chroococcum*), P solubilizer (*B. megatherium*) and K solubilizer (*B. mucilaginosus*) and AM fungi (*G. mosseae* and *G. intradices*) on soil properties and maize growth. The greenhouse study revealed that co-inoculation of mycorrhizal fungus and three species of bacteria significantly increase the growth of maize.

Han et al. (2006) conducted greenhouse experiments to evaluate the potential of co-inoculation with PSB (*B. megatherium* var. *phosphaticum*) and KSB (*B. mucilaginosus*) on mobilization of P and K from rock minerals and their effect on nutrient uptake and growth of pepper and cucumber using an Inceptisol from Korea. The results revealed that rock P and K applied either singly or in combination did not significantly enhance soil availability of P and K, indicating their unsuitability for direct application. They also found that soil inoculation with PSB or KSB significantly increased NPK uptake by roots and shoots in pepper (Table 19.6). The results suggest that the integrated use of bacteria inoculation with PSB and KSB and rock P and rock K materials can improve crop growth and higher nutrient uptake in P- and K-deficient soils.

Han and Lee (2005) reported the synergistic effects of soil fertilization with rock P and K materials and co-inoculation with *B. megaterium* and *B. mucilaginosus* KCTC3870 on the improvements of P and K uptake by eggplant

Table 19.6 Effects of PSB and KSB strains on shoot and root nutrient uptake (mg/plant) of pepper; the values are the mean of four replications

	Shoot			Root		
Treatments	N	P	K	N	P	K
Control	13.6	1.61	19.2	3.69	0.46	5.92
R-P	14.3	1.72	19.9	3.76	0.48	6.19
R-K	14.1	1.68	20.3	3.71	0.47	6.17
R-(P+K)	14.2	1.75	20.9	3.87	0.50	6.68
PSB	14.5	1.88	21.1	3.90	0.50	6.48
KSB	14.5	1.75	21.7	3.82	0.49	6.68
(P+K)SB	14.7	1.99	23.4	3.92	0.54	6.95
Rock P+PSB	15.0	1.91	21.7	3.95	0.53	6.70
Rock P+KSB	14.8	1.84	22.4	3.90	0.51	6.93
Rock-(P+K)+ (P+K)SB	16.5	2.11	25.7	4.27	0.61	7.46
LSD (5 %)	1.01	0.18	2.81	0.43	0.07	0.60

R-P rock phosphate, *R-K* rock K (illite), *PSB* phosphate-solubilizing bacteria, *KSB* K-solubilizing bacteria

(*Solanum torvum* L. NIVOT) grown under limited P and K soil in the greenhouse. They found that soil inoculation with PSB or KSB did not significantly increase the yield and uptake of NPK by the eggplant. Thus, combined application of bacteria that solubilize (fixed form of P and K) they might provide a faster and continuous supply of P and K for optimal plant growth. The use of plant growth–promoting rhizobacteria (PGPRs), including PSB and KSB as biofertilizers, was suggested as a sustainable solution to improve plant nutrition and production (Dotaniya and Meena 2013; Vessey 2003). The important K-releasing bacterial strain *B. edaphicus* NBT was examined for plant growth–promoting effects and nutrient uptake on cotton and rape in K-deficient soil in pot experiments (Sheng 2005). Inoculation with bacterial strain *B. edaphicus* NBT was found to increase root and shoot growth of cotton and rape. Strain NBT was able to mobilize K efficiency in both plants when illite was added to the soil. In cotton and rape growing in soils treated with insoluble K and inoculated with strain NBT, K content was increased by 30 % and 26 %, respectively.

An experiment was conducted to study the dynamics of K release from waste mica inoculated with KSMs to investigate its effectiveness as potassic fertilizer using sudan grass (*Sorghum vulgare* Pers.) var *Sudanensis* as a test crop grown under two Alfisols (Basak and Biswas 2009). Results revealed that application of mica significantly enhanced biomass yield and K uptake by sudan grass in comparison with control (Fig. 19.6). Biomass yield and K uptake increased further when the mica was inoculated with a bacterial strain in both the soils in comparison with uninoculated mica. Alfisol from Hazaribag recorded higher yield, uptake and K recoveries than Alfisol from Bhubaneswar. Basak and Biswas (2010) concluded that co-inoculation of *B. mucilaginosus* and *A. chroococcum* A-41) bacteria in K-bearing mineral (waste mica) could be a promising and alternative option for supplying K and N to crops. Meena et al. (2015b) isolated 12 potassium-solubilizing rhizobacteria (KSR), which actively participate in K release and pH dynamics under clay minerals, i.e. muscovite and biotite. All the strains that performed efficiently for solubilization of insoluble K could be used as novel K biofertilizers. The KSR strains reduced the pH, which is the main mechanism of acidolysis, and promoted K solubilization from waste mica. All the KSR strains performed diversely, among them *A. tumefaciens* OPVS 11 followed by *R. pusense* OPVS 6 performed better. The application of KSR increased the crop productivity of maize, sugarcane, pigeon pea, potato and tobacco.

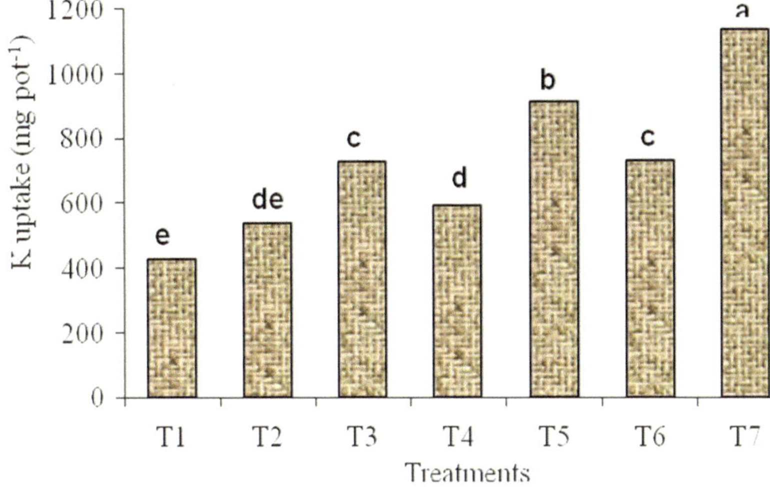

Fig. 19.6 Effect of waste mica co-inoculated with K-solubilizing (*Bacillus mucilaginous*) and N-fixing (*Azotobacter chroococcum* A-41) bacteria on potassium uptake (mg pot^{-1}) by Sudan grass (sum of five cuttings) in an Alfisol. T_1: control (no-mica, no-microbial culture); T_2: mica @ 100 mg K kg^{-1} soil; T_3: *B. mucilaginosus* (K solubilizer); T_4: *A. chroococcum* A-41 (N fixer); T_5: mica @ 100 mg K kg^{-1} soil + *B. mucilaginosus* (K solubilizer); T_6: mica @ 100 mg K kg^{-1} soil + *A. chroococcum* A-41 (N fixer); T_7: mica @ 100 mg K kg^{-1} soil + *B. mucilaginosus* (K solubilizer) + *A. chroococcum* A-41 (N fixer)

19.8 Factors Affecting Potassium Availability in Soil

K availability is affected by various factors: soil texture, soil pH, soil depth and liming material.

19.8.1 Soil Texture

This influences the status of both available and non-exchangeable K. Fine-textured soils generally possess larger amounts of both forms of K compared with coarse textured soils. The effect of texture on K status is more conspicuous in alluvial (illite-dominant) soils than in black (smectite-dominant) or red (kaolinite-dominant) soils because of the presence of K-bearing minerals in the finer fractions of soil. The amount of available K will be dependent on the dominant clay type present in the soil. For example, for soils dominated by illitic clays, it would be difficult to increase soil-test K levels by applying potash. Most soils have a mixture of clay types; therefore, soils vary in their ability to hold and trap applied K.

19.8.2 Soil Depth

The K availability factor determines the K concentration in soil. In most Indian soils, K content decreases with increasing depth in available or non-exchangeable forms. The Indo-Gangetic plain of India shows more plant-available K in surface soil compared with the sub-surface. In Vertisols both available and non-exchangeable K decreases with increasing depth. A long-term experiment showed that application of K fertilizers for balancing showed a significant amount of K leaching in sandy soils (0–20 cm). But the effect was less noticeable in clay soil (Rosolem et al. 2010). K not bound to organic skeletons is washed directly to the soil system by rainfall, independent of the point of organic matter mineralization (Calonego et al. 2005; Rosolem et al. 2007).

19.8.3 Soil pH

The soil pH affects the nutrient availability to plants and chemical and biological processes in soil. It determines the plant nutrient behaviour and affects the K concentration in crop plants. K bears positive charges and it is antagonistically affected by the positive cations in acid soil, i.e. H^+ and Al^{3+}; and Ca^{2+} and Mg^{2+} in higher-pH soil. The competition is mainly for the fixation sites on clay. In low-pH conditions the presence of hydrogen ions reduces the susceptibility to fixing. But by the means of amendments increasing pH, H^+ is neutralized and K ions have more chances to move closer to the soil colloidal surface, where they can be susceptible to fixation. Liming of acidic soil increases the cation exchange capacity of soil, which positively affects the K fixation in colloidal sites, and K availability decreases in soil solution. This is one of the important precautions during the amelioration of acidic soil by lime. Apart from this, increasing the Ca^{2+} ion concentration in soil due to the application of lime reduces the K uptake by plants especially in soils containing high amounts of $CaCO_3$. The fixation of applied K reduces the plant-available K in soil solution, particularly in soils containing K-depleted micaceous and vermiculites minerals (Bertsch and Thomas 1985). The K fixation process reduces the K use efficiency of fertilizers, but it is a favourable factor for limiting K leaching. Consequently, fixed K is also indicative of probable high K release. The conversion of non-exchangeable K to labile K is a reversible reaction. It is controlled by kinetic mechanisms. Therefore, K fixation and release both are symmetric, with the proportionality coefficient of change in non-exchangeable K and the initial K constraint influencing the soil's capacity for K release and fixation in soil (Meyer and Jungk 1993).

19.8.4 Liming

The liming of acid soils increases the cation exchange capacity of soil, which results in increased K adsorption by the soil colloids and decreases the K level in soil solution. The high calcium concentration reduces the K uptake from the soil solution.

19.9 Conclusion and Future Prospects

Traditionally, K fertilization strategies are designed to bring the soil to a target available K content, above which no increase in yield could be anticipated. For the growth of sustainable agriculture, it is vital to understand the conditions in which non-exchangeable K will take part in the replenishment of soil solution K and uptake by crop plants. Use of best agronomic practices and K-recycling sources can be better options for minimizing the negative balance of K in agricultural crop production systems. The best KSM strains can be selected on the basis of their ability to dissolve silicate minerals or release K from silicate minerals. There is an opportunity to isolate some indigenous KSM strains, which may be more suitable in local conditions than alien species. It is essential that the two disciplines of biology and mineralogy have equal weight in experiments of this type, as expertise in both is needed. Future studies should concentrate on use of some low-grade K minerals with suitable KSMs under field conditions, using different crops to develop an alternative and cheaper source of K in place of costly K fertilizers, which will be more appropriate in a specific farming system, especially in organic farming.

Acknowledgments The authors are highly thankful to Dr. Kuldeep Kumar, Scientist, ICAR-Indian Institute of Soil and Water Conservation Research and Training, Dehradun, India, for help during the writing of this manuscript.

References

Alexander M (1977) Introduction to soil microbiology, 1st edn. Wiley, New York

Argelis DT, Gonzala DA, Vizcaino C, Gartia MT (1993) Biochemical mechanism of stone alteration carried out by filamentous fungi living in monuments. Biogeochemistry 19:129–147

Badar MA, Shafei AM, Sharaf El-Deen SH (2006) The dissolution of K and phosphorus bearing minerals by silicate dissolving bacteria and their effect on sorghum growth. Res J Agric Biol Sci 2:5–11

Bajpai PD, Sundara R (1971) Phosphate solubilizing bacteria, solubilization of phosphate in liquid culture by selected bacteria as affected by different pH values. Soil Sci Plant Nutr 17:41–43

Banik S, Dey BK (1982) Available phosphate content of an alluvial soil as influenced by inoculation of some isolated phosphate solubilizing microorganisms. Plant Soil 69:353–364

Barker WW, Welch SA, Chu S, Banfield JF (1998) Experimental observations of the effects of bacteria on aluminosilicate weathering. Am Mineral 83:1551–1563

Basak BB, Biswas DR (2009) Influence of potassium solubilising microorganism (*Bacillus mucilaginosus*) and waste mica on potassium uptake dynamics by sudan grass (*Sorghum vulgare* Pers.) grown under two Alfisols. Plant Soil 317:235–255

Basak BB, Biswas DR (2010) Co-inoculation of potassium solubilizing and nitrogen fixing bacteria on solubilization of waste mica and their effect on growth promotion and nutrient acquisition by a forage crop. Biol Fert Soils. doi:10.1007/s00374-010-0456-x

Beckett PHT (1964) The immediate O/I relations of labile potassium in the soil. J Soil Sci 15:9–23

Bennett PC, Choi WJ, Rogera JR (1998) Microbial destruction of feldspars. Mineral Manag 8(62A):149–150

Bennett PC, Rpgers JR, Choi WJ (2001) Silicates, silicate weathering, and microbial ecology. Geomicrobiol J 18:3–19

Berthelin J (1983) Microbial weathering processes. In: Krumbein WE (ed) Microbial geochemistry. Blackwell, Oxford, pp 223–262

Bertsch PM, Thomas GW (1985) Potassium status of temperate region soils. In: Munson RD (ed) Potassium in agriculture. ASA, Madison, pp 131–162

Brady NC (1990) The nature and properties of soils, 10th edn. Prentice Hall of India Pvt Ltd, New Delhi

Calonego JC, Foloni JSS, Rosolem CA (2005) Lixiviac¸a˜o de pota'ssio da palha de plantas de cobertura em diferentes esta'dios de senesce^ncia apo's a dessecac¸a˜o qui'mica. Rev Bras Cienc Solo 29:99–108. doi:10.1590/S0100-06832005000100011

Claassen N (1990) Naehrstoffaufnahme hoeherer Pflanzen aus dem Boden. Ergebnis von Verfuegbarkeit und Aneignungsvermoegen. Habilitationsschrift, Severin Verlag, Goettingen

Denoroy P, Dubrulle P, Villette C, Colomb B, Fayet G, Schoeser M, Marin-Laflèche A, Pellerin F, Pellerin S, Boiffin J (2004) REGIFERT: Interpréter les résultats des analyses de terre. INRA, Paris

Diep CN, Hieu TN (2013) Phosphate and potassium solubilizing bacteria from weathered materials of denatured rock mountain, Ha Tien, Kiên Giang province Vietnam. Am J Life Sci 1(3):88–92

Dotaniya ML, Meena VD (2013) Rhizosphere effect on nutrient availability in soil and its uptake by plants -A review. Proc Natl Acad Sci India Sec B: Biol Sci 85 (1):1–12. doi:10.1007/s40011-013-0297-0

Dotaniya ML, Sharma MM, Kumar K, Singh PP (2013) Impact of crop residue management on nutrient balance in rice–wheat cropping system in an Aquic hapludoll. J Rural Agric Res 13(1):122–123

Foloni JSS, Tiritan CS, Calonego JC, Alves Junior J (2008) Aplicacaodefosfato natural e reciclagem de fo'sforo por milheto, braquia'ria, milho e soja. Rev Bras Cienc Solo 32:1147–1155. doi:10.1590/S0100-06832008000300023

Friedrich S, Platonova NP, Karavaiko GI, Stichel E, Glombitza F (1991) Chemical and microbiological solubilization of silicates. Acta Biotechnol 11:187–196

Garcia RA, Crusciol CAC, Calonego JC, Rosolem CA (2008) Potassium cycling in a corn-brachiaria cropping system. Eur J Agron 28:579–585. doi:10.1016/j.eja.2008.01.002

Han HS, Lee KD (2005) Phosphate and potassium solubilizing bacteria effect on mineral uptake, soil availability and growth of eggplant. Res J Agric Biol Sci 1(2):176–180

Han HS, Supanjani E, Lee KD (2006) Effect of co-inoculation with phosphate and potassium solubilizing bacteria on mineral uptake and growth of pepper and cucumber. Plant Soil Environ 52(3):130–136

Hu X, Chen J, Guo J (2006) Two phosphate- and potassium-solubilizing bacteria isolated from Tianmu Mountain, Zhejiang, China. World J Microbiol Biotechnol 22:983–990

Huang WH, Keller WD (1972) Organic acids as agents of chemical weathering of silicate minerals. Nature (Physical Sci) 239:149–151

Kloepper JW, Zablowicz RM, Tipping EM, Lifshitz R (1991) Plant growth mediated by bacterial rhizosphere colonizers. In: Keister DL, Gregan B (eds) The rhizosphere and plant growth. Kluwer Academic Publishing, Dordrecht, pp 315–326

Krishnamurthy HA (1989) Effect of pesticides on phosphate solubilizing microorganisms, M.Sc. (Agric.) thesis, University of Agricultural Sciences, Dharwad

Kumar A, Bahadur I, Maurya BR, Raghuwanshi R, Meena VS, Singh DK, Dixit J (2015) Does a plant growth–promoting rhizobacteria enhance agricultural sustainability? J Pure Appl Microbiol 9(1):715–724

Leyval C, Berthelin J (1989) Interaction between *Laccaria laccata*, *Agrobacterium radiobacter* and beech roots: influence on P, K, Mg and Fe mobilization from minerals and plant growth. Plant Soil 117:103–110

Lian B, Fu PQ, Mo DM, Liu CQ (2002) A comprehensive review of the mechanism of potas-sium release by silicate bacteria. Acta Mineral Sin 22:179

Lin QM, Rao ZH, Sun YX, Yao J, Xing LJ (2002) Identification and practical application of silicate-dissolving bacteria. Agric Sci China 1:81–85

Liu D, Lian B, Dong H (2012) Isolation of *Paenibacillus sp.* and assessment of its potential for enhancing mineral weathering. Geomicrobiol J 29(5):413–421

Maurya BR, Meena VS, Meena OP (2014) Influence of Inceptisol and Alfisol's potassium solubilizing bacteria (KSB) isolates on release of K from waste mica. Vegetos 27(1):181–187

Meena OP, Maurya BR, Meena VS (2013) Influence of K-solubilizing bacteria on release of potassium from waste mica. Agric Sustain Dev 1(1):53–56

Meena VS, Maurya BR, Bahadur I (2014a) Potassium solubilization by bacterial strain in waste mica. Bangladesh J Bot 43(2):235–237

Meena VS, Maurya BR, Verma JP (2014b) Does a rhizospheric microorganism enhance K+ availability in agricultural soils? Microbiol Res 169:337–347

Meena RK, Singh RK, Singh NP, Meena SK, Meena VS (2015a) Isolation of low temperature surviving plant growth-promoting rhizobacteria (PGPR) from pea (*Pisum sativum* L.) and documentation of their plant growth promoting traits. Biocatal Agric Biotechnol. doi:10.1016/j.bcab.2015.08.006

Meena VS, Maurya BR, Verma JP, Aeron A, Kumar A, Kim K, Bajpai VK (2015a) Potassium solubilizing rhizobacteria (KSR): isolation, identification, and K-release dynamics from waste mica. Ecol Eng 81:340–347

Meyer D, Jungk A (1993) A new approach to quantify the utilization of non-exchangeable soil potassium by plants. Plant Soil 149:235–243

Mikhailouskaya N, Tchernysh A (2005) K – mobilizing bacteria and their effect on wheat yield. Latv J Agron 8:147–150

Parmar P, Sindhu SS (2013) Potassium solubilization by rhizosphere bacteria: influence of nutritional and environmental conditions. J Microbiol Res 3(1):25–31

Rao AS, Brar MS (2002) Essential nutrients and their uptake by plants. Fundamental of Soil Science, ISSS, New Delhi, pp 368–380

Romheld V, Kirkby EA (2010) Research on potassium in agriculture: needs and prospects. Plant Soil 335:155–180

Rosolem CA (1982) Soybean mineral nutrition and fertilization, 2nd edn. Instituto Internacional da Potassa e Instituto da Potassae Fosfato, Piracicaba

Rosolem CA, Sgariboldi T, Garcia RA, Calonego JC (2010) Potassium leaching as affected by soil texture and residual fertilization in tropical soils. Commun Soil Sci Plant Anal 41(16):1934–1943. doi:10.1080/00103624.2010.495804

Rosolem CA, Calonego JC, Foloni JSS, Garcia RA (2007) Pota'ssio lixiviado da palha de aveia-preta e milheto apo's adessecac¸a˜o quı'mica. Pesqui Agrop Bras 42:1169–1175. doi:10.1590/S0100-204X2007000800014

Sheng XF (2005) Growth promotion and increased potassium uptake of cotton and rape by a potassium releasing strain of *Bacillus edaphicus*. Soil Biol Biochem 37:1918–1922

Sheng XF, He LY (2006) Solubilization of potassium-bearing minerals by a wildtype strain of *Bacillus edaphicus* and its mutants and increased potassium uptake by wheat. Can J Microbiol 52(1):66–72

Sheng XF, Xia JJ, Chen J (2003) Mutagenesis of the Bacillus edphicaus strain NBT and its effect on growth of chili and cotton. Agric Sci China 2:40–41

Simonsson M, Andersson S, Andrist-Rangel Y, Hillier S, Mattsson L, ÖOborn I (2007) Potassium release and fixation as a function of fertilizer application rate and soil parent material. Geoderma 140:188–198

Singh RB, Paroda RS (1994) Sustainability and productivity of rice–wheat systems in Asia-Pacific region: research and technology development needs. In: Paroda RS, Woodhead T, Singh RB (eds) Sustainability of rice–wheat production systems in Asia. FAO, Bangkok, pp 1–35

Singh NP, Singh RK, Meena VS, Meena RK (2015) Can we use maize (*Zea mays*) rhizobacteria as plant growth promoter? Vegetos 28(1):86–99

Sparks DL, Huang PM (1985) Physical chemistry of soil potassium. In: Munson RD (ed) Potassium in agriculture. Wisc, ASA, Madison, pp 201–276

Sperberg JI (1958) The incidence of apatite-solubilizing organisms in the rhizosphere and soil. Aust J Agric Res Econ 9:778

Taha SM, Mahmod SAZ, Halim El-Damaty A, Hafez AM (1969) Activity of phosphate dissolving bacteria in Egyptian soils. Plant Soil 31:149–160

Tandon HLS (1991) Sulphur research and agriculture production in India. The Sulphur Institute, Washington, DC, pp 6–17

Tiwari KN (2000) Potash in Indian farming. Potash & Phosphate Institute of Canada - India Programme, Gurgaon, p 48

Vessey JK (2003) Plant growth promoting rhizobacteria as biofertilizers. Plant Soil 255:571–586

Williams SM, Weil RR (2004) Crop cover root channels mayalleviate soil compaction effects on soybean crop. Soil Sci Soc Am J 68:1403–1409. doi:10.2136/sssaj2004.1403

Wu SC, Cao ZH, Li ZG, Cheung KC, Wong MH (2005) Effects of biofertilizer containing N-fixer, P and K solubilizers and AM fungi on maize growth: a greenhouse trail. Geoderma 125:155–166

Xue-Zhiyong (1996) Effect of silicate bacterium on release of potassium in different soils and on increase yield of sweet potato. Soil Fertil 1:26–28

Yadav RL, Prasad K, Gangwar KS (1998) Analysis of eco-regional production constraints in rice-wheat cropping system. PDCSR Bulletin No. 98-2. Project Directorate for Cropping Systems Research, Modipuram, India, pp 68

Yadav RL, Singh SR, Prasad K, Dwivedi BS, Batta RK, Singh AK, Patil NG, Chaudhary SK (2000) Management of irrigated ecosystem. In: Yadav JSP, Singh GB (eds) Natural resource management for agricultural production in India. Indian Society of Soil Science, New Delhi, pp 775–870

Zhang Z, Liao L, Xia Z (2010) Ultrasound-assisted preparation and characterization of anionic surfactant modified montmorillonites. Appl Clay Sci 50:576–581

Potassium-Solubilizing Microorganisms and Their Role in Enhancing Soil Fertility and Health

20

Ipsita Das and Madhusmita Pradhan

Abstract

Potassium, the third major plant nutrient, occurs in potassium-bearing minerals such as feldspars and micas. The soil-available K is usually very low and mostly becomes unavailable to plants. Furthermore, intensive cropping, runoff, leaching and soil erosion lead to soil potassium deficiency, providing stunted growth and limited physiological activities of plant. There are microbes (bacteria, actinomycetes and fungi) which are able to release soluble K from K-bearing minerals. KSMs, the soil microorganisms, could provide an alternative approach to make K available or soluble for the cropping system. These microbes secrete organic acids, which act as chelating agents to dissolve rock and chelate silicon ions releasing the K ions available to crops. These microbial cultures can be used as biofertilizers to make available K from mineral and rocks, ultimately influencing crop growth and quality which is a prerequisite for eco-friendly and sustainable agriculture.

Keywords

Feldspars • Physiological activities • Organic acids • Environmental sustainability

20.1 Introduction

Besides nitrogen (N) and phosphorus (P), potassium (K) is the third macronutrient of plant, which performs a multitude of various biological activities to maintain or even improve crop growth, grain quality, protein synthesis, enzyme activation and photosynthesis (Read et al. 2006). It is needed for the making of starch, controls root growth, regulates the stomata movement in plant cells and also contributes to quality. K-containing minerals in soil are feldspars and micas. The chemical relationships among different forms of soil K can be summarized as below.

I. Das (✉) • M. Pradhan
Department of Soil Science and Agricultural Chemistry, College of Agriculture, OUAT, Bhubaneswar 751003, India
e-mail: ipsita.isha17@gmail.com

V.S. Meena et al. (eds.), *Potassium Solubilizing Microorganisms for Sustainable Agriculture*,
DOI 10.1007/978-81-322-2776-2_20

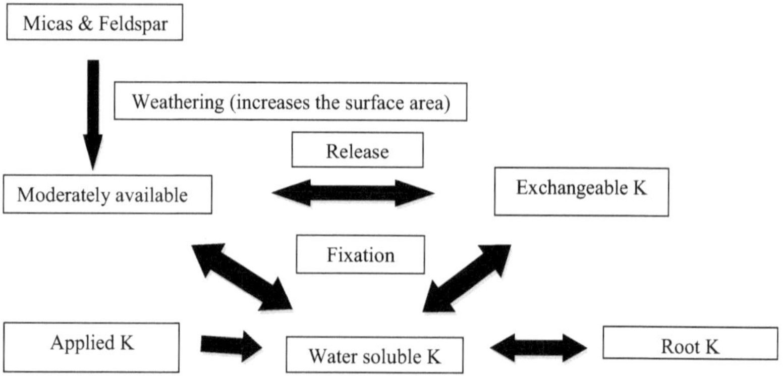

Fig. 20.1 Various forms of K in soil (Bear 1964)

Usually soils carry over 20,000 ppm of total K of which plants can utilize only the exchangeable K on the surface of the soil particles and that dissolved in the soil water which often amounts to less than 100 ppm and comprise only 0.1–2 % of the total K (George and Michael 2002).

Very often, low concentrations of soluble K are found in the soil, and ≥98 % of K exists in the form of insoluble rocks and silicate minerals, viz. microcline, muscovite, orthoclase, biotite, feldspar, etc. (Buchholz and Brown 1993). Out of these forms of K, soil minerals make up ≈90–98 % of soil K which are unavailable for crop uptake (Sparks 1987). The second non-exchangeable form of K makes up approximately 1–10 % of soil K and consists predominantly of interlayer K of non-expanded clay minerals such as illite and lattice K in K-feldspars, which contribute significantly to the plant uptake (Memon et al. 1988; Sharpley 1989) (Fig. 20.1).

Release of non-exchangeable K to the third exchangeable form occurs when level of exchangeable and solution K is decreased by crop removal, runoff, erosion and/or leaching (Sparks 1987; Sheng and Huang 2002). The most commonly found soil components of K, 90–98 %, are feldspar and mica (McAfee 2008). In soil, K occurs in water-soluble (solution K), exchangeable, non-exchangeable and structural or mineral forms. Water-soluble and exchangeable pools directly contribute available K for plant uptake. In some soil types with low levels of exchangeable K, non-exchangeable K can also supply significantly to crop uptake (Memon et al. 1988; Sharpley 1989). Exchangeable, available or soluble K is held by negatively charged clay minerals and organic matter in soils, while non-exchangeable K consists predominantly of interlayer K of non-expanded clay minerals such as illite and lattice K in K minerals such as K-feldspars. The bulk of total soil K is in the mineral fraction (Sparks and Huang 1985; Sparks 1987). A kind of dynamic equilibrium and chemical kinetic reactions was found to exist among the various forms of soil K which affect the level of soil solution K at any particular time and, thus, the amount of readily available K for crop uptake. Soil solution K levels are decided by the equilibria and kinetic reactions occurring among other forms of soil K (Sparks 1987).

The two main factors, viz. rate and direction of reactions, between solution and exchangeable forms of K determine whether the applied K will be leached into lower horizons of soil, taken up by crops or converted into unavailable forms or released into available forms (Sparks 2000). The fate of applied K or fertilizer K in soil is also governed by clay content and clay mineralogy of soil and the nature of crops grown. Decrease in the concentration of exchangeable and solution K either by crop removal or by leaching leads to the conversion of non-exchangeable K to exchangeable forms. Perhaps a vast increase in microbial activities can be one of the measures for this conversion (Sparks 1987). Usually, mineral K forms are slowly available for crop uptake (Sparks and Huang 1985); however, the availability is dependent on a number of factors,

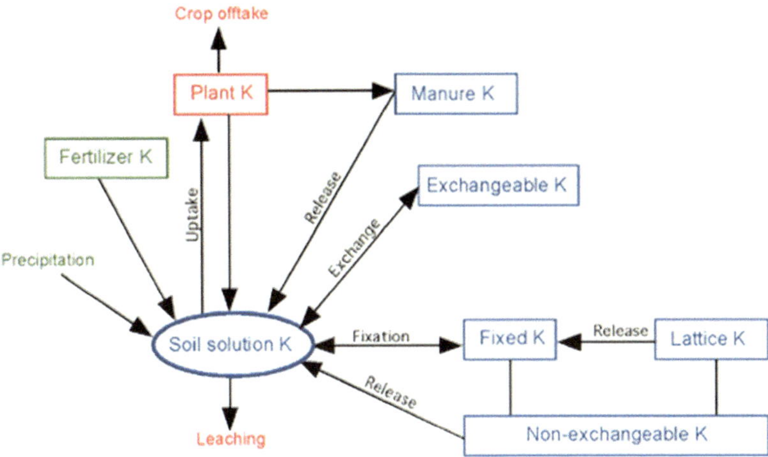

Fig. 20.2 Potassium cycle in the ecosystem (Syers 1998)

which includes the concentrations of other forms of K, such as solution, exchangeable and non-exchangeable and the rate of weathering of the K-bearing minerals like feldspar and micas (Sparks 1987). Optimal nutrition of a crop very often suffers in K-depleted soils. Replenishment of these soil types is mostly affected due to slow release of exchangeable K from the clay minerals. Consequently, for maximal crop growth, soil solution and exchangeable K need to be refilled continually with K by the release of non-exchangeable K through weathering of K reserves (i.e. micas and feldspars) (Sparks and Huang 1985) or the addition of phosphatic fertilizers. Exchangeable K serves as the indicator of soil K status, which occurs in a fast equilibrium with the soil solution K and taken as readily available. Fixed and lattice K can be grouped together to make up the pool of non-exchangeable inorganic K in the soil (Fig. 20.2).

Intensive cropping, runoff, leaching and soil erosion owing to soil K deficiency most of the time result in stunted growth and limited physiological activities of plant. Soil K deficiency inhibits plant to uptake adequate amount of K which in turn results in K deficiency syndromes.

In plants, K deficiency causes yellowing of the leaf edges, giving them a burnt appearance. It can also cause slow growth and incomplete root development. K is involved in numerous biochemical and physiological processes in plants like stomatal regulation for plants. K performs a multitude of functions like activation of enzymes, production of adenosine triphosphate (ATP) and sugar, regulation of rate of photosynthesis and transportation of sugars produced in photosynthesis through the phloem to other parts of the plant for utilization and storage. Plant transport system uses energy in the form of ATP. Therefore, it can be said that K uptake directly influences ATP production. Less is the K uptake then lesser are the amounts of ATP, which slows down the transport system. Transport of water and nutrients in the plant through xylem is also mostly dependent upon the K concentration of plant. The enzyme responsible for synthesis of starch (starch synthase) is activated by K; hence, it plays a crucial role in water and nutrient transport (Usherwood 1985; Doman and Geiger 1979; Marschner 1995; Pettigrew 2008). K promotes root growth, strengthens the stem and increases resistance to cold and water stress. It directly connects with improvement of the quality of crop and reduces pest and disease incidence by enhancing crop resistance as well. Although K is not a constituent of any organic molecule or plant structure, it is involved in numerous biochemical and physiological processes occurring inside the cell, leading to the pivotal role of better plant growth, yield, quality and stress (Cakmak 2005).

K deficiency is severely problematic because soluble K declines easily in soils, while the crops uptake K only from soil. Hence, limited available K of soil often supplied with chemical K-fertilizers such as muriate of potash (KCl) and sulphate of potash (K_2SO_4). Extensive use of these chemical fertilizers has an adverse effect on long-term soil health and quality such as destruction of soil structures, reduction in amount of organic matter in soil, etc. However, the hangover effect of inorganic fertilizer application on soil environment reduces its microbial activity. Therefore, a viable alternative to chemical fertilizers has been a continuous search. Microbial diversity in soil serves a significant role in metabolism of complex molecules, helps in boosting plant nutrition and offers countless new genes, biochemical pathways, antibiotics and other metabolites, important molecules for agronomic productivity.

20.2 Potassium-Solubilizing Microorganisms (KSMs)

Microorganisms play the crucial role in the natural P and K cycles as P- or K-solubilizing microbes in plant rhizosphere (Diep and Hieu 2013). Soil microfloras bear the ability to make available soluble K from K-bearing minerals, such as K-feldspar, mica and illite. These microbes secrete organic acids which act as chelating agents and quickly dissolve rock and silicon ions, ultimately releasing the K ions into the soil (Bennett et al. 1998). Bacteria from genera *Bacillus mucilaginosus* and *B. edaphicus* can release polysaccharide and carboxylic acids, such as tartaric acid and citric acid, to solubilize K compounds that plants can access (Richards and Bates 1989; Lin et al. 2002). The use of KSMs as biofertilizers can increase the concentration of available K^+ in the soil, which will mitigate K deficiency in a more eco-friendly way (Barker et al. 1998). Zhanga and Kong (2014) isolated 27 KSB strains and identified through the homology pattern of 16 s ribosomal DNA. Out of 23 isolates, 17 strains belonged to *Klebsiella variicola* and the rest belonged to genera *Enterobacter cloacae*, *E. asburiae and E. aerogenes*, *Pantoea agglomerans*, *Agrobacterium tumefaciens*, *Microbacterium foliorum*, *Myroides odoratimimus* and *Burkholderia cepacia*. There are also other groups of microorganisms, such as actinomycetes (*Torulaspora globosa*) and fungi, capable of solubilization of K contained in silicate minerals by the process of decomposition (Weed et al. 1969).

Several microorganisms like *Aspergillus niger*, *B. extorquens* and *Clostridium pasteurianum* were able to grow on muscovite, biotite, orthoclase, microcline and micas under in vitro conditions (Reitmeir 1951; Archana et al. 2013). Microbes from the genus *Aspergillus*, *Bacillus* and *Clostridium* are efficient K-solubilizers (Bhardwaj et al. 2014). Further, Prajapati and Modi (2012) reported 14 KSBs, and the best among them was *Enterobacter hormaechei*. Rhizosphere bacteria, viz. *B. mucilaginosus* (Zhao et al. 2008; Basak and Biswas 2009; Raj 2004; Sugumaran and Janarthanam 2007; Zarjani et al. 2013), *B. edaphicus* (Sheng 2005), *B. circulans* (Lian et al. 2002), *Burkholderia*, *A. ferrooxidans* (Sheng and Huang 2002; Sheng and He 2006), *Arthrobacter* sp. (Zarjani et al. 2013), *Enterobacter hormaechei* (KSB-8) (Prajapati et al. 2013), *Paenibacillus mucilaginosus* (Liu et al. 2012; Hu et al. 2006), *P. frequentans*, *Cladosporium* (Argelis et al. 1993), *Aminobacter*, *Sphingomonas*, *Burkholderia* (Uroz et al. 2007) and *Paenibacillus glucanolyticus* (Sangeeth et al. 2012), have the ability to solubilize K from insoluble K-bearing minerals.

KSBs are heterotrophic in nature and derive all their energy and cellular carbon from pre-existing organic material. Thus, in soil, they perform the major functions like the formation of humus, the cycling of other minerals tied up in organic matter and the prevention of buildup of dead organic materials. Besides, KSBs are aerobic bacteria which play the indispensable role in maintaining the soil structure by contributing to formation and stabilization of water-stable soil aggregates. In addition to the above, these bacteria stimulate plant growth and/or show antagonistic effect on soilborne root pathogens.

Moreover, KSBs specifically are well known for its capability to make available rock K mineral (micas, illites and orthoclases). This is done through the production and excretion of organics acids (Han et al. 2006). Therefore, KSBs' function is to increase K availability in soils besides increasing mineral contents in plants. Many scientists reported that *B. mucilaginosus* is a KSB, which releases organic acids and solubilizes K minerals (micas, illites and orthoclases) (Friedrich et al. 1991; Ullman et al. 1996). According to Sheng and He (2006), soil microorganisms produce organic acids (oxalic acid and tartaric acids) and capsular polysaccharides for solubilization of illite and feldspar to release K (Meena et al. 2013; Maurya et al. 2014).

Rosa-Magri et al. (2012) reported that *T. globosa* is capable of solubilizing alkaline ultramafic rock, releasing higher concentrations of K. Therefore, in the bio-solubilization assays with *T. globosa*, the acids secreted chelate the minerals from the rock powder and release ions with an alkalinizing effect, resulting in higher pH and available K. This yeast offers great potential for biotechnological applications, especially production of organic acids and solubilization of rocks, and can be used as biofertilizers. The fact that the yeast *T. globosa* is found in the sugarcane rhizosphere is a positive factor because the application of rock powder in agricultural soils depends greatly on the presence of these microorganisms to convert insoluble nutrients into available elements for crop uptake, thus promoting nutrition and development.

20.3 Mechanism of K-Solubilization

The mechanism for mineral K-solubilization is the production of organic acids which play pivotal role in the mineralization of organic K in soil (Prajapati et al. 2012). Microbes produce the low molecular weight organic acid which primarily serves to release K from minerals during physiological activities and, by its absorption of soluble K, causes an increase in the concentration gradient during mineral hydrolysis. Insoluble K and structurally unavailable forms of K compounds are mobilized and solubilized due to the secretion of various types of organic acids. These acids convert insoluble K (mica, muscovite, biotite feldspar) to the soluble form of K (soil solution form) with the availability of major plant nutrients (Bahadur et al. 2014).

Organic matter after decomposition produces acids like citric acid, formic acid, malic acid and oxalic acid. These organic acids produced enhance the dissolution of K compounds by supplying protons and by complexing Ca^{2+} ions (Shanware et al. 2014). Previous work has shown microbes produce organic compounds such as acetate, citrate and oxalate, which enhance mineral solubilization in soil (Sheng et al. 2003). K-solubilization occurs by buildup of complex between organic acids and metal ions such as Fe^{2+}, Al^{3+} and Ca^{2+} (Styriakova et al. 2003).

In a study, it was reported that KSB like *B. mucilaginosus* is capable of solubilizing rock K mineral powder such as micas, illite and orthoclases through release of organic acids which act as chelating agents (Ullman et al. 1996). Microbe-produced organic ligands include metabolic by-products, extracellular enzymes, chelates and both simple and complex organic acids enhance the dissolution of aluminosilicate mineral or quartz both in field and laboratory experiments (Grandstaff 1986; Surdam and MacGowan 1988). Production of capsular polysaccharides along with organic acid production like tartaric and oxalic acid by the microorganisms leads to solubilization of feldspar and illite to release K (Sheng and He 2006). Another report showed that K was solubilized by release of inorganic and organic acids and due to production of mucilaginous capsules containing of exopolysaccharides by *Bacillus*, *Clostridium* and *Thiobacillus* (Groudev 1987). The weathering ability of the bacteria involves production of protons, organic acids, siderophores and organic ligands, which act as chelating agents and do the process of chelation by making the K mineral free for crop uptake. This was seen in *Cladosporium cladosporioides* and *Penicillium* sp. (Meena et al. 2014a, b; Kumar et al. 2015).

These fungal species have the ability to secrete a variety of organic acids like oxalic, citric and gluconic acids in broth culture, resulting in dissociation of clay silicates, mica and feldspar. The filamentous fungi are known to cause an extensive weathering of stone due to organic acid secretion. The production of gluconate promotes dissolution of silicates like albite, quartz and kaolinite by subsurface bacteria (Argelis et al. 1993). Oxalic acid production caused dissolution of feldspar, while oxalic and tartaric acid were involved in mobilizing illite (Hu et al. 2006).

Organic acids and siderophores could play a crucial role in the solubilization of elements such as K, Si and Fe from the liquid medium containing acid-leached soil, muscovite and biotite. In general, the mechanisms of K-solubilization can be proposed as decomposing ability of the bacteria which involves the production of protons, organic acids, siderophores, capsular polysaccharides and certain organic ligands. Decomposition of silicate minerals such as muscovite and biotite by plant growth-promoting bacteria is dependent on the production of oxalate and citrate and the extent of which polysaccharides absorbed organic acids (Liu et al. 2006). Many rhizospheric microorganisms in soil are able to solubilize unavailable forms of K in silicate minerals, such as micas and orthoclases, by excreting organic acids, which either directly dissolve rock K or chelate silicon ions to bring K into solution (Zarjani et al. 2013).

They convert insoluble K in the soil into a form that plants can access. This is a promising strategy for the improvement of plant absorption of K, thereby reducing the use of chemical fertilizer. Direct application of RP and RK materials may be agronomically more useful and environmentally safer than soluble PK fertilization (Rajan et al. 1996). However, K nutrients are released slowly from the rock materials, and their use as fertilizer often causes significant increase in the yield of crops (Sindhu et al. 2012). Therefore, concerted efforts are made to understand the combined effects of rock material addition. Among the K-bearing silicate minerals, mica was found to weather readily (Tandon and Sekhon 1988). It is reported that KSMs secrete organic acids like citric, oxalic and tartaric acids and due to which they are capable of solubilizing the unavailable forms of K from K-bearing minerals such as micas, illite and orthoclase. These organic acids can facilitate the weathering of minerals by directly dissolving K from rocks or through the formation of metal-organic complexes by chelation with silicon ions to bring the K into solution (Song and Huang 1988; Friedrich et al. 1991; Bennett et al. 1998). Many workers reported KSMs produce carboxylic acids and capsular polysaccharides and help in releasing K from feldspar. KSMs, viz. *B. mucilaginosus* and *B. edaphicus* as biofertilizers, can be used in soil for K-solubilization (Lin et al. 2002; Sheng and Huang 2002).

Malinovskaya et al. (1990) opined that a mixture of polymers and low molecular weight compounds act synergistically on mineral weathering and identified *Bacillus mucilaginosus* capable of secreting these compounds. Furthermore, microorganisms produce various organic ligands during their physiological metabolism. These include metabolic by-products, extracellular enzymes and simple and complex organic acids, which helped in dissolution of feldspar by decreasing the pH of the soil environment (Bennett et al. 2001).

Plant growth promotion as well as K-solubilization activities of *B. mucilaginosus* strain resulted in higher biomass yield in various crops (Sheng 2005; Wu et al. 2005). Certain other hypotheses/mechanisms also involve solubilizing and mobilizing soil K reserve. Formation of biofilms by some specific bacterial strains on the rhizospheric mineral surfaces (Balogh-Brunstad et al. 2008). Biofilms not only protect the microbial community from various environmental factors or threats such as secondary metabolites secreted by other microbes but also help to regulate transport of heavy metals, cations and nutrients to the microbes from bulk soil reserves. There are reports that ectomycorrhizal hyphal networks and root hairs of non-ectomycorrhizal trees, embedded in biofilms (microorganisms surrounded by extracellular polymers) and transfer nutrients to the host. Again, reports suggest that biofilms help to accelerate weathering of minerals like biotite and anorthite, thereby increasing nutrient uptake by the crop. Further, root-derived organic acids have effect on the mobilization of nutrients from the rhizosphere (Jones and Darrah 1994; Jones et al. 1996; Jones 1998).

According to Leyval and Berthelin (1989), ectomycorrhizal fungi solubilize K, Fe, Mg and Al from phlogopite within the rhizosphere. The potential for dissolution of silicate rock powders is enhanced through the removal of nutrients and the addition of organic acids by mycorrhiza. Furthermore, reports revealed that low molecular weight organic anions are being released by some rock-eating fungi (ectomycorrhizal fungi) which forms microscopic tunnels and exude at hyphal tips within minerals like feldspar and hornblende grains in soils, thereby increasing the weathering rates significantly (Van Scholl et al. 2008). Bacterial intervention effects positively on mica and helps in improving the water-soluble, exchangeable and non-exchangeable K pools in soils, thereby influencing the K dynamics in soils into those pools which become readily available to crops. *B. mucilaginosus* is able to solubilize inorganic source of K like muriate of potash and sulphate of potash by means of production of organic acids in order to improve the yield. K-solubilizers were capable of producing citric and oxalic acid, along with other organic acids malic, succinic and tartaric acid, at low concentration (Archana et al. 2013; Singh et al. 2015).

20.4 Effect of KSMs on Soil Fertility and Crop Productivity

In the context of imbalanced fertilization, lower potash application results in a significant depletion of soil potash reserves, yield loss and higher economic risk for farmers. Utilization of microbial inoculants or bioinoculants as K-solubilizers can positively influence crop quality, growth and yield, which can be both economically and environmentally sustainable. N, P and K are the primary nutrients for plant growth and development. Indiscriminate use of synthetic fertilizers for nourishment of plant has caused the contamination of the soil, has polluted the water basins and has destroyed microorganisms, making the soil less fertile. Balanced use of biofertilizers is an environment-friendly approach for supplementation of nutrient to the plant growth and development and crop productivity. These biological agents include PGPMs like N-fixers; mineral solubilizers (P, K, Zn and Fe) by the different microbial genera *Azotobacter* spp. and *Rhizobium* spp. are well known as N-fixers, while *B. megaterium* and *Aspergillus* spp. are the mineral solubilizers. In order to achieve optimum crop productivity and environmental sustainability, K is also required in adequate amount; therefore, we need to exploit more and more KSMs that have ability to solubilize K from its fixed form. Biofertilizer is a good platform to deliver this primary and secondary nutrient by assistance of KSMs. Solubilization of this soil mineral, by fungi and bacteria, is well established, which reflects their use as competent biofertilizers. Isolation, characterization and development of liquid carriers for these potash solubilizers can be the means towards sustainable agriculture development. Apart from the identification of KSMs, there is strong need for field demonstration studies of these isolates along with standardization of the dose for application for various crops so that the technology can be easily adopted by farmers for multiplication at their own level.

Lin et al. (2002) demonstrated significant increases in K and P uptake in tomato plants inoculated with silicate-dissolving bacterium (*B. mucilaginosus*) compared with the non-inoculated plants. Application of *B. mucilaginosus* as a bioinoculant yielded 36.6 % higher biomass over the untreated (Basak and Biswas 2009; Meena et al. 2015c). The microbial inoculant enhanced biomass yield directly by solubilizing soil K reserve as well as indirectly by causing increased growth of the plants. KSB enhanced the plant growth and soil nutrient level significantly (Han and Lee 2005).

Similarly, *Frateuria aurantia* belonging to the family *Pseudomonadaceae* promoted the crop yield (Ramarethinam and Chandra 2006).

In addition, they are also known to secrete amino acids, vitamins and growth-promoting substances like indole-3-acetic acid (IAA) and gibberellic acid (GA_3) which help in increased growth of the plants (Ponmurugan and Gopi 2006). Higher green leaf yield and bush health in terms of number of plucking points, internodal length and physiological and biochemical constituents might be due to the action of KSB by solubilizing K content in the soils (Bagyalakshmi et al. 2012). Prajapati et al. (2013) reported that inoculation with bacterial strain *E. hormaechei* had enhanced the root and shoot length of okra and mobilized K efficiently in plant when feldspar was added to the soil.

KSMs cannot only activate the insoluble K mineral but also change that into available P, Zn and Fe. Thus, the quality of crop can be improved, the effect of environment can be decreased, the physicochemical property can be made better and the cost of production can be reduced (Xue 2005). Silicate bacteria has the ability to decompose soil K minerals, to activate K for the absorption and utilization of plant and to increase crop productivity (Sun and Zhang 2006; Tang and Zhang 2008; Sheng et al. 2001; Meena et al. 2015a, b). *B. mucilaginosus* grows well on K-feldspar-supplemented, nitrogen-free medium, which indicates that it has the ability of N-fixation and decomposition of the minerals.

N-fixers and P- and K-solubilizers influenced crop growth and productivity; these beneficial microbes also enhance the quality of the agricultural products (Wang et al. 2009). The improvement of K nutrient can promote the changes of a series of ~70 enzymatic activities (Ai-min et al. 2013). According to Prajapati et al. (2013), higher nutrient uptake by seedling biopriming with PGPMs leading to production of plant growth regulators (PGRs) at the root interface stimulated root development, resulting in better absorption of water and nutrients from the soil. KSMs can provide less polluting and less energy-consuming approaches (Burgstaller et al. 1992), widely used in improvement of available K for assimilation by plants and removal of impurities from minerals (Nianikova et al. 2002; Meena et al. 2014a, b). KSMs enhance the nutrient availability and crop productivity; it resulted in economic, environmental and agricultural sustainability.

20.5 Conclusions

KSMs play the most important role in the environmental sustainability by providing essential nutrients from primary minerals that are required for plants and to maintain soil health and quality. Furthermore, vigorous research is needed in this field to isolate native strains of microorganisms capable of not only solubilizing K but also secreting plant growth hormones, solubilizing P and inhibiting plant pathogens. This way these microbes can be more beneficial to soil health and crop quality.

References

Ai-min Z, Gang-yong Z, Shuang-feng Z, Rui-ying Z, Bao-cheng Z (2013) Effect of phosphorus and potassium content of plant and soil inoculated with *Paenibacillus kribensis* CX-7 strain antioxidant and antitumor activity of Phyllanthus emblica in colon cancer cell lines. Int J Curr Microbiol Appl Sci 6:273–279

Archana DS, Nandish MS, Savalagi VP, Alagawadi AR (2013) Characterization of potassium solubilizing bacteria (KSB) from rhizosphere. Soil Bioinfolet 10 (1B):248–257

Argelis DT, Gonzala DA, Vizcaino C, Gartia MT (1993) Biochemical mechanism of stone alteration carried out by filamentous fungi living in monuments. Biogeochemistry 19:129–147

Bagyalakshmi B, Ponmurugan P, Marimuthu S (2012) Influence of potassium solubilizing bacteria on crop productivity and quality of tea (*Camellia sinensis*). Afr J Agric Res 7(30):4250–4259

Bahadur I, Meena VS, Kumar S (2014) Importance and application of potassic biofertilizer in Indian agriculture. Int Res J Biol Sci 3(12):80–85

Balogh-Brunstad Z, Keller CK, Gill RA, Bormann BT, Li CY (2008) The effect of bacteria and fungi on chemical weathering and chemical denudation fluxes in pine growth experiments. Biogeochemistry 88:153–167

Barker WW, Welch SA, Chu S, Banfield JF (1998) Experimental observations of the effects of bacteria on aluminosilicate weathering. Am Mineral 83:1551–1563

Basak BB, Biswas DR (2009) Influence of potassium solubilizing microorganism (*Bacillus mucilaginosus*) and waste mica on potassium uptake dynamics by sudan grass (*Sorghum vulgare* Pers.) grown under two Alfisols. Plant Soil 317:235–255

Bear F (1964) Chemistry of the soil. Oxford and IBH Publications, New York, USA, pp 129–130

Bennett PC, Choi WJ, Rogera JR (1998) Microbial destruction of feldspars. Miner Manag 8(62A):149–150

Bennett PC, Rogers JR, Choi WJ (2001) Silicates, silicate weathering, and microbial ecology. Geomicrobiol J 18:3–19

Bhardwaj D, Ansari MW, Sahoo RK, Tuteja N (2014) Biofertilizers function as key player in sustainable agriculture by improving soil fertility, plant tolerance and crop productivity. Microb Cell Factories 13:66

Buchholz DD, Brown JR (1993) Potassium in Missouri soils. Agricultural publication, pp 9–185

Burgstaller ZA, Sreasser H, Wobking H, Shinner F (1992) Solubilization of zinc oxide from filterdust with *Penicillium simplicissimum* bioreactor, leaching and stoichiometry. Environ Sci Technol 26:340–346

Cakmak I (2005) The role of potassium in alleviating detrimental effects of abiotic stresses in plants. J Plant Nutr Soil Sci 168:521–530

Diep CN, Hieu TN (2013) Phosphate and potassium solubilizing bacteria from weathered materials of denatured rock mountain, Ha Tien, Kiên Giang province Vietnam. Am J Life Sci 1(3):88–92

Doman DC, Geiger DR (1979) Effect of exogenously supplied foliar potassium on phloem loading in Beta vulgaris L. Plant Physiol 64:528–533

Friedrich S, Platonova NP, Karavaiko GI, Stichel E, Glombitza F (1991) Chemical and microbiological solubilization of silicates. Acta Biotechnol 11:187–196

George R, Michael S (2002) Potassium for crop production. Communication and Educational Technology Services, University of Minnesota Extension

Grandstaff DE (1986) The dissolution rate of forsteritic olivine from Hawaiian beach sand. In: Colman SM, Dethier DP (eds) Rates of chemical weathering of rocks and minerals. Academic, New York, pp 41–60

Groudev SN (1987) Use of heterotrophic microorganisms in mineral biotechnology. Acta Biotechnol 7:299–306

Han HS, Lee KD (2005) Phosphate and potassium solubilizing bacteria effect on mineral uptake, soil availability and growth of egg plant. Res J Agric Biol Sci 1:176–180

Han HS, Supanjani, Lee KD (2006) Effect of co-inoculation with phosphate and potassium solubilizing bacteria on mineral uptake and growth of pepper and cucumber. Plant Soil Environ 52:130–136

Hu XF, Chen J, Guo JF (2006) Two phosphate and potassium solubilizing bacteria isolated from Tianmu Mountain, Zhejiang, China. World J Microbiol Biotechnol 22:983–990

Jones DL (1998) Organic acids in the rhizosphere—a critical review. Plant Soil 205:25–44

Jones DL, Darrah PR (1994) Role of root derived organic-acids in the mobilization of nutrients from the rhizosphere. Plant Soil 166:247–257

Jones DL, Darrah PR, Kochian LV (1996) Critical-evaluation of organic-acid mediated iron dissolution in the rhizosphere and its potential role in root iron uptake. Plant Soil 180:57–66

Kumar A, Bahadur I, Maurya BR, Raghuwanshi R, Meena VS, Singh DK, Dixit J (2015) Does a plant growth-promoting rhizobacteria enhance agricultural sustainability? J Pure Appl Microbiol 9(1):715–724

Leyval C, Berthelin J (1989) Interactions between Laccaria laccata, Agrobacterium radiobacter and beech roots: influence on P, K, Mg and Fe mobilization from minerals and plant growth. Plant Soil 117:103–110

Lian B, Fu PQ, Mo DM, Liu CQ (2002) A comprehensive review of the mechanism of potassium release by silicate bacteria. Acta Mineral Sin 22:179–183

Lin QM, Rao ZH, Sun YX, Yao J, Xing LJ (2002) Identification and practical application of silicate-dissolving bacteria. Agric Sci China 1:81–85

Liu W, Xu X, Wu S, Yang Q, Luo Y, Christie P (2006) Decomposition of silicate minerals by *Bacillus mucilaginosus* in liquid culture. Environ Geochem Health 28:133–140

Liu D, Lian B, Dong H (2012) Isolation of Paenibacillus sp. and assessment of its potential for enhancing mineral weathering. Geomicrobiol J 29:413–421

Malinovskaya IM, Kosenko LV, Votselko SK, Podgorskii VS (1990) Role of Bacillus mucilaginosus polysaccharide in degradation of silicate minerals. Mikrobiologiya 59:49–55

Marschner H (1995) Functions of mineral nutrients: macronutrients. In: Marschner H (ed) Mineral nutrition of higher plants, 2nd edn. Academic, London, USA, pp 299–312

Maurya BR, Meena VS, Meena OP (2014) Influence of inceptisol and alfisol's potassium solubilizing bacteria (KSB) isolates on release of K from waste mica. Vegetos 27(1):181–187

McAfee J (2008) Potassium, a key nutrient for plant growth. Department of Soil and Crop Sciences. http://jimmcafee.tamu.edu/files/potassium

Meena OP, Maurya BR, Meena VS (2013) Influence of K-solubilizing bacteria on release of potassium from waste mica. Agric Sustain Dev 1(1):53–56

Meena VS, Maurya BR, Bahadur I (2014a) Potassium solubilization by bacterial strain in waste mica. Bangladesh J Bot 43(2):235–237

Meena VS, Maurya BR, Verma JP (2014b) Does a rhizospheric microorganism enhance K+ availability in agricultural soils? Microbiol Res 169:337–347

Meena RK, Singh RK, Singh NP, Meena SK, Meena VS (2015a) Isolation of low temperature surviving plant growth-promoting rhizobacteria (PGPR) from pea (*Pisum sativum* L.) and documentation of their plant growth promoting traits. Biocatal Agric Biotechnol doi:10.1016/j.bcab.2015.08.006

Meena VS, Maurya BR, Verma JP, Aeron A, Kumar A, Kim K, Bajpai VK (2015b) Potassium solubilizing rhizobacteria (KSR): isolation, identification, and K-release dynamics from waste mica. Ecol Eng 81:340–347

Meena VS, Meena SK, Verma JP, Meena RS, Ghosh BN (2015c) The needs of nutrient use efficiency for sustainable agriculture. J Clean Prod 102:562–563

Memon YM, Fergus IF, Hughes JD, Page DW (1988) Utilization of non-exchangeable soil potassium in relation to soil types, plant species and stage of growth. Aust J Soil Res 26:489–496

Nianikova GG, Kuprina EE, Pestova OV, Vodolazhskaia SV (2002) Immobilizing of Bacillus mucilaginosus, a producer of exopolysaccharides, on chitin. Priklalmaia Biokhilia I Mikrobiologiya 38:300–304

Pettigrew WT (2008) Potassium influences on yield and quality production for maize, wheat, soybean and cotton. Physiol Plant 133:670–681

Ponmurugan P, Gopi C (2006) In vitro production of growth regulators and phosphatase activity by phosphate solubilizing bacteria. Afr J Biotechnol 5 (4):348–350

Prajapati KB, Modi HA (2012) Isolation and characterization of potassium solubilizing bacteria from ceramic industry soil. CIBTech J Microbiol 1(2–3):8–14

Prajapati K, Sharma MC, Modi HA (2012) Isolation of two potassium solubilizing fungi from ceramic industry soils. Life Sci Leafl 5:71–75

Prajapati KB, Sharma MC, Modi HA (2013) Growth promoting effect of potassium solubilizing microorganisms on Okra (Abelmoschus esculentus). Int J Agric Sci 3(1):181–188

Raj SA (2004) Solubilization of silicate and concurrent release of phosphorus and potassium in rice ecosystem. In: Book chapter – conference paper biofertilizers technology. Coimbatore, pp 372–378

Rajan SSS, Watkinson JH, Sinclair AG (1996) Phosphate rock for direct application to soils. Adv Agron 57:77–159

Ramarethinam S, Chandra K (2006) Studies on the effect of potash solubilizing bacteria Frateuria aurantia (Symbion-K- liquid formulation) on Brinjal (Solanum melongena L) growth and yield. Pestology 11:35–39

Read JJ, Reddy KR, Jenkins JN (2006) Yield and quality of upland cotton as influenced by nitrogen and phosphorus. Eur J Agron 24:282–290

Reitmeir RF (1951) Soil potassium. In: Norman AG (ed) Advances in agronomy II. Academic, New York, pp 113–164

Richards JE, Bates TE (1989) Studies on the potassium-supplying capacities of southern Ontario soils. III. Measurement of available K. Canadian. J Soil Sci 69 (3):597–610

Rosa-Magri MM, Avansini SH, Lopes-Assad ML, Tauk-Tornisielo SM, Ceccato-Antonini SR (2012) Release of potassium from rock powder by the yeast torulaspora globosa. Braz Arch Biol Technol 55 (4):577–582

Sangeeth KP, Bhai RS, Srinivasan V (2012) Paenibacillus glucanolyticus, a promising potassium solubilizing bacterium isolated from black pepper (*Piper nigrum* L.) rhizosphere. J Spic Aromat Crop 21(2):118–124

Shanware AS, Kalkar SA, Trivedi MM (2014) Potassium solublisers: occurrence, mechanism and their role as competent biofertilizers. Int J Curr Microbiol Appl Sci 3(9):622–629

Sharpley AN (1989) Relationship between soil potassium forms and mineralogy. Soil Sci Soc Am J 52:1023–1028

Sheng XF (2005) Growth promotion and increased potassium uptake of cotton and rape by a potassium releasing strain of Bacillus edaphicus. Soil Biol Biochem 37:1918–1922

Sheng XF, He LY (2006) Solubilization of potassium bearing minerals by a wild type strain of Bacillus edaphicus and its mutants and increased potassium uptake by wheat. Can J Microbiol 52(1):66–72

Sheng XF, Huang WY (2002) Mechanism of potassium release from feldspar affected by the strain NBT of silicate bacterium. Acta Pedol Sin 39:863–871

Sheng XF, Huang WY, Cao XY (2001) Dissolution of feldspar and potassium uptake by the strain NBT of silicate bacterium. Plant Nutr Fertil 7(4):459–466

Sheng XF, Xia JJ, Chen J (2003) Mutagenesis of the Bacillus edaphicus strain NBT and its effect on growth of chili and cotton. Agric Sci China 2:40–412

Sindhu SS, Parmar P, Phour M (2012) Nutrient cycling: potassium solubilization by microorganisms and improvement of crop growth. In: Parmar N, Singh A (eds) Geomicrobiology and biogeochemistry: soil biology. Springer, Wien/New York/Berlin, USA

Singh NP, Singh RK, Meena VS, Meena RK (2015) Can we use maize (*Zea mays*) rhizobacteria as plant growth promoter? Vegetos 28(1):86–99

Song SK, Huang PM (1988) Dynamics of potassium release from potassium-bearing minerals as influenced by oxalic and citric acids. Soil Sci Soc Am J 52:383–390

Sparks DL (1987) Potassium dynamics in soils. Adv Soil Sci 6:1–63

Sparks DL (2000) Bioavailability of soil potassium, D-38-D-52. In: Sumner ME (ed) Handbook of soil science. CRC, Boca Raton

Sparks DL, Huang PM (1985) Physical chemistry of soil potassium. In: Munson RD et al (eds) Potassium in agriculture. ASA, Madison, pp 201–276

Styriakova I, Styriak I, Hradil D, Bezdicka P (2003) The release of iron bearing minerals and dissolution of feldspar by heterotrophic bacteria of Bacillus species. Ceram Silic 47(1):20–26

Sugumaran P, Janarthanam B (2007) Solubilization of potassium containing minerals by bacteria and their effect on plant growth. World J Agric Sci 3:350–355

Sun DS, Zhang Q (2006) Screening of silicate bacteria and bio-leaching silicon from silicate ores. J Xi'an Univ Sci Technol 26(2):235–239

Surdam RC, MacGowan DB (1988) Oil field waters and sandstone diagenesis. Appl Geochem 2(5–6):613–620

Syers JK (1998) Soil and plant potassium in agriculture. In: Proceedings No. 411, The International Fertiliser Society York, UK 32

Tandon HLS, Sekhon GS (1988) Potassium research and agricultural production in India. Fertilizer Development and Consultation Organization, New Delhi, p 144

Tang L, Zhang JZ (2008) Isolation, purification and identification of silicate-dissolving bacterial strains and studies of their biological characteristics. Shandong Agric Sci 1:71–73

Ullman WJ, Kirchman DL, Welch SA (1996) Laboratory evidence for microbially mediated silicate mineral dissolution in nature. Chem Geol 132:11–17

Uroz S, Calvaruso C, Turpault MP, Pierrat JC, Mustin C, Frey-Klett P (2007) Effect of the mycorrhizosphere on the genotypic and metabolic diversity of the bacterial communities involved in mineral weathering in a forest soil. Appl Environ Microbiol 73:3019–3027

Usherwood NR (1985) The role of potassium in crop quality. In: Munson RD (ed) Potassium in agriculture. ASA-CSSA-SSSA, Madison, pp 489–513

Van Schöll L, Kuyper TW, Smits MM, Landeweert R, Hoffland E, van Breemen N (2008) Rock-eating mycorrhizas: their role in plant nutrition and biogeochemical cycles. Plant Soil 303:35–47

Wang W, Li J, Liu JS, Zhu BC (2009) Isolation of the silicate bacteria strain and determination of the activity of releasing silicon and potassium. J Anhui Agric Sci 37(17):7889–7891

Weed SB, Davey CB, Cook MG (1969) Weathering of mica by fungi. Soil Sci Soc Am J 33:702–706

Wu SC, Cao ZH, Li ZG, Cheung KC, Wong MH (2005) Effects of biofertilizer containing N-fixer, P and K solubilizers and AM fungi on maize growth: a greenhouse trail. Geoderma 125:155–166

Xue BG (2005) Screening and identification of the bacterium which have efficiency on resolving phosphorus and potassium and in nitrogen fixation. J Huazhong Agric Univ 24(1):43–48

Zarjani JK, Aliasgharzad N, Oustan S, Emadi M, Ahmadi A (2013) Isolation and characterization of potassium solubilizing bacteria in some Iranian soils. Arch Agron Soil Sci 59(12):1713–1723

Zhanga C, Kong F (2014) Isolation and identification of potassium-solubilizing bacteria from tobacco rhizospheric soil and their effect on tobacco plants. Appl Soil Ecol 82:18–25

Zhao F, Sheng X, Huang Z, He L (2008) Isolation of mineral potassium-solubilizing bacterial strains from agricultural soils in Shandong Province. Biodivers Sci 16:593–600

Potassium-Solubilizing Bacteria and Their Application in Agriculture

21

Maqshoof Ahmad, Sajid Mahmood Nadeem, Muhammad Naveed, and Zahir Ahmad Zahir

Abstract

Potassium (K) is one of the major macronutrients which play an important role in plant growth and development. Total soil potassium reserves are generally large; however, major portion of it exists in insoluble K minerals and very little potassium becomes available to plants. There are certain microorganisms which use a number of biological processes to make potassium available from unavailable forms. These potassium-solubilizing bacteria (KSB) can be used as a promising approach to increase K availability in soils, thus playing an important role for crop establishment under K-limited soils. Owing to naturally available source of potassium in soil and high price of synthetic potassium fertilizers, the importance of KSB is increasing day by day. The use of chemical fertilizers can be decreased by using KSB in agriculture that can lead to sustainable agriculture. A number of workers have demonstrated the role of KSB in crop improvement. The present review highlights the importance of KSB for enhancing crop production. The mechanisms used by KSB for K solubilization have been discussed. The work of various scientists regarding plant growth promotion through KSB has been reviewed in detail, and present constraints and future scope of this technology have also been discussed.

Keywords

Insoluble K • Microorganisms • Biological processes • K solubilization

M. Ahmad
University College of Agriculture and Environmental Sciences, the Islamia University of Bahawalpur, Bahawalpur, Pakistan

S.M. Nadeem
University of Agriculture, Faisalabad, Sub-Campus, Burewala-Vehari, Pakistan

M. Naveed • Z.A. Zahir (✉)
Institute of Soil & Environmental Sciences, University of Agriculture, Faisalabad, Pakistan
e-mail: zazahir@yahoo.com

V.S. Meena et al. (eds.), *Potassium Solubilizing Microorganisms for Sustainable Agriculture*,
DOI 10.1007/978-81-322-2776-2_21

21.1 Introduction

World population is increasing on alarming rate which may rise up to 8.9 billion by 2050 as estimated by the United Nations (Wood 2001). Feeding this burgeoning population will be the major challenge in the near future. This will be more dangerous for the developing countries with fewer resources. In this scenario, understanding of soil-plant-microbe interactions is the only hope that may help the researchers to increase agricultural productivity on sustainable basis through better management of soil biology and agroecosystem. The earth's crust contains minerals that have essential plant nutrients as their structural part. This makes soil a dynamic natural body.

Potassium is the third most important major essential plant nutrient after nitrogen and phosphorus. It has a key role in plant metabolism that triggers most important enzymes involved in the plant physiology. The deficiency of potassium hinders plant physiology leading to poor growth and development and lesser yields (White and Karley 2010). This also disturbs the immune system of plants that increases susceptibility to diseases (Armengaud et al. 2010) and pest attack (Troufflard et al. 2010). Potassium occurs in soil as available, fixed, interlayer, and mineral K. The application of potassium fertilizers is a contemporary practice to supply available K in extensive agricultural systems (Yadegari et al. 2012; Dasan 2012; Zhang et al. 2013). The potash fertilizers that are available in the world market are costly that increase the cost of inputs and decrease the agricultural profitability.

The K-bearing minerals are major source of insoluble potassium in soils and share ≈90–98 % of total potassium (Mengel and Kirkby 2001). The major K-bearing minerals are feldspar, mica, muscovite, biotite, orthoclase, illite, vermiculite, and smectite. These minerals supply slowly available K to plants. The plant availability of K from these minerals can be enhanced by solubilizing these minerals.

The bacteria that are involved in the solubilization of potassium from K-bearing minerals are called potassium-solubilizing bacteria (KSB). They have the ability to convert insoluble/mineral K into available K in soil (Zeng et al. 2012). KSB in the soil and rhizosphere play a central role in the cycling of K (Diep and Hieu 2013). They solubilize the K-bearing minerals by producing acids (Basak and Biswas 2012). Soil-plant-microbe interaction is an important aspect of recent research. The K solubilizers are important part of microbial community in soil especially in the rhizosphere and play an important role in plant growth through solubilization of K-bearing minerals. Most of the farmers only use N and P fertilizers. They are ignoring K fertilizer as they are either unaware of the importance of K fertilizer or the price of these fertilizers are too high that these are out of reach of resource for poor farmers (Mohammadi and Sohrabi 2012). The available K is, therefore, decreasing in soils due to more removal of crops than application of fertilizers. In this situation, the role of KSB is gaining importance in modern agriculture for sustainable crop production. These bacteria release K from insoluble minerals (Basak and Biswas 2009; Archana et al. 2012; Parmar and Sindhu 2013; Prajapati et al. 2013; Gundala et al. 2013). These bacteria are also beneficial for plant growth promotion through providing protection from plant pathogens and protecting them from stress conditions. A considerable number of these bacteria are present in the rhizosphere and improved plant growth by a number of mechanisms (Nadeem et al. 2013; Glick 2014).

The use of chemical fertilizers is becoming difficult due to many reasons such as high cost, timely availability, poor resources, high cost of production, etc. Moreover, the injudicious use of chemical fertilizers is posing hazardous effects on the environment; therefore, there is a dire need to evolve or use some cost-effective and eco-friendly biotechnology for sustainable crop production. The use of KSB may be one of such strategies that can improve agricultural productivity on sustained basis. These bacteria can also be used in degraded lands thus bringing them into crop production system (Basak and Biswas 2012;

Rajawat et al. 2012). Moreover, these bacteria are widely used in mining, metallurgy and feed (Li 2003; Sheng 2005; Zhao et al. 2008). Owing to their ability, KSB used to mobilize potassium from insoluble sources could be very helpful for enhancing potassium availability to crops. The presence of these bacteria in soil environment is beneficial to promote plant growth and decrease the dependence on chemical fertilizers.

21.2 Insoluble Potassium Sources

Potassium is one of the most abundant essential plant nutrients in soil and is one of the seven most abundant elements in earth's crust. All crops need K in higher amounts, so it is a key macronutrient for plant growth and yield. Enzyme activation, osmotic regulation, and charge balance are key functions being performed by potassium in plants. It is the most abundant macronutrient in Pakistani soils. The use of potassium fertilizer has gained little attention as soil test shows sufficient exchangeable K, which is satisfactory in these soils. This could be one of the reasons for more vulnerability of Pakistani soils to potassium deficiency along with intensive cropping system and injudicious use of potash fertilizers (Mengel 2007). It has been observed that application of potassium fertilizer on such soils often gives good response due to hidden hunger (Leaungvutiviroj et al. 2010).

Lithosphere contains ≈ 2.5 % potassium, but actual soil K concentration varies greatly ranging from 0.04 % to 3.0 % (Sparks and Huang 1985). Potassium availability to plant depends on K dynamics along with total potassium concentration in soil. Silicate clay minerals are the major source of insoluble potassium in soils and constitute about 90–98 % of total potassium (Mengel and Kirkby 2001). These minerals include feldspar, mica, muscovite, biotite, orthoclase, illite, vermiculite, and smectite. Clay minerals are selective for K ions and release K slowly from the lattice wedge sites (Mengel and Kirkby 2001). Non-exchangeable K constitutes $\approx$1–10 % of total soil potassium. Total pool of soil potassium is extremely complex and this can be solubilized into plant-available K by potassium-solubilizing microorganisms through their mechanisms (Singh et al. 2010; Basak and Biswas 2012).

The soils are not deficient in total potassium because of the occurrence of mica as dominant source of potassium in parent material. This does not represent the plant-available potassium as the rate of release of potassium from mica is much less than the requirement of the crops. The type and content of soil minerals are responsible for the release and fixation of K (Braunschweigh 1980). It has been observed that smectite-dominant soils have faster release of K than illite soils (Wakeel et al. 2013).

In spite of large reserves of potassium in soils, its amount as well as availability to plant is decreasing day by day. It is generally due to its slow release from the clay minerals and low application of potassium to the soils. Clay minerals have variable efficacy to release potassium in soil environment. This slow K availability can be increased by one or the other way. The use of KSB is one of the effective technologies to fulfill the potassium requirement of the crops.

21.3 Role of KSMs in K Solubilization

Soil microorganisms including bacteria, fungi, and actinomycetes are important in the cycling of mineral elements. Among these microbes, bacteria are the important players in this system. They are involved in solubilization of potassium from insoluble sources and interlayer spaces of potassium clay minerals into plant-available forms of K. The potassium released in this way is available for plant uptake (Li et al. 2006; Gundala et al. 2013). Microbial biotechnology is an eco-friendly and viable approach for the release of potassium from mineral and rocks and is important for improving plant growth and yield. The history of using bacteria for the solubilization of potassium is very old. For example, silicate bacteria were isolated by Aleksandrov et al. (1967) from agricultural soils which were able to solubilize potassium from silicate and aluminosilicates. Muentz (1890) reported the involvement of microbes in the solubilization of potassium from rocks. Gundala et al. (2013) also reported the microbes are able to solubilize K from feldspar and aluminosilicates. Now a wide range of species from the genera *Bacillus*, *Paenibacillus*, *Acidithiobacillus*, *Pseudomonas*, *Burkholderia* and many others have been reported to possess ability of K solubilization (Sheng et al. 2002, 2008; Rajawat et al. 2012; Basak and Biswas 2012; Archana et al. 2013; Syed and Patel 2014). The potassium solubilized by soil bacteria and fungi from soil minerals contributes significantly in the plant-available K in soils (Supanjani et al. 2006; Sindhu et al. 2009). These bacteria were also able to decompose organic matter thus played important roles in nutrient cycling. Silicate mineral-solubilizing bacteria (MSB) have also been reported in bulk soil/non-rhizosphere soils (Lian et al. 2002). According to Mikhailouskaya and Tcherhysh (2005), KSB are also present in the roots of cereal crops grown on soils amended with potassium and silicate. A wide range of soil bacteria are reported as K solubilizers. Some examples of K solubilizers include *Bacillus mucilaginosus* (Basak and Biswas 2009; Zarjani et al. 2013), *Burkholderia* (Sheng and He 2006; Uroz et al. 2007), *Enterobacter hormaechei* (Prajapati et al. 2013), *Arthrobacter* spp. (Zarjani et al. 2013), *Paenibacillus mucilaginosus* (Liu et al. 2012), and *Paenibacillus glucanolyticus* (Sangeeth et al. 2012). These bacteria can be used in mining, feed, and metallurgy in addition to their use as biofertilizers (Li et al. 2006; Zhao et al. 2008). Similarly, species of *rhizobia* have the ability to fix atmospheric nitrogen into ammonia, and *Pseudomonas* have also been reported as K solubilizers (Requena et al. 1997).

There are a number of bacterial strains that have the ability to solubilize potassium. These bacteria are present in a variety of soils and environmental conditions. These bacteria belong to a number of genera and reside in the rhizosphere of economically important crops, therefore playing an important role in plant growth and development.

21.3.1 Mechanisms of Action

Scanty information is available on mechanisms used by bacteria to solubilize potassium present in soils as well as K-containing minerals. The direct mechanisms used by these bacteria (Fig. 21.1) include (i) decreasing the pH by production of organic acids and protons, (ii) production of chelated compounds which make complexes with potassium and enhance its availability, and (iii) acidolysis of the surrounding area of microorganism (Uroz et al. 2009; Zarjani et al. 2013; Parmar and Sindhu 2013). The production of organic acids is the major mechanism used by these bacteria to solubilize minerals. These acids result in the acidification of soil environment and promote solubilization of K from minerals.

The production of different types of organic acids is the primary mechanism, which is used by silicate-solubilizing bacteria for the release of potassium making it bioavailable by solubilization and mobilization of minerals. The availability of nutrients due to mineral solubilization ultimately enhances crop growth and productivity. Acidolysis and complexolysis interchange responses are the primary reactions that are used by bacteria in the transformation of insoluble to soluble form, which is mediated by these organic acids (Uroz et al. 2009). The transformation of fixed forms of potassium, i.e., silicate minerals, by organic acids results in a significant increase in the level of available potassium in soil solution that plants can uptake easily (Sheng et al. 2008). The process of potassium release from minerals by K solubilizers is carried out by the synthesis and discharge of organic acid, and proton action in the rhizosphere results in acidifying the bacterial cells and neighboring environment that eventually results in the discharge of potassium into soil solution (Goldstein 1994). The release of acids by bacterial colonies results in decrease of pH; increase in chelation of

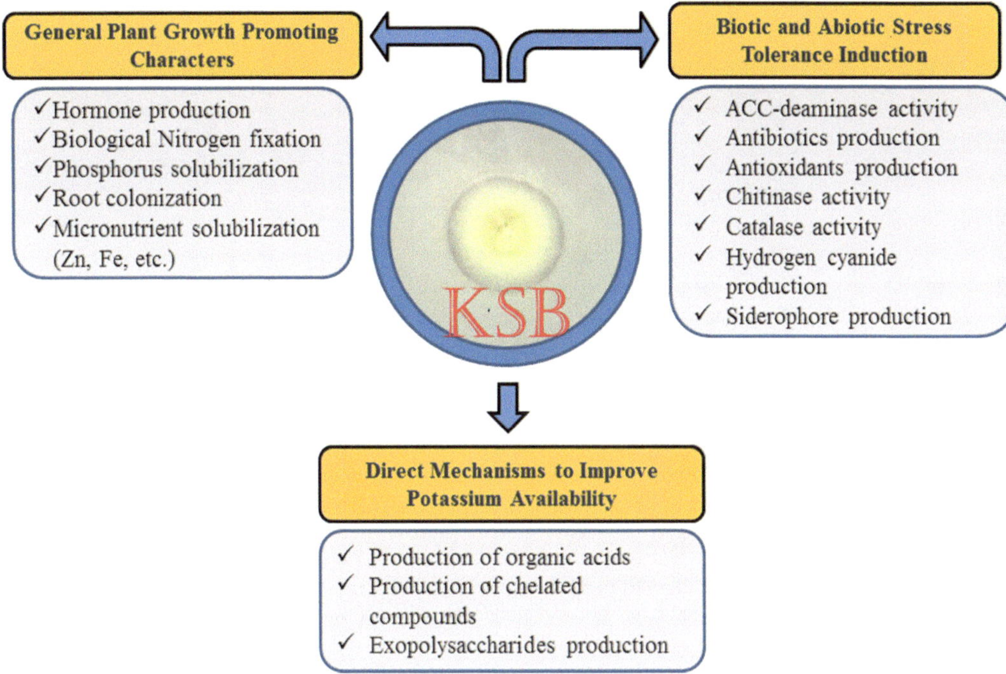

Fig. 21.1 Direct and indirect mechanisms used by potassium-solubilizing bacteria

potassium, aluminum, and iron; and acidolysis of rhizospheric minerals. The decrease in pH of rhizosphere also promotes the liberation protons (Zarjani et al. 2013; Parmar and Sindhu 2013). This acidolysis by KSB via releasing organic acids breaks potassium mineral resulting in slow liberation of readily bioavailable form of potassium in soil solution. Bacterial-mediated acidolysis can also promote the production of chelates of metals linked with potassium minerals (Romheld and Kirkby 2010).

Potassium-solubilizing rhizobacteria (KSR) can also solubilize phlogopite minerals and play an important role in the weathering of these minerals. The process is carried out by Al chelation and acidic solubilization of crystal framework of phlogopite (Abou-el-Seoud and Abdel-Megeed 2012). The solubilization and release of potassium through organic acids by KSB result in plant growth promotion leading to enhanced yield and production. The primary acids produced by the rhizobacteria involved in the dissolution of illite and feldspar are oxalic acid, gluconic acid, tartaric acids, 2-ketogluconic acid, oxalic acid, citric acid, malic acid, and succinic acid (Sheng and He 2006). Among them, tartaric acid is the most abundant in rhizosphere involved in the weathering of potassium minerals (Prajapati et al. 2012; Zarjani et al. 2013). Other types of organic acids produced by K solubilizers as propionic acid, lactic acid, citric acid, tartaric acid, acetic acid, oxalic acid, glycolic acid, succinic acid, malonic acid, fumaric acid, etc., are also important from potassium availability point of view (Wu et al. 2005).

K solubilizers also have other mechanisms that improve plant growth and are indirectly involved to increase potassium availability under normal as well as stressed conditions. The bacteria help exudation of soluble compounds, decomposition of soil organic matter, and mobilization and mineralization of other nutrients (Rajawat et al. 2012; Zeng et al. 2012; Abhilash et al. 2013; Diep and Hieu 2013; Archana et al. 2013). Some of the selected examples about mechanisms used by bacteria to solubilize potassium have been summarized in Table 21.1.

Table 21.1 Mechanisms used by bacteria for potassium solubilization

Bacterial strain	Mechanism used	Source of microorganism	References
Sphingomona, *Burkholderia*	Acidification, complexation	Oak – mycorrhizosphere	Uroz et al. (2007)
B. circulans GY92	Lipo-chitooligosaccharides production	Soybean	Lian et al. (2001)
B. mucilaginosus	Mica through organic acids	Sudan grass	Basak and Biswas (2009)
B. edaphicus	Tartaric acid, oxalic acid	Wheat	Sheng and He (2006)
B. edaphicus	Production of organic acids	Cotton and rapeseed	Sheng (2005)
Pseudomonas spp.	Excretion of organic acids	Wheat rhizosphere	Sheng and Huang (2002)
B. mucilaginosus	Illite solubilization by producing IAA	Eggplant	Han and Lee (2005)
B. mucilaginosus	Organic acids	Pepper and cucumber	Han and Supanjani (2006)
B. glathei	Siderophores, organic ligands	Mycorrhizosphere, bulk soil	Calvaruso et al. (2007)
B. mucilaginosus	Acidification	Maize	Abou-el-Seoud and Abdel-Megeed (2012)
P. glucanolyticus	Organic acids	Black pepper	Sangeeth et al. (2012)
P. mucilaginosus	Tartaric, citric, oxalic acids	Silicate minerals	Liu et al. (2012)
E. hormaechei	Organic acids	Okra	Prajapati et al. (2013)

In addition to K solubilization, bacterial strains also promote plant growth by other mechanisms. For example, ACC deaminase-containing rhizospheric bacteria improved growth of maize and nodulation in mung bean (Shaharoona et al. 2006). In another study, Nadeem et al. (2009) reported that ACC deaminase-containing rhizobacteria improved the uptake of potassium in maize under salinity stress. They observed that these strains were strong colonizers of plant roots. The increased colonization by these bacteria make capable the plants to explore more soil that might have improved the uptake of potassium indirectly. It was observed that these bacteria also have a number of growth-promoting traits such as chitinase activity, nutrient solubilization, root colonization, etc. (Shaharoona et al. 2006; Ahmad et al. 2011). Similarly, the improvement in potassium uptake in mung bean, wheat, and maize has also been reported under pot and field conditions (Nadeem et al. 2010a, b, 2011; Ahmad et al. 2012, 2013a, b, 2014a, b; Aamir et al. 2013). All these studies were conducted under stressed conditions. They observed that bacterial strains improved plant growth due to ACC deaminase activity in addition to hydrogen cyanide (HCN) production, siderophore production, catalase activity, chitinase activity, root colonization ability, etc. Likewise, strains of nitrogen-fixing bacteria, e.g., *Rhizobium phaseoli* and *R. leguminosarum*, have also been reported to increase potassium uptake by crop plants (Iqbal et al. 2012; Ahmad et al. 2012, 2013a).

There are a number of factors that affect the release of potassium from insoluble sources. The amount and type of clay minerals in the parent material greatly affect the availability of potassium. The release of potassium from K minerals like feldspar and mica is also affected by oxygen and pH (Sheng and Huang 2002). The bacterial strains vary greatly in their efficacy to release potassium from minerals. The effectiveness of these strains also varies with the nature of minerals and other climatic factors such as availability of aeration (Uroz et al. 2009). For example, *B. edaphicus* was more effective in solubilization of potassium from illite as compared to feldspar when grown on liquid media (Sheng and He 2006). Precipitated complexes of silicates are being solubilized by silicate-solubilizing bacteria releasing available K, Al, and SiO_2 in solution. In this dissolution of insoluble minerals, H^+ ions play a vital role in the liberation of potassium from insoluble compounds. The dissolution of potassium minerals by rhizobacteria is greatly affected by change in pH, O_2 in soil solution, and strains of bacteria involved (Sheng and Huang 2002). Sheng and He (2006) reported that the inoculation of *B. edaphicus* in soil enhanced the release of potassium up to 84.8–127.9 % as compared to un-inoculated control. Furthermore, they concluded that *B. edaphicus* was more efficient in solubilization of illite as compared to feldspar. At pH 6.5–7.0, maximum release of soluble minerals in soil solution was recorded up to 4.90 mg L^{-1}K (Badr 2006). While in the case of *B. mucilaginosus*, maximum release of potassium was recorded up to 4.29 mg L^{-1}K soil inoculated with muscovite mica (Sugumaran and Janarthanam 2007). The liberation of potassium into solution is also greatly affected by the characteristics of minerals and pH and availability of the O_2 (Chen et al. 2008; Bin et al. 2010).

In the case of structural potassium, a number of studies were conducted under in vitro, greenhouse, and field conditions showing that the release of potassium from these compounds by KSB was a common phenomenon (Sugumaran and Janarthanam 2007; Singh et al. 2010; Prajapati et al. 2012, 2013; Basak and Biswas 2012; Zarjani et al. 2013; Parmar and Sindhu 2013). The release of structural potassium, mediated by rhizobacteria, is carried out by relatively same processes, i.e., organic acid acidolysis and production of chelated molecules (Uroz et al. 2009). Potassium is utilized by rhizobacteria as it performs key processes in bacterial cells. Rhizospheric biomass of bacteria captures a significant amount of released potassium that serves as a potential source of available potassium for plants (Jones et al. 2003; Girgis 2006).

All above-discussed mechanisms support the release of potassium from insoluble sources making it bioavailable; however, the contribution of these mechanisms in weathering of potassium-bearing minerals is variable. These differences in variability are due to bacterial strains, mineral type, and environmental conditions. Man et al. (2014) studied the effect of potassium-solubilizing bacteria on decomposition behavior of potassium-rich shale. They used bacteria-mineral contact/uncontact model to evaluate the efficacy of bacterial strains. They observed that the ability of bacteria to produce acids, proteins, and polysaccharides was high in the bacteria-mineral contact model. According to their view, bacteria-mineral contact method is more profitable for leaching of potassium-rich shale. They observed that biofilm formation promote the release of potassium.

The above discussion shows that bacterial strains have a number of potential mechanisms to solubilize potassium from insoluble sources. Among these mechanisms, the production of organic acids is one of the major mechanisms used by bacteria. It has also been observed from the above discussion that K solubilization by the bacteria is affected by a large number of soil and environment factors.

21.4 Potassium-Solubilizing Bacteria and Soil Type

Soils contain large amount of potassium; however, most of this potassium exists in insoluble form.

KSB solubilize K from insoluble minerals such as illite, mica, and orthoclase (Friedrich et al. 1991) and increase its availability in the soil (Ullman et al. 1996; Bennett et al. 1998). KSR have been found in different soil environments and have variable efficacy regarding their K solubilization potential. A number of K solubilizers have been isolated from various soils (Table 21.2) that show their wide distribution in a variety of soil environments. In spite of their occurrence in a wide variety of soils, the efficiency of KSB varies with variation in chemical composition of minerals (Yakhontova et al. 1987; Liu et al. 2006). Raj (2004) studied the solubilization potential of *Bacillus* spp. on four different types of soils (red soil, clayey soil, forest soil, and sandy soil). He found that efficacy of *Bacillus* spp. regarding K solubilization varied and high quantity of potassium was

Table 21.2 Isolation of potassium-solubilizing bacteria from different soil types

Bacterial sp.	Isolated from	Plant growth promotion impact study on	References
Bacillus sp.	Granite crusher yard	Rice	Raj (2004)
Bacillus sp.	Wheat rhizosphere soil	Wheat	Mikhailouskaya and Tcherhysh (2005)
Bacillus mucilaginosus	Plastic film house area	Eggplant	Han and Lee (2005)
Bacillus mucilaginosus	Illite	ND	Zhou et al. (2006)
Bacillus mucilaginosus	Tianmu mountain	ND	Hu et al. (2006)
B. mucilaginosus	Muscovite mica	Groundnut	Sugumaran and Janarthanam (2007)
Bacillus globisporus	Weathered feldspar	ND	Sheng et al. (2008)
Enterobacter hormaechei	Ceramic industry soil	ND	Prajapati and Modi (2012)
Bacillus spp., *Azotobacter* sp., and *Microbacterium* sp.	Potassium mines	ND	Lang-bo et al. (2012)
Paenibacillus glucanolyticus	Black pepper rhizosphere	Black pepper	Sangeeth et al. (2012)
Pseudomonas putida	Southern Indian tea soils	Tea	Bagyalakshmi et al. (2012a, b)
Bacillus megaterium, *Bacillus coagulans*	Soils/weathered rocks	ND	Diep and Hieu (2013)
Bacillus spp.	Mica mines	ND	Gundala et al. (2013)
Bacillus megaterium, *Arthrobacter* sp.	Iranian soils	ND	Zarjani et al. (2013)
Bacillus spp., *Pseudomonas* spp.	Rhizosphere soil		Archana et al. (2013)
Klebsiella spp., *Enterobacter* spp.	Tobacco rhizosphere soil	Tobacco	Zhang and Kong (2014)
Pseudomonas spp., *Bacillus* sp.	Loamy sand soil	ND	Syed and Patel (2014)

ND Not determined

observed in case of forest soil. He further demonstrated that solubilization was more in unsterile set compared to sterile set. The solubilizing ability of these bacteria in soil may vary due to certain environmental factors. The work of Sheng and Huang (2002) showed that K solubilization was affected by pH, oxygen, and also the bacterial strains used. To evaluate the impact of nutritional and environmental conditions on K solubilization, Parmar and Sindhu (2013) conducted a study by applying two different potassium sources, i.e., KCl and K_2SO_4, to the bacterial strains. They observed that bacterial strain caused maximum K solubilization when KCl was applied as potassium source. More K solubilization was observed in a medium having neutral pH and a decrease in K solubilization occurred with increase in pH (Parmar and Sindhu 2013). Therefore soil pH may also affect K solubilization.

The bacterial strains may have different potential to solubilize K, and K solubilization efficacy of bacterial strains varies with the nature of potassium-bearing minerals and aerobic conditions. Sheng and He (2006) observed that K solubilization potential of *B. edaphicus* was more on illite than feldspar. In another study, Sheng et al. (2008) showed that biotite was the best potassium source for the growth of *B. globisporus* and better growth of this strain was observed on biotite than feldspar and muscovite. These bacteria solubilize K from minerals by excreting organic acids (Sheng and Huang 2002; Girgis 2006). Maurya et al. (2014) evaluated the K-solubilizing bacteria isolated from Inceptisols and Alfisols for their ability to release of K from waste mica. They demonstrated that KSB isolated from Inceptisols showed higher K solubilization potential than isolates from Alfisols. The variable solubilization potential of bacterial strains on different minerals might be due to a variety of acids produced by these strains. As it was determined, the solubilization of feldspar by *B. mucilaginosus* and *B. edaphicus* was associated with the production of citric, tartaric, and oxalic acid (Malinovskaya et al. 1990; Sheng and Huang 2002), while solubilization of silicate minerals was associated with the production of gluconic acid by the bacteria (Sheng et al. 2008). Similarly, Gundala et al. (2013) also reported that gluconic acid was the main organic acid released by *Bacillus* spp. during K solubilization.

Zarjani et al. (2013) tested the ability of six potassium isolates on acid-leached soil, biotite, and muscovite. They observed variable efficacy of these isolates regarding their potential to dissolve K from these soils. Except one strain that showed lower potential for K solubilization in acid-leached soil, all these bacterial strains showed higher potential to solubilize K in biotite than muscovite and acid-leached soil. Sugumaran and Janarthanam (2007) also demonstrated variable efficacy of bacterial strains in acid leached soil. They stated that solubilization potential of bacterial strains varied from 1.90 to 2.26 mg L^{-1}. Basak and Biswas (2009) studied the combined effect of waste mica as insoluble source and *B. mucilaginosus*, a potassium-solubilizing stain on potassium dynamics in two Alfisols. They used Sudan grass (*Sorghum vulgare* Pers.) as test crop. Both the experimental soils were of the same type but collected from two different places, i.e., Hazaribagh and Bhubaneswar. They observed high K recoveries in all forms of K such as exchangeable, non-exchangeable, and water soluble and its uptake. They also observed better growth of Sudan grass in case of Alfisols collected from Hazaribagh. According to them, exchangeable and water-soluble potassium as well as solubilization of K from non-exchangeable to exchangeable pool improved the K status of the soil. The increase in K content and better physicochemical conditions of this soil might be due to higher biomass production. They concluded that the variation in K availability was due to difference in clay type and reported that 2:1 type of clay mineral was more efficient in supplying available K in soils. Similar results have also been reported in former studies of Barre et al. (2008).

In addition to the type of soil, the particle size of the same soil may also affect K solubilization. The work of Meena et al. (2014) supports this observation. They studied the release of K from waste mica (muscovite and biotite) by four KSB. The results showed that high K solubilization was observed with biotite compared to muscovite. They also observed that with decrease in particle size of clay, K solubilization increased. More solubilization occurred with particle size of 1 mm compared to that of 2 mm. Similarly, soil conditions also affect the ability of strains to solubilize K. Under variable and adverse soil conditions, only those strains can provide satisfactory results regarding K solubilization which have the ability to maintain its growth under such conditions. The work of Prajapati and Modi (2012) supports this argument. They isolated KSB from soil samples of ceramic industries by using feldspar (insoluble potassium). They observed that among these isolated strains, one strain had high capacity to grow in saline condition and pH range. This ability of the bacterial strain might be due to its wide range of enzymatic activity and biochemical characteristics that make such strain a good inoculant for its use on commercial basis. Similarly, potassium-solubilizing *Bacillus* spp. isolated from mica cores was capable of growing under aerobic conditions up to pH 12 with optimum growth at pH 10 (Gundala et al. 2013). KSB strain *Frateuria aurantia* can be used as a bio-inoculant that have the ability to solubilize K from almost all types of soils having wide range of pH and temperature (Pindi and Satyanarayana 2012). The former work of Patel

(2011) also highlights the effectiveness of *Frateuria aurantia* for solubilizing and enhancing the availability of potassium for crops.

The above discussion shows that KSB occurred in a variety of soils. These solubilizing bacteria have different potentials to mobilize potassium in the soil. This variable efficacy of bacteria is due to the nature of minerals present in the soils. Certain bacterial strains can mobilize potassium from one type of minerals but cannot perform with similar efficacy with other minerals. In addition to soil type, certain soil factors affect the K solubilization; among them pH is the important one. Similarly, the size of soil particles and unfavorable soil conditions like salinity also affect K solubilization. Therefore, during application of KSB, these soil-related aspects should be kept in mind to have maximum benefit from this beneficial population.

21.5 Potassium-Solubilizing Bacteria as Biofertilizer

For optimum plant growth and sustainable agriculture, maintaining soil fertility is very important. Intensive cropping system causes depletion of soil nutrients. Soil fertility can be restored through integrated soil fertility management approach consisting of natural resource conservation as well as increasing the efficiency of applied inputs (Vlek and Vielhauer 1994). For sustainable agriculture, the use of microorganism and organic material is helpful for reducing the use of inorganic chemicals (Ghorbani et al. 2008). Application of biofertilizers is a very important component of integrated nutrient management system that enhances productivity and sustainability of soil. Biofertilizers having living cells of different types of microorganism are cost-effective and environment-friendly products (Mohammadi and Sohrabi 2012). Such biofertilizers contain a number of microorganisms that increase the availability of nutrients to the plant through their effect on soil processes. Most of the bacteria used in biofertilizers have close relationship with the plant. The application of such microbial inoculants not only increases the nutrient availability to plant roots (Supanjani et al. 2006; Shanware et al. 2014) but also protects the plant from a number of biotic and abiotic stresses through their specific mechanisms (Glick et al. 2007; Zahir et al. 2009; Nadeem et al. 2010b). It has been observed that the biofertilizers can reduce the application rate of chemical fertilizers, because these, in addition to enhancing the solubility of soil nutrients, also increase the efficiency of applied nutrients (Habibi et al. 2011). Some of the common bacteria used in biofertilizers are N-fixing bacteria and P- and K-solubilizing bacteria. Such biofertilizers can be prepared from single bacteria or may be from combination of these. These bacteria increase the solubility of potassium by certain mechanisms as discussed in the previous sections. Some of the important KSB include *B. mucilaginous*, *E. hormaechei*, *B. edaphicus*, *B. circulans*, and *Paenibacillus* spp. (Lian et al. 2002; Sheng and He 2006; Liu et al. 2012; Mohammadi and Sohrabi 2012; Prajapati et al. 2013).

After nitrogen, phosphorus and potassium are two major nutrients required for proper plant growth and development. Although the rock K is a cheaper source of potassium, however, slow release of potassium causes insignificant yield increase (Zapata and Roy 2004). The use of KSB like *B. mucilaginosus* can enhance the potassium content of soil and make it available to plants. Han et al. (Han and Supanjani 2006) studied the impact of biofertilizer containing PSB (*B. megaterium*) and KSB (*B. mucilaginosus*) on the growth of pepper and cucumber. They observed that the sole application of rock P and K did not enhance the P and K content; however, application of biofertilizer containing P and K solubilizers enchained the availability to these nutrients and also improved the growth of cucumber and pepper. Similarly, Supanjani et al. (2006) also evaluated the effectiveness of biofertilizer containing P and K solubilizers on the growth and yield of hot pepper (*Capsicum annuum* L.). The plant was treated with rock phosphate (RP) and potassium rock (KR) for P and K, respectively. They concluded that application of PSB and KSB enhanced the P and K availability to the plant. The plant biomass and yield of the plant were significantly higher than control. They further concluded that application of RP and KR with PSB and KSB was equally effective when compared with chemical fertilizer, so such inoculants could be a suitable supplement of chemical fertilizers. Abou-el-Seoud and Abdel-Megeed (2012) also reported the effectiveness of biofertilizer containing PSB (*B. megaterium* var. phosphaticum) and KSB (*B. mucilaginosus* and *B. subtilis*). Application of biofertilizer improved the maize growth by providing P and K under limited P and K soil conditions.

The KSB can also be used with other microbes like fungus for promoting plant growth. The co-inoculation of KSB and KSF could be beneficial for plant growth owing to number of specific mechanism used by these microbes. It has been observed that the application of KSB with fungus *A. terreus* enhanced root and shoot of okra by increasing soil potassium through K mobilization from insoluble K source, i.e., feldspar (Prajapati et al. 2013). They suggested that improved plant growth was due to K solubilization and release of organic acids by KSMs.

The application of KSMs with other organic materials also proved useful. A significant improvement in growth parameters of *Capsicum*, Hungarian yellow, was observed when a K-solubilizing strain *Frateuria aurantia*-enriched phospho compost was applied (Pindi and Satyanarayana 2012). This growth enhancement was due to enhancing the K availability and release of hormone by the strains. Before this, Patel (2011) also demonstrated that microbial base product containing *F. aurantia* produced plant growth-promoting substance which caused positive effect on plant growth by mobilizing potash and increasing its availability to plants.

Such bio-inoculants are not only useful for enhancing K availability but are also effective for minimizing the impact of pathogens on plant growth. Keeping in view of this dual role, El-Haddad et al. (2011) evaluated the effect of four strains of N-fixing bacteria (*Paenibacillus polymyxa*), three strains of phosphorus-solubilizing bacteria (*B. megaterium*), and three strains of potassium-solubilizing bacteria (*B. circulans*) on tomato plant infested with root-knot nematode. The results showed that these biofertilizers significantly reduced the nematodes population compared to control. They demonstrated that such biofertilizer could be used effectively for improving nutrition and biocontrol agents. Similarly, a biofertilizer containing N-fixing bacteria and P and K solubilizers as well as arbuscular mycorrhizae (AM) caused significant increase in maize growth (Wu et al. 2005). The biofertilizer not only improved plant nutrition but also caused positive effect on soil properties. Similarly, Dasan (2012) also evaluated the effectiveness of a biofertilizer containing PSM (*B. megaterium*) and KSM (*F. aurantia*) against agrochemicals. They observed that these strains were able to grow in the presence of agrochemicals. Therefore these strains can be used as biocontrol agents against plant pathogens.

Although a number of workers have demonstrated the effectiveness of biofertilizers for enhancing plant growth and development, however, the shelf life of such inoculants is still a question. For sustainable effectiveness of biofertilizer, viable microorganism and formulation medium is an important aspect. In certain cases satisfactory results are not obtained from biofertilizers. This might be due to certain factors like low shelf life, effect of storage temperature, and improper packing (Verma et al. 2011). The carrier material that is used for biofertilizer preparation acts as means of transportation for microorganisms. Such materials play an important role in shelf life of biofertilizer prior to its release into the field (Brar et al. 2012). According to Bhattacharyya and Kumar (2000), carrier-based formulation has longer shelf life and can be used without compromising its quality for 6 months. The study conducted by Prajapati and Modi (2014) showed that shelf life of KSB used as biofertilizer had high amount of viable microbial population after 4 months of storage. They tested two biofertilizer (liquid and solid) at two storage temperatures of 4 °C and 30 °C. They observed that liquid broth formulation increased the shelf life of K solubilizer-based biofertilizer. The use of KSB as biofertilizer for improving agriculture productivity can reduce the use of chemical fertilizers thus improving environment quality (Liu et al. 2012; Bagyalakshmi et al. 2012b). So, it could be used as an eco-friendly approach for crop production (Sindhu et al. 2010; Prajapati et al. 2012, 2013; Archana et al. 2013).

From the above review, it has been observed that KSB strains have the potential to be used as biofertilizer. These biofertilizers increase plant growth and development by enhancing K availability to plant and also by number of other direct and indirect mechanisms. There is an opportunity to isolate, screen, and culture these bacteria for the purpose to develop biofertilizer for agricultural improvement in supplementation with chemical fertilizers. This biotechnology is not only beneficial in improving crop yield and quality but also helpful in improving environment quality by decreasing the use of agrochemicals. It has been observed that application of such fertilizers is limited because of problems related to its handling and shelf life. Therefore, for efficient use of these biofertilizers, special emphasis should be given to its shelf life.

21.6 Impact on Plant Growth, Yield, and Nutrient Uptake

KSB have a prominent role for improving crop growth, yield, and quality (Basak and Biswas 2009; Singh et al. 2010). It is well documented that these organisms contribute significantly in improving germination, growth, yield, and nutrient uptake of crop plants grown in gnotobiotic and field conditions (Awasthi et al. 2011; Basak and Biswas 2012; Zhang et al. 2013). Some of the selected studies showing the effectiveness of K solubilizers to improve plant growth, yield, and nutrient uptake have been summarized in Table 21.3. The history of use of silicate-solubilizing bacteria for improving crop yield is very old. Aleksandrov (1985) used silicate-solubilizing bacteria in combination with organo-minerals to improve growth and yield of wheat and maize. Xie (1998) reported that the use of KSMs in maize, wheat, and Sudan grass reduced the use of chemical fertilizers. The improvement in potassium uptake by crop plants due to inoculation with KSMs might be due to stimulation of root growth (Zeng et al. 2012; Sindhu et al. 2012), which contribute to exploration of more soil by plant roots for potassium uptake. A number of studies are available where scientists have isolated KSMs from various crops such as potato (Biswas 2011; Abdel-Salam and Shams 2012), wheat (Zhang et al. 2013), common bean (Kumar et al. 2012), cotton and tomato (Archana et al. 2012), sorghum and chili (Archana et al. 2013), black pepper (Sangeeth et al. 2012), and rice (Muralikannan and Anthomiraj 1998). It has been observed that it can effectively use different K-bearing minerals as K source. These sources include feldspar (Sheng et al. 2008), ceramic industry soil (Prajapati and Modi 2012, mica (Gundala et al. 2013), etc. Such potassium-solubilizing strains show their effectiveness for promoting plant growth and development. Sheng and He (2006) conducted a study under pot conditions and reported that wheat plants inoculated with *B. edaphicus* showed higher nutrient uptake and biomass accumulation as compared to un-inoculated control. In other studies, the K solubilizers have been reported to improve growth of rapeseed and cotton (Sheng 2005) and mung bean (Archana et al. 2008). It has been observed that KSB in addition to solubilizing K also have the potential to improve soil nutrient status by improving N fixation and/or P solubilization (Verma et al. 2010; Basak and Biswas 2012).

The effectiveness of KSB is evident from a number of studies (Table 21.3); however, the strains may have variable efficacy regarding their impact on plant growth. The study of Singh et al. (2010) shows this variable effect when they studied the effectiveness of three KSB strains, i.e., *B. mucilaginosus*, *A. chroococcum*, and *Rhizobium* spp., to improve maize and wheat growth under hydroponics conditions. They used mica as insoluble source and observed a significant improvement in potassium assimilation rate. They reported that

Table 21.3 Effectiveness of potassium-solubilizing bacteria for enhancing plant growth

Crop	Rhizobacterial strain	Experimental conditions	Plant response	References
Sudan grass (*Sorghum vulgar*)	Potassium solubilizer (*Bacillus mucilaginosus*) and nitrogen fixer (*Azotobacter chroococcum*)	Greenhouse pot trial	Co-inoculation enhanced K and N availability and caused an increase in biomass production and nutrient acquisition. *B. mucilaginosus* strain was more effective and potent K solubilizer than *A. chroococcum*	Basak and Biswas (2012)
Maize (*Zea mays*)	Nitrogen fixer (*Azotobacter tropicalis*), phosphate solubilizer (*Burkholderia unamae*), potassium solubilizer (*Bacillus subtilis*)	Greenhouse pot trial	Biofertilizer caused four times increase in corn growth. Growth enhancement was seven times by addition of rock phosphate and K feldspar with biofertilizer	Leaungvutiviroj et al. (2010)
	Potassium solubilizers (*Bacillus mucilaginosus* and *B. subtilis*) and phosphate solubilizer (*Bacillus megaterium*)	Greenhouse pot trial	Co-inoculation increased the availability and uptake of P and K in maize. The shoot and root growth enhanced significantly grown on P- and K-limited soils	Abou-el-Seoud and Abdel-Megeed (2012)
Pepper (*Capsicum annum* L.) and cucumber (*Cucumis sativus* L.)	Phosphate solubilizer (*Bacillus megaterium*) and potassium solubilizer (*Bacillus mucilaginosus*)	Greenhouse pot trial	Co-inoculation resulted in consistently higher P and K availability. The uptake of N, P, and K by shoot and root increased and also improved pepper and cucumber growth	Han and Supanjani (2006)
Tomato (*Solanum lycopersicum*)	*Bacillus cereus*	Field experiment	Inoculation enhanced K uptake and potassium use efficiency. Application of organic K sources considerably improved the soil fertility status	Badr (2006)
	Pseudomonas spp.	Field experiment	Application of P and K solubilizer bacteria significantly improved tomato yield	Lynn et al. (2013)
Peanut (*Arachis hypogaea*) and sesame (*Sesamum indicum*)	*Bacillus pasteurii*	Field experiment	Application of calcium nitrate combined with potassium sulfate, ammonium nitrate, and feldspar; in the presence of K-solubilizing bacteria, it caused significant increase in K availability to peanut and sesame. Highest values of yield components were obtained by inoculation	Youssef et al. (2010)
Potato (*Solanum tuberosum*)	Nitrogen fixer bacteria (*Azotobacter* and *Azospirillum*), phosphate solubilizer (*Bacillus megaterium*), and silicate-dissolving bacteria (*Bacillus circulans*)	Field experiment	Biofertilizer increased tuber size, starch content, leaf area, and total chlorophyll. Improved uptake of macronutrients in tubers and shoots	Abdel-Salam and Shams (2012)

(continued)

Table 21.3 (continued)

Crop	Rhizobacterial strain	Experimental conditions	Plant response	References
Cotton (*Gossypium hirsutum*) and rape (*Brassica napus* L)	*Bacillus edaphicus*	Greenhouse pot trial	Bacteria increased the potassium content by dissolving insoluble potassium. In cotton and rape, 30 and 26 % increase in potassium was recorded, respectively. Enhanced the root and shoot growth of cotton and rape	Sheng (2005)
Maize (*Zea mays*) and wheat (*Triticum aestivum*)	*Bacillus mucilaginosus*, *Azotobacter chroococcum*, and *Rhizobium* spp.	Hydroponic study	Inoculation significantly improved potassium assimilation of maize and wheat. Improved biomass yield, potassium, chlorophyll, and crude protein content of plants. *Bacillus mucilaginosus* significantly caused higher solubilization of potassium than *Azotobacter* and *Rhizobium*	Singh et al. (2010)
Wheat (*Triticum aestivum*)	*Bacillus edaphicus*	Greenhouse pot trial	Inoculation significantly improved the root and shoot growth of wheat. Higher potassium content was observed in plant compared to un-inoculated control	Sheng and He (2006)
	Bacillus spp.	Field experiment	Bacterial inoculation conjugated with organic amendment significantly improved N, P, and K content and wheat yield compared with un-inoculated control	Girgis (2006)
Sorghum (*Sorghum bicolor*)	*Bacillus* cereus	Greenhouse pot trial	Inoculation combined with K- and P-bearing minerals gave 48, 65, and 58 % increase in dry matter yield of sorghum plants grown on clay, sandy, and calcareous soil, respectively. The uptake of K by sorghum plants also increased by 71, 110, and 116 % planted on these soils	Badr et al. (2006)
Okra (*Abelmoschus esculentus*)	*Enterobacter hormaechei*	Greenhouse pot trial	Increased root and shoot growth of okra. In the presence of bacteria, potassium mobilization was efficient in plant when feldspar was added as potassium source in the medium	Prajapati et al. (2013)
Tea (*Camellia sinensis*)	*Pseudomonas putida*	Field experiment	All the tea quality parameters like theaflavin, thearubigin, highly polymerized substances, total liquor color,	Bagyalakshmi et al. (2012b)

(continued)

Table 21.3 (continued)

Crop	Rhizobacterial strain	Experimental conditions	Plant response	References
			caffeine, briskness, color, and flavor indexes were greatly improved in KSB treated plants	
Tobacco (*Nicotiana tabacum*)	*Enterobacter cloacae* and *Klebsiella variicola*	Greenhouse pot trial	Uptake of K and N by tobacco seedlings increased significantly. Improved plant dry weight. *Klebsiella variicola* showed the most pronounced beneficial effect on plant growth and nutrient uptake	Zhang and Kong (2014)
Black pepper (*Piper nigrum* L.)	*Paenibacillus glucanolyticus*	Greenhouse pot trail	Increased tissue dry mass (ranging from 37.0 % to 68.3 %). K uptake in inoculated plants increased by 125–184 % compared to un-inoculated control	Sangeeth et al. (2012)
Eggplant (*Solanum melongena*)	Phosphate solubilizer (*Bacillus megaterium*) and potassium solubilizer (*Bacillus mucilaginosus*)	Greenhouse pot trail	Inoculation increased the availability of P and K in soil by dissolving insoluble P and K. Enhanced N, P, and K uptake and promoted growth of eggplant	Han and Lee (2005)
Hot pepper (*Capsicum annuum* L.)	Phosphate solubilizer (*Bacillus megaterium*) and potassium solubilizer (*Bacillus mucilaginosus*)	Field experiment	Inoculation increased photosynthesis by 16 % and leaf area by 35 % compared to un-inoculated control Similarly, biomass harvest and fruit yield of the treated plants were 23 % and 30 %, respectively, higher as compared with control	Supanjani et al. (2006)
Pine tree (*Pinus sylvestris*)	*Burkholderia glathei*	Greenhouse pot trial	Significantly increased biotite weathering and improved plant growth. Growth-promoting effect of *B. glathei* was mainly resulted from the improved plant nutrition via increased mineral weathering	Calvaruso et al. (2006)
Valencia orange trees (*Citrus sinensis*)	*Bacillus Circulans*	Field study	Feldspar inoculated with *Bacillus circulans* significantly improved leaf K content, fruit weight, yield, and fruit quality	Shaaban et al. (2012)
Khella (*Ammi visnaga*)	Phosphate solubilizer (*Bacillus megaterium*) Potassium solubilizer (*Bacillus circulans*)	Field experiment	Strains had great ability to solubilize P and K from insoluble sources. Inoculation significantly enhanced the growth parameters of the plant	Hassan et al. (2010)

improvement in K uptake resulted in enhanced physiology that ultimately leads to more biomass accumulation. They reported that bacterial strains varied in their effectiveness to mobilize potassium and *B. mucilaginosus* was significantly better than *A. chroococcum* and *Rhizobium* spp. Prior to this, increase in dry matter, P and K uptake, and oil contents of groundnut was observed by inoculation with *B. mucilaginosus* compared to un-inoculated plants (Sugumaran and Janarthanam 2007). The work of Archana et al. (2012) also shows the efficiency of KSB *Bacillus* spp. to improve growth and yield of maize. They reported that *Bacillus* spp. were strong candidates for K solubilization in maize, and the use of these bacteria in combination with chemical fertilizers was significantly better in improving plant growth, yield, and nutrient uptake than the treatment where chemical fertilizers were used.

Although the efficacy of KSB in improving mineral nutrition and plant growth is reported by various workers, however there was an opinion that this efficiency can be further improved if these strains are provided with some supplements. Keeping in view of this, the K-solubilizing ability of bacterial strain was studied by Bagyalakshmi et al. (2012a) in the presence of some supplement. They isolated bacterial strains from tea plant to study the solubilization of murate of potash in combination with ammonium nitrate and glucose as supplements. They observed that K-solubilizing ability of these bacteria was improved when supplemented with ammonium nitrate and glucose. However this aspect need further study because low efficiency in the absence of supplement might be due to some other factors like the nature of potassium source or type of bacterial strain. The recent study of Subhashini (2015) also supports this argument where bacterial strain enhanced the solubilization of sulfate of potash in the absence of supplement. They studied the impact of *Frateuria aurantia*, the KSB strain on growth promotion, and nutrient uptake of tobacco (*Nicotiana tabacum* L.) in a field experiment for two crop seasons. They observed that inoculation increased the biomass, nutrient content, and leaf quality of the tobacco. Bacterial strain enhanced the potassium content of tobacco plant and caused 39 % increase in potassium uptake compared to un-inoculated control. They reported that inoculation enhanced the potassium uptake significantly when sulfate of potash was added in the soil.

The above-discussed review shows the effectiveness of KSB for promoting plant growth and development. This growth improvement was observed by a number of bacterial strains on a variety of crops. It was also observed that this bacterial population is equally effective in controlled as well as natural field conditions. However variable responses are also observed in certain cases. These responses might be due to the type of bacterial strain, nature of source, and environment conditions.

21.7 Conclusion and Future Perspectives

The K-bearing minerals have prominent position in the lithosphere which could contribute to potassium fertilization for crop plants. Due to low application of potassium fertilizer and the presence of insoluble potassium sources, the potassium availability to plants is decreasing. The potassium availability can be enhanced by solubilizing these insoluble minerals. The cycling of potassium between different components is largely controlled by soil microbial community, and KSB are an integral component of this community. They play an important role in making K available to plants. These bacteria have enormous potential to make use of insoluble potassium minerals for fertilization of crop plants in resource-poor developing countries. There are many factors that contribute to the efficiency of these microbes for using them as biofertilizers. These factors may include mechanism used for K solubilization, the strains used, soil nutritional status, type of minerals and soil, environmental factors, etc. There is an opportunity to study all these factors in the future to develop competent biofertilizers for K fertilization of crop plants. There is also a need to study the competency of bacterial strains to use them as consortium in combination with other strains having the ability to improve availability of other essential nutrients such as N, P, Fe, Zn, etc. More important is the induction of these traits in plants to develop genetically modified organisms (GMOs). The demonstration as well as transfer of technology to farmers for commercialization purpose is another aspect that needs to be focused on. The evaluation of these strains on different K sources and under variable environmental conditions is also an important aspect so that promising strains can be selected for biofertilizer preparation that can be used on a wide range of soil and climatic conditions.

References

Aamir M, Aslam A, Khan MY, Jamshaid MU, Ahmad M, Asghar HN, Zahir ZA (2013) Co-inoculation with *Rhizobium* and plant growth promoting rhizobacteria (PGPR) for inducing salinity tolerance in mung bean under field condition of semi-arid climate. Asian J Agric Biol 1:7–12

Abdel-Salam MA, Shams AS (2012) Feldspar-K fertilization of potato (*Solanum tuberosum* L.) augmented by biofertilizer. J Agric Environ Sci 12:694–699

Abhilash PC, Dubey RK, Tripathi V, Srivastava P, Verma JP, Singh HB (2013) Remediation and management of POPs-contaminated soils in a warming climate: challenges and perspectives. Environ Sci Pollut Res 20:5879–5885

Abou-el-Seoud II, Abdel-Megeed A (2012) Impact of rock materials and biofertilizations on P and K availability for maize (*Zea mays*) under calcareous soil conditions. Saudi J Biol Sci 19:55–63

Ahmad M, Zahir ZA, Asghar HN, Asghar M (2011) Inducing salt tolerance in mung bean through coinoculation with *Rhizobium* and PGPR containing ACC deaminase. Can J Microbiol 57:578–589

Ahmad M, Zahir ZA, Asghar HN, Arshad M (2012) The combined application of rhizobial strains and plant growth promoting rhizobacteria improves growth and productivity of mung bean (*Vigna radiata* L.) under salt–stressed conditions. Ann Microbiol 62:1321–1330

Ahmad M, Zahir ZA, Khalid M, Nazli F, Arshad M (2013a) Efficacy of *Rhizobium* and *Pseudomonas* strains to improve physiology, ionic balance and quality of mung bean under salt–affected conditions on farmer's fields. Plant Physiol Biochem 63:170–176

Ahmad M, Zahir ZA, Nadeem SM, Nazli F, Jamil M, Khalid M (2013b) Field evaluation of *Rhizobium* and *Pseudomonas* strains to improve growth, nodulation and yield of mung bean under salt-affected conditions. Soil Environ 32:158–165

Ahmad M, Zahir ZA, Jamil M, Nazli F, Latif M, Akhtar MF (2014a) Integrated use of plant growth promoting rhizobacteria, biogas slurry and chemical nitrogen for sustainable production of maize under salt-affected conditions. Pak J Bot 46:375–382

Ahmad M, Zahir ZA, Nadeem SM, Nazli F, Jamil M, Jamshaid MU (2014b) Physiological response of mung bean to *Rhizobium* and *Pseudomonas* based biofertilizers under salinity stress. Pak J Agric Sci 51:1–8

Aleksandrov VG (1985) Organo-mineral fertilizers and silicate bacteria. Dokl Akad Nauk 7:43–48

Aleksandrov VG, Blagodyr RN, Ilev IP (1967) Liberation of phosphoric acid from apatite by silicate bacteria. Microchem J 29:111–114

Archana DS, Savalgi VP, Alagawadi AR (2008) Effect of potassium solubilizing bacteria on growth and yield of maize. Soil Biol Ecol 28:9–18

Archana DS, Nandish MS, Savalagi VP, Alagawadi AR (2012) Screening of potassium solubilizing bacteria (KSB) for plant growth promotional activity. Bioinfolet 9(4):627–630

Archana DS, Nandish MS, Savalagi VP, Alagawadi AR (2013) Characterization of potassium solubilizing bacteria (KSB) from rhizosphere soil. Bioinfolet 10:248–257

Armengaud P, Breitling R, Amtmann A (2010) Coronatine-intensive 1 (COII) mediates transcriptional responses of *Arabidopsis thaliana* to external potassium supply. Mol Plant 3:390–405

Awasthi R, Tewari R, Nayyar H (2011) Synergy between plants and P-solubilizing microbes in soils: effects on growth and physiology of crops. Int Res J Microbiol 2:484–503

Badr MA (2006) Efficiency of K-feldspar combined with organic materials and silicate dis-solving bacteria on tomato yield. J Appl Sci Res 2:1191–1198

Badr MA, Shafei AM, Sharaf El-Deen SH (2006) The dissolution of K and phosphorus bearing minerals by silicate dissolving bacteria and their effect on sorghum growth. Res J Agric Biol Sci 2:5–11

Bagyalakshmi B, Ponmurugan P, Balamurugan A (2012a) Impact of different temperature, carbon and nitrogen sources on solubilization efficiency of native potassium solubilizing bacteria from tea (*Camellia sinensis*). J Biol Res 3(2):36–42

Bagyalakshmi B, Ponmurugan P, Marimuthu S (2012b) Influence of potassium solubilizing bacteria on crop productivity and quality of tea (*Camellia sinensis*). Afr Agric Res 7:4250–4259

Barre P, Montagnier C, Chenu C, Abbadie L, Velde B (2008) Clay minerals as a soil potassium reservoir: observation and quantification through X-ray diffraction. Plant Soil 302:213–220

Basak BB, Biswas DR (2009) Influence of potassium solubilizing microorganism (*Bacillus mucilaginosus*) and waste mica on potassium uptake dynamics by

Sudan grass (*Sorghum vulgare* Pers.) grown under two Alfisols. Plant Soil 317:235–255

Basak BB, Biswas DR (2012) Modification of waste mica for alternative source of potassium: Evaluation of potassium release in soil from waste mica treated with potassium solubilizing bacteria (KSB). LAP LAMBERT Academic Publishing, India

Bennett PC, Choi WJ, Rogera JR (1998) Microbial destruction of feldspars. Mineral Manag 8:149–150

Bhattacharyya P, Kumar R (2000) Liquid biofertilizer-current knowledge and future prospect National seminar on development and use of biofertilizers, biopesticides and organic manures. Bidhan Krishi Viswavidyalaya, Kalyani, pp 10–12

Bin L, Bin W, Mu P, Liu C, Teng HH (2010) Microbial release of potassium from K-bearing minerals by thermophilic fungus *Aspergillus fumigatus*. Geochim Cosmochim Acta 72:87–98

Biswas DR (2011) Nutrient recycling potential of rock phosphate and waste mica enriched compost on crop productivity and changes in soil fertility under potato-soybean cropping sequence in an Inceptisol of Indo-Gangetic Plains of India. Nutr Cycl Agroecosyst 89:15–30

Brar SK, Sarma SJ, Chaabouni E (2012) Shelf-life of Biofertilizers: an accord between formulations and genetics. J Biofertil Biopesticide 3:1–2

Braunschweigh IC (1980) K^+ availability in relation to clay content. Results of field experiment. Potash Rev 2:1–8

Calvaruso C, Turpault MP, Frey-Klett P (2006) Root-associated bacteria contribute to mineral weathering and to mineral nutrition in trees: a budgeting analysis. Appl Environ Microbiol 72:1258–1266

Calvaruso C, Turpault MP, Leclerc E, Frey-Klett P (2007) Impact of ectomycorrhizosphere on the functional diversity of soil bacterial and fungal communities from a forest stand in relation to nutrient mobilization processes. Microb Ecol 54:567–577

Chen S, Lian B, Liu CQ (2008) *Bacillus mucilaginosus* on weathering of phosphorite and primary analysis of bacterial proteins during weathering. Chin J Geochem 27:209–216

Dasan AS (2012) Compatibility of agrochemicals on the growth of phosphorous mobilizing bacteria *Bacillus megaterium* var. phosphaticum potassium mobilizing bacteria *Frateuria aurantia*. Appl Res Dev Inst J 6:118–134

Diep CN, Hieu TN (2013) Phosphate and potassium solubilizing bacteria from weathered materials of denatured rock mountain, Ha Tien, Kiên Giang province Vietnam. Am J Life Sci 1:88–92

El-Haddad ME, Mustafa I, Selim Sh M, El-Tayeb TS, Mahgoob AE, Abdel-Aziz NH (2011) The nematicidal effect of some bacterial biofertilizers on *Meloidogyne incognita* in sandy soil. Braz J Microbiol 42:105–113

Friedrich S, Platonova NP, Karavaiko GI, Stichel E, Glombitza F (1991) Chemical and microbiological solubilization of silicates. Acta Biotechnol 11:187–196

Ghorbani R, Koocheki A, Jahan M, Asadi GA (2008) Impact of organic amendments and compost extracts on tomato production and storability in agroecological system. Agron J 28:307–311

Girgis MGZ (2006) Response of wheat to inoculation with phosphate and potassium mobilizers and organic amendment. Ann Agric Sci Ain Shams Univ Cairo 51:85–100

Glick BR (2014) Bacteria with ACC deaminase can promote plant growth and help tofeed the world. Microbiol Res 169:30–39

Glick BR, Cheng Z, Czarny J, Cheng Z, Duan J (2007) Promotion of plant growth by ACC deaminase-producing soil bacteria. Eur J Plant Pathol 119:329–339

Goldstein AH (1994) Involvement of the quinoprotein glucose dehydrogenase in the solubilization of exogenous mineral phosphates by gram-negative bacteria. In: Torriani-Gorni A, Yagil E, Silver S (eds) Phosphate in microorganisms: cellular and molecular biology. ASM Press, Washington, DC, pp 197–203

Gundala PB, Chinthala P, Sreenivasulu B (2013) A new facultative alkaliphilic, potassium solubilizing *Bacillus sp.* SVUNM9 isolated from mica cores of Nellore District, Andhra Pradesh, India: research and reviews. J Microbiol Biotechnol 2:1–7

Habibi A, Heidari G, Sohrabi Y, Badakhshan H, Mohammadi K (2011) Influence of bio, organic and chemical fertilizers on medicinal pumpkin traits. J Med Plants Res 5:5590–5597

Han HS, Lee KD (2005) Phosphate and potassium solubilizing bacteria effect on mineral up take, soil availability and growth of eggplant. J Agric Biol Sci 1:176–180

Han HS, Supanjani LKD (2006) Effect of co-inoculation with phosphate and potassium solubilizing bacteria on mineral uptake and growth of pepper and cucumber. Plant Soil Environ 52:130–136

Hassan EA, Hassan EA, Hamad EH (2010) Microbial solubilization of phosphate-potassium rocks and their effect on khella (*Ammi visnaga*) growth. Ann Agric Sci 55:37–53

Hu X, Chen J, Guo J (2006) Two phosphate- and potassium-solubilizing bacteria isolated from Tianmu Mountain, Zhejiang, China. World J Microbiol Biotech 22:983–990

Iqbal MA, Khalid M, Shahzad SM, Ahmad M, Soleman N, Akhtar N (2012) Integrated use of *Rhizobium leguminosarum*, plant growth promoting rhizobacteria and enriched compost for improving growth, nodulation and yield of lentil (*Lens culinaris* Medik.). Chil J Agri Res 72:104–110

Jones DL, Dennis PG, Owen AG, Van-Hees PAW (2003) Organic acid behavior in soils misconceptions and knowledge gaps. Plant Soil 248:31–41

Kumar P, Dubey RC, Maheshwari DK (2012) *Bacillus* strains isolated from rhizosphere showed plant growth promoting and antagonistic activity against phytopathogens. Microbiol Res 67:493–499

Lang-bo YI, Qing-zhong P, Qi-zhuang HE, Qing-jing P (2012) Isolation and identification of potash feldspar-solubilizing bacteria and their potassium-releasing activities. Chin J Microecol 24:773–776

Leaungvutiviroj C, Ruangphisarn P, Hansanimitkul P, Shinkawa H, Sasaki K (2010) Development of a new biofertilizer with a high capacity for N_2 fixation, phosphate and potassium solubilization and auxin production. Biosci Biotechnol Biochem 74:1098–1101

Li DX (2003) Study on the effects of silicate bacteria on the growth and fruit quality of apples. J Fruit Sci 20:64–66

Li FC, Li S, Yang YZ, Cheng LJ (2006) Advances in the study of weathering products of primary silicate minerals, exemplified by mica and feldspar. Acta Petrol Mineral 25:440–448

Lian B, Prithiviraj B, Souleimanova A, Smitha DL (2001) Evidence for the production of chemical compounds analogous to nod factor by the silicate bacterium *Bacillus circulans* GY92. Microbiol Res 156:289–292

Lian B, Fu PQ, Mo DM, Liu CQ (2002) A comprehensive review of the mechanism of potassium release by silicate bacteria. Acta Mineral Sin 22:179–182

Liu W, Xu X, Wu S, Yang Q, Luo Y, Christie P (2006) Decomposition of silicate minerals by *Bacillus mucilaginosus* in liquid culture. Environ Geochem Health 28:133–140

Liu D, Lian B, Dong H (2012) Isolation of *Paenibacillus sp.* and assessment of its potential for enhancing mineral weathering. Geomicrobiol J 29:413–421

Lynn TM, Win HS, Kyaw EP, Latt ZK, Yu SS (2013) Characterization of phosphate solubilizing and potassium decomposing strains and study on their effects on tomato cultivation. Int J Innov Appl Stud 3:959–966

Malinovskaya IM, Kosenko LV, Votselko SK, Podgorskii VS (1990) Role of *Bacillus mucilaginosus* polysaccharide in degradation of silicate minerals. Microbiology 59:49–55

Man LY, Cao XY, Sun DS (2014) Effect of potassium-solubilizing bacteria-mineral contact mode on decomposition behavior of potassium-rich shale. Chin J Nonferrous Metals 24:1099–1109

Maurya BR, Meena VS, Meena OP (2014) Influence of inceptisol and alfisol's potassium solubilizing bacteria (KSB) isolates on release of K from waste mica. An Intl J Plant Res 27:181–187

Meena VS, Maurya BR, Bahadur I (2014) Potassium solubilization by bacterial strain in waste mica. Bangladesh J Bot 43:235–237

Mengel K (2007) Potassium. In: Barker AV, Pilbeam DJ (eds) Handbook of plant nutrition. Taylor & Francis, Boca Raton, pp 91–120

Mengel K, Kirkby EA (2001) Principles of plant nutrition, 5th edn. Kluwer Acad Publishers, Dordrecht, p 849

Mikhailouskaya N, Tcherhysh A (2005) K-mobilizing bacteria and their effect on wheat yield. Latvian J Agron 8:154–157

Mohammadi K, Sohrabi Y (2012) Bacterial biofertilizers for sustainable crop production: a review. ARPN J Agric Biol Sci 7:307–316

Muentz A (1890) Surla decomposition desroches etla formation de la terre arable. C R Acad Sci 110:1370–1372

Muralikannan N, Anthomiraj S (1998) Occurrence of silicate solubilizing bacteria in rice ecosystem. Madras Agric J 85:47–50

Nadeem SM, Zahir ZA, Naveed M, Arshad M (2009) Rhizobacteria containing ACC deaminase confer salt tolerance in maize grown on salt affected soils. Can J Microbiol 55:1302–1309

Nadeem SM, Zahir ZA, Naveed M, Asghar HN, Arshad M (2010a) Rhizobacteria capable of producing ACC-deaminase may mitigate the salt stress in wheat. Soil Sci Soc Am J 74:533–542

Nadeem SM, Zahir ZA, Naveed M, Ashraf M (2010b) Microbial ACC-deaminase: prospects and applications for inducing salt tolerance in plants. Crit Rev Plant Sci 29:360–393

Nadeem SM, Ahmad M, Zahir ZA, Ashraf M (2011) Microbial ACC-deaminase biotechnology: perspectives and applications in stress agriculture. In: Maheshwari DK (ed) Bacteria in agrobiology: stress management. Springer, Heidelberg, pp 141–185

Nadeem SM, Naveed M, Zahir ZA, Asghar HM (2013) Plant–microbe interactions for sustainable agriculture: fundamentals and recent advances. In: Arora NK (ed) Plant microbe symbiosis: fundamentals and advances. Springer, New Delhi

Parmar P, Sindhu SS (2013) Potassium solubilization by rhizosphere bacteria: influence of nutritional and environmental conditions. J Microbiol Res 3:25–31

Patel BC (2011) Advance method of preparation of bacterial formulation using potash mobilizing bacteria that mobilize potash and make it available to crop plant. WIPO Patent Application WO/2011/154961

Pindi PK, Satyanarayana SDV (2012) Liquid microbial consortium- a potential tool for sustainable soil health. J Biofertil Biopesticide 3(4):1–9

Prajapati KB, Modi HA (2012) Isolation and characterization of potassium solubilizing bacteria from ceramic industry soil. CIB Tech J Microbiol 1:8–14

Prajapati KB, Modi HA (2014) The Study of shelf life of potassium solubilizing microorganisms for liquid biofertilizer. Biotechnology 3:13–14

Prajapati K, Sharma MC, Modi HA (2012) Isolation of two potassium solubilizing fungi from ceramic industry soils. Life Sci Leaflets 5:71–75

Prajapati K, Sharma MC, Modi HA (2013) Growth promoting effect of potassium solubilizing microorganisms on *Abelmoscus esculantus*. Int J Agric Sci 3:181–188

Raj SA (2004) Solubilization of silicate and concurrent release of phosphorus and potassium in rice ecosystem. In: Kannaiyan S, Kumar K, Govindarajan K (eds) Biofertilizer technology for rice based cropping system. Scientific Book Center, pp 372–378

Rajawat MVS, Singh S, Singh G, Saxena AK (2012) Isolation and characterization of K-solubilizing bacteria isolated from different rhizospheric soil. In: Proceeding of 53rd annual conference of association of microbiologists of India, 2012, pp 124

Requena BN, Jimenez I, Toro M, Barea JM (1997) Interactions between plant growth promoting rhizobacteria (PGPR), arbuscular mycorrhizal fungi and *Rhizobium* spp. in the rhizosphere of *Anthyllis cytisoides*, a model legume for revegetation in Mediterranean semi-arid ecosystem. New Phytol 136:667–677

Romheld V, Kirkby EA (2010) Research on potassium in agriculture: needs and prospects. Plant Soil 335:155–180

Sangeeth KP, Bhai RS, Srinivasan V (2012) *Paenibacillus glucanolyticus*, a promising potassium solubilizing bacterium isolated from black pepper (*Piper nigrum* L.) rhizosphere. J Spic Aromat Crops 21(2):118–124

Shaaban EA, El-Shamma IMS, El Shazly S, El-Gazzar A, Abdel-Hak RE (2012) Efficiency of rock-feldspar combined with silicate dissolving bacteria on yield and fruit quality of valencia orange fruits in reclaimed soils. J Appl Sci Res 8:4504–4510

Shaharoona B, Arshad M, Zahir ZA (2006) Effect of plant growth promoting rhizobacteria containing ACC-deaminase on maize (*Zea mays* L.) growth under axenic conditions and on nodulation in mung bean (*Vigna radiata* L.). Lett Appl Microbiol 42:155–159

Shanware AS, Kalkar SA, Trivedi MM (2014) Potassium solublisers: occurrence, mechanism and their role as competent biofertilizers. Int J Curr Microbiol App Sci 3:622–629

Sheng XF (2005) Growth promotion and increased potassium uptake of cotton and rape by a potassium releasing strain of *Bacillus edaphicus*. Soil Biol Biochem 37:1918–1922

Sheng XF, He LY (2006) Solubilization of potassium-bearing minerals by a wild type strain of *Bacillus edaphicus* and its mutants and increased potassium uptake by wheat. Can J Microbiol 52:66–72

Sheng XF, Huang WY (2002) Mechanism of potassium release from feldspar affected by the strain NBT of silicate bacterium. Acta Pedol Sin 39:863–871

Sheng XF, He LY, Huang WY (2002) The conditions of releasing potassium by a silicate-dissolving bacterial strain NBT. Agr Sci China 1:662–666

Sheng XF, Zhao F, He H, Qiu G, Chen L (2008) Isolation, characterization of silicate mineral solubilizing *Bacillus globisporus* Q12 from the surface of weathered feldspar. Can J Microbiol 54:1064–1068

Sindhu SS, Verma MK, Suman M (2009) Molecular genetics of phosphate solubilization in rhizosphere bacteria and its role in plant growth promotion. In: Khan MS, Zaidi A (eds) Phosphate solubilizing microbes and crop productivity. Nova Science Publishers, New York, pp 199–228

Sindhu SS, Dua S, Verma MK, Khandelwal A (2010) Growth promotion of legumes by inoculation of rhizosphere bacteria. In: Khan MS, Zaidi A, Musarrat J (eds) Microbes for legume improvement. Springer-Wien, New York, pp 195–235

Sindhu SS, Parmar P, Phour M (2012) Nutrient cycling: potassium solubilization by microorganisms and improvement of crop growth. In: Parmar N, Singh A (eds) Geomicrobiology and biogeochemistry: soil biology. Springer-Wien, New York, pp 27–40

Singh G, Biswas DR, Marwah TS (2010) Mobilization of potassium from waste mica by plant growth promoting rhizobacteria and its assimilation by maize (*Zea mays*) and wheat (*Triticum aestivum* L.). J Plant Nutr 33:1236–1251

Sparks DL, Huang PM (1985) Physical chemistry of soil potassium. In: Munson RD et al (eds) Potassium in agriculture. ASA, Madison, pp 201–276

Subhashini DV (2015) Growth promotion and increased potassium uptake of tobacco by potassium-mobilizing bacterium *Frateuria aurantia* grown at different potassium levels in Vertisols. Commun Soil Sci Plant Anal 46:210–220

Sugumaran P, Janarthanam B (2007) Solubilization of potassium containing minerals by bacteria and their effect on plant growth. World J Agr Sci 3:350–355

Supanjani HHS, Jung SJ, Lee KD (2006) Rock phosphate and potassium rock solubilizing bacteria as alternative sustainable fertilizers. Agro Sust Dev 26:233–340

Syed BA, Patel B (2014) Investigation and correlation of soil biotic and abiotic factors affecting agricultural productivity in semi-arid regions of north Gujarat, India. Int J Res Stud Biosci 2:18–29

Troufflard S, Mullen W, Larson TR, Graham IA, Crozier A, Amtmann A, Armengaud P (2010) Potassium deficiency induced the biosynthesis of oxylipins and glucosinolates in *Arabidopsis thaliana*. Plant Biol 10(1):172–184

Ullman WJ, Kirchman DL, Welch SA, Vandevivere P (1996) Laboratory evidence for microbially mediated silicate mineral dissolution in nature. Chem Geol 132:11–17

Uroz S, Calvaruso C, Turpault MP, Pierrat JC, Mustin C, Frey-Klett P (2007) Effect of the mycorrhizosphere on the genotypic and metabolic diversity of the bacterial communities involved in mineral weathering in a forest soil. Appl Environ Microbiol 73:3019–3027

Uroz S, Calvaruso C, Turpault P, Frey-Klett P (2009) Mineral weathering by bacteria: ecology, actors and mechanisms. Trends Microbiol 17:378–387

Verma JP, Yadav J, Tiwari KN, Lavakush SV (2010) Impact of plant growth promoting rhizobacteria on crop production. Int J Agric Res 5:954–983

Verma M, Sharma S, Prasad R (2011) Liquid biofertilizers: advantages over carrier- based biofertilizers for sustainable crop production. News L Intl Soc Environ Bot 17:2pp

Vlek PLG, Vielhauer K (1994) Nutrient management strategies in stressed environments. In: Virmani SM, Katyal JC, Eswaran H, Abrol IP (eds) Stressed ecosystems and sustainable agriculture. Oxford and IBH Publishing Co., New Delhi, pp 203–229

Wakeel A, Gul M, Sanaullah M (2013) Potassium dynamics in three alluvial soils differing in clay contents. Emirates J Food Agric 25:39–44

White PJ, Karley AJ (2010) Potassium. In: Hell R, Mendel RR (eds) Cell biology of metals and nutrients, plant cell monographs, vol 17. Springer, Berlin, pp 199–224

Wood N (2001) Nodulation by numbers: the role of ethylene in symbiotic nitrogen fixation. Trends Plant Sci 6:501–502

Wu SC, Cao ZH, Li ZG, Cheung KC, Wong MH (2005) Effects of biofertilizer containing N-fixer, P and K solubilizers and AM fungi on maize growth: a greenhouse trial. Geoderma 125:155–166

Xie JC (1998) Present situation and prospects for the world's fertilizer use. Plant Nutri Fertil Sci 4:321–330

Yadegari M, Farahani GHN, Mosadeghzad Z (2012) Biofertilizers effects on quantitative and qualitative yield of Thyme (*Thymus vulgaris*). Afr J Agric Res 7:4716–4723

Yakhontova K, Andreev PI, Ivanova MY, Nesterovich LG (1987) Bacterial decomposition of smecite minerals. Doklady Akademii Nauk, SSR 296:203–206

Youssef GH, Seddik WMA, Osman MA (2010) Efficiency of natural minerals in presence of different nitrogen forms and potassium dissolving bacteria on peanut and sesame yields. J Am Sci 6:647–660

Zahir ZA, Ghani U, NaveedM NSM, Asghar HN (2009) Comparative effectiveness of *Pseudomonas* and *Serratia sp.* containing ACC-deaminase for improving growth and yield of wheat (*Triticum aestivum* L.) under salt-stressed conditions. Arch Microbiol 191:415–424

Zapata F, Roy RN (2004) Use of phosphate rock for sustainable agriculture. FAO and IAEA, Rome

Zarjani JK, Aliasgharzad N, Oustan S, Emadi M, Ahmadi A (2013) Isolation and characterization of potassium solubilizing bacteria in some Iranian soils. Arch Agron Soil Sci 59:1713–1723

Zeng X, Liu X, Tang J, Hu S, Jiang P, Li W et al (2012) Characterization and potassium-solubilizing ability of *Bacillus circulans* Z1-3. Adv Sci Lett 10:173–176

Zhang C, Kong F (2014) Isolation and identification of potassium-solubilizing bacteria from tobacco rhizospheric soil and their effect on tobacco plants. Appl Soil Ecol 82:18–25

Zhang A, Zhao G, Gao T, Wang W, Li J, Zhang S et al (2013) Solubilization of insoluble potassium and phosphate by *Paenibacillus kribensis* CX-7: a soil microorganism with biological control potential. Afr J Microbiol Res 7:41–47

Zhao F, Sheng X, Huang Z, He L (2008) Isolation of mineral potassium-solubilizing bacterial strains from agricultural soils in Shandong Province. Biodiv Sci 16:593–600

Zhou H, Zeng X, Liu F, Qiu G, Hu Y (2006) Screening, identification and desilication of a silicate bacterium. J Cent South Univ Technol 13:337–341

Rhizosphere Microbes: Potassium Solubilization and Crop Productivity – Present and Future Aspects

22

Priyanku Teotia, Vivek Kumar, Manoj Kumar, Neeraj Shrivastava, and Ajit Varma

Abstract

The plant rhizosphere harbor array of potassium-solubilizing microbes (KSMs), which solubilize the insoluble and inaccessible potassium (K) to accessible forms of potassium for plant uptake and transport, is one of the inevitable elements for growth and yield. The process of potassium solubilization is performed by specific rhizosphere microbes, which include bacteria and fungi; the prominent are *Bacillus* sp. (*B. megaterium*, *B. mucilaginosus*, *B. edaphicus*, *B. circulans*, *Acidithiobacillus ferrooxidans*, *Pseudomonas putida*, *Arthrobacter* sp., and *Paenibacillus* sp.) *Aspergillus* spp., and *Aspergillus terreus*. The agricultural soil particulates contain minerals such as orthoclase, illite, biotite, feldspar, and mica which contain potassium, though this is not available to the plants due to its immobilized form. Intermittently, potassium is an important element after N and P in soil chemistry; therefore, the rhizosphere microbes play a significant role in mobilizing the unavailable form of potassium to the plant roots. The potential rhizosphere K-solubilizing microbes such as *Pseudomonas*, *Bacillus*, and *Aspergillus* excrete organic acids, which solubilize the unavailable potassium and make available to plant roots. Till date, most of the work has been done on nitrogen-fixing and phosphate-solubilizing microbes; moreover, the available biofertilizer with solubilized K (readily available) needs more attention at commercial scale. The current chapter addresses the information gaps related to potassium-solubilizing/potassium-mobilizing microorganisms in soil and analyzing current and future aspects of potassium-solubilizing microbes for the crop productivity.

P. Teotia
Department of Botany, CCS University, Meerut, India

V. Kumar (✉) • M. Kumar • N. Shrivastava • A. Varma
Amity Institute of Microbial Technology, Amity University, Noida, Uttar Pradesh, India
e-mail: vivekbps@gmail.com

V.S. Meena et al. (eds.), *Potassium Solubilizing Microorganisms for Sustainable Agriculture*,
DOI 10.1007/978-81-322-2776-2_22

Keywords
Bacteria • Crops • Fungi • Microbes • Potassium • Solubilization

22.1 Introduction

To symbolize potassium, letter K is in use, which is derived from the German word "Kalium." Since long, people used to burn organic matters and wood in pots to make soap. After burning the wood and other materials, the ashes were rinsed with water, and the water was allowed to evaporate, leaving behind the residues, which consisted of potassium salts. The residue was known popularly as "pot ashes" or potash. These residual salts were boiled with animal fat to manufacture soap. A botanist from Connecticut, Samuel William Jackson, analyzed the ash of burned plants. He observed that plant parts consisted of great amounts of potassium, in addition to other micro and macro minerals.

Potassium (K) is a vital and essential nutrient for plant growth. Since huge amounts are absorbed from the root zone in the growth and production of most agronomic crops, therefore, it is classified as a macronutrient. This element plays a crucial role for enzyme activation, protein synthesis, photosynthesis, and quality of production. With the ever-growing interest in rigorous and exhaustive agriculture, the soil potassium levels have declined, owing to crop uptake, leaching, water runoff, and soil erosion (Li et al. 2006). It is, therefore, imperative to develop some alternative sustainable biological means that would effectively minimize this loss. This vital element is associated with movement of water, nutrients, and carbohydrates in plant tissues. If K is scarce or not provided in sufficient amounts, plant growth is stunted and yields are reduced. Ninety-five percent of all potassium fertilizers come in the form of muriate of potash, also known as potassium chloride. For crops unable to tolerate chloride, potassium sulfate, potassium nitrate, and other chloride-free salts are used.

Various research efforts have shown that potassium stimulates early growth; increases protein production; improves the efficiency of water use; is vital for stand persistence, longevity, and winter hardiness of alfalfa; and improves resistance to diseases and insects. This mineral is required in high quantities by plant cells and possesses essential physiological and biochemical functions involving cell osmotic regulation and enzyme activation (Valmorbida and Boaro 2007). The option and prospect of using microbes as a substitute or choice for biological process to persuade the liberation of potassium from potassium-containing rocks and minerals of soils also exist (Rogers et al. 1998). There are several groups of microbes, such as bacteria, fungi, yeast, and actinomycetes, that are proficient in solubilizing the unavailable potassium contained in soil and rock minerals through mineralization (Supanjani et al. 2006; Magril et al. 2012; Meena et al. 2014a). The liberation of potassium from these soil and rock minerals is mainly initiated by the production of microbial organic acids during the process of proliferation and survival in rhizosphere. The rhizosphere microbes produce organic acids such as oxalic acid, formic acid, citric acid, and malic acid. These organic acids augment the solubilization of K ions by providing protons and by complexing Ca^{2+} ions in the soil system. Sheng (2005) has publicized that organic compounds such as acetate, citrate, and oxalate excreted by microbes can boost mineral solubilization in soil. The formation of complex between metal ions (Ca^{++}, Al^{++}, and Fe^{+++}) and organic acids also leads to solubilization of potassium (Styriakova 2003).

The understanding of rhizosphere biology, over the last three decades, has progressed with the finding of a remarkable and particular collection of potential microbes, recognized as plant growth-promoting microbes (PGPMs). These microbes inhabit in and on the plant root system and exercise valuable and positive effects on plant growth promotion using various means (Egamberdiyeva and Hoflich 2004). On the other hand, the use of potassium-mobilizing microbes (KMMs and KSR) as bioinoculants alone or with other microbes has been shown to enhance the plant growth (Meena et al. 2013, 2015b; Maurya et al. 2014; Singh et al. 2015). Singh et al. (2010) have shown that wheat and maize yields increased by KMM (*Bacillus mucilaginosus*, *Azotobacter chroococcum*, and *Rhizobium* spp.), under phytotron growth chamber. Results revealed that KMM significantly improves the assimilation of potassium by both maize and wheat, where waste mica was the sole source of potassium. KMMs have been reported to be important organisms for plant root establishment, root growth prototype, plant competitiveness, and plant nourishment, predominantly under abiotic and biotic stress and nutrient-imbalance conditions (Wu et al. 2005; Meena et al. 2014a, b). Their use in agriculture can errand a decrease in the use of agrochemicals and sustain eco-friendly crop production (Yuan et al. 2000; Pettigrew 2008). Different KMMs including associative bacteria and fungi such as *Paenibacillus*, *Azospirillum*, *Bacillus*, *Pseudomonas*, *Azotobacter*, *Enterobacter*, and *Aspergillus* have been employed for their beneficial outcome on plant growth (Sangeeth et al. 2012; Archana et al. 2013; Diep and Hieu 2013). KMMs improve plant development by various means, but the precise and exact mechanisms involved have not been properly characterized (Lian et al. 2002). The KMMs have been shown to candidly increase plant augmentation by various techniques, solubilization of minerals (Sheng and Huang 2002), and synthesis of phytohormones (Kumar et al. 2012). The direct improvement of plant growth by mineral solubilization and mobilization owing to amplifying the precise ion fluxes by KMM on the root surface has also been reported (Badr et al. 2006; Sheng et al. 2008). Microorganisms play a central role in the natural potassium cycle and K-solubilizing microbes in soil and in plant rhizosphere (Han and Lee 2005; Kumar et al. 2015; Meena et al. 2015a); there are several bacteria species having high ability to solubilize unavailable potassium from insoluble minerals.

22.2 Soil and Potassium

Potassium is a necessary, vital, and indispensable macronutrient found in soil orbed cation that plays an important role in the growth, metabolism, and development of plants. Without adequate potassium, the plants will have poorly developed roots, grow slowly, produce small seeds, and have lower yields. Even though potassium constitutes ~2.52 % of the upper layer of earth's crust, the actual amount of this nutrient in soil varies from 0.04 % to 3.0 % (Baligar 1985; Zorba et al. 2013). The plants avail potassium from the rhizosphere soil, and availability of this nutrient depends upon its amount present in soil and its dynamics. There are three types of potassium available in soil, in readily unavailable form; depending on soil type, 90–98 % of soil potassium exists in this form, and the minerals are feldspar and mica which contain most of the potassium. These minerals are the source of soil potassium, and they release potassium very slowly to the more available forms as they weather and break down. The second type is around 1–10 % of soil potassium and is slowly available form to the plants; this form is the fraction of the inner constitution of the clay minerals forming the colloidal portion of the soil. Slowly available potassium form cannot be restored by usual cation exchange procedure and is known as "non-exchangeable" potassium. The third form is readily available potassium and is comprised of water-soluble potassium and exchangeable potassium in the soil. This form is absorbed on the soil colloid surfaces and is easily accessible to plants (Maathuis and Sanders 1997). In spite of this, higher plants acquire the majority of the potassium from the soil solution part. Out of the three forms of potassium found in the soil, soil minerals make up more than 90–98 % of soil potassium (Memon et al. 1988), and most of it is unavailable for plant uptake. The second non-exchangeable form of potassium makes up approximately 1–10 % of soil potassium and consists predominantly of interlayer K of non-expanded clay

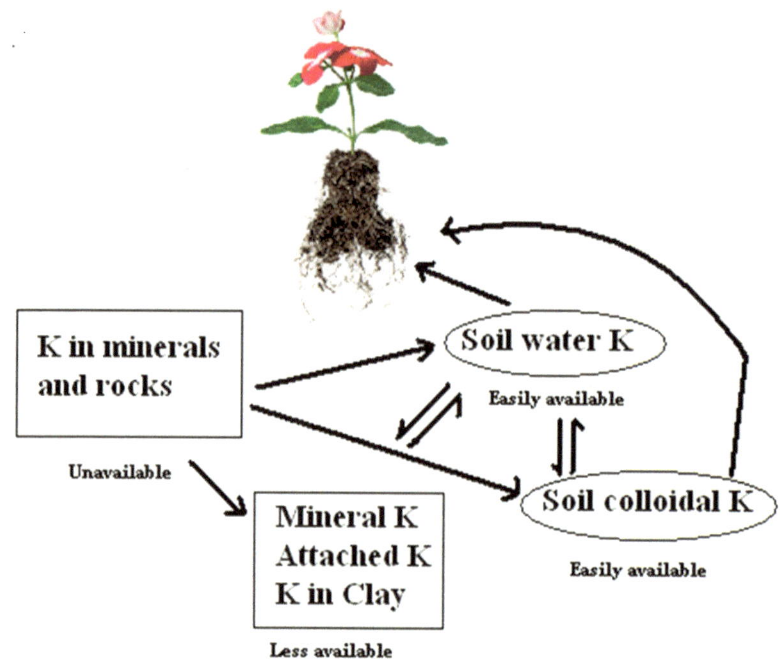

Fig. 22.1 Interrelationship between unavailable, less available, and easily available K in soil and plant system

minerals such as illite and lattice K in K-feldspars, which contribute significantly to the plant uptake (Sheng et al. 2008). Release of non-exchangeable K to the third exchangeable form occurs when level of exchangeable and solution K is decreased by crop removal, runoff, erosion, and/or leaching (Fig. 22.1).

With the introduction of high-yielding crop varieties/hybrids and the progressive intensification of agriculture, the soils are getting depleted in potassium reserve at a faster rate. Moreover, due to imbalanced fertilizer application, potassium deficiency is becoming one of the major constraints in crop production. This emphasized the search to find an alternative indigenous source of K for plant uptake and to maintain K status in soils for sustaining crop production (Supanjani et al. 2006). Soil microbes have been reported to play a key role in the natural K cycle, and, therefore, potassium-solubilizing microorganisms (KSMs) present in the soil could provide an alternative technology to make potassium available for uptake by plants (Lian et al. 2007). Thus, identification of microbial strains capable of solubilizing potassium minerals quickly can conserve our existing resources and avoid environmental pollution hazards caused by heavy application of chemical fertilizers.

Many soil bacteria such as *Burkholderia*, *Pseudomonas*, *Acidithiobacillus ferrooxidans*, *Bacillus mucilaginosus*, *Bacillus edaphicus*, *B. circulans*, and *Paenibacillus* sp. can release and provide potassium to the plants from potassium-bearing minerals (Badr et al. 2006; Han et al. 2006). These KSB have been reported to ooze organic acids, siderophores, and hydrogen ions, resulting in mobilizing potassium from minerals such as illite, feldspar, and micas (Lian et al. 2002; Liu et al. 2012). Inoculation of crops with potassium-solubilizing microbes, such as wheat (Mikhailouskaya and Tcherhysh 2005), sorghum (Badr et al. 2006), cucumber and pepper (Han et al. 2006), eggplant (Han and Lee 2005), soybean and cotton (Pettigrew 2008), and rape and cotton (Xeng 2005), has generated interesting and encouraging results. In the same way, a bacterium identified as *Paenibacillus glucanolyticus* strain IISRBK2 bearing elevated capability to solubilize potash was isolated from the black pepper rhizosphere. This bacterial strain was also assessed for black pepper growth and potassium uptake in soil administered with 0.51 and 1.5 g K kg^{-1} soil in pot, where the wood ash was used as a source of K which contained 53.1 g kg^{-1} K. Consequently, nowadays K-mobilizing microbes are widely employed as bioinoculants in most of the countries, where potassium is deficient or less available to the plants in agricultural soils (Gundala et al. 2013; Zarjani et al. 2013; Diep and Hieu 2013). Hence, appliance of K-mobilizing microbes as bioinoculants for enhanced crop production could mitigate the employment of chemical fertilizer to maintain and sustain the crop production (Sheng 2005; Supanjani et al. 2006).

Currently, little is known about KSMs, their efficacy, and mechanism to solubilize potassium and make it available to plant roots and finally affect plant growth parameters under various agroclimatic conditions. Sheng and Huang (2002) observed that pH, oxygen concentration, and the types of bacterial strains employed affected the potassium release from the minerals. The effectiveness of potassium solubilization by diverse microbes was observed to differ according to the type of environmental conditions and types of minerals. *Bacillus* spp. showed the potassium solubilization in liquid medium. Potassium mineral illite exhibited more growth than feldspar (Sheng and He 2006). Consequently, there could be colossal and immense possibilities for additional augmentation of the crop production by application of K-bearing rock materials and potassium-solubilizing microbes as bioinoculants.

22.3 Potassium and Plant Productivity

Potassium is vital to many plant processes. Its role involves the basic biochemical and physiological activities of plants. Though K does not become an integral part of the chemical structure of plants, it plays numerous imperative regulatory roles in natural growth and development. To carry out any chemical reaction, enzymes serve as catalysts without being utilized and not consumed in the process. The element potassium "activates" at least 60 different enzymes, which are involved in plant growth and production. The potassium changes the physical nature of the enzyme molecule, uncovering the suitable chemically functional sites for reaction. Various organic anions and other organic as well as inorganic compounds are also neutralized by potassium, within the plant cells, resulting in stabilizing the pH between 7 and 8, which is the optimum for most enzyme actions (Meena et al. 2014a). The quantity of potassium present in the cell determines how many of the enzymes can be activated and the rates at which chemical reactions can proceed. Consequently, the rate of a given reaction is governed by the rate at which K ion enters the cell. The opening and closing of plant leaf stomata potassium plays an important role. Stomatal appropriate function is indispensable for photosynthesis process, nutrient and water transport, and also plant cooling. Buildup of K in plant roots generates a gradient of osmotic pressure that absorbs water into the roots. Deficiency of potassium in plants leads to stress conditions and leads to less absorbance of water.

During K deficiency in the plant, the rate of ATP generation and photosynthesis is also declined, and most of the processes dependent on ATP are also slowed down. Carbohydrates produced in photosynthesis process must be transported to other parts of the plant through the phloem for normal growth, utilization, and storage. The plant's transport system uses energy in the form of ATP. If K is inadequate, less ATP is available, and the transport system breaks down. This causes photosynthates to build up in the leaves, and the rate of photosynthesis is reduced. Potassium is required for every major step of protein synthesis. The "reading" of the genetic code in plant cells to produce proteins and enzymes that regulate all growth processes would be impossible without adequate K. When plants are deficient in K, proteins are not synthesized despite an abundance of available nitrogen (N). Instead, protein "raw materials" (precursors) such as amino acids, amides, and nitrate accumulate. Potassium is responsible for starch and protein formation and overall crop quality parameters.

In the plant, potassium is a well-transportable element and is translocated from the older to younger leaves. As a result, potassium

Table 22.1 Potassium fertilizer materials and their percent nutrient content

Elements	Chemical formula	K_2O (%)	N (%)	S (%)
Potassium chloride	KCl	60–62	–	–
Potassium sulfate	K_2SO_4	50–52	–	18
Potassium magnesium sulfate	$K_2SO_4.2MgSO_4$	22	–	22
Potassium nitrate	KNO_3	44	13	–
Potassium sodium nitrate	$KNa(NO_3)_2$	14	15	–
Potassium hydroxide	KOH	83	–	–
Potassium carbonate	$K_2CO_3KHCO_3$	<68	–	–
Potassium orthophosphate	$KH_2PO_4K_2HPO_4$	30–50	–	–
Potassium polyphosphate	$K_4P_2O_7$	22–48	–	–
Potassium metaphosphate	KPO_3	38	–	–

insufficiency symptoms typically happen first on the older (lower) leaves of the plant, and proceed to the younger leaves, depending on the increasing severity of the deficiency. The generally widespread and universal symptoms and indicator of potassium deficiency are the yellow chlorosis or yellow corching, along the entire leaf margin. In acute and critical cases, the yellow dried margin of the leaf may fall down. On the other hand, crops with broad leaves, cotton, soybeans, banana, and the whole leaf, may cast off, which results in premature defoliation of the plant. Under severe potassium efficiency, the crops grow slowly with underdeveloped root systems, the stalks become weak, and the small grain with poor quality is lodged in wheat and other cereals. During insufficient potassium conditions, winter killing of recurrent crops such as alfalfa and grasses could take place (Table 22.1).

22.4 Microbial Mechanism for Potassium Solubilization

The potassium-solubilizing microbes (KSMs) could dissolve potassium from insoluble minerals such as micas, illite, and orthoclases, by producing organic acids. These acids either directly solubilize rock potassium or excrete chelated silicon ions to bring K into available solution form. Microbial organic acids augment the solubilization of potassium compounds by providing protons and also by complexing Ca^{2+} ions in soil. Sheng et al. (2003) reported that organic compounds such as oxalate, citrate, and acetate can enhance mineral dissolution in soil, produced by microorganisms. K solubilization happens by formation of complex between organic acids and metal ions such as iron, calcium, and aluminum (Styriakova et al. 2003).

Microbial-mediated organic ligands and metabolic by-products such as excreted enzymes, small and large molecules of organic acids, and chelates augment solubilization of aluminosilicate mineral or quartz under in vitro and in situ conditions (Zeng et al. 2012). Production of microbial organic acids such as oxalic and tartaric solubilizers feldspar and illite to release potassium (Sheng and He 2006). Another report by Groudev (1987) exhibited solubilization of potassium by inorganic and organic acid production and also due to production of mucilaginous capsules composed of exopolysaccharides produced by bacteria like *Bacillus*, *Clostridium*, and *Thiobacillus* spp.; similar probable method of K solubilization has been reported by Sugumaran and Janarthanam (2007), where *Bacillus mucilaginosus* has been tested for K solubilization. During growth period of bacterium, there was no drop down in pH of medium; therefore, *Bacillus* spp. did not excrete inorganic or organic acids, and slime formation could play a role in K solubilization. The mineral weathering potential of the soil microbes involves production of inorganic and organic acids, protons, and chelates like siderophores and organic ligands. This has been reported in fungus species such as *Cladosporium*, *Aspergillus*, and *Penicillium* spp. which have the potential to excrete huge quantity of citric, gluconic, and oxalic acids in vitro, which leads to solubilization of soil silicates, mica, and feldspar (Lian et al. 2008). Yang et al. (2014) reported the biofilm formation by bacteria onto mineral containing potassium-rich shale and leaching of K from the mineral. These biofilms were composed of acids, protein, and polysaccharides produced by bacteria.

22.5 Present Status of Potassium Solubilization and Crop Production

Plant growth-promoting microbes (PGPMs) found in rhizosphere zone exercise their advantageous effect on nutrient uptake and their mobilization along with diverse mechanisms such as nitrogen fixation, phytohormone and siderophore production, and mobilization of macro- and micronutrient elements such as potassium, phosphorus, zinc, copper, and iron. In India, the total soil potassium reserves are usually enough to support the crop growth, but the distribution and available pattern of element vary from soil to soil and region to region; this makes the unavailability of potassium for plant uptake (Zorb et al. 2014). The release of available and exchangeable form of potassium with the help of a microbe replenishes the root zone environment, which fulfills the basic needs of a plant. This replacement is a must in the rhizosphere zone for optimal crop production, and the microbes inoculated deliberately or naturally present play a significant role (Sheng et al. 2002). Many research works demonstrated that K-solubilizing microbes performed well with many crops under different agroclimatic conditions. Supanjani et al. (2006) showed that potassium-dissolving bacteria (along with phosphate-solubilizing bacteria) increased hot pepper (*Capsicum annuum* L.) photosynthesis by ~16 and leaf area by ~35 % as compared with control plants, and biomass harvest and fruit yield of the treated plants were 23 % and 30 %, respectively, higher as compared with control. The workers found that the treatment with P and K rocks and PK-solubilizing bacterial strains and the treatment with classical, soluble fertilizer have a similar effectiveness. Sheng (2005) worked on potassium-releasing bacterial strain *Bacillus edaphicus* NBT for plant growth-promoting effects and nutrient uptake on cotton and rape in K-deficient soil in pot experiments. Inoculation with bacterial strain *B. edaphicus* NBT was found to increase root and shoot growth of cotton and rape. Strain NBT was able to mobilize potassium efficiently in both plants when illite was added to the soil. In cotton and rape growing in soils treated with insoluble potassium and inoculated with strain NBT, the potassium content was increased by 30 % and 26 %, respectively. Bacterial inoculation also resulted in higher N and P contents of aboveground plant components. The bacterial isolate was also able to colonize and develop in the rhizosphere soil of cotton and rape after root inoculation.

In another study, Basak and Biswas (2008) used to investigate the effectiveness of potassium-solubilizing microorganism (*Bacillus mucilaginosus*) using Sudan grass (*Sorghum vulgare* Pers. var. *sudanense*) as test crop grown under two Alfisols. Results revealed that application of mica significantly enhanced biomass yield, uptake, and percent K recoveries by Sudan grass than control (without potassium). Biomass yield, uptake, and percent K recoveries increased further when mica was inoculated with bacterial strain in both soils than uninoculated mica. Zhang and Konga (2014) isolated 27 - potassium-solubilizing strains; among them, 17 strains belonged to *Klebsiella variicola*, 2 strains belonged to *Enterobacter cloacae*, 2 strains belonged to *Enterobacter asburiae*, and the remaining 6 strains belonged to *Enterobacter aerogenes*, *Pantoea agglomerans*, *Agrobacterium tumefaciens*, *Microbacterium foliorum*, *Myroides odoratimimus*, and *Burkholderia cepacia*, respectively. *K. variicola* occurred at the highest frequency with 18 strains. Four isolates, GL7, JM3, XF4, and XF11, were selected for a greenhouse pot experiment because of their pronounced K-solubilizing capabilities. After being treated with the four KSB strains, plant dry weight and uptake of both K and nitrogen (N) by tobacco seedlings increased significantly. These increases were higher with the combination of KSB inoculation and K-feldspar powder addition. Isolate XF11 showed the most pronounced beneficial effect on plant growth and nutrient uptake by tobacco seedlings. Combining the inoculation of KSB and the addition of K-feldspar powder could be a promising alternative to commercial K fertilizer and may help maintain the availability of soil nutrients.

Recently, Prajapati et al. (2013) isolated potassium-solubilizing bacterium *Enterobacter hormaechei* and a fungal strain *Aspergillus terreus* and studied their effects and nutrient uptake on okra (*Abelmoschus esculentus*) in K-deficient soil using pot experiments. *Enterobacter hormaechei* increased root and shoot growth of plant, and both microbes were able to mobilize potassium in okra plant when feldspar was incorporated into the pot soil. In pot soils amended with insoluble potassium, inoculation of *Enterobacter hormaechei* and *Aspergillus terreus* with the potassium content was increased in okra plants. Similarly, Han et al. (2006) studied the effect of potassium-solubilizing bacteria (KSB) *Bacillus mucilaginosus* on pepper and cucumber. Coinoculation of phosphate-solubilizing bacteria (PSB) and KSB resulted in consistently higher P and K availability than in the control without bacterial inoculum and without rock material fertilizer. Integrated rock P with inoculation of PSB increased the availability of P and K in soil; the uptake of N, P, and K by shoot and root; and the growth of pepper and cucumber. Similar but less pronounced results were obtained when rock K and KSB were added concomitantly. Hassan et al. (2010) studied the efficacy of KSB, *Bacillus circulans* on khella (*Ammi visnaga*) growth. Inoculation of KSB along with feldspar increased plant growth parameters. Sugumaran and Janarthanam (2007) isolated K-solubilizing bacteria (*Bacillus mucilaginosus*); maximum K solubilization was 4.29 mg L^{-1}. Recently, Zhanga and Konga (2014) isolated 27 KSB strains; among them, 17 strains belonged to *Klebsiella variicola*, 2 strains belonged to *Enterobacter cloacae*, 2 strains belonged to *Enterobacter asburiae*, and the remaining 6 strains belonged to *Enterobacter aerogenes*, *Pantoea agglomerans*, *Agrobacterium tumefaciens*, *Microbacterium foliorum*, *Myroides odoratimimus*, and *Burkholderia cepacia*, respectively. The best four KSB strains, when inoculated, plant dry weight, and uptake of both K and nitrogen (N) by tobacco seedlings increased significantly. Isolate XF11 exhibited the most pronounced positive effect on plant augmentation and nutrient uptake by tobacco seedlings. In an interesting research work by Bagyalakshmi et al. (2012), it showed that inoculation of tea plants with K-solubilizing bacteria (*Pseudomonas putida*) leads to the improvement in the productivity and nutrient uptake in plants. Tea quality parameters such as theaflavin, thearubigin, highly polymerized substances, total liquor color, caffeine, briskness, leaf color, and flavor indexes were greatly improved in K-solubilizing bacteria-treated plants.

22.6 Future Aspects

Multifaceted and intricate dealings between potassium-mobilizing bacteria, potential rhizosphere microbes, the plant roots, and the ecosystem are accountable for the solubilization of bound potassium, and their variability uptake influences overall plant growth promotion. Potential and outlook strategies will be to clone genes responsible for potassium solubilization into those microbes having other beneficial functions, such as excellent proliferation and survival in rhizosphere, nitrogen fixation, phosphate solubilization, and phytohormone production. Moreover, the efficiency of K-solubilizing microbes can be augmented by developing the improved cultural techniques and deliverance procedure that support their survival in the rhizospheric zone. The employment of the blending of plant growth-promoting microbes with diverse beneficial functions comprising of K and P solubilization ability could be additional ecologically better tactic since it may lead to enhance survival and proliferation and superior acclimatization to the agroclimatic fluctuations happening all through the plantation period. The other future aspects could be the modern molecular biology approaches and techniques for manipulation of microbial beneficial functions that could lead to develop better potassium solubilization ability, and their commercial employment as K bioinoculants will further augment crop productivity and growth for viable and imperishable agriculture. Additional and expanded studies with concentration on such parallel issues with other available macro- and microelements in soil, especially in rhizosphere zones, and the role of microbial molecular mechanisms on plant nutrition uptake will certainly help in enhancing our knowledge of how to develop better microbial inoculants to enhance the K requirement of a plant.

22.7 Concluding Remarks

Obviously, the use of chemical fertilizers and organic manures cannot be reduced drastically or eliminated at this juncture without considerably decreasing food production. Simultaneously, the deleterious environmental side effects of fertilizer use, such as the expanding dead zones in marine systems all over the world, cannot go unnoticed, and in the long run, it will prove disastrous to our balanced ecosystem. That's why there is an imperative call for integrated nutrient management approach that aims toward less agricultural inputs and also minimizes the unfavorable and unpleasant environmental aftereffects of agricultural fertilizer use and production. Better understanding of the interactions among microbes, fertilizer, and plants is very important. There is a need for more information along the methods earlier mentioned, the application of which is two pronged: firstly, by introducing more applied potassium nutrient into the plant parts with the help of microbial inoculants, because less potassium nutrient is lost to the environment during and after the crop duration. Secondly, the loss of fertilizer may be minimized by increasing plant nutrient uptake efficiency. This could be achieved by applying potential K solubilizers. In both cases, there will be drastic reduction in agro-environmental pollution caused by indiscriminate use of chemical fertilizers. Results have revealed that inoculation of microbial strains of potassium solubilizers along with plant growth-promoting microbes (PGPMs) has resulted in better crop production. For that reason, the employment of this system will be better approach with a plan that integrates consortium of strains. In the meantime, some specific areas do need to be understood better, which includes potassium solubilization under in vitro conditions and then under field conditions. However, data on what quantity of the in vitro potassium solubilization is absorbed by plants under field conditions or consumed by the microbe for its own growth and metabolism is not well understood. This and related information will certainly help in understanding and determining the level of insoluble K and the use of bioinoculants that would be needed for practical purposes under actual field conditions. This is imperative for the reason that the quantity of K solubilized and K need of microbes, root exudation by the individual plant, and the soil conditions (pH, total dissolved solids, total and available K) in real time are among many probable factors that could influence the K solubilization and uptake by the plant.

These results reveal that plant type influences root colonization activity of introduced strains. From the results, we conclude that effective plant growth-promoting bacterium plant systems must be tested and established in controlled vegetation experimental designs with consideration of the specific ecological site conditions of practical applications (soil type, plant type). On the other hand, the degree of stimulus of crop plants by the commercially employed microbial strains and their persistence in the rhizosphere along with their plant growth-promoting ability under authentic field conditions remains ambiguous and vague. Thus, experiments concerning stimulation of cotton and rape must be followed by investigations under field conditions. The potassium-solubilizing microbes we currently use in our cropping system in K-deficient soils or where K is less or not available will certainly help to solve this macroelement predicament, and further research in this area toward developing potential K solubilizers could lead to an alternative mean of K nutrition supplement for use in agriculture.

References

Archana DS, Nandish MS, Savalagi VP, Alagawadi AR (2013) Characterization of potassium solubilizing bacteria (KSB) from rhizosphere soil. Bioinfolet 10:248–257. Availability and growth of eggplant. Res J Agric Boil Sci. 1(2): 176–180

Badr MA, Shafei AM, Sharaf El-Deen SH (2006) The dissolution of K and phosphorus bearing minerals by silicate dissolving bacteria and their effect on sorghum growth. Res J Agric Biol Sci 2:5–11

Bagyalakshmi B, Ponmurugan P, Marimuthu S (2012) Influence of potassium solubilizing bacteria on crop productivity and quality of tea (*Camellia sinensis*). Afr J Agric Res 7(30):4250–4259

Baligar VC (1985) Potassium uptake by plants, as characterized by root density, species and K/Rb ratio. Plant Soil 85:43–53

Basak BB, Biswas DR (2008) Influence of potassium solubilizing microorganism (*Bacillus mucilaginosus*) and waste mica on potassium uptake dynamics by sudan grass (*Sorghum vulgare* Pers.) grown under two Alfisols. Plant Soil 317:235–255

Diep CN, Hieu TN (2013) Phosphate and potassium solubilizing bacteria from weathered materials of denatured rock mountain, Ha Tien, Kiên Giang province, Vietnam. Am J Life Sci 1(3):88–92

Egamberdiyeva D, Hofflich G (2004) Effect of plant growth-promoting bacteria on growth and nutrient uptake of cotton and pea in an arid region of Uzbekistan. J Arid Environ 56:293–301

Groudev SN (1987) Use of heterotrophic microorganisms in mineral biotechnology. Acta Biotechnol 7:299–306

Gundala PB, Chinthala P, Sreenivasulu B (2013) A new facultative alkaliphilic, potassium solubilizing, *Bacillus* Sp. SVUNM9 isolated from mica cores of Nellore District, Andhra Pradesh. India Res Rev J Microbiol Biotechnol 2(1):1–7

Han HS, Lee KD (2005) Phosphate and potassium solubilizing bacteria effect on mineral uptake, soil availability and growth of eggplant. Res J Agric Boil Sci 1(2):176–180

Han HS, Supanjani K, Lee KD (2006) Effect of co-inoculation with phosphate and potassium solubilizing bacteria on mineral uptake and growth of pepper and cucumber. Plant Soil Environ 52:130–136

Hassan EA, Hassan EA, Hamad EH (2010) Microbial solubilization of phosphate – potassium rocks and their effect on khella (*Ammi visnaga*) growth. Ann Agric Sci (Cairo) 55:37–53

Kumar P, Dubey RC, Maheshwari DK (2012) *Bacillus* strains isolated from rhizosphere showed plant growth promoting and antagonistic activity against phytopathogens. Microbiol Res 167:493–499

Kumar A, Bahadur I, Maurya BR, Raghuwanshi R, Meena VS, Singh DK, Dixit J (2015) Does a plant growth-promoting rhizobacteria enhance agricultural sustainability? J Pur Appl Microbiol 9(1):715–724

Li FC, Li S, Yang YZ, Cheng LJ (2006) Advances in the study of weathering products of primary silicate minerals, exemplified by mica and feldspar. Acta Petrol Mineral 25:440–448

Lian B, Fu PQ, Mo DM, Liu CQ (2002) A comprehensive review of the mechanism of potassium release by silicate bacteria. Acta Petrol Sin 22:179–183

Lian B, Wang B, Pan M, Liu C, Teng HH (2007) Microbial release of potassium from K-bearing minerals by thermophilic fungus *Aspergillus fumigatus*. Geochim Cosmochim Acta 72:87–98

Lian B, Wang B, Pan M, Liu C, Teng HH (2008) Microbial release of potassium from K-bearing minerals by thermophilic fungus *Aspergillus fumigatus*. Geochim Cosmochim Acta 72:87–98

Liu D, Lian B, Dong H (2012) Isolation of *Paenibacillus* sp. and assessment of its potential for enhancing mineral weathering. Geomicrobiol J 29:413–421

Maathuis FJM, Sanders D (1997) Regulation of K^+ absorption in plant root cells by external K^+: interplay of different plasma membrane K^+ transporters. J Exp Bot 48:451–458

MagriI MMR, AvansiniI SH, Lopes-AssadI ML, Tauk-TornisieloII SM, Ceccato-Antonini SR (2012) Release of potassium from rock powder by the yeast *Torulaspora globosa*. Braz Arch Biol Technol 55(4):577–582

Maurya BR, Meena VS, Meena OP (2014) Influence of Inceptisol and Alfisol's potassium solubilizing bacteria (KSB) isolates on release of K from waste mica. Vegetos 27(1):181–187

Meena OP, Maurya BR, Meena VS (2013) Influence of K-solubilizing bacteria on release of potassium from waste mica. Agric Sustain Dev 1(1):53–56

Meena VS, Maurya BR, Bahadur I (2014a) Potassium solubilization by bacterial strain in waste mica. Bangladesh J Bot 43(2):235–237

Meena VS, Maurya BR, Verma JP (2014b) Does a rhizospheric microorganism enhance K+ availability in agricultural soils? Microbiol Res 169:337–347

Meena RK, Singh RK, Singh NP, Meena SK, Meena VS (2015a) Isolation of low temperature surviving plant growth-promoting rhizobacteria (PGPR) from pea (*Pisum sativum* L.) and documentation of their plant growth promoting traits. Biocatal Agric Biotechnol. doi:10.1016/j.bcab.2015.08.006

Meena VS, Maurya BR, Verma JP, Aeron A, Kumar A, Kim K, Bajpai VK (2015b) Potassium solubilizing rhizobacteria (KSR): isolation, identification, and K-release dynamics from waste mica. Ecol Eng 81: 340–347

Memon YM, Fergus IF, Hughes JD, Page DW (1988) Utilization of non-exchangeable soil potassium in relation to soil types, plant species and stage of growth. Aust J Soil Res 26:489–496

Mikhailouskaya N, Tcherhysh A (2005) K-mobilizing bacteria and their effect on wheat yield. Lat J Agron 8:154–157

Pettigrew WT (2008) Potassium influences on yield and quality production for maize, wheat, soybean and cotton. Physiol Plant 133:670–681

Prajapati K, Sharma MC, Modi HA (2013) Growth promoting effect of potassium solubilizing microorganisms on okra (*Abelmoschus Esculentus*). Int J Agric Sci 3:181–188

Rogers JR, Bennett PC, Choi WJ (1998) Feldspars as a source of nutrients for microorganisms. Am Miner 83:1532–1540

Sangeeth KP, Bhai RS, Srinivasan V (2012) *Paenibacillus glucanolyticus*, a promising potassium solubilizing bacterium isolated from black pepper (*Piper nigrum* L.) rhizosphere. J Spic Aromat Crops 21:118–124

Sheng XF (2005) Growth promotion and increased potassium uptake of cotton and rape by a potassium releasing strain of *Bacillus edaphicus*. Soil Biol Biochem 37:1918–1922

Sheng XF, He LY (2006) Solubilization of potassium bearing minerals by a wild type strain of *Bacillus edaphicus* and its mutants and increased potassium uptake by wheat. Can J Microbiol 52:66–72

Sheng XF, Huang WY (2002) Study on the conditions of potassium release by strain NBT of silicate bacteria. Sci Agric Sin 35:673–677

Sheng XF, He LY, Huang WY (2002) The conditions of releasing potassium by a silicate-dissolving bacterial strain NBT. Agric Sci China 1:662–666

Sheng XF, Xia JJ, Chen J (2003) Mutagenesis of the *Bacillus edaphicus* strain NBT and its effect on growth of chili and cotton. Agric Sci China 2:40–41

Sheng XF, Zhao F, He LY, Qiu G, Chen L (2008) Isolation and characterization of silicate mineral solubilizing *Bacillus globisporus* Q12 from the surfaces of weathered feldspar. Can J Microbiol 54(5):1064–1068

Singh G, Biswas DR, Marwaha TS (2010) Mobilization of potassium from waste mica by plant growth promoting rhizobacteria and its assimilation by maize (*Zea mays*) and wheat (*Triticum aestivum* l.): a hydroponics study under phytotron growth chamber. J Plant Nutr 33:1236–1251

Singh NP, Singh RK, Meena VS, Meena RK (2015) Can we use maize (*Zea mays*) rhizobacteria as plant growth promoter? Vegetos 28(1):86–99

Styriakova I, Styriak I, Hradil D, Bezdicka P (2003) The release of iron bearing minerals and dissolution of feldspar by heterophic bacteria of Bacillus species. Ceram Silic 47(1):20–26

Sugumaran P, Janarthanam B (2007) Solubilization of potassium containing minerals by bacteria and their effect on plant growth. World J Agric Sci 3:350–355

Supanjani HHS, Jung JS, Lee KD (2006) Rock phosphate-potassium and rock-solubilising bacteria as alternative, sustainable fertilizers. Agron Sustain Dev 26:233–240

Valmorbida J, Boaro CSF (2007) Growth and development of *Mentha piperita* L. in nutrient solution as affected by rates of potassium. Braz Arch Biol Technol 50:379–384

Wu SC, Cao ZH, Li ZG, Cheung KC, Wong MH (2005) Effects of biofertilizer containing N-fixer, P and K solubilizers and AM fungi on maize growth: a greenhouse trial. Geoderma 125:155–166

Xeng XF (2005) Growth promotion and increased potassium uptake of cotton and rape by a potassium releasing strain of *Bacillus edaphicus*. Soil Biol Biochem 37:1918–1922

Yang ML, Yan CX, Si SD (2014) Effect of potassium-solubilizing bacteria-mineral contact mode on decomposition behavior of potassium-rich shale. Chin J Nonferrous Met 24:48–52

Yuan L, Wang Z, Fang D, Shun H, Huang J (2000) Bio-mobilization of potassium from clay minerals: II. By ectomycorrhizal fungi. Pedosphere 10:347–354

Zarjani JK, Aliasgharzad N, Oustan S, Emadi M, Ahmadi A (2013) Isolation and characterization of potassium solubilizing bacteria in some Iranian soils. Arch Agron Soil Sci 59(12):1713–1723

Zeng X, Liu X, Tang J, Hu S, Jiang P, Li W, Xu L (2012) Characterization and potassium solubilizing ability of *Bacillus circulans* Z1-3. Adv Sci Lett 10:173–176

Zhanga C, Konga F (2014) Isolation and identification of potassium-solubilizing bacteria from tobacco rhizospheric soil and their effect on tobacco plants. Appl Soil Ecol 82:18–25

Zörb C, Senbayram M, Peiter E (2014) Potassium in agriculture – status and perspectives. J Plant Physiol 171(9):656–669

Zorba C, Senbayramb M, Peiterc E (2013) Potassium in agriculture -status and perspectives. J Plant Physio. http://dx.doi.org/10.1016/j.jplph.2013.08.008

Global Perspective of Potash for Fertilizer Production

23

Satya Prakash and Jay Prakash Verma

Abstract

Potash is used as a fertilizer to fulfil the requirement of potassium (K) which is one of the essential macronutrients among NPK fertilizers that are widely used in agriculture to increase crop production. Potash is the common name given to a naturally occurring potassium salt, e.g. sylvinite. These salts are water soluble and thus are available for the plant. The sources of potash are sulfosalts found in rock and clay within various geological sets. Potash is mined and processed in few countries (Canada, Belarus, Russia, Germany, Brazil, Chile, Spain, China, the USA, Israel, Jordan, UK and Ukraine). This commodity is consumed in a large number of countries of the world which reflects its importance for discovery of new deposit. The aim of this review is to highlight the various deposits of potash globally to fulfil the current requirement of potassium fertilizers for agricultural production.

Keywords

Potassium • Fertilizers • Potash • Polyhalite • Glauconite

23.1 Introduction

Potash refers to a variety of K-bearing minerals of soil and rocks. Potash is one of the important fertilizers for agricultural lands because it improves water retention and colour and texture of soils and enhances nutrient value, taste and yield as well as disease resistance of food crops. It has wide application to increase the production of fruit and vegetables, rice, wheat and other grains, sugar, corn, soybeans, palm oil and cotton. Potassium (K) is the most important macronutrient for plants. Potassium salts (potassium chloride, sulphate, nitrate, etc.) present in soils which are water soluble separate into the potash cation K^+ and the relevant anions Cl^-, SO_4^-, NO_3^-, etc. Generally, about 70–90 % of potash is present in water in

S. Prakash
Geological Survey of India, Northern Region, Aliganj, Lucknow 226024, Uttar Pradesh, India

J.P. Verma (✉)
Institute of Environment and Sustainable Development, Banaras Hindu University, Varanasi 22100, Uttar Pradesh, India
e-mail: verma_bhu@yahoo.co.in; jpv.iesd@bhu.ac.in

V.S. Meena et al. (eds.), *Potassium Solubilizing Microorganisms for Sustainable Agriculture*,
DOI 10.1007/978-81-322-2776-2_23

soluble form, and the remaining small amounts are bound with the organic material.

Potash from fertilizer and manure operates in the K cycle of soil in a similar manner, and it does not have any distinction in terms of K supply. Potassium fertilizers are not associated with any environmental or health concerns. None of the forms of these materials provide any harmful effects, unless, like any substance, they are used in excess. Farmers should therefore use all potash materials with care and responsibility whatever the farming system is. In particular, excess potash applications can cause scorch at seedling stage, and crops grown artificially under cover require the sulphate form.

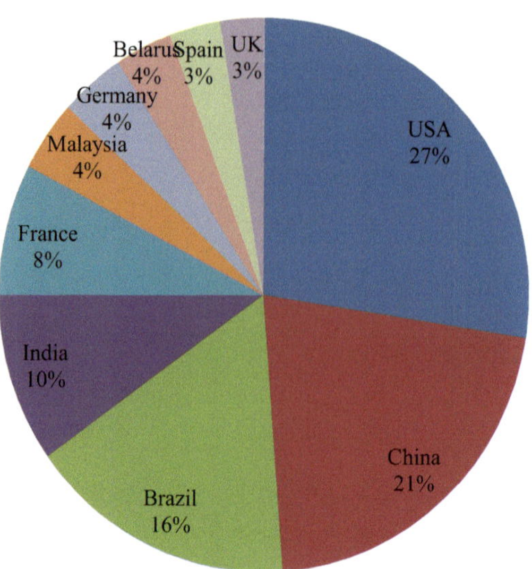

Fig. 23.1 Global consumption of potash in various countries, 2000 (Source: www.fertilizer.org)

23.2 Sources of Potash Fertilizer

The use of potash fertilization in agricultural field started in the nineteenth century after the discoveries made by Justus V. Liebig for the requirement of potassium for plants' need in different proportions and quantities and nutrients such as nitrogen, phosphorus and potassium, in order to build up biomass. The main sources of potassium in the world are the clay minerals of the soils and rocks, in the water of the oceans and in the rock salt deposits containing the crystallized minerals from long dried-up seas. It is supplied commercially as potash, i.e. potassium-bearing minerals, ores and processed products. Potassium salts, chloride, sulphate and nitrate derived from these evaporite rocks and from salt pans which are in current use in certain parts of the world, are the most common forms of fertilizer potash – all of these being naturally water soluble. The oldest form of potash fertilizer is wood ash, but supplies of this material are obviously no longer practical or sustainable. Green wastes, seaweed products and composts also represent alternative minor potash sources. Animal feeds and bedding bought onto farm also represent a supply of nutrient. Various process wastes containing potassium (such as lime kiln dusts) have been considered as sources of K, but the server in K availability can contain undesirable contamination and suffer from irregularity of supply (Fig. 23.1 and Table 23.1).

Table 23.1 Resources of potash in India, 2010 (by grades/states)

By grade	Total resources of potash (in million tonnes)
Glauconite	1968
Polyhalite	16,164
Sylvite	2476
Unclassified	1206
By states	
Madhya Pradesh	1206
Rajasthan	20,419
Uttar Pradesh	190

Source: IMYB-2013

Polyhalite, i.e. potassium chloride (KCl), is the main ore which accounts for most of the K used in world agriculture. It fulfils up to 96 % of the world potash production, and the balance of 4 % comes from potassium sulphate (K_2SO_4), potassium nitrate (KNO_3) and potassium-magnesium salts (Prud'Homme 1997). The K content of KCl fertilizer is frequently expressed on the basis of its oxide equivalent (K_2O), although some countries use the element K. Commercial KCl fertilizers contain between 60 % and 62 % K_2O – the large K content of all the K fertilizers.

The most abundant mineral in commercial deposits is polyhalite (KCl). Polyhalite and halite (NaCl) form the common ore called sylvinite (Garret 1996). Kainite ores (4 KCl·4 $MgSO_4$·11H_2O) are much less common. In brines, carnallite (KCl·$MgCl_2$·6H_2O) is present with halite (Eatock 1985). Bedded marine evaporite deposits and surface and subsurface potash-rich brines are principal sources of potash. The principal ore is sylvinite, a mixture of polyhalite (KCl) and rock salt (NaCl). In India, resources exist mainly as polyhalite, sylvite and glauconite. However, there is no commercial exploitation, and the entire requirements of potassic fertilizers for direct application as well as for production of complex fertilizers are met through imports.

23.3 Occurrence of Potash Deposition

Huge deposits with high grade of potash occurrence are known from Montana and North Dakota as an extension of the Williston Basin deposits in Manitoba and Saskatchewan, Canada. Ore is of high grade with K_2O content varying from 25 % to 30 %. The main ore is sylvinite, some carnallite and clay occurring in uniform thickness of 2.4–3 m mineralization. It lies at depths between 1800 and 3100 m covering about 3110 km^2 area of potash deposit.

Russia-Verkhoyansk deposit is known from the Urals near Solikamsk. The zone of potash ore containing K_2O up to 15 % occurs at depth between 75 and 450 m in 13 potentially minable beds ranging in thickness from 26 to 30 m of sylvinite and 70–80 m of sylvinite and carnallite. Belarus-Starobinski deposit contains 30 potash-rich beds occurring at depth of 350–620 m containing an average K_2O of 11 %. The oldest deposit is the Hessen and Thüringen beds in Southern Germany containing K_2O up to 15–20 % of polyhalite, kieserite and carnallite. Besides these, polyhalite deposits are known from England and sylvinite from Spain. Potassium is extracted mainly from the Dead Sea of Middle East from KCl. In Latin America sylvinite and carnallite in the Sergipe Basin in Brazil KNO_3 in Chile in Atacama Desert, Salar de Atacama, a high-attitude dry lake (brine est. at 120 million tonnes KCl and 80 million K_2SO_4) are other important known deposits of the world. In Asia, carnallite and K-bearing brines are known from Qinghai and Xinjiang area. Thailand, Argentina, Amazon basin in Brazil, Morocco and Poland are other few countries having undeveloped deposits containing polyhalite, kieserite and carnallite.

23.3.1 Resource of Potash in India

In India, glauconite, polyhalite and sylvite from bedded marine evaporite deposits and surface and subsurface potash-rich brines are the principal sources of potash. The whole requirement of potash fertilizers for direct application as well as for production of complex fertilizers is met through imports. Grade wise, polyhalites constitute maximum resources followed by polyhalite and glauconites. A total resource of potash is estimated to be 21,816 million tonnes till 2010, out of which Rajasthan contributes 94 % resources, followed by Madhya Pradesh 5 % and Uttar Pradesh the remaining 1 %. The total consumption of potash in fertilizer industries was about 1.42 million tonnes in 2012–2013, 1.42 million tonnes in 2011–2012 and 1.89 million tonnes in 2010–2011.

23.3.2 Resource of Potash in the World

The world reserves of potash (K_2O) are about 6072 million tonnes (Table 23.2). The main contributing countries of the world are Canada, Russia, Belarus, Germany, Brazil, China, the USA, Israel and Chile. The world production of potash in 2012 was 31.5 million tonnes in terms of K_2O content as against 35 million tonnes in 2011. Canada remained the leading producer of potash with 28 % share in total production in 2012, followed by Russia (17 %), Belarus (15 %), China (8 %), Germany (10 %), Israel (7 %) and Jordan and Chile (3 % each).

Table 23.2 World reserves of potash (by principal countries)

Country	Reserve (in '000 tonnes of K_2O content)
Belarus	3,300,000
Brazil	300,000
Canada	1,000,000
Chile	150,000
China	210,000
Germany	140,000
Israel	40,000
Jordan	40,000
Russia	600,000
Spain	20,000
UK	22,000
USA	200,000
Other countries	50,000

Source: Mineral commodity summaries 2014) (http://www.usgs.gov (USGS)

Table 23.3 Potash consumption in 2000

Countries	K_2O (million tonnes)	World share (%)
USA	4.30	20
China	3.35	15
Brazil	2.56	12
India	1.55	7
France	1.20	5
Malaysia	0.65	3
Germany	0.62	3
Belarus	0.55	3
Spain	0.49	2
UK	0.42	2
World	21.9	–

Source: www.fertilizer.org

23.4 Trend of Production and Consumption of Potash in the World

Records show that the global demands for potassium consumption are continuously increasing for the past 40 years. The consumption of the potassium is increased by 9–22 million tonnes. In the developing countries, the annual growth rate is up to 8.3 % between 1961 and 2000. In the developed countries, the increase in growth rate is 0.6 %. Potash fertilizer reserve and its production are in few countries, but its demand and consumption are noted in 84 countries. Its consumption in the top ten countries of the world is given in year 2000 (Tables 23.3 and 23.4). The consumption of potash fertilizer has been recorded up to 10 % of total consumption of the world (Maene 2007). According to the Food and Agriculture Organization of the United Nations, the world potash balance was about 8,850,000 tonnes in 2013, and it decreased to 8,700,000 tonnes in 2014. The demand for potash seems to increase in the coming years. Its demand is expected to increase by 2.6 % between 2014 and 2018. Reports of FAO also say that potash supply is expected to increase by 4.2 % between 2014 and 2018. Thus, the potential balance of potash will be expected to rise significantly from 8,700,000 tonnes in 2014 to 12,700,000 tonnes in 2018. The potential potash (K_2O) balance as a percentage of global total demand is expected to rise from 25 % in 2014 and expected to reach up to 33 % in 2018.

23.5 Conclusion and Future Prospective

The main source of potash fertilizer is potassium-bearing salt occurring from the world known deposit of few countries. Only these countries are involved with the production of potash; however, their requirements are continuously increasing in most of the countries of the world, leading to imports. At present, no any substitutes are present other than rock and clay that can be used for the production of potash to meet the requirement of agricultural land. The demands of potash are increasing year after year as indicated by the Food and Agriculture Organization of the United Nations. The demand is expected to increase by 2.6 % between 2014 and 2018 as per report. Very limited resources are present to meet the requirement. The potential potash (K_2O) balance as a percentage of global total demand is expected to rise from 25 % in 2014 and expected to reach up to 33 % in 2018. This review report highlighted the potash deposition and their application as a potassium fertilizer for agricultural purpose in various countries. The discovery of potash deposits in the world and in India also needs to be identified for

Table 23.4 World Potash Trade (2000) (www.fertilizer.org)

Exporters	K_2O in MT	World share (%)	Importers	K_2O in MT	World share (%)
Canada	8.5	41	USA	4.8	23
Belarus	2.8	14	China	3.1	15
Russia	2.8	13	India	2.4	12
Germany	2.7	13	Brazil	1.6	8
Israel	2.7	7	Malaysia	1.3	6
Jordan	1.5	6	France	0.8	4
Others	1.1	6	Others	6.6	32

the fulfilment of the future demands for enhancing agricultural production and for providing feed in order to increase the population in a sustainable way.

References

Eatock WH (1985) Advances in potassium mining and refining. In: Munson RD (ed) Potassium in agriculture. ASA/CSSA/SSSA, Madison, pp 29–48

Garret D (1996) Potash: deposits, processing, properties and uses. Chapman & Hall, London

Indian Minerals Year book (2013) Part-III: mineral reviews, 52nd edn (Website: www.ibm.gov.in)

Maene LM (2007) Global potassium fertilizer situation: current use and perspectives. International Fertilizer Industry Association, Paris

Prud'Homme M (1997) World potash supply. IFA Production and International Trade Committee Meeting, 14–15 October 1997, Warsaw (Poland)